INTRODUCTION TO THE PHILOSOPHY OF SCIENCE

A Text by Members of the Department
of the History and Philosophy of Science
of the University of Pittsburgh

Merrilee H. Salmon · *John Earman* · *Clark Glymour*

James G. Lennox · *Peter Machamer* · *J. E. McGuire*

John D. Norton · *Wesley C. Salmon* · *Kenneth F. Schaffner*

PRENTICE HALL

Englewood Cliffs, New Jersey 07632

Library of Congress Cataloging-in-Publication Data

Introduction to the philosophy of science : a text / by members of the
  Department of the History and Philosophy of Science of the
  University of Pittsburgh ; Merrilee H. Salmon . . . [et al.].
    p. cm.
  Includes bibliographical references and index.
  ISBN 0-13-663345-5
  1. Science—Philosophy. I. Salmon, Merrilee H. II. University
of Pittsburgh. Dept. of the History and Philosophy of Science.
Q175.I633 1992
501—dc20   91-29648
                                                     CIP

To the giants
On whose shoulders we stand

Acquisitions editor: *Ted Bolen*
Production editor: *Merrill Peterson*
Interior design: *Joan L. Stone*
Copy editor: *Judith Meiksin*
Cover designer: *John D. Norton/Michele Paccione*
Prepress buyer: *Herb Klein*
Manufacturing buyer: *Patrice Fraccio*
Editorial assistant: *Diane Schaible*

Printed in the United States of America

10 9 8 7 6 5 4 3 2 1

ISBN 0-13-663345-5

PRENTICE-HALL INTERNATIONAL (UK) LIMITED, *London*
PRENTICE-HALL OF AUSTRALIA PTY. LIMITED, *Sydney*
PRENTICE-HALL CANADA INC., *Toronto*
PRENTICE-HALL HISPANOAMERICANA, S.A., *Mexico*
PRENTICE-HALL OF INDIA PRIVATE LIMITED, *New Delhi*
PRENTICE-HALL OF JAPAN, INC., *Tokyo*
SIMON & SCHUSTER ASIA PTE. LTD., *Singapore*
EDITORA PRENTICE-HALL DO BRASIL, LTDA., *Rio de Janeiro*

Contents

Preface   v

Introduction   1

PART ONE: GENERAL TOPICS IN THE PHILOSOPHY OF SCIENCE

1

Scientific Explanation   7

2

The Confirmation of Scientific Hypotheses   42

3

Realism and the Nature of Theories   104

4

Scientific Change: Perspectives and Proposals   132

PART TWO: PHILOSOPHY OF THE PHYSICAL SCIENCES

5

Philosophy of Space and Time   179

6

DETERMINISM IN THE PHYSICAL SCIENCES   232

PART THREE: PHILOSOPHY OF BIOLOGY AND MEDICINE

7

PHILOSOPHY OF BIOLOGY   269

8

PHILOSOPHY OF MEDICINE   310

**PART FOUR: PHILOSOPHY OF BEHAVIORAL
AND SOCIAL SCIENCES**

9

PHILOSOPHY OF PSYCHOLOGY   346

10

ANDROID EPISTEMOLOGY: COMPUTATION, ARTIFICIAL INTELLIGENCE,
AND THE PHILOSOPHY OF SCIENCE   364

11

PHILOSOPHY OF THE SOCIAL SCIENCES   404

BIBLIOGRAPHY   426

INDEX   447

PREFACE

The philosophy of science has become such a far-flung and specialized enterprise that no one person is in a position to write an authoritative survey of the field, even at an introductory level. This is an especially unfortunate situation in view of the felt need for an up-to-date textbook in this field. Our solution is drawn on the combined expertise of members of the History and Philosophy of Science Department of the University of Pittsburgh: John Earman, Clark Glymour, James G. Lennox, J. E. McGuire, Peter Machamer, John D. Norton, Merrilee H. Salmon, Wesley C. Salmon, and Kenneth F. Schaffner. Although individual names are attached to chapters, this volume is the result of a cooperative effort, and to the extent that it succeeds the credit is to be equally divided, except for additional measures for Merrilee Salmon, who conceived the project and prodded us into action, and for John Norton, who drew the figures for Chapters 2, 5, 6, and 9.

The primary audience for our text consists of upper-level undergraduates and beginning graduate students. While it is possible to teach philosophy of science to first and second year students, our experience is that students will benefit more from such a course if they have already mastered some college-level science and /or history of science. We have attempted to reach a compromise between a presentation that is accessible to a wide audience and one that makes contact with current research. This is by no means an easy compromise to achieve, but texts that do not aspire to it are not worth the candle.

The volume contains more material than can be comfortably covered in one semester and certainly more than can be covered in a ten-week quarter. The instructor is thus presented with a number of choices. Those interested mainly in scientific methodology may want to devote the entire course to the general topics in the philosophy of science that are explored in Part One, Chapters 1–4. Alternatively, the first half of the course could be devoted to a survey of the material in Chapters 1–4 and the second to a sampling of the foundational problems in various physical, biological, behavioral and social sciences that are discussed in Chapters 5–11. Or the text could be used as the basis of a two-course sequence.

Each chapter contains a list of suggested readings and study questions. Some of the questions are appropriate for test questions; others can be used as topics for writing assignments.

We have tried to keep technical notation to a minimum, but sometimes its use is unavoidable. Symbols are defined and their use is explained with examples in the contexts in which they first occur. When symbols occur in more than one chapter, first occurrences are noted in the index.

We want to express our deepest thanks to Philip Kitcher, Universiy of California, San Diego, who read the entire manuscript at an early stage and offered extensive and helpful comments on every chapter. We are also grateful to James E. Roper, Michigan State University, Arnold Wilson, University of Cincinnati, and Michael J. Zenzen, Rensselaer Polytechnic Institute, for their critical reading of the manuscript and their good advice. John Beatty, University of Minnesota, Harry Corwin, University of Pittsburgh, and David Kelly, Institute for Objectivist Studies, offered useful suggestions for Chapter 7. Rob Pennock, University of Texas at Austin, offered important suggestions regarding Chapter 1. Kevin Kelly, Carnegie-Mellon University, provided useful suggestions and help with the pictures in Chapters 3 and 10. David Hillman gave valuable help in checking and assembling the bibliography, and Judith Meiksin provided excellent editorial assistance. Madeline Larson compiled the index. The Faculty Proficiency Enhancement Program, University of Pittsburgh, provided the computer system on which the text, graphics, and index were assembled. Special thanks are due to the students in several Honors Philosophy of Science courses at the Universiy of Pittsburgh. They served as willing guinea pigs for our efforts to produce a suitable text for undergraduates, and their lively participation in class discussions helped immensely in refining our pedagogical strategies.

John Earman

INTRODUCTION

Scientific knowledge stands as the supreme intellectual achievement of our society. Governments, private foundations, and businesses support scientific research although it is costly and does not always yield immediate practical benefits. Courses in science are a required part of curricula from grade school through university, and young people are encouraged to undergo the long apprenticeship of study and work that will transform them into scientists. Scientific accomplishments are honored at every level, from awards at local science fairs to Nobel prizes. Major museums in cities all over the western world document and display scientific achievements and inventions. Yet despite the impressive scope of scientific progress and all the attention paid to science and scientists, many questions remain about the nature of science and how it works.

Such questions are not usually raised in the study of specific sciences. Physics is concerned, for example, with providing explanations of why chain reactions occur in certain kinds of materials but not in others; it is not the task of physics to outline an answer to the more general question of what features an explanation must have if it is to be scientifically acceptable. Biologists study populations of fruit flies to draw conclusions about how heredity works. They do not, as biologists, address in a general way the issue of the nature of the relationship between observation and theories. This is not to say that physicists and biologists are incapable of discussing such topics or of clarifying them. When they do so, however, they are speaking philosophically *about* science rather than actually *doing* science.

"Philosophy of science" is the name given to that branch of philosophy that reflects on and critically analyzes science. As a discipline, it tries to understand the aims and methods of science, along with its principles, practices, and achievements. Philosophers try to provide precise answers to very broad questions about science,

1

such as the question just raised about the nature of scientific explanation. Some other questions studied by philosophers of science are as follows:

What are the aims of science?

What is the role of observations and experiments in obtaining scientific knowledge?

How do scientists justify their claims? What is a scientific proof?

What is a scientific law?

Are there methods for making scientific discoveries?

How does scientific knowledge advance and grow?

How do the historical and cultural settings in which scientific work occurs affect the content and quality of such work?

Does science employ or require a special language?

Science itself is made up of many subdisciplines: physics, astronomy, chemistry, biology, psychology, sociology, anthropology, and medicine, to name a few. The presence of so many different fields within science raises interesting questions about what it means to be a science and whether a single method is common to all sciences. Philosophy of science thus addresses also the following sorts of questions:

Is it possible to give a general account of scientific methodology, or are there different methods and forms of explanation for various branches of science?

How do physical, biological, and social sciences differ from one another?

Can some sciences be reduced to others?

Finally, philosophy of science is concerned with specific issues that arise in connection with particular fields of science. For example, while experimentation plays a major role in some sciences, in others, such as astronomy, it does not. Some other discipline-specific questions are these:

Does the existence of free will pose a special problem for a science of human behavior?

Is medicine more an art than a science?

Are statistical techniques useful in anthropology, where sample sizes are very small?

All of the questions raised above are complex and difficult, so it should come as no surprise that the opinions of philosophers of science (and scientists in their philosophical moments) on these topics vary considerably. In the twentieth century, two disparate approaches have been dominant. The earlier tradition, developed by logical positivists (members of the Vienna Circle) and logical empiricists (a similar group from Berlin), set rigorous standards for the conduct of philosophy of science, as close to those of science itself as the subject matter would allow. These philosophers and scientists attempted to provide logical analyses of the nature of scientific concepts, the relation between evidence and theory, and the nature of scientific explanation. In their desire to be precise, they made extensive use of the language and techniques of symbolic logic. Despite many differences in points of view, the logical positivists and logical empiricists generally were concerned with emphasizing such distinctions as

> the demarcation between scientific knowledge and other types of knowledge,
> the difference between facts and values,
> the difference between the language used to state observations and that used to refer to theoretical entities, and
> the difference between how theories are discovered and how they are justified.

Logical empiricists and logical positivists were also concerned with establishing clear meanings for all the terms used in science. Some approached this problem by searching for a verifiability criterion of meaning while others, particularly scientists themselves, tried to formulate operational definitions of scientific terms. These efforts were closely related to their concern with providing a solid foundation for scientific theorizing by linking it firmly to an observational basis. Although they believed that justification rather than discovery was the proper concern of science, they shared an optimism about the ability of science to provide genuine knowledge of the features of an independently existing world.

At the time of World War Two, many of these philosophers left Europe for England and the United States where their works have significantly affected the development of philosophy of science in English-speaking countries. Even at the level of undergraduate education, their influence has been important. Carl G. Hempel, who came to America from Berlin, for example, has literally defined the philosophy of the natural sciences for generations of students who first learned about the subject from his introductory text, *Philosophy of Natural Science* (1966). The power and broad influence of the general approach outlined by Hempel in this work justifies calling it ''the standard view'' of philosophy of science.

During the past twenty-five years, however, many criticisms have been raised against perceived faults of the standard view. (Indeed, Hempel himself has criticized some of its features.) A major objection is that the standard view fails to take account of the bearing of history of science on the philosophy of science. Critics of the standard view cite Thomas Kuhn's *Structure of Scientific Revolutions* (1962, 1970), which argues that most scientific textbooks ignore history and distort the real nature of progress in science by presenting it as a series of accumulations of new discoveries that straightforwardly build on and add to knowledge already attained. Kuhn draws attention to the revolutionary character of science—its replacement of outworn theories by newer ones that are so different from the old that the two do not share the same problems or even a common language. He also draws attention to the ''irrational'' aspects of changes in science, that is to say, those features of scientific change that cannot be accounted for entirely in terms of scientists' allegiance to ''facts'' and logic. Kuhn argues that only a refusal to take seriously the history of science could account for the gross distortion presented in scientific textbooks.

Appealing to Kuhn's account of science, critics of the standard view of philosophy of science say that it embodies and promotes an ahistorical view of scientific activity by emphasizing the logical characteristics of science while ignoring the cultural context of scientific activity, which strongly influences the style of the enterprise and the content of its results. Furthermore, critics say, failure to take account of the rhetorical features of scientific discourse can only lead to a distorted notion of how science really works. The values of society and of individual practitioners of science, they say, influence not only the choice of problems and the amount of effort devoted to their solution, but also the interpretation of the results. They maintain that so-called facts can only be grasped through theories, which are the creations of members of a specific culture, and are never completely free of the values and aspirations of that culture.

Both the standard view and that of its critics have merits and shortcomings. Both views are likewise too complex to state succinctly without distortion and oversimplification; the above brief synopsis is intended only to introduce the reader to the subject. The ensuing chapters will survey many aspects of the dispute and will examine the reasons offered in support of the various positions.

The approach to the philosophy of science exemplified in this work does not fall neatly into either of the two main categories briefly outlined. The authors of this text are all members of a Department of History and Philosophy of Science. The marriage between history and philosophy in the Department is not merely one of convenience between philosophers and historians each of whom happens to be concerned with science. Instead, the Department was founded because the members believe that the study of the philosophy of science must be informed by an understanding of the historical and social context of science, as well as by a grasp of the workings of science itself. At the same time, the general approach of this book disavows the extreme forms of relativism and skepticism that characterize some of the more strident critics of the standard view.

Part One of this book takes up topics requisite for any adequate introduction to the philosophy of science: Explanation; Induction and Confirmation; Realism and the Nature of Scientific Theories; and Scientific Change: Perspectives and Proposals. These four chapters outline and discuss fundamental issues in philosophy of science and form the foundation for discussions in the remaining chapters of the book. In Part One, the reader is introduced to the pertinent history of the topics discussed as well as to the vocabulary, techniques, and most important issues in contemporary philosophy of science. The intention of the authors in each case is to presume no prior knowledge of philosophy of science, but to lead the reader to an appreciation of some of the knottiest problems that concern contemporary philosophers of science. In the first chapter, "Scientific Explanation," Wesley C. Salmon discusses the elements involved in the special kind of understanding of our world and what takes place within it that is provided by the various sciences. In the second chapter, "The Confirmation of Scientific Hypotheses," John Earman and Wesley C. Salmon deal with questions concerning the relationship between empirical evidence and scientific hypotheses, laws, and theories. In the course of the discussion they consider the nature of inductive reasoning and the meanings of the concept of probability. Chapter 3, by Clark Glymour, considers the major traditional arguments against literal belief in the claims of science and a range of responses to those arguments. In the fourth chapter, J. E. McGuire discusses the nature of scientific change and progress in relation to social context and historical development.

In the remaining seven chapters, each of which deals with the philosophy of a special area of science, the authors assume that the reader is familiar with the issues addressed in the first four chapters, though some topics depend less heavily on this than others. The chapters in Parts Two through Four can be read independently of one another, although they do contain references to the materials covered in other chapters.

The philosophy of physical sciences is covered in Part Two (Chapters 5 and 6). In "The Philosophy of Space and Time," John D. Norton introduces questions central to recent work in philosophy of space and time and illustrates how philosophical ideas about verification, conventions, realism, and theory reduction are applied in physical theories of space and time. In "Determinism in the Physical Sciences," John Earman surveys the implications of classical physics, the special and general theories of relativity, and quantum mechanics for the doctrine that the world evolves in a deterministic manner.

Part Three takes up the philosophy of biology and medicine with separate chapters on these topics. In "Philosophy of Biology," James G. Lennox discusses the development of the so-called neo-Darwinian theory of evolution. Lennox shows how this union of Mendelian genetics and Darwin's theory of natural selection provides a powerful tool for explaining evolutionary change and adaptation that operates differently from theories in the physical sciences. In "Philosophy of Medicine," Kenneth F. Schaffner, who has been trained as a physician as well as an historian and philosopher of science, discusses the questions of whether medicine is a science and whether medicine can be reduced to biology. He examines the nature of medicine as an enterprise incorporating ethical principles and the implications of this for medicine's reduction to biology. Part Four, on the behavioral sciences, begins with Peter Machamer's chapter on "Philosophy of Psychology." This chapter briefly surveys the relationships between philosophy and psychology and lays out some of the topics that are and have been important to an understanding of psychology. The bulk of the chapter, however, describes and assesses the nature of psychological theories of perception. The author's intent is to provide a case study of what philosophers who are interested in psychology might do. In Chapter 10, "Android Epistemology," Clark Glymour discusses philosophical issues raised by the exciting new field of Artificial Intelligence. He illustrates the influence of issues in philosophy of science on the design of artificial intelligence and expert systems programs. In the final chapter, "Philosophy of Social Science," Merrilee H. Salmon addresses the issue of whether the so-called social sciences really are entitled to the name, and discusses some of the special problems posed by disciplines that try to explain human behavior by using the same methods that have been so successful in the physical sciences.

Merrilee H. Salmon

One

SCIENTIFIC EXPLANATION

Wesley C. Salmon

The fruits of science are many and various. When science is first mentioned, many people think immediately of high technology. Such items as computers, nuclear energy, genetic engineering, and high-temperature superconductors are likely to be included. These are the fruits of applied science. Evaluating the benefits, hazards, and costs of such technological developments often leads to lively and impassioned debate.

In this chapter, however, we are going to focus on a different aspect of science, namely, the intellectual understanding it gives us of the world we live in. This is the fruit of pure science, and it is one we highly prize. All of us frequently ask the question "Why?" in order to achieve some degree of understanding with regard to various phenomena. This seems to be an expression of a natural human curiosity. Why, during a total lunar eclipse, does the moon take on a coppery color instead of just becoming dark when the earth passes between it and the sun? Because the earth's atmosphere acts like a prism, diffracting the sunlight passing through it in such a way that the light in the red region of the spectrum falls upon the lunar surface. This is a rough sketch of a scientific explanation of that phenomenon, and it imparts at least some degree of scientific understanding.

Our task in this chapter is to try to say with some precision just what scientific explanation consists in. Before we embark on that enterprise, however, some preliminary points of clarification are in order.

1.1 EXPLANATION VS. CONFIRMATION

The first step in clarifying the notion of scientific explanation is to draw a sharp distinction between explaining *why* a particular phenomenon occurs and giving reasons for believing *that* it occurs. My reason for believing *that* the moon turns coppery during total eclipse is that I have observed it with my own eyes. I can also appeal to the testimony of other observers. That is how the proposition that the moon turns coppery during a total eclipse is confirmed,[1] and it is entirely different from explaining *why* it happens. Consider another example. According to contemporary cosmology all of the distant galaxies are receding from us at high velocities. The evidence for this is the fact that the light from them is shifted toward the red end of the spectrum; such evidence confirms the statement that the other galaxies are moving away from our galaxy (the Milky Way). The fact that there is such a red shift does not explain *why* the galaxies are moving in that way; instead, the fact that they are receding explains—in terms of the Doppler effect[2]—why the light is shifted toward the red end of the spectrum. The explanation of the recession lies in the "big bang" with which our universe began several billion years ago; this is what makes all of the galaxies recede from one another and, consequently, makes all of the others move away from us.

1.2 OTHER KINDS OF EXPLANATION

Another preliminary step in clarifying the notion of *scientific* explanation is to recognize that there are many different kinds of explanation in addition to those we classify as scientific. For example, we often encounter explanations of *how* to do something—how to use a new kitchen gadget, or how to find a certain address in a strange city. There are, in addition, explanations of *what*—what an unfamiliar word means, or what is wrong with an automobile. While many, if not all, scientific explanations can be requested by means of why-questions, requests for explanations of these other sorts would not normally be phrased in why-questions; instead, *how-to*-questions and *what*-questions would be natural.

Still other types of explanation exist. Someone might ask for an explanation of the meaning of a painting or a poem; such a request calls for an artistic interpretation. Or, someone might ask for an explanation of a mathematical proof; an appropriate response would be to fill in additional steps to show how one gets from one step to another in the original demonstration. Neither of these qualifies as scientific explanation. Also excluded from our domain of scientific explanation are explanations of formal facts of pure mathematics, such as the infinitude of the set of prime numbers. We are concerned only with explanation in the empirical sciences.

As we understand the concept of *scientific* explanation, such an explanation is an attempt to render understandable or intelligible some particular event (such as the 1986 accident at the Chernobyl nuclear facility) or some general fact (such as the copper color of the moon during total eclipse) by appealing to other particular and/or general facts drawn from one or more branches of empirical science. This formulation

---

[1] Confirmation will be treated in Chapter 2 of this book.

[2] The Doppler effect is the lengthening of waves emitted by a source traveling away from a receiver and the shortening of waves emitted by a source approaching a receiver. This effect occurs in both light and sound, and it can be noticed in the change of pitch of a whistle of a passing train.

is *not* meant as a definition, because such terms as "understandable" and "intelligible" are as much in need of clarification as is the term "explanation." But it should serve as a rough indication of what we are driving at.

In pointing out the distinction between scientific explanations and explanations of other types we do not mean to disparage the others. The aim is only to emphasize the fact that the word "explanation" is extremely broad—it applies to a great many different things. We simply want to be clear on the type of explanation with which our discussion is concerned.

1.3 SCIENTIFIC EXPLANATIONS AND WHY-QUESTIONS

Many scientific explanations are requested by means of why-questions, and even when the request is not actually formulated in that way, it can often be translated into a why-question. For example, "What caused the Chernobyl accident?" or "For what reason did the Chernobyl accident occur?" are equivalent to "Why did the Chernobyl accident occur?" However, not all why-questions are requests for scientific explanations. A woman employee might ask why she received a smaller raise in salary than a male colleague when her job-performance is just as good as his. Such a why-question might be construed as a request for a justification, or, perhaps, simply a request for more pay. A bereaved widow might ask why her husband died even though she fully understands the medical explanation. Such a why-question is a request for consolation, not explanation. Some why-questions are requests for evidence. To the question, "Why should we believe that the distant galaxies are traveling away from us at high velocities?" the answer, briefly, is the red shift. Recall, as we noted in Section 1.1, that the red shift does not explain the recession. The recession explains the red shift; the red shift is evidence for the recession. For the sake of clarity we distinguish *explanation-seeking why-questions* from why-questions that seek such other things as justification, consolation, or evidence.

Can all types of scientific explanation be requested by why-questions? Some authors say "yes" and others say "no." It has been suggested, for example, that some scientific explanations are answers to *how-possibly-questions*. There is an old saying that a cat will always land on its feet (paws), no matter what position it falls from. But remembering the law of conservation of angular momentum we might well ask, "How is it possible for a cat, released (without imparting any angular momentum) from a height of several feet above the ground with its legs pointing upward, to turn over so that its paws are beneath it when it lands? Is this just an unfounded belief with no basis in fact?" The answer is that the cat can (and does) twist its body in ways that enable it to turn over without ever having a total angular momentum other than zero (see Frohlich 1980).

Other requests for explanation may take a *how-actually* form. A simple commonsense example illustrates the point. "How did the prisoner escape?" calls for an explanation of how he did it, not why he did it. The answer to this question might be that he sawed through some metal bars with a hacksaw blade smuggled in by his wife. If we were to ask why, the answer might be his intense desire to be with his wife outside of the prison. For a somewhat more scientific example, consider the question,

"How did large mammals get to New Zealand?" The answer is that they came in boats—the first were humans, and humans brought other large mammals. Or, consider the question, "How is genetic information transmitted from parents to offspring?" The answer to this question involves the structure of the DNA molecule and the genetic code.

In this chapter we will not try to argue one way or other on the issue of whether all scientific explanations can appropriately be requested by means of why-questions. We will leave open the possibility that some explanations cannot suitably be requested by why-questions.

1.4 SOME MATTERS OF TERMINOLOGY

As a further step in preliminary clarification we must establish some matters of terminology. In the first place, any explanation consists of two parts, the *explanandum* and the *explanans*. The explanandum is the fact that is to be explained. This fact may be a *particular* fact, such as the explosion of the Challenger space-shuttle vehicle, or a *general* fact, such as the law of conservation of linear momentum. A statement to the effect that the explanandum obtained is called the *explanandum-statement*. Sometimes, when it is important to contrast the fact-to-be-explained with the statement of the explanandum, we may refer to the explanandum itself as the *explanandum-fact*. When the explanandum is a particular fact we often speak of it as an event or occurrence, and there is no harm in this terminology, provided we are clear on one basic point. By and large, the events that happen in our world are highly complex, and we hardly ever try to explain every aspect of such an occurrence. For example, in explaining the explosion of the Challenger vehicle, we are not concerned to explain the fact that a woman was aboard, the fact that she was a teacher, or the fact that her life had been insured for a million dollars. When we speak of a particular fact, it is to be understood that this term refers to certain limited aspects of the event in question, not to the event in its full richness and complexity.

The other part of an explanation is the explanans. The explanans is that which does the explaining. It consists of whatever facts, particular or general, are summoned to explain the explanandum. When we want to refer to the statements of these facts we may speak of the *explanans-statements*; to contrast the facts with the statements of them we may also speak of the *explanans-facts*.

In the philosophical literature on scientific explanation, the term "explanation" is used ambiguously. Most authors use it to apply to a linguistic entity composed of the explanans-statements and the explanandum-statement. Others use it to refer to the collection of facts consisting of the explanans-facts and the explanandum-fact. In most contexts this ambiguity is harmless and does not lead to any confusion. But we should be aware that it exists.

1.5 DEDUCTION AND INDUCTION

As we will see, one influential philosophical account of explanation regards all bona fide scientific explanations as arguments. An argument is simply a set of statements, one of which is singled out as the conclusion of the argument. The remaining members of the set are premises. There may be one or more premises; no fixed number of premises is required.[3] The premises provide support for the conclusion.

All logically correct arguments fall into two types, deductive and inductive, and these types differ fundamentally from one another. For purposes of this chapter (and later chapters as well) we need a reasonably precise characterization of them. Four characteristics are important for our discussion.

DEDUCTION

1. In a valid deductive argument, all of the content of the conclusion is present, at least implicitly, in the premises. Deduction is *nonampliative*.

2. If the premises are true, the conclusion must be true. Valid deduction is *necessarily truth-preserving*.

3. If new premises are added to a valid deductive argument (and none of the original premises is changed or deleted) the argument remains valid. Deduction is *erosion-proof*.

4. Deductive validity is an *all-or-nothing* matter; validity does not come in degrees. An argument is totally valid or it is invalid.

INDUCTION

1. Induction is *ampliative*. The conclusion of an inductive argument has content that goes beyond the content of its premises.

2. A correct inductive argument may have true premises and a false conclusion. Induction is *not necessarily truth-preserving*.

3. New premises may completely undermine a strong inductive argument. Induction is *not erosion-proof*.

4. Inductive arguments come in different *degrees of strength*. In some inductions the premises support the conclusions more strongly than in others.

These characteristics can be illustrated by means of simple time-honored examples.

(1)   All humans are mortal.
      Socrates is human.
      Socrates is mortal.

Argument (1) is obviously a valid deduction. When we have said that all humans are mortal, we have already said that Socrates is mortal, given that Socrates is human. Thus, it is nonampliative. Because it is nonampliative, it is necessarily truth-preserving. Since nothing is said by the conclusion that is not already stated by the premises, what the conclusion says *must* be true if what the premises assert is true. Moreover, the argument remains nonampliative, and hence, necessarily truth-preserving, if new premises—for example, "Xantippe is human"—are added. You cannot make a valid deduction invalid just by adding premises. Finally, the premises

---

[3] Because of certain logical technicalities, there are valid deductive arguments that have no premises at all, but arguments of this sort will not be involved in our discussion.

support the conclusion totally, not just to some degree; to accept the premises and reject the conclusion would be outright self-contradiction.

(2) <u>All observed ravens have been black.</u>
     All ravens are black.

This argument is obviously ampliative; the premise refers only to ravens that have been observed, while the conclusion makes a statement about all ravens, observed or unobserved. It is not necessarily truth-preserving. Quite possibly there is, was, or will be—at some place or time—a white raven, or one of a different color. It is not erosion-proof; the observation of one non-black raven would undermine it completely. And its strength is a matter of degree. If only a few ravens in one limited environment had been observed, the premise would not support the conclusion very strongly; if vast numbers of ravens have been observed under a wide variety of circumstances, the support would be much stronger. But in neither case would the conclusion be necessitated by the premise.

Deductive validity and inductive correctness do not hinge on the truth of the premises or the conclusion of the argument. A valid deduction may have true premises and a true conclusion, one or more false premises and a false conclusion, and one or more false premises and a true conclusion.[4] When we say that valid deduction is necessarily truth-preserving, we mean that the conclusion would have to be true *if the premises were true*. Thus there cannot be a valid deduction with true premises and a false conclusion. Where correct inductive arguments are concerned, since they are not necessarily truth-preserving, any combination of truth values of premises and conclusion is possible. What we would like to say is that, if the premises are true (and embody all relevant knowledge), the conclusion is probable. As we will see in Chapter 2, however, many profound difficulties arise in attempting to support this claim about inductive arguments.

We have chosen very simple—indeed, apparently trivial—examples in order to illustrate the basic concepts. In actual science, of course, the arguments are much more complex. Most of the deductive arguments found in serious scientific contexts are mathematical derivations, and these can be extremely complicated. Nevertheless, the basic fact remains that all of them fulfill the four characteristics listed above. Although deep and interesting problems arise in the philosophy of mathematics, they are not our primary concern in this book. Our attention is focused on the empirical sciences, which, as we argue in Chapter 2, necessarily involve induction. In that chapter we encounter much more complex and interesting inductive arguments.

1.6 IS THERE ANY SUCH THING AS SCIENTIFIC EXPLANATION?

The idea that science can furnish explanations of various phenomena goes back to Aristotle (4th century B.C.), and it has been reaffirmed by many philosophers and

---

[4] The familiar slogan, "Garbage in, garbage out," does not accurately characterize deductive arguments.

scientists since then. Nevertheless, many other philosophers and scientists have maintained that science must "stick to the facts," and consequently can answer only questions about *what* but not about *why*. To understand "the *why* of things," they felt, it is necessary to appeal to theology or metaphysics. Science can describe natural phenomena and predict future occurrences, but it cannot furnish explanations. This attitude was particularly prevalent in the early decades of the twentieth century. Since it is based upon certain misconceptions regarding scientific explanation, we need to say a bit about it.

It is natural enough, when attempting to find out why a person did something, to seek a conscious (or perhaps unconscious) motive. For example, to the question, "Why did you buy that book?" a satisfactory answer might run, "Because I wanted to read an amusing novel, and I have read several other novels by the same author, all of which I found amusing." This type of explanation is satisfying because we can put ourselves in the place of the subject and *understand* how such motivation works. The concept of *understanding* is critical in this context, for it signifies empathy. If we yearn for that kind of empathetic understanding of nonhuman phenomena, we have to look elsewhere for motivation or purpose. One immediate suggestion is to make the source of purpose supernatural. Thus, prior to Darwin, the variety of species of living things was explained by *special creation*—that is, God's will. Another manifestation of the same viewpoint—held by some, but not all, *vitalists*—was the notion that behind all living phenomena there is a vital force or *entelechy* directing what goes on. These entities—entelechies and vital forces—are not open to empirical investigation.

The insistence that all aspects of nature be explained in human terms is known as *anthropomorphism*. The supposition—common before the rise of modern science—that the universe is a cozy little place, created for our benefit, with humans at its center, is an anthropomorphic conception. The doctrines of special creation and some forms of vitalism are anthropomorphic. So-called "creation science" is anthropomorphic. Teleological explanation of nonhuman phenomena in terms of human-like purposes is anthropomorphic.[5]

Many philosophers and scientists rejected the appeal to anthropomorphic and teleological explanations as an appeal to hypotheses that could not, even in principle, be investigated by empirical science. If this is what is needed for explanation, they said, we want no part of it. Science is simply not concerned with explaining natural phenomena; anyone who wants explanations will have to look for them outside of science. Such scientists and philosophers were eager to make clear that scientific knowledge does not rest on nonempirical metaphysical principles.

Not all philosophers were willing to forgo the claim that science provides explanations of natural phenomena. Karl R. Popper (1935), Carl G. Hempel (1948), R. B. Braithwaite (1953), and Ernest Nagel (1961) published important works in which they maintained that there are such things as legitimate scientific explanations, and that such explanations can be provided without going beyond the bounds of empirical science. They attempted to provide precise characterizations of scientific

---

[5] As James Lennox points out in Chapter 7, teleological explanations are anthropomorphic *only* if they appeal to human-like purposes. In evolutionary biology—and other scientific domains as well—there are teleological explanations that are *not* anthropomorphic.

explanation, and they were, to a very large degree, in agreement with respect to the core of the account. The line of thought they pursued grew into a theory that enjoyed a great deal of acceptance among philosophers of science. We will discuss it at length in later sections of this chapter.

1.7 DOES EXPLANATION INVOLVE REDUCTION TO THE FAMILIAR?

It has sometimes been asserted that explanation consists in reducing the mysterious or unfamiliar to that which is familiar. Before Newton, for example, comets were regarded as mysterious and fearsome objects. Even among educated people, the appearance of a comet signified impending disaster, for example, earthquakes, floods, famines, or epidemic diseases. Newton showed that comets could be understood as planet-like objects that travel around the sun in highly eccentric orbits. For that reason, any given comet spends most of its time far from the sun and well beyond the range of human observation. When one appeared it was a surprise. But when we learned that they behave very much as the familiar planets do, their behavior was explained, and they were no longer objects of dread.

Appealing as the notion of reduction of the unfamiliar to the familiar may be, it is not a satisfactory characterization of scientific explanation. The point can best be made in terms of a famous puzzle known as *Olbers's paradox*—which is named after a nineteenth-century astronomer but was actually formulated by Edmund Halley in 1720—why is the sky dark at night? Nothing could be more familiar than the darkness of the night sky. But Halley and later astronomers realized that, if Newton's conception of the universe were correct, then the whole night sky should shine as brightly as the noonday sun. The question of how to explain the darkness of the sky at night is extremely difficult, and there may be no answer generally accepted by the experts. Among the serious explanations that have been offered, however, appeal is made to such esoteric facts as the non-Euclidean character of space or the mean free path of photons in space. In this case, and in many others as well, a familiar phenomenon is explained by reference to facts that are very unfamiliar indeed.

I suspect that a deep connection exists between the anthropomorphic conception of explanation and the thesis that explanation consists in reduction of the unfamiliar to the familiar. The type of explanation with which we are best acquainted is that in which human action is explained in terms of conscious purposes. If it is possible to explain the phenomena of physics or biology in terms of attempting to realize a goal, that is a striking case of reduction to the familiar. A problem with this approach is, of course, that a great deal of the progress in scientific understanding has resulted in the elimination, not the injection, of purposes.

1.8 THE DEDUCTIVE-NOMOLOGICAL PATTERN OF SCIENTIFIC EXPLANATION

In a classic 1948 paper, Carl G. Hempel and Paul Oppenheim formulated, with great precision, one pattern of scientific explanation that is central to all discussions of the subject. It is known as the *deductive-nomological (D-N) model* of scientific explanation. Stated very simply, an explanation of this type explains by subsuming its explanandum-fact under a general law. This can best be appreciated by looking at an example.

A figure skater with arms outstretched stands balanced on one skate. Propelling herself with her other skate she begins to rotate slowly. She stops propelling herself, but she continues to rotate slowly for a few moments. Suddenly—without propelling herself again and without being propelled by any external object, such as another skater—she begins spinning very rapidly. Why? Because she drew her arms in close to her body, thus concentrating her total body mass closer to the axis of rotation. Because of the law of conservation of angular momentum, her rate of rotation had to increase to compensate for her more compact body configuration.

More technically, the angular momentum of an object is the product of its angular velocity (rate of rotation) and its moment of inertia. The moment of inertia depends upon the mass of the object and the average distance of the mass from the axis of rotation; for a fixed mass, the moment of inertia is smaller the more compactly the mass is distributed about the axis of rotation. The law of conservation of angular momentum says that the angular momentum of a body that is not being propelled or retarded by external forces does not change; hence, since the moment of inertia is decreased, the rate of rotation must increase to keep the value of the product constant.[6]

According to Hempel and Oppenheim, an explanation of the foregoing sort is to be viewed as a deductive argument. It can be set out more formally as follows:

(3) The angular momentum of any body (whose rate of rotation is not being increased or decreased by external forces) remains constant.
The skater is not interacting with any external object in such a way as to alter her angular velocity.
The skater is rotating (her angular momentum is not zero).
The skater reduces her moment of inertia by drawing her arms in close to her body.

The skater's rate of rotation increases.

The explanandum—the increase in the skater's rate of rotation—is the conclusion of the argument. The premises of the argument constitute the explanans. The first premise states a law of nature—the law of conservation of angular momentum. The remaining three premises state the antecedent conditions. The argument is logically correct; the conclusion follows validly from the premises. For purposes of our discussion, we may take the statements of antecedent conditions as true; expert figure skaters do this maneuver frequently. The law of conservation of angular momentum can also be regarded as true since it is a fundamental law of physics which has been confirmed by a vast quantity of empirical data.

Hempel and Oppenheim set forth four conditions of adequacy for D-N explanations:

---

[6] In this example we may ignore the friction of the skate on the ice, and the friction of the skater's body in the surrounding air.

1. The explanandum must be a logical consequence of the explanans; that is, the explanation must be a valid deductive argument.
2. The explanans must contain at least one general law, and it must actually be required for the derivation of the explanandum; in other words, if the law or laws were deleted, without adding any new premises, the argument would no longer be valid.
3. The explanans must have empirical content; it must be capable, at least in principle, of test by experiment or observation.
4. The sentences constituting the explanans must be true.

These conditions are evidently fulfilled by our example. The first three are classified as *logical* conditions of adequacy; the fourth is *empirical*. An argument that fulfills all four conditions is an explanation (for emphasis we sometimes say "true explanation"). An argument that fulfills the first three conditions, without necessarily fulfilling the fourth, is called a *potential explanation*. It is an argument that would be an explanation if its premises were true.[7]

According to Hempel and Oppenheim, it is possible to have D-N explanations, not only of particular occurrences as in argument (3), but also of general laws. For example, in the context of Newtonian mechanics, it is possible to set up the following argument:

(4) $F = ma$ (Newton's second law).
For every action there is an equal and opposite reaction (Newton's third law).

In every interaction, the total linear momentum of the system of interacting bodies remains constant (law of conservation of linear momentum).

This argument is valid, and among its premises are statements of general laws. There are no statements of antecedent conditions, but that is not a problem since the conditions of adequacy do not require them. Because we are not concerned to explain any particular facts, no premises regarding particular facts are needed. Both premises in the explanans are obviously testable, for they have been tested countless times. Thus, argument (4) fulfills the logical conditions of adequacy, and consequently, it qualifies as a potential explanation. Strictly speaking, it does not qualify as a true explanation, for we do not consider Newton's laws of motion literally true, but in many contexts they can be taken as correct because they provide extremely accurate approximations to the truth.

Although Hempel and Oppenheim discussed both deductive explanations of particular facts and deductive explanations of general laws, they offered a precise characterization only of the former, but not of the latter. They declined to attempt to provide a characterization of explanations of general laws because of a problem they recognized but did not know how to solve. Consider Kepler's laws of planetary motion K and Boyle's law of gases B. If, on the one hand, we conjoin the two to form

---

[7] Hempel and Oppenheim provide, in addition to these conditions of adequacy, a precise technical definition of "explanation." In this book we will not deal with these technicalities.

a law *K . B,* we can obviously deduce *K* from it. But this could not be regarded as an explanation of *K*, for it is only a pointless derivation of *K* from itself. On the other hand, the derivation of *K* from Newton's laws of motion and gravitation constitutes an extremely illuminating explanation of Kepler's laws. Hempel and Oppenheim themselves confessed that they were unable to provide any criterion to distinguish the pointless pseudoexplanations from the genuine explanations of laws (see Hempel and Oppenheim 1948 as reprinted in Hempel 1965b, 273, f.n. 33).

Hempel and Oppenheim envisioned two types of D-N explanation, though they were able to provide an account of only one of them. In addition, they remarked that other types of explanation are to be found in the sciences, namely, explanations that appeal, not to universal generalizations, but to statistical laws instead (ibid., 250–251).

Table 1.1 shows the four kinds of explanations to which Hempel and Oppenheim called attention; they furnished an account only for the type found in the upper left-hand box. Some years later Hempel (1962) offered an account of the I-S pattern in the lower left-hand box. In Hempel (1965b) he treated both the I-S and the D-S patterns. In 1948, Hempel and Oppenheim were looking forward to the time when theories of explanation dealing with all four boxes would be available.

TABLE 1.1

Explananda / Laws	Particular Facts	General Regularities
Universal Laws	D-N Deductive-Nomological	D-N Deductive-Nomological
Statistical Laws	I-S Inductive-Statistical	D-S Deductive-Statistical

1.9 WHAT ARE LAWS OF NATURE?

Hempel and Oppenheim emphasized the crucial role played by laws in scientific explanation; in fact, the D-N pattern is often called *the covering-law model.* As we will see, laws play a central part in other conceptions of scientific explanation as well. Roughly speaking, a *law* is a regularity that holds throughout the universe, at all places and all times. A *law-statement* is simply a statement to the effect that such a regularity exists. A problem arises immediately. Some regularities appear to be lawful and others do not. Consider some examples of laws:

(i) All gases, kept in closed containers of fixed size, exert greater pressure when heated.
(ii) In all closed systems the quantity of energy remains constant.
(iii) No signals travel faster than light.

Contrast these with the following:

(iv) All of the apples in my refrigerator are yellow.

(v) All Apache basketry is made by women.

(vi) No golden spheres have masses greater than 100,000 kilograms.

Let us assume, for the sake of argument, that all of the statements (i)–(vi) are true. The first thing to notice about them is their generality. Each of them has the overall form, "All *A* are *B*" or "No *A* are *B*." Statements having these forms are known as *universal generalizations*. They mean, respectively, "*Anything* that is an *A* is also a *B*" and "*Nothing* that is an *A* is also a *B*." Nevertheless, statements (i)–(iii) differ fundamentally from (iv)–(vi). Notice, for example, that none of the statements (i)–(iii) makes any reference to any particular object, event, person, time, or place. In contrast, statement (iv) refers to a particular person (me), a particular object (my refrigerator), and a particular time (now). This statement is not completely general since it singles out certain particular entities to which it refers. The same remark applies to statement (v) since it refers to a particular limited group of people (the Apache).

Laws of nature are generally taken to have two basic capabilities. First, they support counterfactual inferences. A *counterfactual statement* is a conditional statement whose antecedent is false. Suppose, for example, that I cut a branch from a tree and then, immediately, burn it in my fireplace. This piece of wood was never placed in water and never will be. Nevertheless, we are prepared to say, without hesitation, that *if it had been placed in water, it would have floated*. This italicized statement is a counterfactual conditional. Now, a law-statement, such as (i), will support a counterfactual assertion. We can say, regarding a particular sample of some gas, held in a closed container of fixed size but not actually being heated, that *if* it were heated it *would* exert greater pressure. We can assert the counterfactual because we take statement (i) to be a statement of a law of nature.

When we look at statement (iv) we see that it does not support any such counterfactual statement. Holding a red delicious apple in my hand, I cannot claim, on the basis of (iv), that this apple would be yellow if it were in my refrigerator.

A second capability of laws of nature is to support *modal* statements of physical necessity and impossibility. Statement (ii), the first law of thermodynamics, implies that it is impossible to create a perpetual motion machine of the first kind—that is, a machine that does useful work without any input of energy from an external source. In contrast, statement (v) does not support the claim that it is impossible for an Apache basket to be made by a male. It is physically possible that an Apache boy might be taught the art of basket making, and might grow up to make a career of basketry.

When we compare statements (iii) and (vi) more subtle difficulties arise. Unlike statements (iv) and (v), statement (vi) does not make reference to any particular entity or place or time.[8] It seems clear, nevertheless, that statement (vi)—even assuming it to be true—cannot support either modal statements or counterfactual

---

[8] If the occurrence of the kilogram in (vi) seems to make reference to a particular object—the international prototype kilogram kept at the international bureau of standards—the problem can easily be circumvented by defining mass in terms of atomic mass units.

conditionals. Even if we agree that nowhere in the entire history of the universe—past, present, or future—does there exist a gold sphere of mass greater than 100,000 kilograms, we would not be justified in claiming that it is *impossible* to fabricate a gold sphere of such mass. I once made a rough calculation of the amount of gold in the oceans of the earth, and it came to about 1,000,000 kilograms. If an incredibly rich prince were determined to impress a woman passionately devoted to golden spheres it would be physically possible for him to extract a little more than 100,000 kilograms from the sea to create a sphere that massive.

Statement (vi) also lacks the capacity to support counterfactual conditionals. We would not be justified in concluding that, if two golden hemispheres, each of 50,001 kilogram mass, were put together, they would not form a golden sphere of mass greater than 100,000 kilograms. To appreciate the force of this point, consider the following statement:

(vii) No enriched uranium sphere has a mass greater than 100,000 kilograms.

This *is* a lawful generalization, because the critical mass for a nuclear chain reaction is just a few kilograms. If 100,000 kilograms of enriched uranium were to be assembled, we would have a gigantic nuclear explosion. No comparable catastrophe would ensue, as far as we know, if a golden sphere of the same mass were put together.

Philosophers have often claimed that we can distinguish true generalizations that are *lawful* from those that are *accidental*. Even if we grant the truth of (vi), we must conclude that it is an accidental generalization. Moreover, they have maintained that among universal generalizations, regardless of truth, it is possible to distinguish *lawlike generalizations* from those that are not lawlike. A lawlike generalization is one that has all of the qualifications for being a law except, perhaps, being true.

It is relatively easy to point to the characteristic of statements (iv) and (v) that makes them nonlawlike, namely, that they make reference to particular objects, persons, events, places, or times. The nonlawlike character of statement (vi) is harder to diagnose. One obvious suggestion is to apply the criteria of supporting counterfactual and/or modal statements. We have seen that (vi) fails on that score. The problem with that approach is that it runs a serious risk of turning out to be circular. Consider statement (ii). Why do we consider it *physically impossible* to build a perpetual motion machine (of the first type)? Because to do so would violate a law of nature, namely (ii). Consider statement (vi). Why do we consider it *physically possible* to fabricate a golden sphere whose mass exceeds 100,000 kilograms? Because to do so would not violate a law of nature. It appears that the question of what modal statements to accept hinges on the question of what regularities qualify as laws of nature.

A similar point applies to the support of counterfactual conditionals. Consider statement (i). Given a container of gas that is not being heated, we can say that, if it were to be heated, it would exert increased pressure on the walls of its container—sufficient in many cases to burst the container. (I learned my lesson on this as a Boy Scout heating an unopened can of beans in a camp fire.) The reason that we can make such a counterfactual claim is that we can infer from statement (i) what would happen, and (i) states a law of nature. Similarly, from (iii) we can deduce that if something travels faster than light it is not a signal—that is, it cannot transmit information. You might think that this is vacuous because, as the theory of relativity tells us, nothing can travel faster than light. However, this opinion is incorrect. Shadows and various other kinds of "things" can easily be shown to travel faster than light. We can legitimately conclude that, if something does travel faster than light, it is not functioning as a signal, because (iii) is, indeed, a law of nature.

What are the fundamental differences between statement (vi) on the one hand and statements (i)–(iii) and (vii) on the other? The main difference seems to be that (i)–(iii) and (vii) are all deeply embedded in well-developed scientific theories, and that they have been, directly or indirectly, extensively tested. This means that (i)–(iii) and (vii) have a very different status within our body of scientific knowledge than do (iv)–(vi). The question remains, however, whether the regularities described by (i)–(iii) and (vii) have a different status in the physical universe than do (iv)–(vi).

At the very beginning of this chapter, we considered the explanation of the fact that the moon assumes a coppery hue during total eclipse. This is a regularity found in nature, but is it a lawful regularity? Is the statement, "The moon turns a coppery color during total eclipses," a law-statement? The immediate temptation is to respond in the negative, for the statement makes an explicit reference to a particular object, namely, our moon. But if we reject that statement as a lawful generalization, it would seem necessary to reject Kepler's laws of planetary motion as well, for they make explicit reference to our solar system. Galileo's law of falling bodies would also have to go, for it refers to things falling near the surface of the earth. It would be unreasonable to disqualify all of them as laws.

We can, instead, make a distinction between basic and derived laws. Kepler's laws and Galileo's law can be derived from Newton's laws of motion and gravitation, in conjunction with descriptions of the solar system and the bodies that make it up. Newton's laws are completely general and make no reference to any particular person, object, event, place, or time. The statement about the color of the moon during total eclipse can be derived from the laws of optics in conjunction with a description of the earth's atmosphere and the configuration of the sun, moon, and earth when an eclipse occurs. The statement about the color of the moon can also be taken as a derivative law. But what about statements (iv) and (v)? The color of the apples in my refrigerator can in no way be derived from basic laws of nature in conjunction with a description of the refrigerator. No matter how fond I may be of golden delicious apples, there is no physical impossibility of a red delicious getting into my refrigerator. Similarly, there are no laws of nature from which, in conjunction with descriptions of the Apache and their baskets, it would be possible to derive that they can only be made by women.

1.10 PROBLEMS FOR THE D-N PATTERN OF EXPLANATION

Quite remarkably the classic article by Hempel and Oppenheim received virtually no attention for a full decade. Around 1958, however, a barrage of criticism began and a lively controversy ensued. Much of the criticism was brought into sharp focus by means of counterexamples that have, themselves, become classic. These examples fall into two broad categories. The first consists of arguments that fulfill all of the requirements for D-N explanation, yet patently fail to qualify as bona fide explanations. They show that the requirements set forth by Hempel and Oppenheim are not *sufficient* to determine what constitutes an acceptable scientific explanation. The second consists of examples of allegedly bona fide explanations that fail to fulfill the Hempel-Oppenheim requirements. They are meant to show that it is not *necessary* to fulfill those requirements in order to have correct explanations. We must treat this second category with care, for Hempel and Oppenheim never asserted that all correct explanations fit the D-N pattern. They explicitly acknowledged that legitimate statistical explanations can be found in science. So, statistical explanations are not appropriate as counterexamples. However, the attempt has been to find examples that are clearly not statistical, but which fail to fulfill the Hempel-Oppenheim criteria. Let us look at some counterexamples of each type.

CE-1. The flagpole and its shadow.[9] On a flat and level piece of ground stands a flagpole that is 12' tall. The sun, which is at an elevation of 53.13° in the sky, shines brightly. The flagpole casts a shadow that is 9' long. If we ask why the shadow has that length, it is easy to answer. From the elevation of the sun, the height of the flagpole, and the rectilinear propagation of light, we can deduce, with the aid of a bit of trigonometry, the length of the shadow. The result is a D-N explanation that most of us would accept as correct. So far, there is no problem.

If, however, someone asks why the flagpole is 12' tall, we could construct essentially the same argument as before. But instead of deducing the length of the shadow from the height of the flagpole and the elevation of the sun, we would deduce the height of the flagpole from the length of the shadow and the elevation of the sun. Hardly anyone would regard that argument, which satisfies all of the requirements for a D-N explanation, as an adequate explanation of the height of the flagpole.

We can go one step farther. From the length of the shadow and the height of the flagpole, using a similar argument, we can deduce that the sun is at an elevation of 53.13°. It seems most unreasonable to say that the sun is that high in the sky because a 12' flagpole casts a 9' shadow. From the fact that a 12' flagpole casts a 9' shadow we can infer *that* the sun is that high in the sky, but we cannot use those data to explain *why* it is at that elevation. Here we must be sure to remember the distinction between confirmation and explanation (discussed in Section 1.1). The explanation of the elevation rests upon the season of the year and the time of day.

The moral: The reason it is legitimate to explain the length of the shadow in terms of the height of the flagpole and the elevation of the sun is that the shadow is the effect of those two causal factors. We can explain effects by citing their causes. The reason it is illegitimate to explain the height of the flagpole by the length of the shadow is that the length of the shadow is an effect of the height of the flagpole (given the elevation of the sun), but it is no part of the cause of the height of the flagpole. We cannot explain causes in terms of their effects. Furthermore, although the eleva-

---

[9] The counterexample was devised by Sylvain Bromberger, but to the best of my knowledge he never published it.

tion of the sun is a crucial causal factor in the relation between the height of the flagpole and the length of the shadow, the flagpole and its shadow play no causal role in the position of the sun in the sky.

CE-2. The barometer and the storm.   Given a sharp drop in the reading on a properly functioning barometer, we can predict that a storm will shortly occur. Nevertheless, the reading on the barometer does not explain the storm. A sharp drop in atmospheric pressure, which is registered on the barometer, explains both the storm and the barometric reading.

The moral: Many times we find two effects of a common cause that are correlated with one another. In such cases we do not explain one effect by means of the other. The point is illustrated also by diseases. A given illness may have many different symptoms. The disease explains the symptoms; one symptom does not explain another.

CE-3. A solar eclipse.   From the present positions of the earth, moon, and sun, using laws of celestial mechanics, astronomers can predict a future total eclipse of the sun. After the eclipse has occurred, the very same data, laws, and calculations provide a legitimate D-N explanation of the eclipse. So far, so good. However, using the same laws and the same positions of the earth, moon, and sun, astronomers can retrodict the previous occurrence of a solar eclipse. The argument by which this retrodiction is made fulfills the requirements for a D-N explanation just as fully as does the prediction of the eclipse. Nevertheless, most of us would say that, while it is possible to explain an eclipse in terms of antecedent conditions, it is not possible to explain an eclipse in terms of subsequent conditions.

The moral: We invoke earlier conditions to explain subsequent facts; we do not invoke later conditions to explain earlier facts. The reason for this asymmetry seems to lie in the fact that causes, which have explanatory import, precede their effects—they do not follow their effects.

CE-4. The man and the pill.   A man explains *his* failure to become pregnant during the past year on the ground that he has regularly consumed his wife's birth control pills, and that any man who regularly takes oral contraceptives will avoid getting pregnant.

The moral: This example shows that it is possible to construct valid deductive arguments with true premises in which some fact asserted by the premises is actually irrelevant. Since men do not get pregnant regardless, the fact that this man took birth control pills is irrelevant. Nevertheless, it conforms to the D-N pattern.

Counterexamples CE-1–CE-4 are all cases in which an argument that fulfills the Hempel-Oppenheim requirements manifestly fails to constitute a bona fide explanation. They were designed to show that these requirements are too weak to sort out the illegitimate explanations. A natural suggestion would be to strengthen them in ways that would rule out counterexamples of these kinds. For example, CE-1 and CE-2 could be disqualified if we stipulated that the antecedent conditions cited in the explanans must be causes of the explanandum. CE-3 could be eliminated by insisting that the so-called antecedent conditions must actually obtain prior to the explanandum. And CE-4 could be ruled out by stipulating that the antecedent conditions must be relevant to the explanandum. For various reasons Hempel declined to strengthen the requirements for D-N explanation in such ways.

The next counterexample has been offered as a case of a legitimate explanation that does not meet the Hempel-Oppenheim requirements.

CE-5. The ink stain.   On the carpet, near the desk in Professor Jones's office, is an unsightly black stain. How does he explain it? Yesterday, an open bottle of black ink stood on his desk, near the corner. As he went by he accidentally bumped it with his elbow, and it fell to the floor, spilling ink on the carpet. This seems to be a perfectly adequate explanation; nevertheless, it does not incorporate any laws. Defenders of the D-N pattern would say that this is simply an incomplete explanation, and that the laws are tacitly assumed. Michael Scriven, who offered this example, argued that the explanation is clear and complete as it stands, and that any effort to spell out the laws and initial conditions precisely will meet with failure.

The moral: It is possible to have perfectly good explanations without any laws. The covering law conception is not universally correct.

The fifth counterexample raises profound problems concerning the nature of causality. Some philosophers, like Scriven, maintain that one event, such as the bumping of the ink bottle with the elbow, is obviously the cause of another event, such as the bottle falling off of the desk. Moreover, they claim, to identify the cause of an event is all that is needed to explain it. Other philosophers, including Hempel, maintain that a causal relation always involves (sometimes explicitly, sometimes implicitly) a general causal law. In the case of the ink stain, the relevant laws would include the laws of Newtonian mechanics (in explaining the bottle being knocked off the desk and falling to the floor) and some laws of chemistry (in explaining the stain on the carpet as a result of spilled ink).

1.11 TWO PATTERNS OF STATISTICAL EXPLANATION

Anyone who is familiar with any area of science—physical, biological, or social—realizes, as Hempel and Oppenheim had already noted, that not all explanations are of the deductive-nomological variety. Statistical laws play an important role in virtually every branch of contemporary science and statistical explanations—those falling into the two lower boxes in Table 1.1—are frequently given. In 1965b Hempel published a comprehensive essay, "Aspects of Scientific Explanation," in which he offered a theory of statistical explanation encompassing both types.

In the first type of statistical explanation, the *deductive-statistical (D-S) pattern*, statistical regularities are explained by deduction from more comprehensive statistical laws. Many examples can be found in contemporary science. For instance, archaeologists use the radiocarbon dating technique to ascertain the ages of pieces of wood or charcoal discovered in archaeological sites. If a piece of wood is found to have a concentration of C^{14} (a radioactive isotope of carbon) equal to one-fourth that of newly cut wood, it is inferred to be 11,460 years old. The reason is that the half-life of C^{14} is 5730 years, and in two half-lives it is extremely probable that about three-fourths of the C^{14} atoms will have decayed. Living trees replenish their supplies of C^{14} from the atmosphere; wood that has been cut cannot do so. Here is the D-S explanation:

(5) Every C^{14} atom (that is not exposed to external radiation[10]) has a probability of ½ of disintegrating within any period of 5730 years.

In any large collection of C^{14} atoms (that are not exposed to external radiation) approximately three-fourths will *very probably* decay within 11,460 years.

This derivation constitutes a deductive explanation of the probabilistic generalization that stands as its conclusion.

Deductive-statistical explanations are very similar, logically, to D-N explanations of generalizations. The only difference is that the explanation is a statistical law and the explanans must contain at least one statistical law. Universal laws have the form "All A are B" or "No A are B"; statistical laws say that a certain proportion of A are B.[11] Accordingly, the problem that plagued D-N explanations of universal generalizations also infects D-S explanations of statistical generalizations. Consider, for instance, one of the statistical generalizations in the preceding example—namely, that the half-life of C^{14} is 5730 years. There is a bona fide explanation of this generalization from the basic laws of quantum mechanics in conjunction with a description of the C^{14} nucleus. However, this statistical generalization can also be deduced from the conjunction of itself with Kepler's laws of planetary motion. This deduction would not qualify as any kind of legitimate explanation; like the case cited in Section 1.8, it would simply be a pointless derivation of the generalization about the half-life of C^{14} from itself.

Following the 1948 article, Hempel never returned to this problem concerning explanations of laws; he did not address it in Hempel (1965a), which contains characterizations of all four types of explanation represented in Table 1.1. This leaves both boxes on the right-hand side of Table 1.1 in a highly problematic status. Nevertheless, it seems clear that many sound explanations of both of these types can be found in the various sciences.

The second type of statistical explanation—the *inductive-statistical (I-S) pattern*—explains particular occurrences by subsuming them under statistical laws, much as D-N explanations subsume particular events under universal laws. Let us look at one of Hempel's famous examples. If we ask why Jane Jones recovered rapidly from her streptococcus infection, the answer is that she was given a dose of penicillin, and almost all strep infections clear up quickly upon administration of penicillin. More formally:

(6) Almost all cases of streptococcus infection clear up quickly after the administration of penicillin.
Jane Jones had a streptococcus infection.

[10] This qualification is required to assure that the disintegration is spontaneous and not induced by external radiation.

[11] As James Lennox remarks in Chapter 7 on philosophy of biology, Darwin's principle of natural selection is an example of a statistical law.

Jane Jones received treatment with penicillin.
$$===[r]$$
Jane Jones recovered quickly.

This explanation is an argument that has three premises (the explanans); the first premise states a statistical regularity—a statistical law—while the other two state antecedent conditions. The conclusion (the explanandum) states the fact-to-be-explained. However, a crucial difference exists between explanations (3) and (6): D-N explanations subsume the events to be explained deductively, while I-S explanations subsume them inductively. The single line separating the premises from the conclusion in (3) signifies a relation of deductive entailment between the premises and conclusion. The double line in (6) represents a relationship of inductive support, and the attached variable r stands for the strength of that support. This strength of support may be expressed exactly, as a numerical value of a probability, or vaguely, by means of such phrases as "very probably" or "almost certainly."

An explanation of either of these two kinds can be described as an argument to the effect that *the event to be explained was to be expected by virtue of certain explanatory facts*. In a D-N explanation, the event to be explained is deductively certain, given the explanatory facts; in an I-S explanation the event to be explained has high inductive probability relative to the explanatory facts. This feature of expectability is closely related to the *explanation-prediction symmetry thesis* for explanations of particular facts. According to this thesis any acceptable explanation of a particular fact is an argument, deductive or inductive, that could have been used to predict the fact in question if the facts stated in the explanans had been available prior to its occurrence.[12] As we shall see, this symmetry thesis met with serious opposition.

Hempel was not by any means the only philosopher in the early 1960s to notice that statistical explanations play a significant role in modern science. He was, however, the first to present a detailed account of the nature of statistical explanation, and the first to bring out a fundamental problem concerning statistical explanations of particular facts. The case of Jane Jones and her quick recovery can be used as an illustration. It is well known that certain strains of the streptococcus bacterium are penicillin-resistant, and if Jones's infection were of that type, the probability of her quick recovery after treatment with penicillin would be small. We could, in fact, set up the following inductive argument:

(7) Almost no cases of penicillin-resistant streptococcus infection clear up quickly after the administration of penicillin.
Jane Jones had a penicillin-resistant streptococcus infection.
Jane Jones received treatment with penicillin.
$$===[q]$$
Jane Jones did not recover quickly.

The remarkable fact about arguments (6) and (7) is that their premises are mutually compatible—they could all be true. Nevertheless, their conclusions contra-

---

[12] This thesis was advanced for D-N explanation in Hempel-Oppenheim (1948, 249), and reiterated, with some qualifications, for D-N and I-S explanations in Hempel (1965a, Sections 2.4, 3.5).

dict one another. This is a situation that can never occur with deductive arguments. Given two valid deductions with incompatible conclusions, their premises must also be incompatible. Thus, the problem that has arisen in connection with I-S explanations has no analog in D-N explanations. Hempel called this *the problem of ambiguity of I-S explanation.*

The source of the problem of ambiguity is a simple and fundamental difference between universal laws and statistical laws. Given the proposition that all *A* are *B*, it follows immediately that all things that are both *A* and *C* are *B*. If all humans are mortal, then all people who are over six feet tall are mortal. However, even if almost all humans who are alive now will be alive five years from now, *it does not follow* that almost all living humans with advanced cases of pancreatic cancer will be alive five years hence. As we noted in Section 1.5, there is a parallel fact about arguments. Given a valid deductive argument, the argument will remain valid if additional premises are supplied, as long as none of the original premises is taken away. Deduction is erosion-proof. Given a strong inductive argument—one that supports its conclusion with a high degree of probability—the addition of one more premise may undermine it completely. For centuries Europeans had a great body of inductive evidence to support the proposition that all swans are white, but one true report of a black swan in Australia completely refuted that conclusion. Induction is not erosion-proof.

Hempel sought to resolve the problem of ambiguity by means of his *requirement of maximal specificity (RMS)*. It is extremely tricky to state RMS with precision, but the basic idea is fairly simple. In constructing I-S explanations we must include all relevant knowledge we have that would have been available, in principle, prior to the explanandum-fact. If the information that Jones's infection is of the penicillin-resistant type is available to us, argument (6) would not qualify as an acceptable I-S explanation.[13]

In Section 1.8 we stated Hempel and Oppenheim's four conditions of adequacy for D-N explanations. We can now generalize these conditions so that they apply both to D-N and I-S explanations as follows:

1. The explanation must be an argument having correct (deductive or inductive) logical form.
2. The explanans must contain at least one general law (universal or statistical), and this law must actually be required for the derivation of the explanandum.
3. The explanans must have empirical content; it must be capable, at least in principle, of test by experiment or observation.
4. The sentences constituting the explanans must be true.
5. The explanation must satisfy the requirement of maximal specificity.[14]

---

[13] Nor would (6) qualify as an acceptable I-S explanation if we had found that Jones's infection was of the non-penicillin-resistant variety, for the probability of quick recovery among people with that type of infection is different from the probability of quick recovery among those who have an unspecified type of streptococcus infection.

[14] D-N explanations of particular facts automatically satisfy this requirement. If all *A* are *B*, the probability that an *A* is a *B* is one. Under those circumstances, the probability that an *A* which is also a *C* is a *B* is also one. Therefore, no partition of *A* is relevant to *B*.

The theory of scientific explanation developed by Hempel in his "Aspects" essay won rather wide approval among philosophers of science. During the mid-to-late 1960s and early 1970s it could appropriately be considered *the received view of scientific explanation*. According to this view, every legitimate scientific explanation must fit the pattern corresponding to one or another of the four boxes in Table 1.1.

1.12 CRITICISMS OF THE I-S PATTERN OF SCIENTIFIC EXPLANATION

We noticed in Section 1.10 that major criticisms of the D-N pattern of scientific explanation can be posed by means of well-known counterexamples. The same situation arises in connection with the I-S pattern. Consider the following:

CE-6. Psychotherapy. Suppose that Bruce Brown has a troublesome neurotic symptom. He undergoes psychotherapy and his symptom disappears. Can we explain his recovery in terms of the treatment he has undergone? We could set out the following inductive argument, in analogy with argument (6):

(8) Most people who have a neurotic symptom of type N and who undergo psychotherapy experience relief from that symptom.
Bruce Brown had a symptom of type N and he underwent psychotherapy.
$$\overline{\hspace{9cm}}[r]$$
Bruce Brown experienced relief from his symptom.

Before attempting to evaluate this proffered explanation we should take account of the fact that there is a fairly high spontaneous remission rate—that is, many people who suffer from that sort of symptom get better regardless of treatment. No matter how large the number r, if the rate of recovery for people who undergo psychotherapy is no larger than the spontaneous remission rate, it would be a mistake to consider argument (8) a legitimate explanation. A high probability is not *sufficient* for a correct explanation. If, however, the number r is not very large, but is greater than the spontaneous remission rate, the fact that the patient underwent psychotherapy has at least some degree of explanatory force. A high probability is not *necessary* for a sound explanation.

Another example reinforces the same point.

CE-7. Vitamin C and the common cold.[15] Suppose someone were to claim that large doses of vitamin C would produce rapid cures for the common cold. To ascertain the efficacy of vitamin C in producing rapid recovery from colds, we should note, it is *not* sufficient to establish that most people recover quickly; most colds disappear within a few days regardless of treatment. What is required is a double-

---

[15] Around the time Hempel was working out his theory of I-S explanation, Linus Pauling's claims about the value of massive doses of vitamin C in the prevention of common colds was receiving a great deal of attention. Although Pauling made no claims about the ability of vitamin C to cure colds, it occurred to me that a fictitious example of this sort could be concocted.

blind controlled experiment[16] in which the rate of quick recovery for those who take vitamin C is compared with the rate of quick recovery for those who receive only a placebo. If there is a significant difference in the probability of quick recovery for those who take vitamin C and for those who do not, we may conclude that vitamin C has some degree of causal efficacy in lessening the duration of colds. If, however, there is no difference between the two groups, then it would be a mistake to try to explain a person's quick recovery from a cold by constructing an argument analogous to (6) in which that result is attributed to treatment with vitamin C.

The moral: CE-6 and CE-7 call attention to the same point as CE-4 (the man and the pill). All of them show that something must be done to exclude irrelevancies from scientific explanations. If the rate of pregnancy among men who consume oral contraceptives is the same as for men who do not, then the use of birth control pills is causally and explanatorily irrelevant to pregnancy among males. Likewise, if the rate of relief from neurotic symptoms is the same for those who undergo psychotherapy as it is for those who do not, then psychotherapy is causally and explanatorily irrelevant to the relief from neurotic symptoms. Again, if the rate of rapid recovery from common colds is the same for those who do and those who do not take massive doses of vitamin C, then consumption of massive doses of vitamin C is causally and explanatorily irrelevant to rapid recovery from colds.[17] Hempel's requirement of maximal specificity was designed to insure that *all* relevant information (of a suitable sort) is included in I-S explanations. What is needed in addition is a requirement insuring that *only* relevant information is included in D-N *or* I-S explanations.

CE-8. Syphilis and paresis.   Paresis is a form of tertiary syphilis which can be contracted only by people who go through the primary, secondary, and latent forms of syphilis without treatment with penicillin. If one should ask why a particular person suffers from paresis, a correct answer is that he or she was a victim of untreated latent syphilis. Nevertheless, only a small proportion of those with untreated latent syphilis—about 25%—actually contract paresis. Given a randomly selected member of the class of victims of untreated latent syphilis, one should predict that that person *will not* develop paresis.

The moral: there are legitimate I-S explanations in which the explanans *does not* render the explanandum highly probable. CE-8 responds to the explanation-prediction symmetry thesis—the claim that an explanation is an argument of such a sort that it could have been used to predict the explanandum if it had been available prior to the

---

[16] In a controlled experiment there are two groups of subjects, the experimental group and the control group. These groups should be as similar to one another as possible. The members of the experimental group receive the substance being tested, vitamin C. The members of the control group receive a placebo, that is, an inert substance such as a sugar pill that is known to have no effect on the common cold. In a blind experiment the subjects do not know whether they are receiving vitamin C or the placebo. This is important, for if the subjects knew which treatment they were receiving, the power of suggestion might skew the results. An experiment is double-blind if neither the person who hands out the pills nor the subjects know which subject is getting which type of pill. If the experiment is not double-blind, the person administering the pills might, in spite of every effort not to, convey some hint to the subject.

[17] It should be carefully noted that I am claiming *neither* that psychotherapy is irrelevant to remission of neurotic symptoms *nor* that vitamin C is irrelevant to rate of recovery from colds. I *am* saying that that is the point at issue so far as I-S explanation is concerned.

fact-to-be-explained. It is worth noting, in relation to CE-6 and CE-7, that untreated latent syphilis is highly relevant to the occurrence of paresis, although it does not make paresis highly probable, or even more probable than not.

CE-9. The biased coin.   Suppose that a coin is being tossed, and that it is highly biased for heads—in fact, on any given toss, the probability of getting heads is 0.95, while the probability of tails is 0.05. The coin is tossed and comes up heads. We can readily construct an I-S explanation fitting all of the requirements. But suppose it comes up tails. In this case an I-S explanation is out of the question. Nevertheless, to the degree that we understand the mechanism involved, and consequently the probable outcome of heads, to that same degree we understand the improbable outcome, even though it occurs less frequently.

The moral: If we are in a position to construct statistical explanations of events that are highly probable, then we also possess the capability of framing statistical explanations of events that are extremely improbable.

1.13 DETERMINISM, INDETERMINISM, AND STATISTICAL EXPLANATION

When we look at an I-S explanation such as (6), there is a strong temptation to regard it as incomplete. It may, to be sure, incorporate all of the relevant knowledge we happen to possess. Nevertheless, we may feel, it is altogether possible that medical science will discover enough about streptococcus infections and about penicillin treatment to be able to determine precisely which individuals with strep infections will recover quickly upon treatment with penicillin and which individuals will not. When that degree of knowledge is available we will not have to settle for I-S explanations of rapid recoveries from strep infections; we will be able to provide D-N explanations instead. Similar remarks can also be made about several of the counterexamples—in particular, examples CE-6–CE-9.

Consider CE-8, the syphilis-paresis example. As remarked above, with our present state of knowledge we can predict that about 25% of all victims of untreated latent syphilis contract paresis, but we do not know how to distinguish those who will develop paresis from those who will not. Suppose Sam Smith develops paresis. At this stage of our knowledge the best we can do by way of an I-S explanation of Smith's paresis is the following:

(9)   25% of all victims of untreated latent syphilis develop paresis.
       Smith had untreated latent syphilis.
       ===[.25]
       Smith contracted paresis.

This could not be accepted as an I-S explanation because of the weakness of the relation of inductive support.

Suppose that further research on the causes of paresis reveals a factor in the blood—call it the *P*-factor—which enables us to pick out, with fair reliability—say

95%—those who will develop paresis. Given that Smith has the *P*-factor, we can construct the following argument:

(10) 95% of all victims of untreated latent syphilis who have the *P*-factor develop paresis.
      Smith had untreated latent syphilis.
      Smith had the *P*-factor.
      ===[.95]
      Smith developed paresis.

In the knowledge situation just described, this would count as a pretty good I-S explanation, for 0.95 is fairly close to 1.

Let us now suppose further that additional medical research reveals that, among those victims of untreated latent syphilis who have the *P*-factor, those whose spinal fluid contains another factor *Q* invariably develop paresis. Given that information, and the fact that Smith has the *Q*-factor, we can set up the following explanation:

(11) All victims of untreated latent syphilis who have the *P*-factor and the *Q*-factor develop paresis.
      Smith had untreated latent syphilis.
      Smith had the *P*-factor.
      Smith had the *Q*-factor.
      _____

      Smith developed paresis.

If the suppositions about the *P*-factor and the *Q*-factor were true, this argument would qualify as a correct D-N explanation. We accepted (10) as a correct explanation of Smith's paresis only because we were lacking the information that enabled us to set up (11).

Determinism is the doctrine that says that everything that happens in our universe is completely determined by prior conditions.[18] If this thesis is correct, then each and every event in the history of the universe—past, present, or future—is, in principle, deductively explainable. If determinism is true, then every sound I-S explanation is merely an incomplete D-N explanation. Under these circumstances, the I-S pattern is not really a stand-alone type of explanation; all fully correct explanations fit the D-N pattern. The lower left-hand box of Table 1.1 would be empty. This does *not* mean that I-S explanations—that is, incomplete D-N explanations—are useless, only that they are incomplete.

Is determinism true? We will not take a stand on that issue in this chapter. Modern physics—quantum mechanics in particular—seems to offer strong reasons to believe that determinism is false, but not everyone agrees with this interpretation. However, we will take the position that indeterminism *may* be true, and see what the consequences are with respect to statistical explanation.

According to most physicists and philosophers of physics, the spontaneous disintegration of the nucleus of an atom of a radioactive substance is a genuinely indeterministic happening. Radioactive decay is governed by laws, but they are

---

[18] Determinism is discussed in detail in Chapter 6.

fundamentally and irreducibly statistical. Any C^{14} atom has a fifty-fifty chance of spontaneously disintegrating within the next 5730 years and a fifty-fifty chance of not doing so. Given a collection of C^{14} atoms, the probability is overwhelming that some will decay and some will not in the next 5730 years. However, no way exists, even in principle, to select in advance those that will. No D-N explanation of the decay of any such atom can possibly be constructed; however, I-S explanations can be formulated. For example, in a sample containing 1 milligram of C^{14} there are approximately 4×10^{19} atoms. If, in a period of 5730 years, precisely half of them decayed, approximately 2×10^{19} would remain intact. It is *extremely unlikely* that *exactly* half of them would disintegrate in that period, but it is *extremely likely* that *approximately* half would decay. The following argument—which differs from (5) by referring to one particular sample *S*—would be a strong I-S explanation:

(12) *S* is a sample of C^{14} that contained one milligram 5730 years ago.
  S has not been exposed to external radiation.[19]
  The half-life of C^{14} is 5730 years.
  ═══[*r*]
  S now contains one-half milligram ($\pm 1\%$) of C^{14}.

In this example, *r* differs from 1 by an incredibly tiny margin, but is not literally equal to 1. In a world that is not deterministic, I-S explanations that are not merely incomplete D-N explanations can be formulated.

1.14 THE STATISTICAL RELEVANCE (S-R) MODEL OF EXPLANATION

According to the received view, scientific explanations are arguments; each type of explanation in Table 1.1 is some type of argument satisfying certain conditions. For this reason, we can classify the received view as an *inferential conception* of scientific explanation. Because of certain difficulties, associated primarily with I-S explanation, another pattern for statistical explanations of particular occurrences was developed. A fundamental feature of this model of explanation is that it *does not* construe explanations as arguments.

One of the earliest objections to the I-S pattern of explanation—as shown by CE-6 (psychotherapy) and CE-7 (vitamin C and the common cold)—is that statistical relevance rather than high probability is the crucial relationship in statistical explanations. Statistical relevance involves a relationship between two different probabilities. Consider the psychotherapy example. Bruce Brown is a member of the class of people who have a neurotic symptom of type *N*. Within that class, regardless of what the person does in the way of treatment or nontreatment, there is a certain probability of relief from the symptom (*R*). That is the *prior probability* of recovery; let us symbolize it as "*Pr(R/N)*." Then there is a probability of recovery in the class of

---

[19] This qualification is required to assure that the disintegrations have been spontaneous and not induced by external radiation.

people with that symptom who undergo psychotherapy (P); it can be symbolized as "$Pr(R/N \cdot P)$." If

$$Pr(R/N \cdot P) > Pr(R/N)$$

then psychotherapy is positively relevant to recovery, and if

$$Pr(R/N \cdot P) < Pr(R/N)$$

then psychotherapy is negatively relevant to recovery. If

$$Pr(R/N \cdot P) = Pr(R/N)$$

then psychotherapy is irrelevant to recovery. Suppose psychotherapy is positively relevant to recovery. If someone then asks why Bruce Brown, who suffered with neurotic symptom N, recovered from his symptom, we can say that it was because he underwent psychotherapy. That is at least an important part of the explanation.

Consider another example. Suppose that Grace Green, an American woman, suffered a serious heart attack. In order to explain why this happened we search for factors that are relevant to serious heart attacks—for example, smoking, high cholesterol level, and body weight. If we find that she was a heavy cigarette smoker, had a serum cholesterol level above 300, and was seriously overweight, we have at least a good part of an explanation, for all of those factors are positively relevant to serious heart attacks. There are, of course, other relevant factors, but these three will do for purposes of illustration.

More formally, if we ask why this member of the class A (American women) has characteristic H (serious heart attack), we can take the original reference class A and subdivide or *partition* it in terms of such relevant factors as we have mentioned: S (heavy cigarette smokers), C (high cholesterol level), and W (overweight). This will give us a partition with eight cells (where the dot signifies conjunction and the tilde "$\sim$" signifies negation):

$S \cdot C \cdot W$	$\sim S \cdot C \cdot W$
$S \cdot C \cdot \sim W$	$\sim S \cdot C \cdot \sim W$
$S \cdot \sim C \cdot W$	$\sim S \cdot \sim C \cdot W$
$S \cdot \sim C \cdot \sim W$	$\sim S \cdot \sim C \cdot \sim W$

An S-R explanation of Green's heart attack has three parts:

1. The prior probability of H, namely, $Pr(H/A)$.
2. The posterior probabilities of H with respect to each of the eight cells, $Pr(H/ S \cdot C \cdot W)$, $Pr(H/S \cdot C \cdot \sim W)$, $\ldots$, $Pr(H/ \sim S \cdot \sim C \cdot \sim W)$.
3. The statement that Green is a member of $S \cdot C \cdot W$.

It is stipulated that the partition of the reference class must be made in terms of all and only the factors relevant to serious heart attacks.

Clearly, an explanation of that sort is not an argument; it has neither premises nor conclusion. It does, of course, consist of an explanans and an explanandum.

Items 1–3 constitute the explanans; the explanandum is Green's heart attack. Moreover, no restrictions are placed on the size of the probabilities—they can be high, middling, or low. All that is required is that these probabilities differ from one another in various ways, because we are centrally concerned with relations of statistical relevance.

Although the S-R pattern of scientific explanation provides some improvements over the I-S model, it suffers from a fundamental inadequacy. It focuses on statistical relevance rather than causal relevance. It may, as a result, tend to foster a confusion of causes and correlations. In the vitamin C example, for instance, we want a controlled experiment to find out whether taking massive doses of vitamin C is *causally relevant* to quick recovery from colds. We attempt to find out whether taking vitamin C is *statistically relevant* to rapid relief because the statistical relevance relation is evidence regarding the presence or absence of *causal relevance*. It is causal relevance that has genuine explanatory import. The same remark applies to other examples as well. In the psychotherapy example we try to find out whether such treatment is statistically relevant to relief from neurotic symptoms in order to tell whether it is causally relevant. In the case of the heart attack, many clinical studies have tried to find statistical relevance relations as a basis for determining what is causally relevant to the occurrence of serious heart attacks.

1.15 TWO GRAND TRADITIONS

We have been looking at the development of the received view, and at some of the criticisms that have been leveled against it. The strongest intuitive appeal of that view comes much more from explanations of laws than from explanations of particular facts. One great example is the *Newtonian synthesis*. Prior to Newton we had a miscellaneous collection of laws including Kepler's three laws of planetary motion and Galileo's laws of falling objects, inertia, projectile motion, and pendulums. By invoking three simple laws of motion and one law of gravitation, Newton was able to explain these laws—and in some cases correct them. In addition, he was able to explain many other regularities, such as the behavior of comets and tides, as well. Later on, the molecular-kinetic theory provided a Newtonian explanation of many laws pertaining to gases. Quite possibly the most important feature of the Newtonian synthesis was the extent to which it systematized our knowledge of the physical world by subsuming all sorts of regularities under a small number of very simple laws. Another excellent historical example is the explanation of light by subsumption under Maxwell's theory of electromagnetic radiation.

The watchword in these beautiful historical examples is *unification*. A large number of specific regularities are unified in one theory with a small number of assumptions or postulates. This theme was elaborated by Michael Friedman (1974) who asserted that our comprehension of the universe is increased as the number of independently acceptable assumptions we require is reduced. I would be inclined to add that this sort of systematic unification of our scientific knowledge provides a comprehensive world picture or worldview. This, I think, represents one major aspect of scientific explanation—it is the notion that we understand what goes on in the world if we can fit it into a comprehensive worldview. As Friedman points out, this is a *global* conception of explanation. The value of explanation lies in fitting things into a universal pattern, or a pattern that covers major segments of the universe.[20]

As we look at many of the criticisms that have been directed against the received view, it becomes clear that causality is a major focus. Scriven offered his ink stain example, CE-5, to support the claim that finding the explanation amounts, in many cases, simply to finding the causes. This is clearly explanation on a very *local* level. All we need to do, according to Scriven, is to get a handle on events in an extremely limited spacetime region that led up, causally, to the stain on the carpet, and we have adequate understanding of that particular fact. In this connection, we should also recall CE-1 and CE-2. In the first of these we sought a local causal explanation for the length of a shadow, and in the second we wanted a causal explanation for a particular storm. Closely related noncausal ''explanations'' were patently unacceptable. In such cases as the Chernobyl accident and the Challenger space-shuttle explosion we also seek causal explanations, partly in order to try to avoid such tragedies in the future. Scientific explanation has its practical as well as its purely intellectual value.

It often happens, when we try to find causal explanations for various occurrences, that we have to appeal to entities that are not directly observable with the unaided human senses. For example, to understand AIDS (Acquired Immunodeficiency Syndrome), we must deal with viruses and cells. To understand the transmission of traits from parents to offspring, we become involved with the structure of the DNA molecule. To explain a large range of phenomena associated with the nuclear accident at Three Mile Island, we must deal with atoms and subatomic particles. When we try to construct causal explanations we are attempting to discover the mechanisms—often hidden mechanisms—that bring about the facts we seek to understand. The search for causal explanations, and the associated attempt to expose the hidden workings of nature, represent a second grand tradition regarding scientific explanation. We can refer to it as *the causal-mechanical tradition*.

Having contrasted the two major traditions, we should call attention to an important respect in which they overlap. When the search for hidden mechanisms is successful, the result is often to reveal a small number of basic mechanisms that underlie wide ranges of phenomena. The explanation of diverse phenomena in terms of the same mechanisms constitutes theoretical unification. For instance, the kinetic-molecular theory of gases unified thermodynamic phenomena with Newtonian particle mechanics. The discovery of the double-helical structure of DNA, for another example, produced a major unification of biology and chemistry.

Each of the two grand traditions faces certain fundamental problems. The tradition of explanation as unification—associated with the received view—still faces the problem concerning explanations of laws that was pointed out in 1948 by Hempel and Oppenheim. It was never solved by Hempel in any of his subsequent work on scientific explanation. If the technical details of Friedman's theory of unification were satisfactory, it would provide a solution to that problem. Unfortunately, it appears to encounter serious technical difficulties (see Kitcher 1976 and Salmon 1989).

---

[20] The unification approach has been dramatically extended and improved by Philip Kitcher (1976, 1981, and 1989).

The causal-mechanical tradition faces a longstanding philosophical difficulty concerning the nature of causality that had been posed by David Hume in the eighteenth century. The problem—stated extremely concisely—is that we seem unable to identify the *connection* between cause and effect, or to find the *secret power* by which the cause brings about the effect. Hume is able to find certain *constant conjunctions*— for instance, between fire and heat—but he is unable to find the connection. He is able to see the spatial contiguity of events we identify as cause and effect, and the temporal priority of the cause to the effect—as in collisions of billiard balls, for instance—but still no *necessary connection*. In the end he locates the connection in the human imagination—in the psychological expectation we feel with regard to the effect when we observe the cause.[21]

Hume's problem regarding causality is one of the most recalcitrant in the whole history of philosophy. Some philosophers of science have tried to provide a more objective and robust concept of causality, but none has enjoyed widespread success. One of the main reasons the received view was reticent about incorporating causal considerations in the analysis of scientific explanation was an acute sense of uneasiness about Hume's problem. One of the weaknesses of the causal view, as it is handled by many philosophers who espouse it, is the absence of any satisfactory theory of causality.[22]

1.16 THE PRAGMATICS OF EXPLANATION

As we noted in Section 1.4, the term ''explanation'' refers sometimes to linguistic entities—that is, collections of statements of facts—and sometimes to nonlinguistic entities—namely, those very facts. When we think in terms of the human activity of explaining something to some person or group of people, we are considering linguistic behavior. Explaining something to someone involves uttering or writing statements. In this section we look at some aspects of this *process of explaining*. In this chapter, up to this point, we have dealt mainly with the *product* resulting from this activity, that is, the explanation that was offered in the process of explaining.

When philosophers discuss language they customarily divide the study into three parts: syntax, semantics, and pragmatics. Syntax is concerned only with relationships among the symbols, without reference to the meanings of the symbols or the people who use them. Roughly speaking, syntax is pure grammar; it deals with the conventions governing combinations and manipulations of symbols. Semantics deals with the relationships between symbols and the things to which the symbols refer. Meaning and truth are the major semantical concepts. Pragmatics deals with the relationships among symbols, what they refer to, and the users of language. Of particular interest for our discussion is the treatment of the context in which language is used.

The 1948 Hempel-Oppenheim essay offered a highly formalized account of D-N explanations of particular facts, and it characterized such explanations in syntactical

---

[21] Hume's analysis of causation is discussed in greater detail in Chapter 2, Part II.

[22] I have tried to make some progress in this direction in Salmon (1984, Chapters 5–7).

and semantical terms alone. Pragmatic considerations were not dealt with. Hempel's later characterization of the other types of explanations were given mainly in syntactical and semantical terms, although I-S explanations are, as we noted, relativized to knowledge situations. Knowledge situations are aspects of the human contexts in which explanations are sought and given. Such contexts have other aspects as well.

One way to look at the pragmatic dimensions of explanation is to start with the question by which an explanation is sought. In Section 1.3 we touched briefly on this matter. We noted that many, if not all, explanations can properly be requested by *explanation-seeking why-questions*. In many cases, the first pragmatic step is to clarify the question being asked; often the sentence uttered by the questioner depends upon contextual clues for its interpretation. As Bas van Fraassen, one of the most important contributors to the study of the pragmatics of explanation, has shown, the emphasis with which a speaker poses a question may play a crucial role in determining just what question is being asked. He goes to the Biblical story of the Garden of Eden to illustrate. Consider the following three questions:

(i) Why did Adam eat *the apple?*

(ii) Why did *Adam* eat the apple?

(iii) Why did Adam *eat* the apple?

Although the words are the same—and in the same order—in each, they pose three very different questions. This can be shown by considering what van Fraassen calls the *contrast class*. Sentence (i) asks why Adam ate the apple instead of a pear, a banana, or a pomegranate. Sentence (ii) asks why Adam, instead of Eve, the serpent, or a goat, ate the apple. Sentence (iii) asks why Adam ate the apple instead of throwing it away, feeding it to a goat, or hiding it somewhere. Unless we become clear on the question being asked, we can hardly expect to furnish appropriate answers.

Another pragmatic feature of explanation concerns the knowledge and intellectual ability of the person or group requesting the explanation. On the one hand, there is usually no point in including in an explanation matters that are obvious to all concerned. Returning to (3)—our prime example of a D-N explanation of a particular fact—one person requesting an explanation of the sudden dramatic increase in the skater's rate of rotation might have been well aware of the fact that she drew her arms in close to her body, but unfamiliar with the law of conservation of angular momentum. For this questioner, knowledge of the law of conservation of angular momentum is required in order to understand the explanandum-fact. Another person might have been fully aware of the law of conservation of angular momentum, but failed to notice what the skater did with her arms. This person needs to be informed of the skater's arm maneuver. Still another person might have noticed the arm maneuver, and might also be aware of the law of conservation of angular momentum, but failed to notice that this law applies to the skater's movement. This person needs to be shown how to apply the law in the case in question.

On the other hand, there is no point in including material in an explanation that is beyond the listeners' ability to comprehend. To most schoolchildren, for example, an explanation of the darkness of the night sky that made reference to the non-Euclidean structure of space or the mean free path of a photon would be inappropriate. Many of the explanations we encounter in real-life situations are incomplete on account of the explainer's view of the background knowledge of the audience.

A further pragmatic consideration concerns the interests of the audience. A scientist giving an explanation of a serious accident to a congressional investigating committee may tell the members of Congress far more than they want to know about the scientific details. In learning why an airplane crashed, the committee might be very interested to find that it was because of an accumulation of ice on the wing, but totally bored by the scientific reason why ice-accumulations cause airplanes to crash.

Peter Railton (1981) has offered a distinction that helps considerably in understanding the role of pragmatics in scientific explanation. First, he introduces the notion of an *ideal explanatory text*. An ideal explanatory text contains *all* of the facts and *all* of the laws that are relevant to the explanandum-fact. It details *all* of the causal connections among those facts and *all* of the hidden mechanisms. In most cases the ideal explanatory text is huge and complex. Consider, for example, an explanation of an automobile accident. The *full* details of such items as the behavior of both drivers, the operations of both autos, the condition of the highway surface, the dirt on the windshields, and the weather, would be unbelievably complicated. That does not really matter, for the ideal explanatory text is seldom, if ever, spelled out fully. What is important is to have the ability to illuminate portions of the ideal text as they are wanted or needed. When we do provide knowledge to fill in some aspect of the ideal text we are furnishing *explanatory information*.

A request for a scientific explanation of a given fact is almost always—if not literally always—a request, not for the ideal explanatory text, but for explanatory information. The ideal text contains all of the facts and laws pertaining to the explanandum-fact. These are the completely objective and nonpragmatic aspects of the explanation. If explanatory information is to count as legitimate it must correspond to the objective features of the ideal text. The ideal text determines what is *relevant* to the explanandum-fact. Since, however, we cannot provide the whole ideal text, nor do we want to, a selection of information to be supplied must be made. This depends on the knowledge and interests of those requesting and those furnishing explanations. The information that satisfies the request in terms of the interests and knowledge of the audience is *salient* information. The pragmatics of explanation determines salience—that is, what aspects of the ideal explanatory text are appropriate for an explanation in a particular context.

1.17 CONCLUSION

Several years ago, a friend and colleague—whom I will call *the friendly physicist*—was sitting on a jet airplane awaiting takeoff. Directly across the aisle was a young boy holding a helium-filled balloon by a string. In an effort to pique the child's curiosity, the friendly physicist asked him what he thought the balloon would do when the plane accelerated for takeoff. After a moment's thought the boy said that it would move toward the back of the plane. The friendly physicist replied that *he* thought it would move toward the front of the cabin. Several adults in the vicinity became interested in the conversation, and they insisted that the friendly physicist was wrong. A flight attendant offered to wager a miniature bottle of Scotch that he was mistaken—a bet that he was quite willing to accept. Soon thereafter the plane accelerated, the balloon moved forward, and the friendly physicist enjoyed a free drink.[23]

Why did the balloon move toward the front of the cabin? Two explanations can be offered, both of which are correct. First, one can tell a story about the behavior of the molecules that made up the air in the cabin, explaining how the rear wall collided with nearby molecules when it began its forward motion, thus creating a pressure gradient from the back to the front of the cabin. This pressure gradient imposed an unbalanced force on the back side of the balloon, causing it to move forward with respect to the walls of the cabin.[24] Second, one can cite an extremely general physical principle—Einstein's *principle of equivalence*—according to which an acceleration is physically equivalent, from the standpoint of the occupants of the cabin, to a gravitational field. Since helium-filled balloons tend to rise in the atmosphere in the earth's gravitational field, they will move forward when the airplane accelerates, reacting just as they would if a massive object were suddenly placed behind the rear wall.

The first of these explanations is causal-mechanical. It appeals to unobservable entities, describing the causal processes and causal interactions involved in the explanandum phenomenon. When we are made aware of these explanatory facts we understand how the phenomenon came about. This is the kind of explanation that advocates of the causal-mechanical tradition find congenial. The second explanation illustrates the unification approach. By appealing to an extremely general physical principle, it shows how this odd little occurrence fits into the universal scheme of things. It does not refer to the detailed mechanisms. This explanation provides a different kind of understanding of the same fact.

Which of these explanations is correct? Both are. Both of them are embedded in the ideal explanatory text. Each of them furnishes valuable explanatory information. It would be a serious error to suppose that any phenomenon has only one explanation. It is a mistake, I believe, to ask for *the* explanation of any occurrence. Each of these explanations confers a kind of scientific understanding. Pragmatic considerations might dictate the choice of one rather than the other in a given context. For example, the explanation in terms of the equivalence principle would be unsuitable for a ten-year-old child. The same explanation might be just right in an undergraduate physics course. But both are bona fide explanations.

As we noted in Section 1.10, the 1948 Hempel-Oppenheim essay attracted almost no attention for about a decade after its publication. Around 1959 it became the focus of intense controversy, much of it stemming from those who saw causality as central to scientific explanation. The subsequent thirty years have seen a strong opposition between the advocates of the received view and the proponents of causal explanation. Each of the two major approaches has evolved considerably during this

[23] This little story was previously published in Salmon (1980). I did not offer an explanation of the phenomenon in that article.

[24] Objects that are denser than air do not move toward the front of the cabin because the pressure difference is insufficient to overcome their inertia.

period—indeed, they have developed to the point that they can peacefully coexist as two distinct aspects of scientific explanation. Scientific understanding is, after all, a complicated affair; we should not be surprised to learn that it has many different aspects. Exposing underlying mechanisms and fitting phenomena into comprehensive pictures of the world seem to constitute two important aspects. Moreover, as remarked above, we should remember that these two types of understanding frequently overlap. When we find that the same mechanisms underlie diverse types of natural phenomena this ipso facto constitutes a theoretical unification.

On one basic thesis there is nearly complete consensus. Recall that in the early decades of the twentieth century many scientists and philosophers denied that there can be any such thing as scientific explanation. Explanation is to be found, according to this view, only in the realms of theology and metaphysics. At present it seems virtually unanimously agreed that, however it may be explicated, there is such a thing as scientific explanation. Science *can* provide deep understanding of our world. We do *not* need to appeal to supernatural agencies to achieve understanding. Equally importantly, we can contrast the objectively based explanations of contemporary science with the pseudounderstanding offered by such flagrantly unscientific approaches as astrology, creation science, and scientology. These are points worth remembering in an age of rampant pseudoscience.

QUESTIONS

1. Must every scientific explanation contain a law of nature? According to philosophers who support "the received view" the answer is affirmative. Other philosophers have answered in the negative. Discuss critically the arguments pro and con. Give your own answer, supported by reasons.

2. Are there any inductive or statistical explanations of particular facts? In their classic 1948 paper Hempel and Oppenheim say that there are such explanations, but do not offer any explication of their nature. Later attempts to work out the details ran into many difficulties. Discuss these problems and say whether you think they are insuperable. Give your reasons.

3. According to the *explanation-prediction symmetry thesis*, every satisfactory scientific explanation could (in some suitable context) serve as a scientific prediction, and every scientific prediction could (in some suitable context) serve as a scientific explanation. Critically discuss both parts of this symmetry thesis. Give your reasons for accepting or rejecting each part.

4. Are there any *fundamental* differences between explanations in the natural sciences and explanations in the social sciences? (See Merrilee H. Salmon's chapter on philosophy of the social sciences.) Are there basic differences between human behavior and the behavior of other kinds of physical objects that make one kind more amenable to explanation than the other? Is explanation of human behavior that involves conscious deliberation and free choice possible? Discuss critically.

5. In this chapter it was suggested that "No gold sphere has a mass greater than 100,000 kg" is not a lawlike statement, whereas "No enriched uranium sphere has a mass greater than 100,000 kg" is a lawlike statement. Discuss the distinction between lawlike and accidental generalizations. Explain as clearly as possible why one is lawlike and the other is not.

6. Discuss the role of causality in scientific explanation. Do all legitimate scientific explanations make reference to causal relations? Is causality essentially irrelevant to scientific explanation? Are some good explanations causal and other good explanations not? Discuss critically.

7. Choose an actual example of a scientific explanation from a magazine such as *Scientific American, Science, Nature, American Scientist*, or from a textbook you have used in a science course. Give a concise summary of this explanation, and analyze it in terms of the models (such as D-N, I-S, D-S, S-R) and concepts (such as covering law, causal-mechanical, unification) introduced in this chapter. Evaluate the explanation in terms of these models and/or concepts.

8. In Section 1.9 it was claimed that

(i) All gases, kept in closed containers of fixed size, exert greater pressure when heated

is a general statement, whereas

(v) All Apache basketry is made by women

is not completely general because it refers specifically to a particular group of people. But, it might be objected, (i) refers to physical objects of a specific type, namely, gases in closed containers, so it is not completely general either. Moreover, (v) is a general statement about the Apache. Discuss this objection. Hint: Statement (i) can be formulated as follows: "If *anything* is a gas in a closed container that is heated, it will expand." But: Statement (v) can likewise be reformulated as follows: "If anything is an Apache basket, it was made by a woman." Is there a fundamental logical difference between the two statements as reformulated?

SUGGESTED READINGS

FRIEDMAN, MICHAEL (1974), "Explanation and Scientific Understanding," *Journal of Philosophy 71*: 5–19. Reprinted in Pitt (1988). This is the original statement of the unification approach to scientific explanation. Although brief highly technical parts appear near the end, the article contains interesting general discussion of the basic issues.

HEMPEL, CARL G. (1942), "The Function of General Laws in History," *Journal of Philosophy 39*: 35–48. Reprinted in Hempel (1965b). Hempel's original essay on explanation in history.

———. (1959), "The Logic of Functional Analysis," in Llewellyn Gross (ed.), *Symposium on Sociological Theory*. New York: Harper & Row, pp. 271–307. Reprinted in Hempel (1965b). Hempel's original essay on functional explanation.

———. (1962), "Explanation in Science and in History," in Robert G. Colodny (ed.), *Frontiers of Science and Philosophy*. Pittsburgh: University of Pittsburgh Press, pp. 7–33. A lucid and highly readable brief presentation of Hempel's basic views.

———. (1965a), "Aspects of Scientific Explanation," in Hempel (1965b), pp. 331–496. Hempel's magisterial comprehensive monograph on scientific explanation.

———. (1965b), *Aspects of Scientific Explanation and Other Essays in the Philosophy of Science*. New York: The Free Press. Part 4 contains four classic articles on scientific explanation.

———. (1966), *Philosophy of Natural Science*. Englewood Cliffs, NJ: Prentice-Hall. Chapters 5–8 offer an extremely elementary and highly readable introduction to Hempel's views.

HEMPEL, CARL G. and PAUL OPPENHEIM (1948), "Studies in the Logic of Explanation," *Philosophy of Science 15:* 135–175. Reprinted in Hempel (1965b) and in Pitt (1988). This is *the* modern classic on scientific explanation. Parts 1 and 3 are especially important.

HUMPHREYS, PAUL (1981), "Aleatory Explanation," *Synthese 48:* 225–232. An introductory account of a new approach to statistical explanation.

JEFFREY, RICHARD C. (1969), "Statistical Explanation vs. Statistical Inference," in Nicholas Rescher (ed.), *Essays in Honor of Carl G. Hempel*. Dordrecht: Reidel, pp. 104–113. Reprinted in Wesley C. Salmon and others (1970) *Statistical Explanation and Statistical Relevance*. Pittsburgh: University of Pittsburgh Press. This philosophical gem explicitly raises the question of whether explanations are arguments.

PITT, JOSEPH C. (ed.) (1988), *Theories of Explanation*. New York: Oxford University Press. An anthology containing a number of important articles.

RAILTON, PETER (1981), "Probability, Explanation, and Information," *Synthese 48:* 233–256. A clear and careful elementary exposition of a mechanistic approach to probabilistic explanation.

SALMON, WESLEY C. (1978), "Why Ask, 'Why'?—An Inquiry Concerning Scientific Explanation," *Proceedings and Addresses of the American Philosophical Association 51:* 683–705. An introductory discussion of explanation by means of unobservable entities.

———. (1982), "Comets, Pollen, and Dreams: Some Reflections on Scientific Explanation," in Robert McLaughlin (ed.), *What? Where? When? Why?* Dordrecht: Reidel, pp. 155–178. A popularized discussion of some of the basic issues concerning scientific explanation.

———. (1990), *Four Decades of Scientific Explanation*. Minneapolis: University of Minnesota Press. A historical survey of philosophical developments since the classic Hempel-Oppenheim (1948) essay.

SCRIVEN, MICHAEL (1959), "Explanation and Prediction in Evolutionary Theory," *Science 130:* 477–482. One of the first strong challenges to Hempel's account.

———. (1962), "Explanations, Predictions, and Laws," in Herbert Feigl and Grover Maxwell (eds.), *Minnesota Studies in the Philosophy of Science.* Vol. 3, *Scientific Explanation, Space and Time.* Minneapolis: University of Minnesota Press, pp. 170–230. Another strong attack on the received view of scientific explanation.

VAN FRAASSEN, BAS C. (1980), *The Scientific Image*. Oxford: Clarendon Press. Chapter 5 presents an influential treatment of the pragmatics of explanation.

WRIGHT, LARRY (1976), *Teleological Explanations*. Berkeley and Los Angeles: University of California Press. An important treatment of teleological and functional explanation.

Two

THE CONFIRMATION
OF SCIENTIFIC HYPOTHESES

John Earman and Wesley C. Salmon

In Chapter 1 we considered the nature and importance of scientific explanation. If we are to be able to provide an explanation of any fact, particular or general, we must be able to establish the statements that constitute its explanans. We have seen in the Introduction that many of the statements that function as explanans cannot be established in the sense of being conclusively verified. Nevertheless, these statements can be supported or confirmed to some degree that falls short of absolute certainty. Thus, we want to learn what is involved in *confirming* the kinds of statements used in explanations, and in other scientific contexts as well.

This chapter falls into four parts. Part I (Sections 2.1–2.4) introduces the problem of confirmation and discusses some attempts to explicate the qualitative concept of support. Part II (2.5–2.6) reviews Hume's problem of induction and some attempted resolutions. Part III (2.7–2.8) develops the mathematical theory of probability and discusses various interpretations of the probability concept. Finally, Part IV (2.9–2.10) shows how the probability apparatus can be used to illuminate various issues in confirmation theory.

Parts I, II, and III can each stand alone as a basic introduction to the topic with which it deals. These three parts, taken together, provide a solid introduction to the basic issues in confirmation, induction, and probability. Part IV covers more advanced topics. Readers who prefer not to bring up Hume's problem of induction can omit Part II without loss of continuity.

Part I: Qualitative Confirmation

2.1 EMPIRICAL EVIDENCE

The physical, biological, and behavioral sciences are all empirical. This means that their assertions must ultimately face the test of observation. Some scientific statements face the observational evidence directly; for example, "All swans are white," was supported by many observations of European swans, all of which were white, but it was refuted by the observation of black swans in Australia. Other scientific statements confront the observational evidence in indirect ways; for instance, "Every proton contains three quarks," can be checked observationally only by looking at the results of exceedingly complex experiments. Innumerable cases, of course, fall between these two extremes.

Human beings are medium-sized objects; we are much larger than atoms and much smaller than galaxies. Our environment is full of other medium-sized things—for example, insects, frisbees, automobiles, and skyscrapers. These can be observed with normal unaided human senses. Other things, such as microbes, are too small to be seen directly; in these cases we can use instruments of observation—microscopes—to extend our powers of observation. Similarly, telescopes are extensions of our senses that enable us to see things that are too far away to be observed directly. Our senses of hearing and touch can also be enhanced by various kinds of instruments. Ordinary eyeglasses—in contrast to microscopes and telescopes—are not extensions of normal human senses; they are devices that provide more normal sight for those whose vision is somewhat impaired.

An observation that correctly reveals the features—such as size, shape, color, and texture—of what we are observing is called *veridical*. Observations that are not veridical are *illusory*. Among the illusory observations are hallucinations, afterimages, optical illusions, and experiences that occur in dreams. Philosophical arguments going back to antiquity show that we cannot be absolutely certain that our direct observations are veridical. It is impossible to prove conclusively, for example, that any given observation is not a dream experience. That point must be conceded. We can, however, adopt the attitude that our observations of ordinary middle-sized physical objects are reasonably reliable, and that, even though we cannot achieve certainty, we can take measures to check on the veridicality of our observations and make corrections as required (see Chapter 4 for further discussion of the topics of skepticism and antirealism).

We can make a rough and ready distinction among three kinds of entities: (i) those that can be observed directly with normal unaided human senses; (ii) those that can be observed only indirectly by using some instrument that extends the normal human senses; and (iii) those that cannot be observed either directly or indirectly, whose existence and nature can be established only by some sort of theoretical inference. We do not claim that these distinctions are precise; that will not matter for our subsequent discussion. We say much more about category (iii) and the kinds of inferences that are involved as this chapter develops.

Our scientific languages should also be noted to contain terms of two types. We have an *observational vocabulary* that contains expressions referring to entities, properties, and relations that we can observe. "Tree," "airplane," "green," "soft," and "is taller than" are familiar examples. We also have a *theoretical vocabulary* containing expressions referring to entities, properties, and relations that we cannot observe. "Microbe," "quark," "electrically charged," "ionized," and "contains more protons than" exemplify this category. The terms of the theoretical vocabulary tend to be associated with the unobservable entities of type (iii) of the preceding paragraph, but this relationship is by no means precise. The distinction between observational terms and theoretical terms—like the distinction among the three kinds of entities—is useful, but it is not altogether clear and unambiguous. One further point of terminology. Philosophers often use the expression "theoretical entity," but it would be better to avoid that term and to speak either of *theoretical terms* or *unobservable entities*.

At this point a fundamental moral concerning the nature of scientific knowledge can be drawn. It is generally conceded that scientific knowledge is not confined to what we have observed. Science provides predictions of future occurrences—such as the burnout of our sun when all of its hydrogen has been consumed in the synthesis of helium—that have not yet been observed and that may never be observed by any human. Science provides knowledge of events in the remote past—such as the extinction of the dinosaurs—before any human observers existed. Science provides knowledge of other parts of the universe—such as planets orbiting distant stars—that we are unable to observe at present. This means that much of our scientific knowledge depends upon inference as well as observation. Since, however, deductive reasoning is nonampliative (see Chapter 1, Section 1.5), observations plus deduction cannot provide knowledge of the unobserved. Some other mode of inference is required to account for the full scope of our scientific knowledge.

2.2 THE HYPOTHETICO-DEDUCTIVE METHOD

As we have seen, science contains some statements that are reports of direct observation, and others that are not. When we ask how statements of this latter type are to meet the test of experience, the answer often given is the *hypothetico-deductive (H-D) method;* indeed, the H-D method is sometimes offered as *the* method of scientific inference. We must examine its logic.

The term *hypothesis* can appropriately be applied to any statement that is intended for evaluation in terms of its consequences. The idea is to articulate some statement, particular or general, from which observational consquences can be drawn. An *observational consequence* is a statement—one that might be true or might be false—whose truth or falsity can be established by making observations. These observational consequences are then checked by observation to determine whether they are true or false. If the observational consequence turns out to be true, that is said to *confirm* the hypothesis to some degree. If it turns out to be false, that is said to *disconfirm* the hypothesis.

Let us begin by taking a look at the H-D testing of hypotheses having the form of universal generalizations. For a very simple example, consider Boyle's law of gases, which says that, for any gas kept at a constant temperature T, the pressure P is inversely proportional to the volume V,[1] that is,

$$P \times V = \text{constant (at constant } T).$$

This implies, for example, that doubling the pressure on a gas will reduce its volume by a half. Suppose we have a sample of gas in a cylinder with a movable piston, and that the pressure of the gas is equal to the pressure exerted by the atmosphere—about 15 pounds per square inch. It occupies a certain volume, say, 1 cubic foot. We now apply an additional pressure of 1 atmosphere, making the total pressure 2 atmospheres. The volume of the gas decreases to ½ cubic foot. This constitutes a hypothetico-deductive confirmation of Boyle's law. It can be schematized as follows:

(1)   At constant temperature, the pressure of a gas is inversely proportional to its volume (Boyle's law).
The initial volume of the gas is 1 cubic ft.
The initial pressure is 1 atm.
The pressure is increased to 2 atm.
The temperature remains constant.
_____
The volume decreases to ½ cubic ft.

Argument (1) is a valid deduction. The first premise is the *hypothesis* that is being tested, namely, Boyle's law. It should be carefully noted, however, that Boyle's law is *not* the only premise of this argument. *From the hypothesis alone it is impossible to deduce any observational prediction;* other premises are required. The following four premises state the *initial conditions* under which the test is performed. The conclusion is the *observational prediction* that is derived from the hypothesis and the initial conditions. Since the temperature, pressure, and volume can be directly measured, let us assume for the moment that we need have no serious doubts about the truth of the statements of initial conditions. The argument can be schematized as follows:

(2) H (test hypothesis)
I (initial conditions)
_____
O (observational prediction)

When the experiment is performed we observe that the observational prediction is true.

As we noted in Chapter 1, it is entirely possible for a valid deductive argument to have one or more false premises and a true conclusion; consequently, the fact that (1) has a true conclusion does not prove that its premises are true. More specifically, we cannot validly conclude that our hypothesis, Boyle's law, is true just because the observational prediction turned out to be true. In (1) the argument from premises to

_____

[1] This relationship does not hold for temperatures and pressures close to the point at which the gas in question condenses into a liquid or solid state.

conclusion is a valid deduction but the argument *from* the conclusion *to* the premises is not. If it has any merit at all, it must be as an inductive argument.

Let us reconstruct the argument *from* the observational prediction *to* the hypothesis as follows:

(3)　The initial volume of the gas is 1 cubic ft.
　　 The initial pressure is 1 atm.
　　 The pressure is increased to 2 atm.
　　 The temperature remains constant.
　　 The volume decreases to ½ cubic ft.

At constant temperature, the pressure of a gas is inversely proportional to its volume (Boyle's law).

No one would seriously suppose that (3) establishes Boyle's law conclusively, or even that it renders the law highly probable. At best, it provides a *tiny bit* of inductive support. If we want to provide solid inductive support for Boyle's law it is necessary to make repeated tests of this gas, at the same temperature, for different pressures and volumes, and to make other tests at other temperatures. In addition, other kinds of gases must be tested in a similar manner.

In one respect, at least, our treatment of the test of Boyle's law has been oversimplified. In carrying out the test we do not directly observe—say by feeling the container—that the initial and final temperatures of the gas are the same. Some type of thermometer is used; what we observe directly is not the temperature of the gas but the reading on the thermometer. We are therefore relying on an *auxiliary hypothesis* to the effect that the thermometer is a reliable instrument for the measurement of temperature. On the basis of an additional hypothesis of this sort we claim that we can observe the temperature indirectly. Similarly, we do not observe the pressures directly, by feeling the force against our hands; instead, we use some sort of pressure gauge. Again, we need an auxiliary hypothesis stating that the instrument is a reliable indicator.

The need for auxiliary hypotheses is not peculiar to the example we have chosen. In the vast majority of cases—if not in every case—auxiliary hypotheses are required. In biological and medical experiments, for example, microscopes of various types are employed—from the simple optical type to the tunneling scanning electron microscope, each of which requires a different set of auxiliary hypotheses. Likewise, in astronomical work telescopes—refracting and reflecting optical, infrared, radio, X-ray, as well as cameras are used. The optical theory of the telescope and the chemical theory of photographic emulsions are therefore required as auxiliary hypotheses. In sophisticated physical experiments using particle accelerators, an elaborate set of auxiliary hypotheses concerning the operation of all of the various sorts of equipment is needed. In view of this fact, schema (2) should be expanded:

(4)　*H* (test hypothesis)
　　 A (auxiliary hypotheses)
　　 I (initial conditions)
　　 O (observational prediction)

Up to this point we have considered the case in which the observational prediction turns out to be true. The question arises, what if the observational prediction happens to be false? To deal with this case we need a different example.

At the beginning of the nineteenth century a serious controversy existed about the nature of light. Two major hypotheses were in contention. According to one theory light consists of tiny particles; according to the other, light consists of waves. If the corpuscular theory is true, a circular object such as a coin or ball bearing, if brightly illuminated, will cast a uniformly dark circular shadow. The following H-D test was performed:

(5) Light consists of corpuscles that travel in straight lines.[2]
     A circular object is brightly illuminated.
     _____
     The object casts a uniform circular shadow.

Surprisingly, when the experiment was performed, it turned out that the shadow had a bright spot in its center. Thus, the result of the test was negative; the observational prediction was false.

Argument (5) is a valid deduction; accordingly, if its premises are true its conclusion must also be true. But the conclusion is not true. Hence, at least one of the premises must be false. Since the second premise was known to be true on the basis of direct observation, the first premise—the corpuscular hypothesis—must be false.

We have examined two examples of H-D tests of hypotheses. In the first, Boyle's law, the outcome was positive—the observational prediction was found to be true. We saw that, even assuming the truth of the other premises in argument (1), the positive outcome could, at best, lend a small bit of support to the hypothesis. In the second, the corpuscular theory of light, the outcome was negative—the observational prediction was found to be false. In that case, assuming the truth of the other premise, the hypothesis was conclusively refuted.

The negative outcome of an H-D test is often less straightforward than the example just discussed. For example, astronomers who used Newtonian mechanics to predict the orbit of the planet Uranus found that their observational predictions were incorrect. In their calculations they had, of course, taken account only of the gravitational influences of the planets that were known at the time. Instead of taking the negative result of the H-D test as a refutation of Newtonian mechanics, they postulated the existence of another planet that had not previously been observed. That planet, Neptune, was observed shortly thereafter. An auxiliary hypothesis concerning the constitution of the solar system was rejected rather than Newtonian mechanics.

It is interesting to compare the Uranus example with that of Mercury. Mercury also moves in a path that differs from the orbit calculated on the basis of Newtonian mechanics. This irregularity, however, could not be successfully explained by postulating another planet, though this strategy was tried. As it turned out, the perturbation of Mercury's orbit became one of three primary pieces of evidence supporting Einstein's general theory of relativity—the theory that has replaced Newtonian me-

_____
[2] Except when they pass from one medium (e.g., air) to another medium (e.g., glass or water).

chanics in the twentieth century. The moral is that negative outcomes of H-D tests sometimes do, and sometimes do not, result in the refutation of the test hypothesis. Since auxiliary hypotheses are almost always present in H-D tests, we must face the possibility that an auxiliary hypothesis, rather than the test hypothesis, is responsible for the negative outcome.

2.3 PROBLEMS WITH THE HYPOTHETICO-DEDUCTIVE METHOD

The H-D method has two serious shortcomings that must be taken into account. The first of these might well be called *the problem of alternative hypotheses*. Let us reconsider the case of Boyle's law. If we represent that law graphically, it says that a plot of pressures against volumes is a smooth curve, as shown in Figure 2.1.

The result of the test, schematized in argument (1), is that we have two points (indicated by arrows) on this curve—one corresponding to a pressure of 1 atmosphere and a volume of 1 cubic foot, the other corresponding to a pressure of 2 atmospheres and a volume of ½ cubic foot. While these two points conform to the solid curve shown in the figure, they agree with infinitely many other curves as well—for example, the dashed straight line through those two points. If we perform another test, with a pressure of 3 atmospheres, we will find that it yields a volume of ⅓ cubic foot. This is incompatible with the straight line curve, but the three points we now have are still compatible with infinitely many curves, such as the dotted one, that go through these three. Obviously we can make only a finite number of tests; thus, it is clear that, no matter how many tests we make, the results will be compatible with infinitely many different curves.

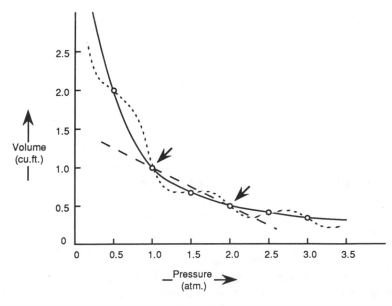

Figure 2.1

This fact poses a profound problem for the hypothetico-deductive method. *Whenever an observational result of an H-D test confirms a given hypothesis, it also confirms infinitely many other hypotheses that are incompatible with the given one.* In that case, how can we maintain that the test confirms our test hypothesis in preference to an infinite number of other possible hypotheses? This is the problem of alternative hypotheses. The answer often given is that we should prefer the *simplest* hypothesis compatible with the results of the tests. The question then becomes, what has simplicity got to do with this matter? Why are simpler hypotheses preferable to more complex ones? The H-D method, as such, does not address these questions.

The second fundamental problem for the H-D method concerns cases in which observational predictions cannot be *deduced*. The situation arises typically where statistical hypotheses are concerned. This problem may be called *the problem of statistical hypotheses*. Suppose, to return to an example cited in Chapter 1, that we want to ascertain whether massive doses of vitamin C tend to shorten the duration of colds. If this hypothesis is correct, the probability of a quick recovery is increased for people who take the drug. (As noted in Chapter 1, this is a fictitious example; the genuine question is whether vitamin C lessens the frequency of colds.) As suggested in that chapter, we can conduct a double-blind controlled experiment. However, we *cannot deduce* that the average duration of colds among people taking the drug will be smaller than the average for those in the control group. We can only conclude that, if the hypothesis is true, *it is probable* that the average duration in the experimental group will be smaller than it is in the control group. If we predict that the average duration in the experimental group will be smaller, the inference is *inductive*. The H-D method leaves no room for arguments of this sort. Because of the pervasiveness of the testing of statistical hypotheses in modern science, this limitation constitutes a severe shortcoming of the H-D method.

2.4 OTHER APPROACHES TO QUALITATIVE CONFIRMATION

The best known alternative to the H-D method is an account of qualitative confirmation developed by Carl G. Hempel (1945). The leading idea of Hempel's approach is that hypotheses are confirmed by their positive instances. Although seemingly simple and straightforward, this intuitive idea turns out to be difficult to pin down. Consider, for example, Nicod's attempt to explicate the idea for universal conditionals; for example:

$$H: (x) (Rx \supset Bx) \text{ (All ravens are black).}$$

(The symbol "(x)" is the so-called universal quantifier, which can be read, "for every object x"; "$\supset$" is the sign of material implication, which can be read *very roughly* "if . . . then . . . .") Although this statement is too simpleminded to qualify as a serious scientific hypothesis, the logical considerations that will be raised apply to all universal generalizations in science, no matter how sophisticated—see Section

2.11 of this chapter. According to Nicod, *E* Nicod-confirms such an *H* just in case *E* implies that some object is an instance of the hypothesis in the sense that it satisfies both the antecedent and the consequent, for example, *E* is *Ra.Ba* (the dot means "and"; it is the symbol for conjunction). To see why this intuitive idea runs into trouble, consider a plausible constraint on qualitative confirmation.

Equivalence condition: If *E* confirms *H* and $\vdash H \equiv H'$, then *E* confirms *H'*.

(The triple bar "$\equiv$" is the symbol for material equivalence; it can be translated *very roughly* as "if and only if," which is often abbreviated "iff." The turnstile "$\vdash$" preceding a formula means that the formula is a truth of logic.) The failure of this condition would lead to awkward situations since then confirmation would depend upon the mode of presentation of the hypothesis. Now consider

$$H' : (x) (\sim Bx \supset \sim Rx).$$

(The tilde "$\sim$" signifies negation; it is read simply as "not.") *H'* is logically equivalent to *H*. But

$$E: Ra.Ba$$

does not Nicod-confirm *H'* although it does Nicod-confirm *H*. Or consider

$$H'': (x) [(Rx.\sim Bx) \supset (Px.\sim Px)].$$

Again *H''* is logically equivalent to *H*. But by logic alone, nothing can satisfy the consequent of *H''* and if *H* is true nothing can satisfy the antecedent. So if *H* is true nothing can Nicod-confirm *H''*.[3]

After rejecting the Nicod account because of these and other shortcomings, Hempel's next step was to lay down what he regarded as conditions of adequacy for qualitative confirmation—that is, conditions that should be satisfied by any adequate definition of qualitative confirmation. In addition to the equivalence condition there are (among others) the following:

Entailment condition: If $E \vdash H$, the *E* confirms *H*.

(When the turnstile is preceded by a formula ("*E*" in "$E \vdash H$"), it means that whatever comes before the turnstile *logically entails* that which follows the turnstile—*E* logically entails *H*.)

---

[3] Such examples might lead one to try to build the equivalence condition into the definition of Nicod-confirmation along the following lines:

(N') *E* Nicod-confirms *H* just in case there is an *H'* such that $\vdash H \equiv H'$ and such that *E* implies that the objects mentioned satisfy both the antecedent and consequent of *H'*.

But as the following example due to Hempel shows, (N') leads to confirmation where it is not wanted in the case of multiply quantified hypotheses. Consider

$$H: (x) (y) Rxy$$
$$H': (x) (y)[\sim(Rxy . Ryx) \supset (Rxy.\sim Ryx)]$$
$$E: Rab. \sim Rba$$

E implies that the pair *a*, *b* satisfies both the antecedent and the consequent of *H'*, and *H'* is logically equivalent to *H*. So by (N') *E* Nicod-confirms *H*. But this is an unacceptable result since *E* contradicts *H*.

Special consequence condition: If E confirms H and $H \vdash H'$ then E confirms H'.

Consistency condition: If E confirms H and also confirms H' then H and H' are logically consistent.

As a result, he rejects

Converse consequence condition: If E confirms H and $H' \vdash H$ then E confirms H'.

For to accept the converse consequence condition along with the entailment and special consequence conditions would lead to the disastrous result that any E confirms any H. (Proof of this statement is one of the exercises at the end of this chapter.) Note that the H-D account satisfies the converse consequence condition but neither the special consequence condition nor the consistency condition.

Hempel provided a definition of confirmation that satisfies all of his adequacy conditions. The key idea of his definition is that of the *development, dev_I(H)*, of a hypothesis H for a set I of individuals. Intuitively, $dev_I(H)$ is what H says about a domain that contains exactly the individuals of I. Formally, universal quantifiers are replaced by conjunctions and existential quantifiers are replaced by disjunctions. For example, let $I = \{a, b\}$, and take

$$H: (x) \; Bx \; \text{(Everything is beautiful)}$$

then

$$dev_I(H) = Ba.Bb.$$

Or take

$$H': (\exists x) \; Rx \; \text{(Something is rotten)}$$

then

$$dev_I(H') = Ra \lor Rb.$$

(The wedge "∨" symbolizes the *inclusive disjunction;* it means "and/or"—that is, "one, or the other, or both.") Or take

$$H'': (x) \; (\exists y) \; Lxy \; \text{(Everybody loves somebody)};$$

then

$$dev_I(H'') = (Laa \lor Lab).(Lba \lor Lbb).^4$$

Using this notion we can now state the main definitions:

---

[4] In formulas like H'' that have mixed quantifiers, we proceed in two steps, working from the inside out. In the first step we replace the existential quantifier by a disjunction, which yields

$$(x) \; (Lxa \lor Lxb).$$

In the next step we replace the universal quantifier with a conjunction, which yields $dev_I(H'')$.

Def. E directly-Hempel-confirms H just in case $E \vdash dev_I(H)$ for the class I of individuals menioned in E.

Def. E Hempel-confirms H just in case E directly confirms every member of a set of sentences K such that $K \vdash H$.

To illustrate the difference between the two definitions, note that *Ra.Ba* does not directly-Hempel-confirm $Rb \supset Bb$ but it does Hempel-confirm it. Finally, disconfirmation can be handled in the following manner.

Def. E Hempel-disconfirms H just in case E confirms $\sim H$.

Despite its initial attractiveness, there are a number of disquieting features of Hempel's attempt to explicate the qualitative concept of confirmation. The discussion of these features can be grouped under two queries. First, is Hempel's definition too stringent in some respects? Second, is it too liberal in other respects? To motivate the first worry consider

$$H: (x) \, Rxy.$$

(The expression "*Rxy*"means "*x* bears relation *R* to *y*.") *H* is Hempel-confirmed by

$$E: Raa.Rab.Rbb.Rba.$$

But it is not confirmed by

$$E' : Raa.Rab.Rbb$$

even though intuitively the latter evidence does support H. Or consider the compound hypothesis

$$(x) \, (\exists y) \, Rxy.(x) \sim Rxx.(x) \, (y) \, (z) \, [(Rxy.Ryz) \supset Rxz],$$

which is true, for example, if we take the quantifiers to range over the natural numbers and interpret *Rxy* to mean that *y* is greater than *x*. (Thus interpreted, the formula says that for any number whatever, there exists another number that is larger. Although this statement is true for the whole collection of natural numbers, it is obviously false for any finite set of integers.) This hypothesis cannot be Hempel-confirmed by any consistent evidence statement since its development for any finite I is inconsistent. Finally, if H is formulated in the theoretical vocabulary, then, except in very special and uninteresting cases, H cannot be Hempel-confirmed by evidence E stated entirely in the observational vocabulary. Thus, Hempel's account is silent on how statements drawn from such sciences as theoretical physics—for example, all protons contain three quarks—can be confirmed by evidence gained by observation and experiment. This silence is a high price to pay for overcoming some of the defects of the more vocal H-D account.

This last problem is the starting point for Glymour's (1980) so-called *bootstrapping* account of confirmation. Glymour sought to preserve the Hempelian idea that hypotheses are confirmed by deducing instances of them from evidence statements, but in the case of a theoretical hypothesis he allowed that the deduction of instances can proceed with the help of auxiliary hypotheses. Thus, for Glymour the basic confirmation relation is three-place—E confirms H relative to H'—rather than two-place. In the main intended application we are dealing with a scientific theory T which consists of a network of hypotheses, from which H and H' are both drawn. If T is finitely axiomatizable—that is, if T consists of the set of logical consequences of a finite set of hypotheses, $H_1, H_2, \ldots, H_n$—we can say that T is bootstrap-confirmed if for each H_i there is an H_j such that E confirms H_i relative to H_j. These ideas are most easily illustrated for the case of hypotheses consisting of simple linear equations.

Consider a theory consisting of the following four hypotheses (and all of their deductive consequences):

$$H_1: O_1 = X$$
$$H_2: O_2 = Y + Z$$
$$H_3: O_3 = Y + X$$
$$H_4: O_4 = Z$$

The Os are supposed to be observable quantities while the Xs and Ys are theoretical.

For purposes of a concrete example, suppose that we have samples of four different gases in separate containers. All of the containers have the same volume, and they are at the same pressure and temperature. According to Avogadro's law, then, each sample contains the same number of molecules. Observable quantities O_1–O_4 are simply the weights of the four samples:

$$O_1 = 28 \text{ g}, O_2 = 44 \text{ g}, O_3 = 44 \text{ g}, O_4 = 28 \text{ g}.$$

Our hypotheses say

H_1: The first sample consists solely of molecular nitrogen—N_2—molecular weight 28; X is the weight of a mole of N_2 (28 g).

H_2: The second sample consists of carbon dioxide—CO_2—molecular weight 44; Y is the weight of a mole of atomic oxygen O (16 g), Z is the weight of a mole of carbon monoxide CO (28 g).

H_3: The third sample consists of nitrous oxide—N_2O—molecular weight 44; Y is the weight of a mole of atomic oxygen O (16 g) and X is the weight of a mole of molecular nitrogen (28 g).

H_4: The fourth sample consists of carbon monoxide—CO—molecular weight 28; Z is the weight of a mole of CO (28 g).

(The integral values for atomic and molecular weights are not precisely correct, but they furnish a good approximation for this example.)

To show how H_1 can be bootstrap-confirmed relative to the other three hypotheses, suppose that an experiment has determined values O_1, O_2, O_3, O_4, for the observables. From the values for O_2 and O_4 we can, using H_2 and H_4, compute values for $Y + Z$ and for Z. Together these determine a value for Y. Then from the value for O_3 we can, using H_3, compute a value for $Y + X$. Then from these latter two values we get a value for X. Finally, we compare this computed value for X with the observed value for O_1. If they are equal, H_1 is confirmed. Although this simple example may seem a bit contrived, it is in principle similar to the kinds of measurements and reasoning actually used by chemists in the nineteenth century to establish molecular and atomic weights.

If we want the bootstrap procedure to constitute a test in the sense that it carries with it the potential for falsification, then we should also require that there are possible values for the observables such that, using these values and the very same bootstrap calculations that led to a confirmatory instance, values for the theoretical quantities are produced that contradict the hypothesis in question. This requirement is met in the present example.

In Glymour's original formalization of the bootstrapping idea, macho bootstrapping was allowed; that is, in deducing instances of H, it was allowed that H itself could be used as an auxiliary assumption. To illustrate, consider again the earlier example of the perfect gas law P(ressure) $\times$ V(olume) $= K \times T$(emperature), and suppose P, V, T to be observable quantities while the gas constant K is theoretical. We proceed to bootstrap-test this law relative to itself by measuring the observables on two different occasions and then comparing the values k_1 and k_2 for K deduced from the law itself and the two sets of observation values p_1, v_1, t_1 and p_2, v_2, t_2. However, macho bootstrapping can lead to unwanted results, and in any case it may be unnecessary since, for instance, in the gas law example it is possible to analyze the logic of the test without using the very hypothesis being tested as an auxiliary assumption in the bootstrap calculation (see Edidin 1983 and van Fraassen 1983). These and other questions about bootstrap testing are currently under discussion in the philosophy journals. (The original account of bootstrapping, Glymour 1980, is open to various counterexamples discussed in Christensen 1983; see also Glymour 1983.)

Let us now return to Hempel's account of confirmation to ask whether it is too liberal. Two reasons for giving a positive answer are contained in the following paradoxes.

Paradox of the ravens.   Consider again the hypothesis that all ravens are black: $(x) (Rx \supset Bx)$. Which of the following evidence statements Hempel-confirm the ravens hypothesis?

$$E_1: Ra_1.Ba_1$$
$$E_2: \sim Ra_2$$
$$E_3: Ba_3$$
$$E_4: \sim Ra_4.\sim Ba_4$$
$$E_5: \sim Ra_5.Ba_5$$
$$E_6: Ra_6.\sim Ba_6$$

The answer is that E_1–E_5 all confirm the hypothesis. Only the evidence E_6 that refutes the hypothesis fails to confirm it. The indoor ornithology of some of these Hempel-confirmation relations—the confirmation of the ravens hypothesis, say, by the evidence that an individual is a piece of white chalk—has seemed to many to be too easy to be true.

Goodman's paradox.   If anything seems safe in this area it is that the evidence $Ra.Ba$ that a is a black raven confirms the ravens hypothesis $(x) (Rx \supset Bx)$. But on

Hempel's approach nothing rides on the interpretation of the predicates Rx and Bx. Thus, Hempel confirmation would still obtain if we interpreted Bx to mean that x is blite, where "blite" is so defined that an object is blite if it is examined on or before December 31, 2000, and is black or else is examined afterwards and found to be white. Thus, by the special consequence condition, the evidence that a is a black raven confirms the prediction that if b is a raven examined after 2000, it will be white, which is counterintuitive to say the least.

Part II: Hume's Problem of Induction

2.5 THE PROBLEM OF JUSTIFYING INDUCTION

Puzzles of the sort just mentioned—involving blite ravens and grue emeralds (an object is grue if it is examined on or before December 31, 2000 and is green, or it is examined thereafter and is blue)—were presented in Nelson Goodman (1955) under the rubric of *the new riddle of induction*. Goodman sought the basis of our apparent willingness to generalize inductively with respect to such predicates as "black," "white," "green," and "blue," but not with respect to "blite" and "grue." To mark this distinction he spoke of *projectible predicates* and *unprojectible predicates*, and he supposed that there are predicates of each of these types. The problem is to find grounds for deciding which are which.

There is, however, a difficulty that is both historically and logically prior. In his *Treatise of Human Nature* ([1739–1740] 1978) and his *Enquiry Concerning Human Understanding* (1748) David Hume called into serious question the thesis that we have any logical or rational basis for any inductive generalizations—that is, for considering any predicate to be projectible.

Hume divided all reasoning into two types, reasoning concerning *relations of ideas* and reasoning concerning *matters of fact and existence*. All of the deductive arguments of pure mathematics and logic fall into the first category; they are unproblematic. In modern terminology we say that they are necessarily truth-preserving because they are nonampliative (see Chapter 1, Section 1.5). If the premises of any such argument are true its conclusion must also be true because the conclusion says nothing that was not said, at least implicitly, by the premises.

Not all scientific reasoning belongs to the first category. Whenever we make inferences from observed facts to the unobserved we are clearly reasoning ampliatively—that is, the content of the conclusion goes beyond the content of the premises. When we predict future occurrences, when we retrodict past occurrences, when we make inferences about what is happening elsewhere, and when we establish generalizations that apply to all times and places we are engaged in reasoning concerning matters of fact and existence. In connection with reasoning of the second type Hume directly poses the question: What is the foundation of our inferences from the observed to the unobserved? He readily concludes that such reasoning is based upon relations of cause and effect. When we see lightning nearby (cause) we infer that the sound of thunder (effect) will ensue. When we see human footprints in the sand (effect) we infer that a person recently walked there (cause). When we hear a knock and a familiar voice saying "Anybody home?" (effect) we infer the presence of a friend (cause) outside the door.

The next question arises automatically: How can we establish knowledge of the cause-effect relations to which we appeal in making inferences from the observed to the unobserved? Hume canvasses several possibilities. Do we have a priori knowledge of causal relations? Can we look at an effect and deduce the nature of the cause? He answers emphatically in the negative. For a person who has had no experience of diamonds or of ice—which are very similar in appearance—there is no way to infer that intense heat and pressure can produce the former but would destroy the latter. Observing the effect, we have no way to deduce the cause. Likewise, for a person who has had no experience of fire or snow, there is no way to infer that the former will feel hot while the latter will feel cold. Observing the cause, we have no way to deduce the effect. All of our knowledge of causal relations must, Hume argues, be based upon experience.

When one event causes another event, we might suppose that three factors are present—namely, the cause, the effect, and the causal connection between them. However, in scrutinizing such situations Hume fails to find the third item—the causal connection itself. Suppose that one billiard ball lies at rest on a table while another moves rapidly toward it. They collide. The ball that was at rest begins to move. What we observe, Hume notes, is the initial motion of the one ball and its collision with the other. We observe the subsequent motion of the other. This is, he says, as perfect a case of cause and effect as we will ever see. We notice three things about the situation. The first is *temporal priority;* the cause comes before the effect. The second is spatiotemporal *proximity;* the cause and effect are close together in space and time. The third is *constant conjunction;* if we repeat the experiment many times we find that the result is just the same as the first time. The ball that was at rest always moves away after the collision.

Our great familiarity with situations similar to the case of the billiard balls may give us the impression that "it stands to reason" that the moving ball will produce motion in the one at rest, but Hume is careful to point out that *a priori reasoning* cannot support any such conclusion. We can, without contradiction, imagine many possibilities: When they collide the two balls might vanish in a puff of smoke; the moving ball might jump right over the one at rest; or the ball that is initially at rest might remain fixed while the moving ball returns in the direction from which it came. Moreover, no matter how closely we examine the situation, the thing we cannot see, Hume maintains, is the causal connection itself—the "secret power" by which the cause brings about the effect. If we observe two events in spatiotemporal proximity, one of which follows right after the other, *just once,* we cannot tell whether it is a mere coincidence or a genuine causal connection. Hans Reichenbach reported an incident that occurred in a theater in California as he was watching a movie. Just as a large explosion was depicted on the screen the theatre began to tremble. An individual's first instinct was to link them as cause and effect, but, in fact, by sheer coincidence, a minor earthquake occurred at precisely that moment. Returning to Hume's billiard ball example, on the first observation of such a collision we would not know whether the motion of the ball originally at rest occurred by coincidence or as a result of the collision with the moving ball. It is only after repeated observations of such events that we are warranted in concluding that a genuine causal relation exists. This fact shows that the *causal connection* itself is not an observable feature of the situation. If it were an observable feature we would not need to observe repetitions of the sequence of events, for we would be able to observe it in the first instance.[5]

What, then, is the basis for our judgements about causal relations? Hume answers that it is a matter of custom or habit. We observe, on one occasion, an event of type C and observe that it is followed by an event of type E. On another occasion we observe a similar event of type C followed by a similar event of type E. This happens repeatedly. Thereafter, when we notice an event of type C we expect that it will be followed by an event of type E. This is merely a fact about human psychology; we form a habit, we become conditioned to expect E whenever C occurs. There is no logical necessity in all of this.

Indeed, Hume uncovered a logical circle. We began by asking for the basis on which inferences from the observed to the unobserved are founded. The answer was that all such reasoning is based upon relations of cause and effect. We then asked how we can establish knowledge of cause-effect relations. The answer was that we assume—or psychologically anticipate—that future cases of events of type C will be followed by events of type E, just as in past cases events of type C were followed by events of type E. In other words, we assume that nature is uniform—that the future will be like the past—that regularities that have been observed to hold up to now will continue to hold in the future.

But what *reason* do we have for supposing that nature is uniform? If you say that nature's uniformity has been established on the basis of past observations, then to suppose that it will *continue* to be uniform is simply to suppose that the future will be like the past. That is flagrantly circular reasoning. If you say that science proceeds on the presumption that nature is uniform, and that science has been extremely successful in predicting future occurrences, Hume's retort is the same. To assume that future scientific endeavors will succeed because science has a record of *past* success is, again, to suppose that the future will be like the past. Furthermore, Hume points out, it is entirely possible that nature will *not* be uniform in the future—that the future *need not* be like the past—for we can consistently imagine all sorts of other possibilities. There is no contradiction in supposing that, at some future time, a substance resembling snow should fall from the heavens, but that it would feel like fire and taste like salt. There is no contradiction in supposing that the sun will not rise tomorrow morning. One can consistently imagine that a lead ball, released from the hand, would rise rather than fall. We do not expect such outlandish occurrences, but that is a result of our psychological makeup. It is not a matter of logic.

---

[5] The English philosopher John Locke had claimed that in one sort of situation we do observe the actual power of one event to bring about another, namely, in cases in which a person has a volition or desire to perform some act and does so as a result. We might, for example, wish to raise our arms, and then do so. According to Locke we would be aware of our power to produce motion in a part of our body. Hume gave careful consideration to Locke's claim, and argued that it is incorrect. He points out, among other things, that there is a complex relationship of which we are not directly aware—involving transmission of impulses along nerves and the contractions of various muscles—between the volition originating in the brain and the actual motion of the arm. Hume's critique effectively cut the ground from under Locke's claim.

What applies to Hume's commonsense examples applies equally to scientific laws, no matter how sophisticated they may be. We have never observed an exception to the law of conservation of angular momentum; nevertheless, tomorrow it may fail. Within our experience, the half-life of C^{14} has been 5730 years; tomorrow it could be 10 minutes. We have never found a signal that could be propagated faster than light; tomorrow we may find one. There is no guarantee that the chemistry of the DNA molecule will be the same in the future as it has been up to now. The possibilities are endless.

We should be clear about the depth and scope of Hume's arguments. Hume is *not* merely saying that we cannot be *certain* about the results of science—about scientific predictions, for example. That point had been recognized by the ancient skeptics many centuries before Hume's time. Hume's point is that we have no logical basis for placing *any* confidence in *any* scientific prediction. From this moment on, for all we can know *every* scientific prediction might fail. We cannot say even that scientific predictions are probable (the concept of probability will be examined in detail in Part III). We have no rational basis for placing more confidence in the predictions of science than in the predictions of fortune tellers or in wild guesses. The basis of our inferences from the observed to the unobserved is, to use Hume's terms, *custom and habit*.

2.6 ANSWERS TO HUME

Hume's critique of inductive reasoning struck at the very foundations of empirical science. It can be formulated as a dilemma. Science involves ampliative inference in an essential way. If we ask for the warrant or justification of any sort of ampliative inference, two responses seem possible. We could, on the one hand, attempt to offer a deductive argument to show that the conclusion follows from the premises—that the conclusion will be true if the premises are—but if any such argument could be given it would transform induction into deduction, and we would be left without any sort of ampliative inference. We could, on the other hand, try to offer an inductive justification, but any such justification would be circular—it would involve the use of induction itself to justify induction. The result is that, on either alternative, it is impossible to provide a suitable justification for the kinds of reasoning indispensable to science—and to common sense as well. It is this situation that led Broad (1926) to remark that induction is the glory of science and the scandal of philosophy.

It goes almost without saying that philosophers have adopted a variety of strategies to deal with Hume's dilemma. We consider some of the more appealing and/or influential ones (a number of these approaches are discussed in detail in Salmon 1967, Chapter 2).

1. The success of science. In spite of Hume's clear arguments concerning the circularity of justifying induction by using induction, it is difficult to escape the feeling that the most basic reason for relying on the methods of science is the remarkable success they have achieved in enabling us to explain natural phenomena and predict future events. What better basis could there be for judging the worth of a method than its track record up to now? And certainly no method of astrology, crystal gazing, divination, entrail reading, fortune telling, guessing, palmistry, or prophesy can begin to match the success of science. It would seem absurd to give up a highly successful method in exchange for one whose record is patently inferior.

Suppose, however, that a scientist—either an actual practitioner of science or anyone else who believes in the scientific method—is challenged by a crystal gazer. The scientist disparages crystal gazing as a method for predicting the future on the ground that it has not in the past been a very successful method, while the scientific method has, on the whole, worked well. The crystal gazer might correctly accuse the scientist of using the scientific method to justify the scientific method. The method of science is based, after all, on projecting past regularities into the future. To predict that the scientific method will continue to be successful in the future because it has been successful in the past is flagrantly circular. "If you are going to use your method to judge your method," the crystal gazer might remark, "then I have every right to use my method to judge my method." After looking into the crystal ball, the crystal gazer announces that the method of crystal gazing (in spite of its past lack of success) is about to become a very reliable method of prediction. "Furthermore," the crystal gazer might add, "since you used your method to cast aspersions on my method, I will use my method to judge yours: I see in my crystal ball that the scientific method is going to have a run of really bad luck in its forthcoming use as a method of prediction."

As Hume's argument regarding circularity had clearly shown, it is difficult to see anything wrong with the logic of the crystal gazer.[6]

2. Ordinary language dissolution.   Perhaps the most widely adopted approach to Hume's problem of induction is the attempt to dissolve it—rather than trying to solve it—by showing that it was not a genuine problem in the first place. One way to state the argument is this. If we ask what it means to be reasonable, the obvious answer is that it means to fashion our beliefs in terms of the available evidence. But what is the meaning of the concept of evidence? There are two forms of evidence, corresponding to two kinds of arguments—namely, deductive and inductive. To say that a proposition is supported by deductive evidence means that it can be deduced from propositions that we are willing to accept. If, for example, we accept the postulates of Euclidean geometry, then we are permitted to accept the Pythagorean theorem, inasmuch as it follows deductively from those postulates. Similarly—so the argument goes—if we accept the vast body of empirical evidence that is available, we should be prepared to accept the law of conservation of angular momentum (recall the figure skater in Chapter 1). This evidence inductively supports the claim that angular momentum is always conserved, and hence, that it will continue to be conserved tomorrow, next week, next month, next year and so on. To ask—in the spirit of Hume—whether we are justified in believing that angular momentum will be conserved tomorrow is to ask whether it is reasonable to base our beliefs on the available evidence, which, in this case, is inductive evidence. But basing our beliefs on evidence is just what it *means* to be rational. To ask whether we should believe on the basis of inductive evidence is tantamount to asking whether it is reasonable to be

[6] Max Black and R. B. Braithwaite both argued that inductive justifications of induction could escape circularity. The arguments of Black are criticized in detail in Salmon (1967, 12–17); Braithwaite's arguments are open to analogous criticism.

reasonable (two classic statements of this view are given by Ayer 1956, 71–75 and Strawson 1952, Chapter 9). The problem vanishes when we achieve a clear understanding of such terms as "evidence" and "rationality."

The foregoing argument is often reinforced by another consideration. Suppose someone continues to demand a justification for the fundamental principles of induction, for example, that past regularities can be projected into the future. The question then becomes, to what principle may we appeal in order to supply any such justification? Since the basic principles of inductive reasoning, like those of deductive reasoning, are ultimate, it is impossible to find anything more basic in terms of which to formulate a justification. Thus, the demand for justification of our most basic principles is misplaced, for such principles define the concept of justification itself.

In spite of its popular appeal among philosophers, this attempt to dispose of Hume's problem of justification of induction is open to serious objection. It can be formulated in terms of a useful distinction, drawn by Herbert Feigl (1950), between two kinds of justification—*validation* and *vindication*. A validation of a principle consists in a derivation of that principle from other, more basic, principles that we accept. For example, we often try to validate moral and/or legal principles. Some people argue that abortion is wrong, and should be outlawed, because it is wrong to take human life (except in certain extreme circumstances) and human life begins at the time of conception. Others (in America) argue, by appealing to certain rights they take to be guaranteed by the Constitution of the United States, that abortion should be permitted. What counts as a validation for any individual obviously depends upon the fundamental principles that person adopts.

Validation also occurs in mathematics and logic. The derivation of the Pythagorean theorem from the postulates of Euclidean geometry constitutes a good mathematical example. In logic, the inference rule modus tollens

$$(6) \quad \begin{array}{c} p \supset q \\ \sim q \\ \hline \sim p \end{array}$$

can be validated by appealing to modus ponens

$$(7) \quad \begin{array}{c} p \supset q \\ p \\ \hline q \end{array}$$

and contraposition

$$(p \supset q) \equiv (\sim q \supset \sim p).^{7}$$

[7] A less trivial example in deductive logic is the validation of the rule of conditional proof by means of the deduction theorem. The deduction theorem shows that any conclusion that can be established by means of conditional proof can be derived using standard basic deductive rules without appeal to conditional proof. Conditional proof greatly simplifies many derivations, but it does not allow the derivation of any conclusion that cannot be derived without it.

The Confirmation of Scientific Hypotheses

To *vindicate* a rule or a procedure involves showing that the rule or procedure in question serves some purpose for which it is designed. We vindicate the basic rules of deductive logic by showing that they are truth-preserving—that it is impossible to derive false conclusions from true premises when these rules are followed. This is a vindication because we want to be guaranteed that by using deductive rules we will never introduce a false conclusion by deduction from true premises.

Where induction is concerned we know that—because it is ampliative—truth preservation cannot be guaranteed; we sometimes get false conclusions from true premises. We would like to be able to guarantee that we will *usually* get true conclusions from true premises, but Hume's arguments show that this goal cannot be guaranteed either. As we will see, Reichenbach tried to give a different kind of vindication, but that is not the issue right now. The point is that, of the two kinds of justification, only one—*validation*—requires appeal to more basic principles; *vindication* does not. Vindications appeal to purposes and goals. When it is noted—as in the foregoing argument—that there is no principle more basic in terms of which induction can be justified, that shows that induction cannot be validated; it does *not* follow that induction cannot be vindicated.

If we keep clearly in mind the distinction between validation and vindication, we can see that the ordinary language dissolution fails. When we pose the question, "Is it reasonable to be reasonable?" it is easy to be fooled by an equivocation. Two senses of the word "reasonable" correspond to the two senses of "justification." One sense of "reasonable" ("reasonable$_1$") corresponds to vindication; in this sense, to ask whether something is reasonable is to ask whether it is a good means for achieving some desired goal. Where induction is concerned, that goal may be described roughly as getting true conclusions or making correct predictions as often as possible. The other sense of "reasonable" ("reasonable$_2$") corresponds to validation. In this sense, being reasonable includes adopting the generally accepted basic principles of inductive inference. If we now ask, "Is it reasonable$_1$ to be reasonable$_2$?" the question is far from trivial; it now means, "Does it serve our goal of predicting correctly as often as possible (reasonable$_1$) to use the accepted rules of inductive inference (reasonable$_2$)?" This is just another way of phrasing the fundamental question Hume raised concerning the justifiability of induction; the basic problem has not been dissolved, but only reformulated.

3. Inductive intuition. When Goodman posed *the new riddle of induction*, he made some sweeping claims about the nature of justification of logical principles. These claims applied both to deduction and to induction. In both cases, he said, we must confront the basic principles we hold dear with the kinds of arguments we are prepared to accept as valid or logically correct. (The term "valid" is often defined so that it characterizes logically correct deductive arguments only; if it is so construed, we need another term, such as "logically correct" to characterize inductive arguments that conform to appropriate logical principles.) When an argument that we want to retain conflicts with a principle we do not want to relinquish, some adjustment must be made. Speaking of deduction, Goodman says:

> The point is that rules and particular inferences alike are justified by being brought into agreement with each other. *A rule is amended if it yields an inference we are unwilling to*

accept; an inference is rejected if it violates a rule we are unwilling to amend. The process of justification is the delicate one of making mutual adjustments between rules and accepted inferences; and in the agreement achieved lies the only justification needed for either. (1955, 67, italics in the original)

He continues:

All this applies equally well to induction. An inductive inference, too, is justified by conformity to general rules, and a general rule by conformity to accepted inductive inferences. Predictions are justified if they conform to valid canons of induction; and the canons are valid if they accurately codify accepted inductive practice. (Ibid.)

Rudolf Carnap, whose theory of probability will be examined in item 6 in Section 2.8, seems to have had a similar point in mind when he said that the basic justification for the axioms of inductive logic rests on our inductive intuitions (Schilpp 1963, 978).

Goodman's claim about deductive logic is difficult to defend. We reject, as fallacious, the form of affirming the consequent

$$(8) \quad p \supset q$$
$$q$$
$$\overline{\qquad}$$
$$p$$

because it is easy to provide a *general* proof that it is not necessarily truth-preserving. The rejection is *not* the result of a delicate adjustment between particular arguments and general rules; it is based upon a demonstration that the form lacks one of the main features demanded of deductive rules. Other argument forms, such as modus ponens and modus tollens, are accepted because we can demonstrate *generally* that they are necessarily truth-preserving. (Going beyond truth-functional logic, there are general proofs of the consistency and completeness of standard first-order logic.)

The situation in inductive logic is complicated. As we will see when we study the various proposed interpretations of the concept of probability, there is an enormous plethora of possible rules of inference. We can illustrate by looking at three simple rules as applied to a highly artificial example. Suppose we have a large urn containing an extremely large number of marbles, all of which are known beforehand to be either red, yellow, or blue. We do not know beforehand what proportion of the marbles is constituted by each color; in fact, we try to learn the color constitution of the population of marbles in the urn by removing samples and observing the colors of the marbles in the samples.

Suppose, now, that the contents of the urn are thoroughly mixed, and that we draw out a sample containing n marbles, of which m are red. Consider three possible rules for inferring (or estimating) the percentage of the marbles in the urn that are red:

Induction by enumeration: if m/n of the marbles in the sample are red, infer that approximately m/n of all marbles in the urn are red.

A priori rule: regardless of the makeup of the observed sample, infer that approximately ⅓ of all marbles in the urn are red. (The fraction ⅓ is chosen because three colors occur in the total population of marbles in the urn.)

Counterinductive rule: if m/n of the marbles in the sample are red, infer that approximately $(n - m)/n$ of the marbles in the urn are red.

Certain characteristics of these rules can be established by *general arguments*. The counterinductive rule is so called because it uses observed evidence in a negative way. If we observe the proportion of red marbles in a sample, this rule instructs us to project that the proportion of red in the whole population is approximately equal to the proportion that are *not* red in the sample. Use of this rule would rapidly land us in an outright contradiction. Suppose, for the sake of simplicity, that our observed sample contains ⅓ red, ⅓ yellow, and ⅓ blue. Using the counterinductive rule for each of the colors would yield the conclusion that ⅔ of the marbles in the urn are red, and ⅔ of the marbles in the urn are yellow, and ⅔ of the marbles in the urn are blue. This is logically impossible; clearly, the counterinductive rule is unsatisfactory.

Suppose we use the a priori rule. Then, even if 98 percent of our observed sample were red, 1 percent yellow, and 1 percent blue, the rule would direct us to ignore that empirical evidence and infer that only about ⅓ of the marbles in the urn are red. Because the a priori rule makes observation irrelevant to prediction, it, too, should be rejected.

The rule of induction by enumeration does not have either of the foregoing defects, and it has some virtues. One virtue is that if it is used persistently on larger and larger samples, it must eventually yield inferences that are approximately correct. If we are unlucky, and begin by drawing unrepresentative samples, it will take a long time to start giving accurate results; if we are lucky and draw mainly representative samples, the accurate results will come much sooner. (Some philosophers have derived considerable comfort from the fact that the vast majority of samples that could be drawn are very nearly representative. See Williams 1947.)

Obviously many—indeed, infinitely many—possible rules exist for making inductive inferences. The problem of deciding which of these rules to use is complicated and difficult. We have seen, nevertheless, that general considerations can be brought to bear on the choice. It is not just a matter of consulting our intuitions regarding the acceptability or nonacceptability of particular inductive inferences. This is not to deny, however, that intuitive consideration of particular inferences has a great deal of heuristic value.

Although we have been skeptical about Goodman's success in dismissing the old riddle of induction, we must remark on the importance of his new riddle. First, Hume never explicitly took account of the fact that some forms of constant conjunction do not give rise to habits of expectation. Such Goodmanian predicates as "blite" and "grue" call attention vividly to this point. Second, Goodman's examples provide another way of showing that there can be no noncircular justification of induction by means of a uniformity principle. There are many uniformities, and the question of which ones will extend into the future is the problem of induction all over again.

4. Deductivism. One influential philosopher, Sir Karl Popper, has attacked Hume's problem by denying that science involves any use of induction. He takes

Hume to have proved decisively that induction cannot be justified, and he concludes that science—if it is to be a rational enterprise—must do without it. The only logic of science, he maintains, is deduction.

Popper characterizes the method of science as *trial and error*, as *conjecture and refutation*. The scientist formulates bold explanatory hypotheses, and then subjects them to severe testing. This test procedure is very much like hypothetico-deductive testing, but there is an absolutely crucial difference. According to the H-D theory, when the observational prediction turns out to be true, that *confirms* the hypothesis to some degree. Popper denies that there is any such thing as confirmation. If, however, the observational prediction turns out to be false, modus tollens can be used to conclude deductively that some premise is false. If we are confident of the initial conditions and auxiliary hypotheses, then we reject the hypothesis. The hypothesis was a *conjecture*; the test provided a *refutation*. Hypotheses that are refuted must be rejected.

If a bold hypothesis is subjected to severe testing and is not refuted, it is said to be *corroborated*. Popper emphatically denies that corroboration is any brand of confirmation. H-D theorists regard confirmation as a process that increases to some degree the probability of the hypothesis and, by implication, the probability that the hypothesis will yield correct predictions. Corroboration, in contrast, says nothing whatever about the future predictive success of the hypothesis; it is, instead, a report exclusively on the past performance of the hypothesis. The corroboration-rating is a statement of the past success of the hypothesis as an explanatory theory. The corroboration report is not contaminated with any inductive elements.

Even if we were to grant Popper's dubious claim that *theoretical science* is concerned only with explanation, and not with prediction, it would be necessary to recognize that we use scientific knowledge in making practical decisions. If we wish to put an artificial satellite into an orbit around the earth, we use Newtonian mechanics to compute the trajectory, and we confidently expect the satellite to perform as predicted. An inductivist would claim that we base such expectations on the fact that, within certain well-defined limits, Newtonian mechanics is a well-confirmed theory. Popper maintains that, for purposes of practical prediction, using well-corroborated theories is advisable, for nothing could be more rational.

The crucial question is, however, whether anything could be less rational than to use the corroboration-rating of a theory as a basis for choosing it for predictive purposes. Recalling that Popper has emphatically stated that the corroboration-rating refers only to past performance, and not to future performance, the corroboration-rating would seem to be totally irrelevant to the predictive virtues of the theory. The use of highly corroborated theories for prediction has no greater claim to rationality than do the predictions of fortune-tellers or sheer blind guessing. The price for banishing all inductive elements from science is to render science useless for prediction and practical decision making (see Salmon 1981).

5. Pragmatic vindication. Reichenbach fully accepted Hume's conclusion about the impossibility of proving that nature is uniform. He agreed that we have no way of knowing whether past uniformities will extend into the future. He recognized that, for all we can know, every inductive inference we make in the future may lead to a false prediction. Nevertheless, he attempted to construct a practical decision-theoretic justification for the use of induction.

Given our inability to know whether nature is uniform, we can consider what happens in either case. Hume showed convincingly that, if nature is uniform, inductive reasoning will work very well, whereas, if nature is not uniform, inductive reasoning will fail. This much is pretty easy to see. Reichenbach suggested, however, that we should consider other options besides the use of induction for purposes of trying to predict the future. Suppose we try consulting a crystal gaze to get our predictions. We cannot say a priori that we will get correct predictions, even if nature turns out to be uniform, but we cannot say a priori that we won't. We just don't know. Let us set up a chart:

TABLE 2.1

	Nature is uniform	Nature is not uniform
We use induction	Success	Failure
We don't use induction	Success or Failure	Failure

The crucial entry is in the lower right-hand box. What if nature is not uniform and we do not use induction? One possibility is simply not to make any predictions at all; whether nature is uniform or not, that obviously does not result in successful predictions. Another possibility is that we adopt a noninductive method such as crystal gazing. Any method—including wild guessing—may yield a true prediction once in a while by chance, whether nature is uniform or not. But suppose that crystal gazing were to work consistently. Then, that would be an important uniformity, and it could be established inductively—that is, on the basis of the observed record of the crystal gazer in making successful predictions we could infer inductively that crystal gazing will be successful in making correct predictions in the future. Thus, if crystal gazing can produce consistent successful predictions so can the use of induction. What has just been said about crystal gazing obviously applies to any noninductive method. Reichenbach therefore concluded that if any method will succeed consistently, then induction will succeed consistently. The same conclusion can be reformulated (by contraposition) as follows: If induction does not work, then no other method will work. We therefore have everything to gain and nothing to lose—so far as predicting the future is concerned—by adopting the inductive method. No other method can make an analogous claim. Reichenbach's argument is an attempt at *vindication* of induction. He is trying to show that—even acknowledging Hume's skeptical arguments—induction is better suited to the goal of predicting the future than any other methods that might be adopted.

Although Reichenbach's pragmatic justification may seem promising at first glance, it does face serious difficulties on closer inspection. The greatest problem with the foregoing formulation is that it suffers from severe vagueness. What do we mean by speaking of *the uniformity of nature*? Nature is not completely uniform; things do change. At the same time—up to the present at any rate—nature has exhibited certain kinds of uniformity. What degree of uniformity do we need in order for the argument to succeed? We should be much more precise on this point. Likewise, when we spoke about noninductive methods we did not carefully survey all of the available options. When the argument is tightened sufficiently, it turns out, it does not vindicate just one rule of inductive inference; instead, it equally justifies an infinite class of rules. Serious efforts—up to this time—to find a satisfactory basis for selecting a unique rule have been unsuccessful, (the technical details are discussed in Salmon 1967, Chapter 6).

Where do things stand now—250 years after the publication of Hume's *Treatise of Human Nature*—with respect to the problem we have inherited from him? Although many ingenious attempts have been made to solve or dissolve it there is still no consensus. It still stands as an item of "unfinished business" for philosophy of science (see Salmon 1978a). The problem may, perhaps, best be summarized by a passage from Hume himself:

> Let the course of things be allowed hitherto ever so regular, that alone, without some new argument or inference, proves not that for the future it will continue so. In vain do you pretend to have learned the nature of bodies from your past experience. Their secret nature, and consequently all their effects and influence, may change without any change in their sensible qualities. This happens sometimes, and with regard to some objects: Why may it not happen always, and with regard to all objects? What logic, what process or argument secures you against this supposition? My practice, you say, refutes my doubts. But you mistake the purport of my question. As an agent, I am quite satisfied in the point; but as a philosopher . . . I want to learn the foundation of this inference. (1748, Section 4)

As Hume makes abundantly clear, however, life—and science—go on in spite of these troubling philosophical doubts.

Part III: Probability

2.7 THE MATHEMATICAL THEORY OF PROBABILITY

Our discussion up to this point has been carried on without the aid of a powerful tool-—the calculus of probability. The time has come to invoke it. The defects of the qualitative approaches to confirmation discussed in Sections 2.3 and 2.4 suggest that an adequate account of the confirmation of scientific statements must resort to quantitative or probabilistic methods. In support of this suggestion, recall that we have already come across the concept of probability in the discussion of the qualitative approaches. In our discussion of the H-D method, for instance, we encountered the concept of probability in at least two ways. First, noting that a positive result of an H-D test does not conclusively establish a hypothesis, we remarked that it might render the hypothesis a little more probable than it was before the test. Second, in dealing with the problem of statistical hypotheses, we saw that only probabilistic observational predictions can be derived from such test hypotheses. In order to pursue our investigation of the issues that have been raised we must take a closer look at the concept or concepts of probability.

The modern theory of probability had its origins in the seventeenth century. Legend has it that a famous gentleman, the Chevalier de Méré, posed some questions about games of chance to the philosopher-mathematician Blaise Pascal. Pascal communicated the problems to the mathematician Pierre de Fermat, and that was how it all began. Be that as it may, the serious study of mathematical probability theory began around 1660, and Pascal and Fermat, along with Christian Huygens, played crucial roles in that development, (for an historical account see Hacking 1975 and Stigler 1986).

In order to introduce the theory of probability, we take probability to be a relationship between events of two different types—for example, between tossing a standard die and getting a six, or drawing from a standard bridge deck and getting a king. We designate probabilities by means of the following notation:

$Pr(B|A)$ is the probability of a result of the type B given an event of the type A.

If A is a toss of a standard die and B is getting a three, then "$Pr(B|A)$" stands for the probability of getting a three if you toss a standard die. As the theory of probability is seen today, all of the elementary rules of probability can be derived from a few simple axioms. The meanings of these axioms and rules can be made intuitively obvious by citing examples from games of chance that use such devices as cards and dice. After some elementary features of the mathematical calculus of probability have been introduced in this section, we look in the following section at a variety of interpretations of probability that have been proposed.

AXIOMS (BASIC RULES)

Axiom (rule) 1: Every probability is a unique real number between zero and one inclusive; that is,

$$O \leq Pr(B|A) \leq 1.$$

Axiom (rule) 2: If A logically entails B, then $Pr(B|A) = 1$.

Definition: Events of types B and C are *mutually exclusive* if it is impossible for both B and C to happen on any given occasion. Thus, for example, on any draw from a standard deck, drawing a heart and drawing a spade are mutually exclusive, for no card is both a heart and a spade.

Axiom (rule) 3: If B and C are mutually exclusive, then

$$Pr(B \vee C|A) = Pr(B|A) + Pr(C|A).$$

This axiom is also known as the *special addition rule*.

Example: The probability of drawing a heart or a spade equals the probability of drawing a heart plus the probability of drawing a spade.

Axiom (rule) 4: The probability of a *joint occurrence*—that is, of a conjunction of B and C—is equal to the probability of the first multiplied by the probability of the second given that the first has occurred:

$$Pr(B.C|A) = Pr(B|A) \times Pr(C|A.B).$$

This axiom is also known as the *general multiplication rule*.

Example: If you make two draws *without replacement* from a standard deck, what is the probability of getting two aces? The probability of getting an ace on the first draw is 4/52; the probability of getting an ace on the second draw *if you have already drawn an ace on the first draw* is 3/51, because there are only 51 cards left in the deck and only 3 of them are aces. Thus, the probability of getting two aces is

$$4/52 \times 3/51 = 12/2652 = 1/221.$$

SOME DERIVED RULES

From the four axioms (basic rules) just stated, several other rules are easy to derive that are extremely useful in calculating probabilities. First, we need a definition:

Definition: The events B and C are *independent* if and only if

$$Pr(C|A.B) = Pr(C|A).$$

When the events B and C are *independent* of one another, the multiplication rule (axiom 4) takes on a very simple form:

Rule 5: If B and C are independent, given A, then

$$Pr(B.C|A) = Pr(B|A) \times Pr(C|A).$$

This rule is known as the *special multiplication rule*. (Proofs, sketches of proofs, and other technical items will be placed in boxes. They can be omitted on first reading.)

Proof of Rule 5: Substitute $Pr(C|A)$ for $Pr(B.C|A)$ in Axiom 4.

Example: What is the probability of getting double 6 (''boxcars'') when a standard pair of dice is thrown? Since the outcomes on the two dice are independent, and the probability of 6 on each die is 1/6, the probability of double 6 is

$$1/6 \times 1/6 = 1/36$$

Example: What is the probability of drawing two spades on two consecutive draws when the drawing is done *with replacement*? The probability of getting a spade on the first draw is $13/52 = 1/4$. After the first card is drawn, whether it is a spade or not, it is put back in the deck and the deck is reshuffled. Then the second card is drawn. Because of the replacement, the outcome of the second draw is independent of the outcome of the first draw. Therefore, the probability of getting a spade on the second draw is just the same as it was on the first draw. Thus, the probability of getting two spades on two consecutive draws is

$$1/4 \times 1/4 = 1/16$$

NOTE CAREFULLY. If the drawing is done *without replacement*, the special multiplication rule cannot be used because the outcomes are *not independent*. In that case Rule 4 must be used.

Rule 6: $Pr(\sim B|A) = 1 - Pr(B|A)$.

This simple rule is known as the *negation rule*. It is very useful.

Example: Suppose you would like to know the probability of getting at least one 6 if you toss a standard die three times.[8] That means you want to know the probability of getting a 6 on the first toss *or* on the second toss *or* on the third toss, where this is an *inclusive or*. Thus, the outcomes are *not* mutually exclusive, so you cannot use the special addition rule (Axiom 3). We can approach this problem via the negation. To fail to get at least one 6 in three tosses means to get non-6 on the first toss *and* non-6 on the second toss *and* non-6 on the third toss. Since the probability of 6 is 1/6, the negation rule tells us that the probability of non-6 is 5/6. Because the outcomes on the three tosses are *independent*, we can use Rule 5 to obtain the probability of non-6 on all three tosses as

$$5/6 \times 5/6 \times 5/6 = 125/216.$$

The probability of getting at least one 6, which is the negation of not getting any 6, is therefore

$$1 - 125/216 = 91/216.$$

NOTE CAREFULLY: The probability of getting at least one 6 in three tosses is *not* 1/2. It is equal to 91/216, which is approximately 0.42.

Proof of Rule 6: Obviously every A is either a B or not a B. Therefore, by Axiom 2,

$$Pr(B \vee \sim B|A) = 1.$$

Since B and $\sim B$ are mutually exclusive, Axiom 3 yields

$$Pr(B|A) + Pr(\sim B|A) = 1.$$

Rule 6 results from subtracting $Pr(B|A)$ from both sides.

Rule 7: $Pr(B \vee C|A) = Pr(B|A) + Pr(C|A) - Pr(B.C|A)$.

This is the *general addition rule*. Unlike Rule 3, this rule applies to outcomes B and C even if they are not mutually exclusive.

Example: What is the probability of getting a spade or a face card in a draw from a standard deck? These two alternatives are not mutually exclusive, for there are three

---

[8] This example is closely related to one of the problems posed by the Chevalier de Méré. How many tosses of a pair of dice, he asked, are required to have at least a fifty-fifty chance of getting at least one double 6? It seems that a common opinion among gamblers at the time was that 24 tosses would be sufficient. The Chevalier doubted that answer, and it turned out that he was right. One needs 25 tosses to have at least a fifty-fifty chance.

cards—king, queen, and jack of spades—that are both face cards and spades. Since there are 12 face cards and 13 spades, the probability of a spade or a face card is

$$12/52 + 13/52 - 3/52 = 22/52$$

It is easy to see why this rule has the form that it does. If B and C are not mutually exclusive, then some outcomes may be both B and C. Any such items will be counted twice—once when we count the Bs and again when we count the Cs. (In the foregoing example, the king of spades is counted once as a face card and again as a spade. The same goes for the queen and jack of spades.) Thus, we must subtract the number of items that are both B and C, in order that they be counted only once.

How to prove Rule 7. First, we note that the class of things that are B or C in the *inclusive* sense consists of those things that are $B.C$ or $\sim B.C$ or $B.\sim C$, where these latter three classes are *mutually exclusive*. Thus, Rule 3 can be applied, giving

$$Pr(B \vee C|A) = Pr(B.C|A) + Pr(\sim B.C|A) + Pr(B.\sim C|A).$$

Rule 4 is applied to each of the three terms on the right-hand side, and then Rule 6 is used to get rid of the negations inside of the parentheses. A bit of simple algebra yields Rule 7.

Rule 8: $Pr(C|A) = Pr(B|A) \times Pr(C|A.B) + Pr(\sim B|A) \times Pr(C|A.\sim B)$.

This is the *rule of total probability*. It can be illustrated as follows:

Example: Imagine a factory that produces frisbees. The factory contains just two machines, a new machine **B** that produces 800 frisbees each day, and an old machine $\sim$**B** that produces 200 frisbees per day. Among the frisbees produced by the new machine, 1% are defective; among the frisbees produced by the old machine, 2% are defective. Let A stand for the frisbees produced in a given day at that factory. Let B stand for the frisbees produced by the new machine; $\sim B$ then stands for those produced by the old machine. Let C stand for defective frisbees. Then,

$Pr(B|A)$ = the probability that a frisbee is produced by machine **B** = 0.8

$Pr(\sim B|A)$ = the probability that a frisbee is produced by machine $\sim$**B** = 0.2

$Pr(C|A.B)$ = the probability that a frisbee produced by machine **B** is defective = 0.01

$Pr(C|A.\sim B)$ = the probability that a frisbee produced by machine $\sim$**B** is defective = 0.02

Therefore, the probability that a frisbee is defective =

$$0.8 \times 0.01 + 0.2 \times 0.02 = 0.012$$

As can be seen from this artificial example, the rule of total probability can be used to calculate the probability of an outcome that can occur in either of two ways, either by the occurrence of some intermediate event *B* or by the nonoccurrence of *B*. The situation can be shown in a diagram:

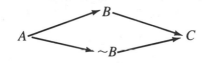

> Proof of Rule 8: Since every *C* is either a *B* or not a *B*, the class *C* is identical to the class $B.C \vee \sim B.C$; moreover, since nothing is both a *B* and not a *B*, the classes $B.C$ and $\sim B.C$ are mutually exclusive. Hence,
> $$Pr(C|A) = Pr(C.[B \vee \sim B] |A)$$
> $$= Pr([B.C \vee \sim B.C]|A)$$
> $$= Pr(B.C|A) + Pr(\sim B.C|A) \qquad \text{by Rule 3}$$
> $$= Pr(B|A) \times Pr(C|A.B) + Pr(\sim B|A) \times Pr(C|A.\sim B)$$
> $$\text{by Rule 4 applied twice}$$

We now come to the rule of probability that has special application to the problem of confirmation of hypotheses

Rule 9: $$Pr(B|A.C) = \frac{Pr(B|A) \times Pr(C|A.B)}{P(C|A)}$$

$$= \frac{Pr(B|A) \times Pr(C|A.B)}{Pr(B|A) \times Pr(C|A.B) + Pr(\sim B|A) \times Pr(C|A.\sim B)}$$

provided that $Pr(C|A) \neq 0$. The fact that these two forms are equivalent follows immediately from the rule of total probability (Rule 8), which shows that the denominators of the right-hand sides are equal to one another.

Rule 9 is known as *Bayes's rule*; it has extremely important applications. For purposes of illustration, however, let us go back to the trivial example of the frisbee factory that was used to illustrate the rule of total probability.

Example: Suppose we have chosen a frisbee at random from the day's production and it turns out to be defective. We did not see which machine produced it. What is the probability—$Pr(B|A \cdot C)$—that it was produced by the new machine? Bayes's rule gives the answer:

$$\frac{0.8 \times 0.01}{0.8 \times 0.01 + 0.2 \times 0.02} = \frac{0.008}{0.012} = 2/3$$

The really important fact about Bayes's rule is that it tells us a great deal about the confirmation of hypotheses. The frisbee example illustrates this point. We have a frisbee produced at this factory (*A*) that turns out, on inspection, to be defective (*C*), and we wonder whether it was produced (caused) by the new machine (*B*). In other

Proof of Bayes's rule: Bayes's rule has two forms as given above; we show how to prove both. We begin by writing Rule 4 twice; in the second case we interchange B and C.

$$Pr(B.C|A) = Pr(B|A) \times Pr(C|A.B)$$

$$Pr(C.B|A) = Pr(C|A) \times Pr(B|A./C)$$

Since the class $B.C$ is obviously identical to the class $C.B$ we can equate the right-hand sides of the two equations:

$$Pr(C|A) \times Pr(B|A.C) = Pr(B|A\ ) \times Pr(C|A.B)$$

Assuming that $P(C|A) \neq 0$, we divide both sides by that quantity:

$$Pr(B|A.C) = \frac{Pr(B|A) \times Pr(C|A.B)}{P(C|A)}$$

This is the first form. Using Rule 8, the rule of total probability, we replace the denominator, yielding the second form:

$$Pr(B|A.C) = \frac{Pr(B|A) \times Pr(C|A.B)}{Pr(B|A) \times Pr(C|A.B) + Pr(\sim B|A) \times Pr(C|A.\sim B)}$$

words, we are evaluating the hypothesis that the new machine produced this defective frisbee. As we have just seen, the probability is 2/3.

Inasmuch as we are changing our viewpoint from talking about types of objects and events $A, B, C, \ldots$ to talking about hypotheses, let us make a small change in notation to help in the transition. Instead of using "A" to stand for the day's production of frisbees, we shall use "K" to stand for our background *knowledge* about the situation in that factory. Instead of using "B" to stand for the frisbees produced by the new machine **B**, we shall use "H" to stand for the *hypothesis* that a given frisbee was produced by machine **B**. And instead of using "C" to stand for defective frisbees, we shall use "E" to stand for the *evidence* that the given frisbee is defective. Now Bayes's rule reads as follows:

$$\text{Rule 9: } Pr(H|K.E) = \frac{Pr(H|K) \times Pr(E\ |K.H)}{P(E|K)}$$

$$= \frac{Pr(H|K) \times Pr(E|K.H)}{Pr(H|K) \times Pr(E|K.H) + Pr(\sim H|K) \times Pr(E|K.\sim H)}$$

Changing the letters in the formula (always replacing the same old letter for the same new letter) obviously makes no difference to the significance of the rule. If the axioms are rewritten making the same changes in variables, Rule 9 would follow from them in exactly the same way. And inasmuch as we are still talking about *probabilities*—albeit the probabilities of hypotheses instead of the probabilities of events—we still need the same rules.

We can now think of the probability expressions that occur in Bayes's rule in the following terms:

$Pr(H|K)$ is the *prior probability* of hypothesis H just on the basis of our background knowledge K without taking into account the specific new evidence E. (In our example, it is the probability that a given frisbee was produced by machine **B**.) $Pr(\sim H|K)$ is the *prior probability* that our hypothesis H is false. (In our example, it is the probability that a given frisbee was produced by machine $\sim$**B**.) Notice that H and $\sim H$ must exhaust all of the possibilities.

By the negation rule (Rule 6), these two prior probabilities must add up to 1; hence, if one of them is known the other can immediately be calculated.

$Pr(E|K.H)$ is the probability that evidence E would obtain given the truth of hypothesis H in addition to our background knowledge K. (In our example, it is the probability that a particular frisbee is defective, given that it was produced by machine **B**.) This probability is known as a *likelihood*.
$Pr(E|K.\sim H)$ is the probability that evidence E would obtain if our hypothesis H is false. (In our example, it is the probability that a particular frisbee is defective if it was not produced by machine **B**.) This probability is also a *likelihood*.

The two likelihoods—in sharp contrast to the prior probabilities—are independent of one another. Given only the value of one of them, it is *impossible* to calculate the value of the other.

$P(E|K)$ is the probability that our evidence E would obtain, regardless of whether hypothesis H is true or false. (In our example, it is the probability that a given frisbee is defective, regardless of which machine produced it.) This probability is often called the *expectedness* of the evidence.[9]
$Pr(H|K . E)$ is the probability of our hypothesis, judged in terms of our background knowledge K and the specific evidence E. It is known as the *posterior probability*. This is the probability we are trying to ascertain. (In our example, it is the probability that the frisbee was produced by the new machine. Since the posterior probability of H is different from the prior probability of H, the fact that the frisbee is defective is evidence relevant to that hypothesis.)

[9] Expectedness is the opposite of *surprisingness*. If the expectedness of the evidence is small the evidence is surprising. Since the expectedness occurs in the denominator of the fraction, the smaller the expectedness, the greater the value of the fraction. Surprising evidence confirms hypotheses more than evidence that is to be expected regardless of the hypothesis.

Notice that, although the *likelihood* of a defective product is twice as great for the old machine (0.02) as for the new (0.01), the *posterior probability* that a defective frisbee was produced by the new machine (2/3) is twice as great as the probability that it was produced by the old one (1/3).

In Section 2.9 we return to the problem of assigning probabilities to hypotheses, which is the main subject of this chapter.

2.8 THE MEANING OF PROBABILITY

In the preceding section we discussed the notion of probability in a formal manner. That is, we introduced a symbol, "*Pr*(|)," to stand for probability, and we laid down some formal rules governing the use of that symbol. We illustrated the rules with concrete examples, to give an intuitive feel for them, but we never tried to say what the word "probability" or the symbol "*Pr*" means. That is the task of this section.

As we discuss various suggested meanings of this term, it is important to recall that we laid down certain basic rules (axioms). If a proposed definition of "probability" satisfies the basic rules—and, consequently, the derived rules, since they are deduced from the basic rules—we say that the suggested definition provides an *admissible interpretation* of the probability concept. If a proposed interpretation violates those rules, we consider it a serious drawback.

1. The classical interpretation.   One famous attempt to define the concept of probability was given by the philosopher-scientist Pierre Simon de Laplace ([1814] 1951). It is known as the *classical interpretation*. According to this definition, *the probability of an outcome is the ratio of favorable cases to the number of equally possible cases.* Consider a simple example. A standard die (singular of "dice") has six faces numbered 1–6. When it is tossed in the standard way there are six *possible* outcomes. If we want to know the probability of getting a 6, the answer is 1/6, for only one possible outcome is *favorable.* The probability of getting an even number is 3/6, for three of the possible outcomes (2, 4, 6) are favorable.

Laplace was fully aware of a fundamental problem with this definition. The definition refers not just to possible outcomes, but to *equally possible* outcomes. Consider another example. Suppose two standard coins are flipped simultaneously. What is the probability of getting two heads? Someone might say it is 1/3, for there are three possible outcomes, two heads, one head and one tail, or two tails. We see immediately that this answer is incorrect, for these possible outcomes are not equally possible. That is because one head and one tail can occur in two different ways—head on coin #1 and tail on coin #2, or tail on coin #1 and head on coin #2. Hence, we should say that there are four *equally possible* cases, so the probability of two heads is 1/4.

In order to clarify his definition Laplace needed to say what is meant by "equally possible," and he endeavored to do so by offering the famous *principle of indifference.* According to this principle, two outcomes are equally possible—we might as well say "equally probable"—*if we have no reason to prefer one to the other.*

Compare the coin example with the following from modern physics. Suppose you have two helium-4 atoms in a box. Each one has a fifty-fifty chance of being in the left-hand side of the box at any given time. What is the probability of both atoms being in the left-hand side at a particular time? The answer is 1/3. Since the two atoms are in principle indistinguishable—unlike the coins, which are obviously distinguishable—we cannot regard atom #1 in the left-hand side and atom #2 in the right-hand side as a case distinct from atom #1 in the right-hand side and atom #2 in the left-hand side. Indeed, it does not even make sense to talk about atom #1 and atom #2 since we have no way, even in principle, of telling which is which.

Suppose, for example, that we examine a coin very carefully and find that it is perfectly symmetrical. Any reason one might give to suppose it will come up heads can be matched by an equally good reason to suppose it will land tails up. We say that the two sides are equally possible, and we conclude that the probability of heads is 1/2. If, however, we toss the coin a large number of times and find that it lands heads up in about 3/4 of all tosses and tails up in about 1/4 of all tosses, we *do have* good reason to prefer one outcome to the other, so we would *not* declare them equally possible. The basic idea behind the principle of indifference is this: when we have no *reason* to consider one outcome more probable than another, we should not *arbitrarily* choose one outcome to favor over another. This seems like a sound principle of probabilistic reasoning.

There is, however, a profound difficulty connected with the principle of indifference; its use can lead to outright inconsistency. The problem is that it can be applied in different ways to the same situation, yielding incompatible values for a particular probability. Again, consider an example, namely, the case of Joe, the sloppy bartender. When a customer orders a 3:1 martini (3 parts of gin to 1 part of dry vermouth), Joe may mix anything from a 2:1 to a 4:1 martini, and there is no further information to tell us where in that range the mix may lie. According to the principle of indifference, then, we may say that there is a fifty-fifty chance that the mix will be between 2:1 and 3:1, and an equal chance that it will be between 3:1 and 4:1. Fair enough. But there is another way to look at the same situation. A 2:1 martini contains 1/3 vermouth, and a 4:1 martini contains 1/5 vermouth. Since we have no further information about the proportion of vermouth we can apply the principle of indifference once more. Since $1/3 = 20/60$ and $1/5 = 12/60$, we can say that there is a fifty-fifty chance that the proportion of vermouth is between 20/60 and 16/60 and an equal chance that it is between 16/60 and 12/60. So far, so good?

Unfortunately, no. We have just contradicted ourselves. A 3:1 martini contains 25 percent vermouth, which is equal to 15/60, *not* 16/60. The principle of indifference has told us *both* that there is a fifty-fifty chance that the proportion of vermouth is between 20/60 and 16/60, *and also* that there is a fifty-fifty chance that it is between

20/60 and 15/60. The situation is shown graphically in Figure 2.2. As the graph shows, the same result occurs for those who prefer their martinis drier; the numbers are, however, not as easy to handle.

We must recall, at this point, our first axiom, which states, in part, that the probability of a given outcome under specified conditions is a *unique* real number. As we have just seen, the classical interpretation of probability does not furnish unique results; we have just found two different probabilities for the same outcome. Thus, it turns out, the classical interpretation is *not* an admissible interpretation of probability.

You might be tempted to think the case of the sloppy bartender is an isolated and inconsequential fictitious example. Nothing could be farther from the truth. This example illustrates a broad range of cases in which the principle of indifference leads to contradiction. The source of the difficulty lies in the fact that we have two quantities—the ratio of gin to vermouth and the proportion of vermouth—that are interdefinable; if you know one you can calculate the other. However, as Figure 2.2 clearly shows, the definitional relation is not linear; the graph is not a straight line. We can state generally: Whenever there is a nonlinear definitional relationship between two quantities, the principle of indifference can lead to a similar contradiction. To convince yourself of this point, work out the details of another example. Suppose there is a square piece of metal inside of a closed box. You cannot see it. But you are told that its area is somewhere between 1 square inch and 4 square inches, but nothing else is known about the area. First apply the principle of indifference to the area of the square, and then apply it to the length of the side which is, of course, directly

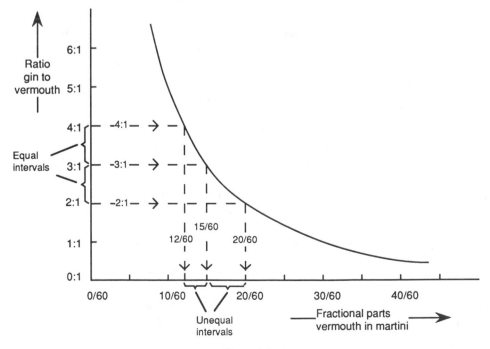

Figure 2.2

ascertainable from the area. (For another example, involving a car on a racetrack, see Salmon 1967, 66–67.)

Although the classical interpretation fails to provide a satisfactory basic definition of the probability concept, that does not mean that the idea of the ratio of favorable to equiprobable possible outcomes is useless. The trouble lies with the principle of indifference, and its aim of transforming ignorance of probabilities into values of probabilities. However, in situations in which we have positive knowledge that we *are* dealing with alternatives that have equal probabilities, the strategy of counting equiprobable favorable cases and forming the ratio of favorable to equiprobable possible cases is often handy for facilitating computations.

2. The frequency interpretation. The frequency interpretation has a venerable history, going all the way back to Aristotle (4th century B.C.), who said that the probable is that which happens often. It was first elaborated with precision and in detail by the English logician John Venn (1866, [1888] 1962). The basic idea is easily illustrated. Consider an ordinary coin that is being flipped in the standard way. As it is flipped repeatedly a sequence of outcomes is generated:

H T H T T T H H T T H T T T T T H T H T T T H H H H . . .[10]

We can associate with this sequence of results a sequence of *relative frequencies*— that is, the proportion of tosses that have resulted in heads up to a given point in the sequence—as follows:

1/1, 1/2, 2/3, 2/4, 2/5, 2/6, 3/7, 4/8, 4/9, 4/10, 5/11, 5/12, 5/13/ 5/14, 5/15, 6/16, 6/17, 7/18, 7/19, 7/20, 7/21, 8/22, 9/23, 10/24, 11/25, . . .

The denominator in each fraction represents the number of tosses made up to that point; the numerator represents the number of heads up to that point. We could, of course, continue flipping the coin, recording the results, and tabulating the associated relative frequencies. We are reasonably convinced that this coin is fair and that it was flipped in an unbiased manner. Thus, we believe that the probability of heads is 1/2. If that belief is correct, then, as the number of tosses increases, the relative frequencies will become and remain close to 1/2. The situation is shown graphically in Figure 2.3. There is no particular number of tosses at which the fraction of heads is and remains precisely 1/2; indeed, in an odd number of tosses the ratio cannot possibly equal 1/2. Moreover, if, at some point in the sequence, the relative frequency does equal precisely 1/2, it will necessarily differ from that value on the next flip. Instead of saying that the relative frequency must equal 1/2 in any particular number of throws, we say that it approaches 1/2 *in the long run*.

Although we know that no coin can ever be flipped an infinite number of times, it is useful, as a mathematical idealization, to think in terms of a *potentially infinite* sequence of tosses. That is, we imagine that, no matter how many throws have been

[10] These are the results of 25 flips made in an actual trial by the authors.

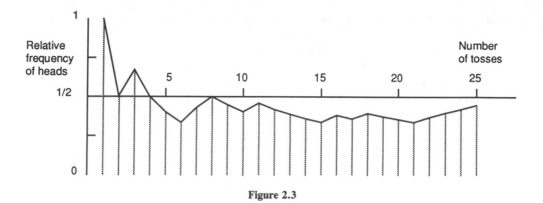

Figure 2.3

made, it is still possible to make more; that is, there is no particular finite number N at which point the sequence of tosses is considered complete. Then we can say that the *limit of the sequence* of relative frequencies equals the probability; this is the meaning of the statement that the *probability* of a particular sort of occurrence is, by definition, its long run relative frequency.

What is the meaning of the phrase "limit of the relative frequency"? Let $f_1, f_2, f_3, \ldots$ be the successive terms of the sequence of relative frequencies. In the example above, $f_1 = 1, f_2 = 1/2, f_3 = 2/3$, and so on. Suppose that p is the limit of the relative frequency. This means that the values of f_n become and remain arbitrarily close to p as n becomes larger and larger. More precisely, let δ be any small number greater than 0. Then, there exists some finite integer N such that, for any $n > N, f_n$ does not differ from p by more than δ.

Many objections have been lodged against the frequency interpretation of probability. One of the least significant is that mentioned above, namely, the finitude of all actual sequences of events, at least within the scope of human experience. The reason this does not carry much weight is the fact that science is full of similar sorts of idealizations. In applying geometry to the physical world we deal with ideal straight lines and perfect circles. In using the infinitesimal calculus we assume that certain quantities—such as electric charge—can vary continuously, when we know that they are actually discrete. Such practices carry no danger provided we are clearly aware of the idealizations we are using. Dealing with infinite sequences is technically easier than dealing with finite sequences having huge numbers of members.

A much more serious problem arises when we ask how we are supposed to ascertain the values of these limiting frequencies. It seems that we observe some limited portion of such a sequence and then extrapolate on the basis of what has been observed. We may not want to judge the probability of heads for a certain coin on the basis of 25 flips, but we might well be willing to do so on the basis of several hundred. Nevertheless, there are several logical problems with this procedure. First, no matter how many flips we have observed, it is always *possible* for a long run of heads to occur that would raise the relative frequency of heads well above 1/2. Similarly, a long run of future tails could reduce the relative frequency far below 1/2.

Another way to see the same point is this. Suppose that, for each n, m/n is the fraction of heads to tosses as of the nth toss. Suppose also that f_n does have the

The Confirmation of Scientific Hypotheses

limiting value p. Let a and b be any two fixed positive integers where $a \leq b$. If we add the constant a to every value of m and the constant b to every value of n, the resulting sequence $(m + a)/(n + b)$ will converge to the very same value p. That means that you could attach any sequence of b tosses, a of which are heads, to the beginning of your sequence, without changing the limiting value of the relative frequency. Moreover, you can chop off any finite number b of members, a of which are heads, from the beginning of your sequence without changing the limiting frequency p. As m and n get very large, the addition or subtraction of fixed numbers a and b has less and less effect on the value of the fraction. This seems to mean that *the observed relative frequency in any finite sample is irrelevant to the limiting frequency*. How, then, are we supposed to find out what these limiting frequencies— probabilities—are?

It would seem that things could not get much worse for the frequency interpretation of probability, but they do. For any sequence, such as our sequence of coin tosses, there is no guarantee that any limit of the relative frequency even exists. It is logically possible that long runs of heads followed by longer runs of tails followed by still longer runs of heads, and so on, might make the relative frequency of heads fluctuate between widely separated extremes throughout the infinite remainder of the sequence. If no limit exists *there is no such thing as the probability of a head when this coin is tossed*.

In spite of these difficulties, the frequency concept of probability seems to be used widely in the sciences. In Chapter 1, for instance, we mentioned the spontaneous decay of C^{14} atoms, commenting that the half-life is 5730 years. That is the rate at which atoms of this type have decayed in the past; we confidently predict that they will continue to do so. The relative frequency of disintegration of C^{14} atoms within 5730 years is 1/2. This type of example is of considerable interest to archaeologists, physicists, and geophysicists. In the biological sciences it has been noted, for example, that there is a very stable excess of human male births over human female births, and that is expected to continue. Social scientists note, however, that human females, on average, live longer than human males. This frequency is also extrapolated.

It is easy to prove that the frequency interpretation satisfies the axioms of probability laid down in the preceding section. This interpretation is, therefore, admissible. Its main difficulty lies in the area of *ascertainability*. How are we to establish values of probabilities of this sort? This question again raises Hume's problem of justification of induction.

A further problem remains. Probabilities of the frequency variety are used in two ways. On the one hand, they appear in statistical laws, such as the law of radioactive decay of unstable species of nuclei. On the other hand, they are often applied in making predictions of single events, or finite classes of events. Pollsters, for example, predict outcomes of single elections on the basis of interviews with samples of voters. If, however, probability is defined as a limiting frequency in a potentially infinite sequence of events, it does not seem to make any sense to talk about probabilities of single occurrences. The *problem of the single case* raises a problem about the *applicability* of the frequency interpretation of probability.

Before we leave the frequency interpretation a word of caution is in order. The frequency interpretation and the classical interpretations are completely different from one another, and they should not be confused. When the classical interpretation refers to possible outcomes and favorable outcomes it is referring to types or classes of events—for example, the class of all cases in which heads comes up is *one* possible outcome; the class of cases in which tails comes up is *one* other possible outcome. In this example there are only two possible outcomes. These *classes—not their members*—are what you count for purposes of the classical interpretation. In the frequency interpretation, it is the *members* of these classes that are counted. If the coin is tossed a large number of times there are many heads and many tails. In the frequency interpretation, the numbers of items of which ratios are formed keep changing as the number of individual events increases. In the classical interpretation, the probability does not depend in any way on how many heads or tails actually occur.

3. The propensity interpretation. The propensity interpretation is a relatively recent innovation in the theory of probability. Although suggested earlier, particularly by Charles Saunders Peirce, it was first clearly articulated by Popper (1957b, 1960). It was introduced specifically to deal with the problem of the single case.

The sort of situation Popper originally envisaged was a potentially infinite sequence of tosses of a loaded die that was biased in such a way that side 6 had a probability of 1/4. The limiting frequency of 6 in this sequence is, of course, 1/4. Suppose, however, that three of the tosses were *not* made with the biased die, but rather with a fair die. Whatever the outcomes of these three throws, they would have no effect on the limiting frequency. Nevertheless, Popper maintained, we surely want to say that the probability of 6 on those three tosses was 1/6—*not* 1/4. Popper argued that the appropriate way to deal with such cases is to associate the probability with the *chance setup* that produces the outcome, rather than to define it in terms of the sequence of outcomes themselves. Thus, he claims, each time the fair die is thrown, the mechanism—consisting of the die and the thrower—has a causal tendency or *propensity* of 1/6 to produce the outcome 6. Similarly, each time the loaded die is tossed, the mechanism has a propensity of 1/4 to produce the outcome 6.

Although this idea of propensity—probabilistic causal tendency—is important and valuable, it does not provide an admissible interpretation of the probability calculus. This can easily be seen in terms of the case of the frisbee factory introduced in the preceding section. That example, we recall, consisted of two machines, each of which had a certain propensity or tendency to produce defective frisbees. For the new machine the propensity was 0.01; for the old machine it was 0.02. Using the rule of total probability we calculated the propensity of the factory to produce faulty frisbees; it was 0.012. So far, so good.

The problem arises in connection with Bayes's rule. Having picked a defective frisbee at random from the day's production, we asked for the probability that it was produced by the new machine; the answer was 2/3. This is a perfectly legitimate probability, but it cannot be construed as a propensity. It makes no sense to say that this frisbee has a propensity of 2/3 to have been produced by the new machine. Either it was produced by the new machine or by the old. It does not have a tendency of 1/3 to have been produced by the old machine and a tendency of 2/3 to have been produced by the new one. The basic point is that causes precede their effects and causes produce their effects, even if the causal relationship has probabilistic aspects. We can speak meaningfully of the causal tendency of a machine to produce a faulty product. Effects do not produce their causes. It does *not* make sense to talk about the causal tendency of the effect to have been produced by one cause or another.

Bayes's rule enables us to compute what are sometimes called *inverse probabilities*. Whereas the rule of total probability enables us to calculate the *forward* probability of an effect, given suitable information about antecedent causal factors, Bayes's rule allows us to compute the inverse probability that a given effect was produced by a particular cause. These inverse probabilities are an integral part of the mathematical calculus of probability, but no propensities correspond to them. For this reason the propensity interpretation is not an admissible interpretation of the probability calculus.

4. The subjective interpretation.   Both the frequency interpretation and the propensity interpretation are regarded by their proponents as types of *physical probabilities*. They are objective features of the real world. But probability seems to many philosophers and mathematicians to have a subjective side as well. This aspect has something to do with the degree of conviction with which an individual believes in one proposition or another. For instance, Mary Smith is sure that it will be cold in Montana next winter—that is, in some place in that state the temperature will fall below 50 degrees Fahrenheit between 21 December and 21 March. Her subjective probability for this event is extremely close to 1. Also, she disbelieves completely that Antarctica will be hot any time during its summer—that is, she is sure that the temperature will not rise above 100 degrees Fahrenheit between 21 December and 21 March. Her subjective probability for real heat in Antarctica in summer is very close to 0. She neither believes in rain in Pittsburgh tomorrow, nor disbelieves in rain in Pittsburgh tomorrow; her conviction for either one of these alternatives is just as strong as for the other. Her subjective probability for rain tomorrow in Pittsburgh is just about 1/2. As she runs through the various propositions in which she might believe or disbelieve she finds a range of degrees of conviction spanning the whole scale from 0 to 1. Other people will, of course, have different degrees of conviction in these same propositions.

It is easy to see immediately that subjective degrees of commitment do not provide an admissible interpretation of the probability calculus. Take a simple example. Many people believe that the probability of getting a 6 with a fair die is 1/6, and that the outcomes of successive tosses are independent of one another. They also believe that we have a fifty-fifty chance of getting 6 at least once in three throws. As we saw in the previous section, however, that probability is significantly below 1/2. Therefore, the preceding set of degrees of conviction violate the mathematical calculus of probability. Of course, not everyone makes that particular mistake, but extensive empirical research has shown that most of us do make various kinds of mistakes in dealing with probabilities. In general, a given individual's degrees of conviction fail to satisfy the mathematical calculus.

5. Personal probabilities.   What if there were a person whose degrees of conviction did not violate the probability calculus? That person's subjective probabilities would constitute an admissible interpretation. Whether there actually is any such person, we can think of such an organization of our degrees of conviction as an ideal.

Compare this situation with deductive logic. One of its main functions is to help us avoid certain types of logical errors. Anyone who believes, for example, that all humans are mortal and Socrates is human, but that Socrates is immortal, is guilty of self-contradiction. Whoever wants to believe only what is true must try to avoid contradictions, for contradictions cannot possibly be true. In this example, among the three statements, "All humans are mortal," "Socrates is human," and "Socrates is immortal," at least one must be false. Logic does not tell us which statement is false, but it does tell us to make some change in our set of beliefs if we do not want to believe falsehoods. A person who avoids logical contradictions—inconsistencies—has a consistent set of beliefs.

A set of degrees of conviction that violate the calculus of probability is said to be *incoherent*. Anyone who holds a degree of conviction of 1/6 that a fair die, when tossed, will come up 6, and who also considers successive tosses independent (whose degree of conviction in 6 on the next toss is not affected by the outcome of previous tosses), and who is convinced to the degree 1/2 that 6 will come up at least once in three tosses, is being incoherent. So also is anyone who assigns two different values to the probability that a martini mixed by Joe, the sloppy bartender, is between 3:1 and 4:1.

A serious penalty results from being incoherent. A person who has an incoherent set of degrees of conviction is vulnerable to a *Dutch book*. A Dutch book is a set of bets such that, no matter what the outcome of the event on which the bets are made, the subject loses. Conside a very simple example. The negation rule of the probability calculus tells us that $Pr(B|A)$ and $Pr(\sim B|A)$ must add up to 1. Suppose someone has a degree of conviction of 2/3 that the next toss of a particular coin will result in heads, and also a degree of conviction of 2/3 that it will result in tails. This person should be willing to bet at odds of 2 to 1 that the coin will come up heads, and also at odds of 2 to 1 that it will come up tails. These bets constitute a Dutch book because, if the coin comes up heads the subject wins $1 but loses $2, and if it comes up tails the subject loses $2 and wins $1. Since these are the only possible outcomes, the subject loses $1 no matter what happens.

It has been proved in general that a person is subject to a Dutch book if and only if that person holds an incoherent set of degrees of conviction. Thus, we can look at the probability calculus as a kind of system of logic—the logic of degrees of conviction. Conforming to the rules of the probability calculus enables us to avoid certain kinds of blunders in probabilistic reasoning, namely, the type of error that makes one subject to a Dutch book. In light of these considerations, *personal probabilities* have been defined as *coherent sets of degrees of conviction*. It follows immediately that personal probabilities constitute an admissible interpretation of the probability calculus, for they have been defined in just that way.

One of the major motivations of those who accept the personalist interpretation of probability lies in the use of Bayes's rule; indeed, those who adhere to personal probabilities are often called "Bayesians." To see why, let us take another look at Bayes's rule (Rule 9):

$$Pr(H|K.E) = \frac{Pr(H|K) \times Pr(E|K.H)}{Pr(H|K) \times Pr(E|K.H) + Pr(\sim H|K) \times Pr(E|K.\sim H)}$$

provided that $Pr(E|K) \neq 0$.

Consider the following simple example. Suppose that someone in the next room is flipping a penny, and that we receive a reliable report of the outcome after each toss. We cannot inspect the penny, but for some reason we suspect that it is a two-headed coin. To keep the arithmetic simple, let us assume that the coin is either two-headed or fair. Let K stand for our background knowledge and opinion, H for the hypothesis that the coin is two-headed, and E for the results of the flips. For any given individual $Pr(H|K)$—the *prior probability*—represents that person's antecedent degree of conviction that the coin is two-headed before any of the outcomes have been reported. Probability $Pr(E|K.H)$—one of the *likelihoods*—is the probability of the outcome reported to us, given that the coin being flipped is two-headed. If an outcome of tails is reported, that probability obviously equals zero, and the hypothesis H is refuted.

If one or more heads are reported, that probability clearly equals 1. The probability $Pr(\sim H|K)$ is the prior probability that the coin is not two-headed—that is, that it is fair. On pain of incoherence, this probability must equal $1 - Pr(H|K)$. The probability $Pr(E|\sim H.K)$ is also a likelihood; it is the probability of the outcomes reported to us given that the coin is not two-headed. The probability $Pr(H|K.E)$—the *posterior probability*—is the probability that the coin is two-headed given both our background knowledge and knowledge of the results of the tosses. That probability represents an assessment of the hypothesis in the light of the observational evidence (reported reliably to us).

Suppose that John's prior personal probability that the coin is two-headed is 1/100. The result of the first toss is reported, and it is a head. Using Bayes's rule, he computes the posterior probability as follows:

$$\frac{1/100 \times 1}{1/100 \times 1 + 99/100 \times 1/2} = 2/101 \approx 0.02.$$

After two heads the result would be

$$\frac{1/100 \times 1}{1/100 \times 1 + 99/100 \times 1/4} = 4/103 \approx 0.04.$$

After ten heads the result would be

$$\frac{1/100 \times 1}{1/100 \times 1 + 99/100 \times 1/1024} = 1024/1123 \approx 0.91.$$

Suppose Wes's personal prior probability, before any outcomes are known, is much higher than John's; Wes has a prior conviction of 1/2 that the coin is two-headed. After the first head, he makes the following calculation:

$$\frac{1/2 \times 1}{1/2 \times 1 + 1/2 \times 1/2} = 2/3 \approx 0.67.$$

After the second head, he has

$$\frac{1/2 \times 1}{1/2 \times 1 + 1/2 \times 1/4} = 4/5 = 0.80.$$

After ten heads, he has

$$\frac{1/2 \times 1}{1/2 \times 1 + 1/2 \times 1/1024} = 1024/1025 > 0.99.$$

These calculations show two things. First, they show how Bayes's rule can be used to ascertain the probability of a hypothesis *if we have values for the prior probabilities*. If we employ *personal probabilities* the prior probabilities become available. They are simply a person's degrees of conviction in the hypothesis prior to receipt of the observational evidence. In this kind of example the likelihoods can be calculated from assumptions we share concerning the behavior of fair and two-headed coins.

Second, these calculations illustrate a phenomenon known as *washing out of the priors* or *swamping of the priors*. Notice that we did two sets of calculations—one for John and one for Wes. We started with widely divergent degrees of conviction in the hypothesis; Wes's was 1/2 and John's was 1/100. As the evidence accumulated our degrees of conviction became closer and closer. After ten heads, Wes's degree of conviction is approximately 0.99 and John's is approximately 0.91. As more heads occur our agreement becomes even stronger. This illustrates a general feature of Bayes's rule. Suppose there are two people with differing prior probabilities—as far apart as you like provided neither has an extreme value of 0 or 1. Then, if they agree on the likelihoods and if they share the same observational evidence, their posterior probabilities will get closer and closer together as the evidence accumulates. The influence of the prior probabilities on the posterior probabilities decreases as more evidence becomes available. This phenomenon of washing out of the priors should help to ease the worry we might have about appealing to admittedly subjective degrees of conviction in our evaluations of scientific hypotheses.

Still, profound problems are associated with the personalistic interpretation of probability. The only restriction imposed by this interpretation on the values of probabilities is that they be coherent—that they satisfy the rules of mathematical probability. This is a very weak constraint. If we look at the rules of probability we note that, with a couple of trivial exceptions, the mathematical calculus of probability does not by itself furnish us with any values of probabilities. The exceptions are that a logically necessary proposition must have probability 1 and a contradiction must have probability 0. In all other cases, the rules of probability enable us to calculate some probability values from others. You plug in some probability values, turn the crank, and others come out. This means that there need be little contact between our personal probabilities and what goes on in the external world. For example, it is possible for a person to have a degree of conviction of 9/10 that the next toss of a coin will result in heads even though the coin has been tossed hundreds of times and has come up tails on the vast majority of these tosses. By suitably adjusting one's other probabilities one can have such personal probabilities as these without becoming incoherent. If our probabilities are to represent *reasonable* degrees of conviction some stronger restrictions surely appear to be needed.

6. The logical interpretation.   One of the most ambitious twentieth-century attempts to deal with the problems of probability and confirmation was the construction of a theory of *logical probability* by Rudolf Carnap. Carnap was not the first to make efforts in that direction, but his was the most systematic and precise. In fact, Carnap maintained that there are two important and legitimate concepts of probability—relative frequencies and logical probabilities—but his main work was directed toward the latter. He referred to logical probability as *degree of confirmation*. Many philosophers refer to logical probability as *inductive probability*. The three terms are essentially interchangeable.

Carnap's program was straightforward in intent. He believed that it is possible to develop a formal inductive logic along much the same lines as formal deductive logic. In fact, he constructed a basic logical language in which both deductive and inductive relations would reside. In deductive logic, if a statement E entails another statement *H*, *E* supports *H* completely—if *E* is true *H* must also be true. In inductive logic, if *E* is evidence for a hypothesis *H*, *E* provides some sort of *partial support* for *H;* indeed, this type of partial support is often referred to as *partial entailment.*

The easiest way to understand what Carnap did is to work out the details of a simple and highly artificial example. Let us construct a language which deals with a universe containing only three entities, and each of these entities has or lacks one property. We let *a, b,* and *c* denote the three individuals, and we use *F* to designate the property. To make the example concrete, we can think of the individuals as three balls and the property as red. The notation *Fa* says that the first ball is red; ~*Fa* says that it is not red. We need a few other basic logical symbols. We use *x* and *y* as variables for individuals, and (*x*), which is known as the *universal quantifier*, means "for every *x*." The notation ($\exists x$), which is known as the *existential quantifier*, means "there exists at least one *x* such that." A dot "." is used for the conjunction *and;* a wedge "v" for the disjunction *or*. That is about all of the logical equipment we will need.

The model universe we are discussing is a very simple place, and we can describe it completely; indeed, we can describe every *logically possible* state of this universe. Any such complete description of a possible state is a *state description;* there are eight:

1. *Fa.Fb.Fc*	5. *Fa.~Fb.~Fc*
2. *Fa.Fb.~Fc*	6. *~Fa.Fb.~Fc*
3. *Fa.~Fb.Fc*	7. *~Fa.~Fb.Fc*
4. *~Fa.Fb.Fc*	8. *~Fa.~Fb.~Fc*

Any consistent statement that we can form in this miniature language can be expressed by means of these state descriptions. For example, (*x*)*Fx*, which says that every ball is red, is equivalent to state description 1. The statement ($\exists x$)*Fx*, which says that at least one ball is red, is equivalent to the disjunction of state descriptions 1–7; that is, it says that either state description 1 or 2 or 3 or 4 or 5 or 6 or 7 is true. *Fa* is equivalent to the disjunction of state descriptions 1, 2, 3, and 5. *Fa.Fb* is equivalent to the disjunction of state descriptions 1 and 2. If we agree to admit—just for the sake of convenience—that there can be disjunctions with only one term, we can say that every consistent statement is equivalent to some disjunction of state descriptions. The state descriptions in any such disjunction constitute the *range* of that statement. A contradictory statement is equivalent to the denial of all eight of the state descriptions. Its range is empty.

In the following discussion, H is any statement that is being taken as a hypothesis and E any statement that is being taken as evidence. In this discussion any consistent statement that can be formulated in our language can serve as a statement of evidence, and any statement—consistent or inconsistent—can serve as a hypothesis. Now, consider the hypothesis $(\exists x)Fx$ and evidence Fc. Clearly this evidence deductively entails this hypothesis; if the third ball is red at least one *must* be red. If we look at the ranges of this evidence and this hypothesis, we see that the range of Fc (state descriptions 1, 3, 4, 7) is entirely included in the range of $(\exists x)Fx$ (state descriptions 1–7). This situation always holds. If one statement entails another, the range of the first is included within the range of the second. This means that every possible state of the universe in which the first is true is a possible state of the universe in which the second is true. If two statements have identical ranges, they are logically equivalent, and each one entails the other. If two statements are logically incompatible with one another, their ranges do not overlap at all—that is, there is no possible state of the universe in which they can both be true. We see, then, that deductive relationships can be represented as relationships among the ranges of the statements involved.

Let us now turn to inductive relationships. Consider the hypothesis $(x)Fx$ and the evidence Fa. This evidence obviously does not entail the hypothesis, but it seems reasonable to suppose that it provides some degree of inductive support or confirmation. The range of the evidence (1, 2, 3, 5) is not completely included in the range of the hypothesis (1), but it does overlap that range—the two ranges have state description 1 in common. What we need is a way of expressing the idea of confirmation in terms of the overlapping of ranges. When we take any statement E as evidence, we are accepting it as true; in so doing we are ruling out all possible states of the universe that are incompatible with the evidence E. Having ruled out all of those, we want to know to what degree the possible states in which the evidence holds true are possible states in which the hypothesis also holds true. This can be expressed in the form of a ratio, range $(E.H)$/range (E), and this is the basic idea behind the concept of *degree of confirmation*.

Consider the range of $(x)Fx$; this hypothesis holds in one state description out of eight. If, however, we learn that Fa is true, we rule out four of the state descriptions, leaving only four as possibilities. Now the hypothesis holds in one out of four. If we now discover that Fb is also true, our combined evidence $Fa.Fb$ holds in only two state descriptions, and our hypothesis holds in one of the two. It looks reasonable to say that our hypothesis had a probability of 1/8 on the basis of no evidence, a probability of 1/4 on the basis of the first bit of evidence, and a probability of 1/2 on the two pieces of evidence. (This suggestion was offered by Wittgenstein 1922). But appearances are deceiving in this case.

If we were to adopt this suggestion as it stands, Carnap realized, we would rule out altogether the possibility of learning from experience; we would have no basis at all for predicting future occurrences. Consider, instead of $(x)Fx$, the hypothesis Fc. By itself, this hypothesis holds in four (1, 3, 4, 7) out of eight state descriptions. Suppose we find as evidence that Fa. The range of this evidence is four state descriptions (1, 2, 3, 5), and the hypothesis holds in two of them. But $4/8 = 2/4$, so the evidence has done nothing to support the hypothesis. Moreover, if we learn that Fb is true our new evidence is $Fa.Fb$, which holds in two state descriptions (1, 2), and our hypothesis holds in only one of them, giving us a ratio of 1/2. Hence, according to this way of defining confirmation, what we observe in the past and present has no bearing on what will occur in the future. This is an unacceptable consequence. When we examined the hypothesis $(x)Fx$ in the preceding paragraph we appeared to be achieving genuine confirmation, but that was not happening at all. The hypothesis $(x)Fx$ simply states that a, b, and c all have property F. When we find out by observing the first ball that it is red, we have simply reduced the predictive content of h. At first it predicted the color of three balls; after we examine the first ball it predicts the color of only two balls. After we observe the second ball, the hypothesis predicts the color of only one ball. If we were to examine the third ball and find it to be red, our hypothesis would have no predictive content at all. Instead of confirming our hypothesis we were actually simply reducing its predictive import.

In order to get around the foregoing difficulty, Carnap proposed a different way of evaluating the ranges of statements. The method adopted by Wittgenstein amounts to assigning equal weights to all of the state descriptions. Carnap suggested assigning unequal weights on the following basis. Let us take another look at our list of state descriptions in Table 2.2:

TABLE 2.2

State Description	Weight	Structure Description	Weight
1. $Fa.Fb.Fc$	1/4	All F	1/4
2. $Fa.Fb.{\sim}Fc$	1/12		
3. $Fa.{\sim}Fb.Fc$	1/12	2 F, 1 ${\sim}F$	1/4
4. ${\sim}Fa.Fb.Fc$	1/12		
5. $Fa.{\sim}Fb.{\sim}Fc$	1/12		
6. ${\sim}Fa.Fb.{\sim}Fc$	1/12	1 F, 2 ${\sim}F$	1/4
7. ${\sim}Fa.{\sim}Fb.Fc$	1/12		
8. ${\sim}Fa.{\sim}Fb.{\sim}Fc$	1/4	No F	1/4

Carnap noticed that state descriptions 2, 3, and 4 make similar statements about our miniature universe; they say that two entities have property F and one lacks it. Taken together, they describe a certain structure. They differ from one another in identifying the ball that is not red, but Carnap suggests that that is a secondary consideration. Similarly, state descriptions 5, 6, and 7, taken together describe a certain structure, namely, a universe in which one individual has property F and two lack it. Again, they differ in identifying the object that has this property. In contrast, state description 1, all by itself, describes a particular structure, namely, all three entities have property F. Similarly, state description 8 describes the structure in which no object has that property.

Having identified the structure descriptions, Carnap proceeds to assign equal weights to them (each gets 1/4); he then assigns equal weights to the state descriptions within each structure description. The resulting system of weights is shown above. These weights are then used as a measure of the ranges of statements;[11] this system of measures is called m^*. A confirmation function c^* is defined as follows:[12]

$$c^*(H|E) = m^*(H.E)/m^*(E).$$

To see how it works, let us reconsider the hypothesis Fc in the light of different bits of evidence. First, the range of Fc consists of state description 1, which has weight 1/4, and 3, 4, and 7, each of which has weight 1/12. The sum of all of them is 1/2; that is, the probability of our hypothesis before we have any evidence. Now, we find that Fa; its measure is 1/2. The range of $Fa.Fc$ is state descriptions 1 and 3, whose weights are, respectively, 1/4 and 1/12, for a total of 1/3. We can now calculate the degree of confirmation of our hypothesis on this evidence:

$$c^*(H|E) = m^*(E.H)/m^*(E) = 1/3 \div 1/2 = 2/3.$$

Carrying out the same sort of calculation for evidence $Fa.Fb$ we find that our hypothesis has degree of confirmation 3/4. If, however, our first bit of evidence had been $\sim Fa$, the degree of confirmation of our hypothesis would have been 1/3. If our second bit of evidence had been $\sim Fb$, that would have reduced its degree of confirmation to 1/4. The confirmation function c^* seems to do the right sorts of things. When the evidence is what we normally consider to be positive, the degree of confirmation goes up. When the evidence is what we usually take to be negative, the degree of confirmation goes down. Clearly, c^* allows for learning from experience.

A serious philosophical problem arises, however. Once we start playing the game of assigning weights to state descriptions, we face a huge plethora of possibilities. In setting up the machinery of state descriptions and weights, Carnap demands only that the weights for all of the state descriptions add up to 1, and that each state description have a weight greater than 0. These conditions are sufficient to guarantee an admissible interpretation of the probability calculus. Carnap recognized the obvious fact that infinitely many confirmation functions satisfying this basic requirement are possible. The question is how to make an appropriate choice. It can easily be shown that choosing a confirmation function is precisely the same as assigning prior probabilities to all of the hypotheses that can be stated in the given language.

Consider the following possibility for a measure function:

---

[11] The measure of the range of any statement H can be identified with the prior probability of that statement in the absence of any background knowledge K. It is an *a priori* prior probability.

[12] Wittgenstein's measure function assigns the weight 1/8 to each state description; the confirmation function based upon it is designated $c\dagger$.

The Confirmation of Scientific Hypotheses

TABLE 2.3

State Description	Weight	Structure Description	Weight
1. *Fa.Fb.Fc*	1/20	All *F*	1/20
2. *Fa.Fb.~Fc*	3/20		
3. *Fa.~Fb.Fc*	3/20	2 *F*, 1 ~*F*	9/20
4. *~Fa.Fb.Fc*	3/20		
5. *Fa.~Fb.~Fc*	3/20		
6. *~Fa.Fb.~Fc*	3/20	1 *F*, 2 ~*F*	9/20
7. *~Fa.~Fb.Fc*	3/20		
8. *~Fa.~Fb.~Fc*	1/20	No *F*	1/20

(The idea of a confirmation function of this type was given in Burks 1953; the philosophical issues are further discussed in Burks 1977, Chapter 3.) This method of weighting, which may be designated $m^\diamond$, yields a confirmation function $C^\diamond$, which is a sort of counterinductive method. Whereas $m*$ places higher weights on the first and last state descriptions, which are state descriptions for universes with a great deal of uniformity (either every object has the property, or none has it), $m^\diamond$ places lower weights on descriptions of uniform universes. Like $c*$, $c^\diamond$ allows for "learning from experience," but it is a funny kind of anti-inductive "learning." Before we reject $m^\diamond$ out of hand, however, we should ask ourselves if we have any a priori guarantee that our universe is uniform. Can we select a suitable confirmation function without being totally arbitrary about it? This is the basic problem with the logical interpretation of probability.

Part IV: Confirmation and Probability

2.9 THE BAYESIAN ANALYSIS OF CONFIRMATION

We now turn to the task of illustrating how the probabilistic apparatus developed above can be used to illuminate various issues concerning the confirmation of scientific statements. Bayes's theorem (Rule 9) will appear again and again in these illustrations, justifying the appellation of Bayesian confirmation theory.

Various ways are available to connect the probabilistic concept of confirmation back to the qualitative concept, but perhaps the most widely followed route utilizes an incremental notion of confirmation: E confirms H relative to the background knowledge K just in case the addition of E to K raises the probability of H, that is, $Pr(H|E.K) >) Pr(H|K)$.[13] Hempel's study of instance confirmation in terms of a

---

[13] Sometimes, when we say that a hypothesis has been confirmed, we mean that it has been rendered highly probable by the evidence. This is a *high probability* or *absolute* concept of confirmation, and it should be carefully distinguished from the *incremental* concept now under discussion (see Carnap 1962, Salmon 1973, and Salmon 1975). Salmon (1973) is the most elementary discussion.

two-place relation can be taken to be directed at the special case where K contains no information. Alternatively, we can suppose that K has been absorbed into the probability function in the sense that $Pr(K) = 1$,[14] in which case the condition for incremental confirmation reduces to $Pr(H|E) > Pr(H)$. (The unconditional probability $Pr(H)$ can be understood as the conditional probability $Pr(H|T)$, where T is a vacuous statement, for example, a tautology. The axioms of Section 2.7 apply only to conditional probabilities.)

It is easy to see that on the incremental version of confirmation, Hempel's consistency condition is violated as is

Conjunction condition: If E confirms H and also H' then E confirms $H.H'$.

It takes a bit more work to construct a counterexample to the special consequence condition. (This example is taken from Carnap 1950 and Salmon 1975, the latter of which contains a detailed discussion of Hempel's adequacy conditions in the light of the incremental notion of confirmation.) Towards this end take the background knowledge to contain the following information. Ten players participate in a chess tournament in Pittsburgh; some are locals, some are from out of town; some are juniors, some are seniors; and some are men (M), some are women (W). Their distribution is given by

TABLE 2.4

	Locals	Out-of-towners
Juniors	M, W, W	M, M
Seniors	M, M	W, W, W

And finally, each player initially has an equal chance of winning. Now consider the hypotheses H: an out-of-towner wins, and H': a senior wins, and the evidence E: a woman wins. We find that

$$Pr(H|E) = 3/5 > Pr(H) = 1/2$$

so E confirms H. But

$$Pr(H \lor H'|E) = 3/5 <(Pr(H \lor H') = 7/10.$$

So E does not confirm $H \lor H'$; in fact E confirms $\sim(H \lor H')$ and so disconfirms $H \lor H'$ even though $H \lor H'$ is a consequence of H.

The upshot is that on the incremental conception of confirmation, Hempel's adequacy conditions and, hence, his definition of qualitative confirmation, are inadequate. However, his adequacy conditions fare better on the high probability conception of confirmation according to which E confirms H relative to K just in case $Pr(H|E.K) > r$, where r is some number greater than 0.5. But this notion of

[14] As would be the case if learning from experience is modeled as change of probability function through conditionalization; that is, when K is learned, Pr_{old} is placed by $Pr_{new}(\ ) = Pr_{old}(\ |\ K)$. From this point of view, Bayes's theorem (Rule 9) describes how probability changes when a new fact is learned.

confirmation cannot be what Hempel has in mind; for he wants to say that the observation of a single black raven (E) confirms the hypothesis that all ravens are black (H), although for typical K, $Pr(H|E.K)$ will surely not be as great as 0.5. Thus, in what follows we continue to work with the incremental concept.

The probabilistic approach to confirmation coupled with a simple application of Bayes's theorem also serves to reveal a kernel of truth in the H-D method. Suppose that the following conditions hold:

(i) H, $K \vdash E$; (ii) $1 > Pr(H|K) > 0$; and (iii) $1 > Pr(E|K) > 0$.

Condition (i) is the basic H-D condition. Conditions (ii) and (iii) say that neither H nor E is known on the basis of the background information K to be almost surely false or almost surely true. Then on the incremental conception it follows, as the H-D methodology would have it, that E confirms H on the basis of K. By Bayes's theorem

$$Pr(H|E.K) = \frac{Pr(H|K)}{Pr(E|K)}$$

since by (i),

$$Pr(E|H.K) = 1.$$

It then follows from (ii) and (iii) that

$$Pr(H|E.K) > Pr(H|K).$$

Notice also that the smaller $Pr(E|K)$ is, the greater the incremental confirmation afforded by E. This helps to ground the intuition that "surprising" evidence gives better confirmational value. However, this observation is really double-edged as will be seen in Section 2.10.

The Bayesian analysis also affords a means of handling a disquieting feature of the H-D method, sometimes called the problem of irrelevant conjunction. If the H-D condition (i) holds for H, then it also holds for $H.X$ where X is anything you like, including conjuncts to which E is intuitively irrelevant. In one sense the problem is mirrored in the Bayesian approach, for assuming that $1 > Pr(H.X|K) > 0$, it follows that E incrementally confirms $H.X$. But since the special consequence condition does not hold in the Bayesian approach, we cannot infer that E confirms the consequence X of $H.X$. Moreover, under the H-D condition (i), the incremental confirmation of a hypothesis is directly proportional to its prior probability. Since $Pr(H|K) \geq Pr(H.X|K)$, with strict inequality holding in typical cases, the incremental confirmation for H will be greater than for $H.X$.

Bayesian methods are flexible enough to overcome various of the shortcomings of Hempel's account. Nothing, for example, prevents the explication of confirmation in terms of a Pr-function which allows observational evidence to boost the probability of theoretical hypotheses. In addition the Bayesian approach illuminates the paradoxes of the ravens and Goodman's paradox.

In the case of the ravens paradox we may grant that the evidence that the individual a is a piece of white chalk can confirm the hypothesis that "All ravens are black" since, to put it crudely, this evidence exhausts part of the content of the hypothesis. Nevertheless, as Suppes (1966) has noted, if we are interested in subjecting the hypothesis to a sharp test, it may be preferable to do outdoor ornithology and sample from the class of ravens rather than sampling from the class of nonblack things. Let a denote a randomly chosen object and let

$$Pr(Ra.Ba) = p_1, \qquad Pr(Ra.{\sim}Ba) = p_2$$
$$Pr({\sim}Ra.Ba) = p_3, \qquad Pr({\sim}Ra.{\sim}Ba) = p_4.$$

Then

$$Pr({\sim}Ba|Ra) = p_2 \neq (p_1 + p_2)$$
$$Pr(Ra|{\sim}Ba) = p_2 \neq (p_2 + p_4)$$

Thus, $Pr({\sim}Ba|Ra) > Pr(Ra|{\sim}Ba)$ just in case $p_4 > p_1$. In our world it certainly seems true that $p_4 > p_1$. Thus, Suppes concludes that sampling ravens is more likely to produce a counterinstance to the ravens hypothesis than is sampling the class of nonblack things.

There are two problems here. The first is that it is not clear how the last statement follows since a was supposed to be an object drawn at random from the universe at large. With that understanding, how does it follow that $Pr({\sim}Ba|Ra)$ is the probability that an object drawn at random from the class of ravens is nonblack? Second, it is the anti-inductivists such as Popper (see item 4 in Section 2.8 above and 2.10 below) who are concerned with attempts to falsify hypotheses. It would seem that the Bayesian should concentrate on strategies that enhance absolute and incremental probabilities. An approach due to Gaifman (1979) and Horwich (1982) combines both of these points.

Let us make it part of the background information K that a is an object drawn at random from the class of ravens while b is an object drawn at random from the class of nonblack things. Then an application of Bayes's theorem shows that

$$Pr(H|Ra.Ba.K) > Pr(H|{\sim}Rb.{\sim}Bb.K)$$

just in case

$$1 > Pr({\sim}Rb|K) > Pr(Ba|K).$$

To explore the meaning of the latter inequality, use the principle of total probability to find that

$$Pr(Ba|K) = Pr(Ba|H.K) \cdot Pr(H|K) + Pr(Ba|{\sim}H.K) \cdot Pr({\sim}H|K)$$
$$= Pr(H|K) + Pr(Ba|{\sim}H.K) \cdot Pr({\sim}H|K)$$

and that

$$Pr({\sim}Rb|K) = Pr(H|K) + Pr({\sim}Rb|{\sim}H.K) \cdot Pr({\sim}H|K).$$

So the inequality in question holds just in case

$$1 > Pr({\sim}Rb|{\sim}H.K) > Pr(Ba|{\sim}H.K),$$

or

$$Pr({\sim}Ba|{\sim}H.K) > Pr(Rb|{\sim}H.K) > 0,$$

which is presumably true in our universe. For supposing that some ravens are non-black, a random sample from the class of ravens is more apt to produce such a bird than is a random sample from the class of nonblack things since the class of nonblack things is much larger than the class of ravens. Thus, under the assumption of the stated sampling procedures, the evidence $Ra.Ba$ does raise the probability of the ravens hypothesis more than the evidence $\sim Rb.\sim Bb$ does. The reason for this is precisely the differential propensities of the two sampling procedures to produce counterexamples, as Suppes originally suggested.

The Bayesian analysis also casts light on the problems of induction, old and new, Humean and Goodmanian. Russell (1948) formulated two categories of induction by enumeration:

> Induction by simple enumeration is the following principle: "Given a number n of α's which have been found to be β's, and no α which has been found to be not a β, then the two statements: (a) 'the next α will be a β,' (b) 'all α's are β's,' both have a probability which increases as n increases, and approaches certainty as a limit as n approaches infinity."
>
> I shall call (a) "particular induction" and (b) "general induction." (1948, 401)

Between Russell's "particular induction" and his "general induction" we can interpolate another type, as the following definitions show (note that Russell's "α" and "β" refer to properties, not to individual things):

Def. Relative to K, the predicate "P" is *weakly projectible* over the sequence of individuals $a_1, a_2, \ldots$ just in case[15]

$$\lim_{n \to \infty} Pr(Pa_{n+1}|Pa_1. \ldots .Pa_n.K) = 1.$$

Def. Relative to K, "P" is *strongly projectible* over $a_1, a_2, \ldots$ just in case

$$\lim_{n, m \to \infty} Pr(Pa_{n+1}. \ldots .Pa_{n+m}| Pa_1. \ldots .Pa_n.K) = 1.$$

(The notation $\lim_{m, n \to \infty}$ indicates the limit as m and n both tend to infinity in any manner you like.) A sufficient condition for both weak and strong probability is that the general hypothesis H: $(i)Pa_i$ receives a nonzero prior probability. To see that it is sufficient for weak projectibility, we follow Jeffreys's (1957) proof. By Bayes's theorem

$$Pr(H|Pa_1. \ldots .Pa_{n+1} . K) = \frac{Pr(Pa_1. \ldots .Pa_{n+1}|H.K) \cdot Pr(H|K)}{Pr(Pa_1. \ldots .Pa_{n+1}|K)}$$

$$= \frac{Pr(H|K)}{Pr(Pa_1|K) \cdot Pr(Pa_2|Pa_1.K) \cdot \ldots \cdot Pr(Pa_{n+1}|Pa_1. \ldots .Pa_n \cdot K)}$$

---

[15] Equation $\lim_{n \to \infty} x_n = L$ means that, for any real number $\epsilon > 0$, there is an integer $N > 0$ such that, for all $n > N$, $|x_n - L| < \epsilon$.

Unless $Pr(Pa_{n+1}|Pa_1. \ldots .Pa_n.K)$ goes to 1 as $n \to \infty$, the denominator on the right-hand side of the second equality will eventually become less than $Pr(H|K)$, contradicting the truth of probability that the left-hand side is no greater than 1.

The posit that

$$(P) \ Pr([(i)Pa_i|K] > 0$$

is not necessary for weak projectibility. Carnap's systems of inductive logic (see item 6 in Section 2.8 above) are relevant examples since in these systems (P) fails in a universe with an infinite number of individuals although weak projectibility can hold in these systems.[16] But if we impose the requirement of countable additivity

$$(CA) \ \lim_{n \to \infty} Pr(Pa_i. \ldots .Pa_n|K) = Pr[(i) Pa_i|K]$$

then (P) is necessary as well as sufficient for strong projectibility.

Also assuming (CA), (P) is sufficient to generate a version of Russell's "general induction," namely

$$(G) \ \lim_{n \to \infty} Pr[(i)Pa_i|Pa_1. \ldots .Pa_n.K] = 1.$$

(Russell 1948 lays down a number of empirical postulates he thought were necessary for induction to work. From the present point of view these postulates can be interpreted as being directed to the question of which universal hypotheses should be given nonzero priors.)

Humean skeptics who regiment their beliefs according to the axioms of probability cannot remain skeptical about the next instance or the universal generalization in the face of ever-increasing positive instances (and no negative instances) unless they assign a zero prior to the universal generalization. But

$$Pr[(i)Pa_i|K] = 0$$

implies that

$$Pr[(\exists i) \sim Pa_i|K] = 1,$$

which says that there is certainty that a counterinstance exists, which does not seem like a very skeptical attitude.

---

[16] A nonzero prior for the general hypothesis is a necessary condition for strong projectibility but not for weak projectibility. The point can be illustrated by using de Finetti's representation theorem, which says that if P is exchangeable over $a_1, a_2, \ldots$ (which means roughly that the probability does not depend on the order) then:

$$Pr(Pa_1.Pa_2. \ldots .Pa_n \mid K) = \int_0^1 \theta^n \, d\mu(\theta)$$

where $\mu(\theta)$ is a uniquely determined measure on the unit interval $0 \le \theta \le 1$. For the uniform measure $d\mu(\theta) = d(\theta)$ we have

$$Pr(Pa_{n+1}|Pa_1. \ldots . Pa_n. K) = n + 1/n + 2$$

and

$$Pr(Pa_{n+1}. \ldots . Pa_{n+m}|Pa_1. \ldots . Pa_n.K) = m + 1/n + m + 1.$$

Note also that the above results on instance induction hold whether "P" is a normal or a Goodmanized predicate—for example, they hold just as well for $P*a_i$ which is defined as

$$[(i \leq 2000).Pa_i] \vee [(i > 2000).{\sim}Pa_i)],$$

where Pa_i means that a_i is purple. But this fact just goes to show how weak the results are; in particular, they hold only in the limit as $n \to \infty$ and they give no information about how rapidly the limit is approached.

Another way to bring out the weakness is to note that (P) does not guarantee even a weak form of Hume projectibility.

> *Def.* Relative to K, "P" is *weakly Hume projectible* over the doubly infinite sequence . . . , $a_{-2}, a_{-1}, a_0, a_1, a_2, \ldots$ just in case for any n,
> $\lim\limits_{k \to \infty} Pr(Pa_n|Pa_{n-1}. \ldots .Pa_{n-k} . K) = 1.$

(To illustrate the difference between the Humean and non-Humean versions of projectibility, let Pa_n mean that the sun rises on day n. The non-Humean form of projectibility requires that if you see the sun rise on day 1, on day 2, and so on, then for any $\varepsilon > 0$ there will come a day N when your probability that the sun will rise on day $N + 1$ will be at least $1 - \varepsilon$. By contrast, Hume projectibility requires that if you saw the sun rise yesterday, the day before yesterday, and so on into the past, then eventually your confidence that the sun will rise tomorrow approaches certainty.)

If (P) were sufficient for Hume projectibility we could assign nonzero priors to both $(i)Pa_i$ and $(i)P*a_i$, with the result that as the past instances accumulate, the probabilities for Pa_{2001} and for $P*a_{2001}$ both approach 1, which is a contradiction.

A sufficient condition for Hume projectibility is exchangeability.

> *Def.* Relative to K, "P" is *exchangeable* for Pr over the a_is just in case for any n and m
>
> $$Pr(\pm Pa_n. \ldots .\pm Pa_{n + m}|K) = Pr(\pm Pa_{n'}. \ldots .\pm Pa_{n' + m}|K)$$

where $\pm$ indicates that either P or its negation may be chosen and $[a_{i'}]$ is any permutation of the a_is in which all but a finite number are left fixed. Should we then use a Pr-function for which the predicate "purple" is exchangeable rather than the Goodmanized version of "purple"? Bayesianism per se does not give the answer anymore than it gives the answer to who will win the presidential election in the year 2000. But it does permit us to identify the assumptions needed to guarantee the validity of one form or another of induction.

Having touted the virtues of the Bayesian approach to confirmation, it is now only fair to acknowledge that it is subject to some serious challenges. If it can rise to these challenges, it becomes all the more attractive.

2.10 CHALLENGES TO BAYESIANISM

1. Nonzero priors. Popper (1959) claims that "in an infinite universe . . . the probability of any (non-tautological) universal law will be zero." If Popper were right and universal generalizations could not be probabilified, then Bayesianism would be worthless as applied to theories of the advanced sciences, and we would presumably have to resort to Popper's method of corroboration (see item 4 in Section 2.8 above).

To establish Popper's main negative claim it would suffice to show that the prior probability of a universal generalization must be zero. Consider again $H: (i)Pa_i$. Since for any n

$$H \vdash Pa_1. Pa_2. \dots .Pa_n,$$
$$Pr(H|K) \leq \lim_{n \to \infty} Pr(Pa_1. \dots .Pa_n|K).$$

Now suppose that

(I)   For all n, $Pr(Pa_1. \dots .Pa_n|K) = Pr(Pa_1|K) \cdot \dots \cdot Pr(Pa_n|K)$

and that

(E)   For any m and n, $Pr(Pa_m|K) = Pr(Pa_n|K)$.

Then except for the uninteresting case that $Pr(Pa_n|K) = 1$ for each n, it follows that

$$\lim_{n \to \infty} Pr(Pa_1. \dots .Pa_n|K) = 0$$

and thus that $Pr(H|K) = 0$.

Popper's argument can be attacked in various places. Condition (E) is a form of exchangeability, and we have seen above that it cannot be expected to hold for all predicates. But Popper can respond that if (E) does fail then so will various forms of inductivism (e.g., Hume projectibility). The main place the inductivist will attack is at the assumption (I) of the independence of instances. Popper's response is that the rejection of (I) amounts to the postulation of something like a causal connection between instances. But this a red herring since the inductivist can postulate a probabilistic dependence among instances without presupposing that the instances are cemented together by some sort of causal glue.

In another attempt to show that probabilistic methods are ensnared in inconsistencies, Popper cites Jeffreys's proof sketched above that a non-zero prior for $(i)Pa_i$ guarantees that

$$\lim_{n \to \infty} Pr(Pa_{n+1}|Pa_1. \dots .Pa_n. K) = 1.$$

But, Popper urges, what is sauce for the goose is sauce for the gander. For we can do the same for a Goodmanized P^*, and from the limit statements we can conclude that for some $r > 0.5$ there is a sufficiently large N such that for any $N' > N$, the probabilities for $P_{a_{N'}}$ and for $P^*_{a_{N'}}$ are both greater than r, which is a contradiction for appropriately chosen P^*. But the reasoning here is fallacious and there is in fact no contradiction lurking in Jeffreys's limit theorem since the convergence is not supposed to be uniform over different predicates—indeed, Popper's reasoning shows that it cannot be.

Of course, none of this helps with the difficult questions of which hypotheses should be assigned nonzero priors and how large the priors should be. The example from item 5 in Section 2.8 above suggests that the latter question can be ignored to some extent since the accumulation of evidence tends to swamp differences in priors and force merger of posterior opinion. Some powerful results from advanced probability theory show that such merger takes place in a very general setting (on this matter see Gaifman and Snir 1982).

2. Probabilification vs. inductive support.   Popper and Miller (1983) have argued that even if it is conceded that universal hypotheses may have nonzero priors and thus can be probabilified further and further by the accumulation of positive evidence, the increase in probability cannot be equated with genuine inductive support. This contention is based on the application of two lemmas from the probability calculus:

Lemma 1. $Pr(\sim H|E.K) \times Pr(\sim E|K) = Pr(H \ \lor \ \sim E|K) - Pr(H \ \lor \ \sim E|E.K)$.

Lemma 1 leads easily to

Lemma 2. If $Pr(H|E.K) < 1$ and $Pr(E|K) < 1$ then

$$Pr(H \lor \sim E|E.K) < Pr(H \lor \sim E|K).$$

Let us apply Lemma 2 to the case discussed above where Bayesianism was used to show that under certain conditions the H-D method does lead to incremental confirmation. Recall that we assumed that

$$H, K \vdash E; \ 1 > Pr(E|K) > 0; \ \text{and} \ 1 > Pr(H|K) > 0$$

and then showed that

$$Pr(H|E.K) > Pr(H|K),$$

which the inductivists want to interpret as saying that E inductively supports H on the basis of K. Against this interpretation, Popper and Miller note that H is logically equivalent to $(H \lor E).(H \lor \sim E)$. The first conjunct is deductively implied by E, leading Popper and Miller to identify the second conjunct as the part of H that goes beyond the evidence. But by Lemma 2 this part is *countersupported* by E, except in the uninteresting case that $E.K$ makes H probabilistically certain.

Jeffrey (1984) has objected to the identification of $H \lor \sim E$ as the part of H that goes beyond the evidence. To see the basis of his objection, take the case where

$$H: (i)Pa_i \ \text{and} \ E: Pa_1. \ \ldots .Pa_n.$$

Intuitively, the part of H that goes beyond this evidence is $(i) [(i > n) \ldots Pa_i]$ and not the Popper-Miller $(i)Pa_1 \lor \sim(Pa_1. \ \ldots .Pa_n)$.

Gillies (1986) restated the Popper-Miller argument using a measure of inductive support based on the incremental model of confirmation: (leaving aside K) the support given by E to H is $S(H, E) = Pr(H|E) - Pr(H)$. We can then show that

Lemma 3. $S(H, E) = S(H \lor E, E) + S(H \lor \sim E, E)$.

Gillies suggested that $S(H \lor EE,)$ be identified as the deductive support given H by E and $S(H \lor {\sim}E, E)$ as the inductive support. And as we have already seen, in the interesting cases the latter is negative. Dunn and Hellman (1986) responded by dualizing. Hypothesis H is logically equivalent to $(H.E) \lor (H.{\sim}E)$ and $S(H, E) = S(H.E, E) + S(H.{\sim}E, E)$. Identify the second component as the deductive countersupport. Since this is negative, any positive support must be contributed by the first component which is a measure of the nondeductive support.

3. The problem of old evidence.

In the Bayesian identification of the valid kernel of the H-D method we assumed that $Pr(E|K) < 1$, that is, there was some surprise to the evidence E. But this is often not the case in important historical examples. When Einstein proposed his general theory of relativity (H) at the close of 1915 the anomalous advance of the perihelion of Mercury (E) was old news, that is, $Pr(E|K) = 1$. Thus, $Pr(H|E.K) = Pr(H|K)$, and so on the incremental conception of confirmation, Mercury's perihelion does not confirm Einstein's theory, a result that flies in the face of the fact that the resolution of the perihelion problem was widely regarded as one of the major triumphs of general relativity. Of course, one could seek to explain the triumph in nonconfirmational terms, but that would be a desperate move.

Garber (1983) and Jeffrey (1983) have suggested that Bayesianism be given a more human face. Actual Bayesian agents are not logically omniscient, and Einstein for all his genius was no exception. When he proposed his general theory he did not initially know that it did in fact resolve the perihelion anomaly, and he had to go through an elaborate derivation to show that it did indeed entail the missing 43" of arc per century. Actual flesh and blood scientists learn not only empirical facts but logicomathematical facts as well, and if we take the new evidence to consist in such facts we can hope to preserve the incremental model of confirmation. To illustrate, let us make the following assumptions about Einstein's degrees of belief in 1915:

(a) $Pr(H|K) > 0$ (Einstein assigned a nonzero prior to his general theory.)

(b) $Pr(E|K) = 1$ (The perihelion advance was old evidence.)

(c) $Pr(H \vdash E|K) < 1$ (Einstein was not logically omniscient and did not invent his theory so as to guarantee that it entailed the 43".)

(d) $Pr[(H \vdash E) \lor (H \vdash {\sim}E)|K] = 1$ (Einstein knew that his theory entailed a definite result for the perihhelion motion.)

(e) $Pr[H.(H \vdash {\sim}E)|K] = Pr[H.(H \vdash {\sim}E).{\sim}E|K]$ (Constraint on interpreting $\vdash$ as logical implication.)

From (a)–(e) it can be shown that $Pr[H|(H \vdash E).K]. > Pr(H|K)$. So learning that his theory entailed the happy result served to increase Einstein's confidence in the theory.

Although the Garber-Jeffrey approach does have the virtue of making Bayesian agents more human and, therefore, more realistic, it avoids the question of whether the perihelion phenomena did in fact confirm the general theory of relativity in favor of focusing on Einstein's personal psychology. Nor is it adequate to dismiss this concern with the remark that the personalist form of Bayesianism is concerned precisely with psychology of particular agents, for even if we are concerned principally with Einstein himself, the above calculations seem to miss the mark. We now believe that for Einstein in 1915 the perihelion phenomena provided a strong confirmation of his general theory. And contrary to what the Garber-Jeffrey approach would suggest, we would not change our minds if historians of science discovered a manuscript showing that as Einstein was writing down his field equations he saw in a flash of mathematical insight that $H \vdash E$ or alternatively that he consciously constructed his field equations so as to guarantee that they entailed E. "Did E confirm H for Einstein?" and "Did learning that $H \vdash E$ increase Einstein's confidence in H?" are two distinct questions with possibly different answers. (In addition, the fact that agents are allowed to assign $Pr\ (H \vdash E|K) < 1$ means that the Dutch book justification for the probability axioms has to be abandoned. This is anathema for orthodox Bayesian personalists who identify with the betting quotient definition of probability.)

A different approach to the problem of old evidence is to apply the incremental model of confirmation to the counterfactual degrees of belief that would have obtained had E not been known. Readers are invited to explore the prospects and problems of this approach for themselves. (For further discussion of the problem of old evidence, see Howson 1985, Eells 1985, and van Fraassen 1988.)

2.11 CONCLUSION

The topic of this chapter has been *the logic of science*. We have been trying to characterize and understand the patterns of inference that are considered legitimate in establishing scientific results—in particular, in providing support for the hypotheses that become part of the corpus of one science or another. We began by examining some extremely simple and basic modes of reasoning—the hypothetico-deductive method, instance confirmation, and induction by enumeration. Certainly (pace Popper) all of them are frequently employed in actual scientific work.

We find—both in contemporary science and in the history of science—that scientists do advance hypotheses from which (with the aid of initial conditions and auxiliary hypotheses) they deduce observational predictions. The test of Einstein's theory of relativity in terms of the bending of starlight passing close to the sun during a total solar eclipse is an oft-cited example. Others were given in this chapter. Whether the example is as complex as general relativity or as simple as Boyle's law, the logical problems are the same. Although the H-D method contains a valid kernel—as shown by Bayes's rule—it must be considered a serious oversimplification of what actually is involved in scientific confirmation. Indeed, Bayes's rule itself seems to offer a schema far more adequate than the H-D method. But—as we have seen—it, too, is open to serious objections (such as the problem of old evidence).

When we looked at Hempel's theory of instance confirmation, we discussed an example that has been widely cited in the philosophical literature—namely, the generalization "All ravens are black." If this is a scientific generalization, it is certainly at a low level, but it is not scientifically irrelevant. More complex examples raise the same logical problems. At present, practicing scientists are concerned with—and excited by—such generalizations as, "All substances having the chemical structure given by the formula $YBa_2Cu_3O_7$ are superconductors at 70 kelvins." As if indoor ornithology weren't bad enough, we see, by Hempel's analysis, that we can confirm this latter-day generalization by observing black crows. It seems that observations by birdwatchers can confirm hypotheses of solid state physics. (We realize that bird-lovers would disapprove of the kind of test that would need to be performed to establish that a raven is not a superconductor at 70°K.) We have also noted, however, the extreme limitations of the kind of evidence that can be gathered in any such fashion.

Although induction by enumeration is used to establish universal generalizations, its most conspicuous use in contemporary science is connected with statistical generalizations. An early example is found in Rutherford's counting of the frequencies with which alpha particles bombarding a gold foil were scattered backward (more or less in the direction from which they came). The counting of instances led to a statistical hypothesis attributing stable frequencies to such events. A more recent example—employing highly sophisticated experiments—involves the detection of neutrinos emitted by the sun. Physicists are puzzled by the fact that they are detecting a much smaller frequency than current theory predicts. (Obviously probabilities of the type characterized as frequencies are involved in examples of the sort mentioned here.) In each of these cases an inductive extrapolation is drawn from observed frequencies. In our examination of induction by enumeration, however, we have found that it is plagued by Hume's old riddle and Goodman's new one.

One development of overwhelming importance in twentieth-century philosophy of science has been the widespread questioning of whether there is any such thing as a *logic* of science. Thomas Kuhn's influential work, *The Structure of Scientific Revolutions* (1962, 1970), asserted that the choice of scientific theories (or hypotheses) involves factors that go beyond observation and logic—including judgement, persuasion, and various psychological and sociological influences. There is, however, a strong possibility that, when he wrote about going beyond the bounds of observation and logic, the kind of logic he had in mind was the highly inadequate H-D schema, (see Salmon 1989 for an extended discussion of this question, and for an analysis of Kuhn's views in the light of Bayes's rule). The issues raised by the Kuhnian approach to philosophy of science are discussed at length in Chapter 4 of this book.

Among the problems we have discussed there are—obviously—many to which we do not have adequate solutions. Profound philosophical difficulties remain. But the deep and extensive work done by twentieth-century philosophers of science in these areas has cast a good deal of light on the nature of the problems. It is an area in which important research is currently going on and in which significant new results are to be expected.

DISCUSSION QUESTIONS

1. Select a science with which you are familiar and find a case in which a hypothesis or theory is taken to be confirmed by some item of evidence. Try to characterize the relationship between the evidence and hypothesis or theory confirmed in terms of the schemas discussed here. If none of them is applicable, can you find a new schema that is?

2. If the prior probability of every universal hypothesis is zero how would you have to rate the probability of the statement that unicorns (at least one) exist? Explain your answer.

3. Show that accepting the combination of the entailment condition, the special consequence condition, and the converse consequence condition (see Section 2.4) entails that any E confirms any H.

4. Consider a population that consists of all of the adult population of some particular district. We want to test the hypothesis that all voters are literate,

$$(x)(Vx \supset Lx),$$

which is, of course, equivalent to

$$(x)(\sim Vx \supset \sim Lx).$$

Suppose that approximately 75 percent of the population are literate voters, approximately 15 percent are literate nonvoters, approximately 5 percent are illiterate nonvoters, and approximately 5 percent are illiterate voters—but this does not preclude the possibility that no voters are illiterate. Would it be best to sample the class of voters or the class of illiterate people? Explain your answer. (This example is given in Suppes 1966, 201.)

5. Goodman's examples challenge the idea that hypotheses are confirmed by their instances. Goodman holds that the distinction between those hypotheses that are and those that are not projectable on the basis of their instances is to be drawn in terms of *entrenchment*. Predicates become entrenched as antecedents or consequents by playing those roles in universal conditionals that are actually projected. Call a hypothesis *admissible* just in case it has some positive instances, no negative instances, and is not exhausted. Say that *H overrrides H'* just in case *H* and *H'* conflict, *H* is admissible and is better entrenched than *H'* (i.e., has a better entrenched antecedent and equally well entrenched consequent or vice versa), and *H* is not in conflict with some still better entrenched admissible hypothesis. Critically discuss the idea that *H* is projectable on the basis of its positive instances just in case it is admissible but not overridden.

6. Show that

$$H: (x) (\exists y) Rxy.(x) \sim Rxx.(x) (y) (z) [(Rxy.Ryz) \supset Rxz]$$

cannot be Hempel-confirmed by any consistent E.

7. It is often assumed in philosophy of science that if one is going to represent numerically the degree to which evidence E supports hypothesis H with respect to background B, then the numbers so produced — $P(H|E.B)$ — must obey the probability calculus. What are the prospects of alternative calculi? (Hint: Consider each of the axioms in turn and ask under what circumstances each axiom could be violated in the context of a confirmation theory. What alternative axiom might you choose?)

8. If Bayes's rule is taken as a schema for confirmation of scientific hypotheses, it is necessary to decide on an interpretation of probability that is suitable for that context. It is especially crucial to think about how the prior probabilities are to be understood. Discuss this problem in the light of the admissible interpretations offered in this chapter.

9. William Tell gave his young cousin Wesley a two-week intensive archery course. At its completion, William tested Wes's skill by asking him to shoot arrows at a round target, ten feet in radius with a centered bull's-eye, five feet in radius.

"You have learned *no* control at all," scolded William after the test. "Of those arrows that hit the target, five are within five feet of dead center and five more between five and ten feet from dead center." "Not so," replied Wes, who had been distracted from archery practice by his newfound love of geometry. "That five out of ten arrows on the target hit the bull's-eye shows I *do* have control. The bullseye is only one quarter the total area of the target."

Adjudicate this dispute in the light of the issues raised in the chapter. Note that an alternative form of Bayes's rule which applies when one considers the relative confirmation accrued by two hypotheses H_1 and H_2 by evidence E with respect to background B is:

$$\frac{Pr(H_1|E.B)}{Pr(H_2|E.B)} = \frac{Pr(E|H_1.B)}{Pr(E|H_2.B)} \cdot \frac{Pr(H_1|B)}{Pr(H_2|B)}$$

10. Let $\{H_1, H_2, \ldots, H_n\}$ be a set of competing hypotheses. Say that E selectively Hempel-confirms some H_j just in case it Hempel-confirms H_j but fails to confirm the alternative Hs. Use this notion of selective confirmation to discuss the relative confirmatory powers of black ravens versus nonblack nonravens for alternative hypotheses about the color of ravens.

11. Prove Lemmas 1, 2, and 3 of Section 2.10.

12. Discuss the prospects of resolving the problem of old evidence by using counterfactual degrees of belief, that is, the degrees of belief that would have obtained had the evidence E not been known.

13. Work out the details of the following example, which was mentioned in Section 2.8. There is a square piece of metal in a closed box. You cannot see it. But you are told that its area is somewhere between 1 square inch and 4 square inches. Show how the use of the principle of indifference can lead to conflicting probability values.

14. Suppose there is a chest with two drawers. In each drawer are two coins; one drawer contains two gold coins, the other contains one gold coin and one silver coin. A coin will be drawn from one of these drawers. Suppose, further, that you know (without appealing to the principle of indifference) that each drawer has an equal chance of being chosen for the draw, and that, within each drawer, each coin has an equal chance of being chosen. When the coin is drawn it turns out to be gold. What is the probability that the other coin in the same drawer is gold? Explain how you arrived at your answer.

15. Discuss the problem of ascertaining limits of relative frequencies on the basis of observed frequencies in initial sections of sequences of events. This topic is especially suitable for those who have studied David Hume's problem regarding the justification of inductive inference in Part II of this chapter.

16. When scientists are considering new hypotheses they often appeal to plausibility arguments. As a possible justification for this procedure, it has been suggested that plausibility arguments are attempts at establishing prior probabilities. Discuss this suggestion, using concrete illustrations from the history of science or contemporary science.

17. Analyze the bootstrap confirmation of the perfect gas law in such a way that no "macho" bootstrapping is used, that is, the gas law itself is not used as an auxiliary to deduce instances of itself.

SUGGESTED READINGS

GLYMOUR, CLARK (1980), *Theory and Evidence*. Princeton: Princeton University Press. This book, which is rather technical in parts, contains the original presentation of bootstrap confirmation.

GOODMAN, NELSON (1955), *Fact, Fiction, and Forecast*. 1st ed. Cambridge, MA: Harvard University Press. This twentieth-century classic is now in its 4th edition. Chapter 3 contains Goodman's dissolution of "the old riddle of induction" and presents his "new riddle of induction"—the grue-bleen paradox. Chapter 4 gives Goodman's solution of the new riddle in terms of projectibility.

Hempel, Carl G. (1945), "Studies in the Logic of Confirmation," *Mind 54:* 1–26, 97–121. Reprinted in Hempel (1965, see Bibliography), with a 1964 Postscript added. This classic essay contains Hempel's analysis of the Nicod criterion of confirmation, and it presents Hempel's famous paradox of the ravens, along with his analysis of it.

———— (1966), *Philosophy of Natural Science*. Englewood Cliffs, NJ: Prentice-Hall. Chapters 2–4 provide an extremely elementary and readable introduction to the concept of scientific confirmation.

Hume, David (1748), *An Enquiry Concerning Human Understanding*. Many editions available. This is *the* philosophical classic on the problem of induction, and it is highly readable. Sections 4–7 deal with induction, causality, probability, necessary connection, and the uniformity of nature.

Popper, Karl R. (1972), "Conjectural Knowledge: My Solution of the Problem of Induction," in *Objective Knowledge*. Oxford: Clarendon Press, pp. 1–31. This is an introductory presentation of his deductivist point of view.

Salmon, Wesley C. (1967), *The Foundations of Scientific Inference*. Pittsburgh: University of Pittsburgh. This book provides an introductory, but moderately thorough, survey of many of the issues connected with confirmation, probability, and induction.

Strawson, P. F. (1952), *Introduction to Logical Theory*. London: Methuen. Chapter 9 presents the ordinary language dissolution of the problem of induction.

Three

REALISM
AND THE NATURE
OF THEORIES

Clark Glymour

When we read about the results of scientific discoveries, we usually understand these accounts as descriptions of nature. If we pick up an issue of *Scientific American,* for example, we may find an article about the way in which some feature of the immune system works, or about the origin of the planets. It seems strangely trivial to note that we understand such articles to be telling us about the immune system, or how the planets came to be. What else would they be about except what their words say they are about? The same applies to articles in scientific journals and books, and to scientific lectures and even to conversations on scientific matters. Obviously they are about what they seem to be talking about.

On many occasions throughout the history of modern science this apparently obvious view of what scientific papers, books, lectures and conversations are about has been emphatically denied. Many of the naysayers have been among the greatest of scientists, or have attached their denials to great scientific works. The preface to Copernicus's great work, *De Revolutionibus* (1952), was not written by Copernicus himself, but it announced that despite what the words in the book appeared to mean, Copernicus's theory was not really about how the planets move; instead, so the preface claimed, the book merely presented a mathematical device for computing the positions of the planets on the celestial sphere. Copernican theory, according to the preface, was an *instrument,* not a description. In the 1830s different methods of determining the atomic masses of the elements gave different and conflicting results. Methods using chemical analogies gave different masses from methods using the law of Dulong and Petit which determine atomic masses from heat capacities. The latter methods gave still different masses from methods that used vapor density measurements. Jean Marie Dumas, the leading French chemist of the time, concluded that a different atomic mass should apply according to how the atomic mass is determined. Thus, said Dumas, atoms don't have a single mass, they have a mass associated with vapor density measurements, another mass associated with heat capacity measurements, and so on. Contrary to appearances, in Dumas's view science does not show us how to measure one and the same property of things in different ways; instead each kind of measurement operation has a distinct property. In the twentieth century, after the theory of relativity had replaced the Newtonian concept of mass with two distinct notions—rest mass and proper mass—Percy Bridgeman (1927), a distinguished physicist, proposed that every distinct physical operation determines a distinct property, a view he called *operationalism*. Late in the nineteenth century many consistent alternatives to Euclidean geometry had been developed, and Henri Poincaré, one of the greatest mathematicians and mathematical physicists of the time, realized that by various changes in physical dynamics any of these alternative geometries could be made consistent with scientific observations. Poincaré argued that the geometry of space is not something the world forces upon us; instead, we force geometry on the world. We, in effect, adopt the *convention* that we will measure things in such a way, and formulate our physics in such a way, that Euclidean geometry is true.

Many, many other examples like these could be given. The question of just how scientific theories should be understood becomes a burning issue whenever it seems that the content of our theories is uncertain, and especially whenever it seems that science is faced with alternative theories that could equally account for all possible observations. The threat that the content of science might in some respects be underdetermined by all evidence we might ever have often brings about proposals by scientists and by philosophers to reinterpret the meaning of that content.

In this chapter we consider the evolution of modern philosophical and scientific debates over *scientific realism*. These debates concern both the nature and content of scientific theories, and whether we can have any real justification for believing the claims of science.

3.1 METAPHYSICAL SKEPTICISM AND INDUCTIVE SKEPTICISM

In the seventeenth century, mechanical explanations flourished in chemistry, in physics, in the study of gases, and in many other areas of science. The emerging new sciences tried to explain phenomena by the motions, masses, shapes and collisions of component bodies—the atoms or "corpuscles" that were thought to make up all matter. English philosophers of science in the seventeenth and eighteenth centuries tried to make philosophical sense of the great revolution in the sciences that had taken hold in the seventeenth century. One of the most immediate philosophical problems was the difference between the world we experience and the world described by seventeenth-century mechanical theories of nature. The world of our experience is filled with colors, tastes, smells, sounds, heats, and more that does not appear among the basic properties in Newtonian mechanical science. One fundamental philosophical question therefore had to do with the connection between the world we experience and the purely mechanical world that the new sciences postulated: How are the colors, tastes, odors, and heats we observe in the world produced by mechanical actions, and how can we know how this comes about?

John Locke's *Essay Concerning Human Understanding* ([1690] 1924), engaged the question by supposing that the world we experience is entirely mental, a world of *ideas*. The simple ideas in us are created entirely by the mechanical operation of corpuscles in the world outside us; the corpuscles act mechanically on our bodies, which are also mechanical systems, and somehow the result is that mental events take place in us: we see bodies, smell odors, hear sounds, feel heat, and so forth. From these simple ideas our minds create more complex ideas, Locke held, by forming combinations of simple ideas. We have no ideas whatsoever except those that arise by combining ideas that come either from sensation or from observing in ourselves the operations of our own minds. Some of our simple ideas correspond to qualities that external bodies really have; our ideas of solidity, motion, shape, and number of bodies, for example, correspond to features of external bodies. Locke called these properties "primary qualities." Other ideas, such as those of heat, taste, odor and color do not, according to Locke, correspond to properties that external bodies really have. External bodies are not colored, for example, but they may have the power (because of their arrangements of primary qualities) to produce the idea of color in us.

Locke's picture of how we fit into the world is not very different from the pictures that some people might give using contemporary physics rather than the mechanical physics with which Locke was familiar. But no matter whether we use Newtonian mechanics or modern physics, such pictures raise a fundamental philosophical problem to which Locke had no solution: If Locke is right that all we experience are our "ideas," not things themselves, then how can we have any *knowledge* of things themselves? How, for instance, could Locke possibly know that our ideas of corpuscles, solidity, figure, and motion really do correspond to qualities of bodies themselves, but our ideas of color and odor do not? How could anyone know that the claims of physics are true? Our experience would be the same, it seems, if bodies really did have colors and odors, and not just shapes and masses and velocities. Still more fantastically, our experience might be the same if the external world we see really didn't exist at all, and things in themselves had no mass, no shape and no velocity. The problem with Locke's picture is that it postulates a scientific description of how things are in themselves but it seems to forbid any method by which we could obtain relevant evidence as to the truth of that description.

The problem Locke's picture poses is referred to as the problem of *metaphysical skepticism*. Metaphysical skepticism arises whenever it seems that two or more ways are possible in which the world could be constituted and *not even all possible evidence* would be sufficient to determine in which of these ways the world actually is constituted. In that case, we might reasonably doubt that anyone could really know (or even have good reason to think) that the world is one way rather than the other. Locke's veil of ideas is only one way in which metaphysical skepticism can be posed. In the nineteenth century the Reverend Phillip Gosse invented a novel version of metaphysical skepticism in order to dispute the fossil evidence indicating that the age of the Earth is much greater than is suggested by scripture. God, according to Reverend Gosse, created the world less than 6,000 years ago complete with geological strata and fossils. Evidently, barring time machines, no observations that could ever be obtained would refute this theory. For a modern version of metaphysical skepticism, consider the possibility that you are a brain in a vat and that all of what

Realism and the Nature of Theories

you see, hear, taste, smell and feel is unreal; your sensations are produced by stimulating your sensory nerves according to a plan that is executed by a sophisticated computer. No matter what you may observe in the course of your life, your observations will be consistent both with the hypothesis that you are a brain in a vat and with the hypothesis that you are not.

To such metaphysical perplexities about science and knowledge, David Hume added another in the eighteenth century. Hume's problem of induction has already been considered (see Chapter 2). Hume phrased it as a problem about our knowledge of any *necessary connection* between cause and effect. He argued that we do not perceive any necessary connection between a cause and its effect, and therefore the idea of necessary connection cannot come from our experience of the external world. Hume proposed that the source—and therefore the content—of the idea of necessary connection is our observation within ourselves of the force of habit that leads us to expect a familiar effect when we observe a familiar cause. From this analysis, Hume argued that neither experience nor reason can guarantee that future examples of any familiar cause will be followed by the familiar effect. For all that reason or experience can establish, bread will not nourish people tomorrow, or tomorrow daytime will not be followed by night.

Hume framed his argument in terms that were appropriate to the "idea" idea and to the philosophical language of the eighteenth century, but his *inductive skepticism* is almost as old as philosophy. A related point is made very clearly, and in some respects more clearly than in Hume, by Plato (1900–1903) in his dialogue, *The Meno*. Plato's protagonists, Socrates and Meno, try to establish a general, universal hypothesis by looking for examples or counterexamples. The general hypotheses that Plato considers in the dialogue are proposed conditions for virtue, but Plato's philosophical point applies as well to any general scientific hypotheses, for example hypotheses about the melting point of water.

Imaginably, all pure water could melt at one and the same temperature, or different samples of water could melt at different temperatures. It is perfectly *consistent* to imagine a world in which before this time all samples of water melted at 0 degrees celsius, but after this time water will melt at 10 degrees celsius. We trust that we don't actually live in such a world, but we can imagine one well enough. Suppose we conjecture on the basis of our experience that all pure water melts at 0 degrees celsius and suppose in fact we are right about the actual world we inhabit. In the future we will therefore continue to be correct about the melting point of water. A modern version of Plato's question is this: Even if we are correct right now about what the melting point of water is for all cases everywhere, in the past and in the future, how can we *know right now* that we are correct?

We can understand the logical structure of Hume's and Plato's problems if we imagine a game a scientist must play against a demon bent on deceiving the scientist. We will suppose the two players can live forever. The demon has available a collection of possible worlds. Both the demon and the scientist know which worlds are in the collection. The scientist and the demon agree on a proposition that is true in some of these possible worlds and false in others. The proposition might be "bread always nourishes" or "the melting point of water is zero degrees celsius" or any other proposition of interest. The demon gets to choose a possible world, and he gives the scientist facts about that world one at a time in any order the demon pleases. But the demon is not allowed to lie about the facts and is not allowed to withhold any relevant fact; every fact about the world the demon has chosen must eventually be put in evidence for the scientist. The scientist must guess whether or not the proposition is true in the world the demon has chosen. If he wishes the scientist can change his guess whenever a new fact is received.

What does it mean to win this game? Winning is coming to know. For Plato (and, less clearly, for Hume) the scientist knows the truth or falsity of the proposition if and only if no matter which world the demon chooses, after some finite number of facts have been received, the scientist can announce the conjecture that will always henceforth be given about the proposition, and that conjecture is correct. For Plato, knowledge requires *true* belief acquired by a *reliable* method that permits the knower to *know that she knows*.

The simple but devastating logical fact is that except for the very trivial propositions or very restricted sets of alternative worlds, the scientist can never win this sort of game. The conditions Plato and Hume require for knowledge can almost never be met unless the alternative circumstances are restricted a priori, a supposition Plato endorsed and Hume denied.

In one form or another the problems of metaphysical skepticism and inductive skepticism have been central to theories of knowledge and philosophies of science throughout the history of philosophy, and the same remains true of philosophy of science today. The two problems are alike in calling into question our ability to have scientific knowledge by considering circumstances in which alternative hypotheses are consistent with the evidence we might have. The problems differ in detail. Inductive skepticism trades on the fact that we are finite beings who only have a finite amount of data available at any time. Metaphysical skepticism trades on the assumption of a separation between the kinds of things we have evidence about and the kinds of things we wish to know about; even if we had infinite knowledge of things of the first kind, it would not suffice to determine the truth about things of the second kind. These problems are at the bottom of many disputes over the nature of scientific claims, the structure of scientific theories, and the appropriateness of belief in the claims of modern science.

3.2 THE KANTIAN SOLUTION

Immanuel Kant, an eighteenth-century physicist and philosopher, offered a novel solution to both the problem of metaphysical skepticism and the problem of inductive skepticism. His ideas had an extraordinary influence on both science and philosophy from the nineteenth century into the twentieth. It is almost impossible to understand twentieth-century debates about the structure of science without some grasp of Kant's views.

Kant drew a radical conclusion from the perplexities of English philosophers such as Locke and Hume. Kant's conclusion is that *we can have no knowledge whatsoever of things in themselves*. But why should we want to? Kant knew that the world that affects each of us, the world we care about and want to predict and control and understand, is the world we experience, not the world of things in themselves. In the world we experience we have tables, chairs, houses and other people; in the world we experience, bodies move (approximately) according to the law of inertia, events occur one after another, things have shapes, the properties of figures and shapes are described by geometrical principles, and familiar causes produce familiar effects. This is the world science tries to describe, not the world of things in themselves.

In an important sense, Kant rejected the notion that what we perceive are our own ideas. We perceive things in the world of experience, not our own ideas. In Kant's view it is true enough that the structure of our experience depends in part on the structure of our minds. The objects we see, touch, hear, feel or smell are in some respects constituted by our own minds: The unknowable things in themselves together with our minds produce the world of experience. But exactly *what it is* to be an object of experience—an ordinary chair for example—is to be an entity constructed in this way.

While this is one way out of metaphysical skepticism, what about inductive skepticism? Kant had an ingenious solution there as well, and it is much more intricate and requires a further look at Kant's philosophical project. Kant asked the following sorts of questions: In order for there to be any experience of things in space and time and any experience of events occurring one after the other, what conditions are necessary? We can interpret this sort of question psychologically—how must the mind work in order for one to have experience of the sort we do?—or logically—what principles are logically presupposed by the assumption that we have experience at all? (Kant's words suggest both a psychological and a logical reading of his project, and scholars have argued about the matter ever since.) Kant called these sorts of questions *transcendental,* and to this day philosophers call "transcendental" any arguments that try to establish that something must be true because it is (or is claimed to be) a necessary condition for human knowledge or for human experience.

What separates transcendental arguments from ordinary empirical arguments of the kind given in the sciences is not entirely clear. Some empirical arguments take as premises particular facts of our experience (for example, that particular samples of sodium melted at particular temperatures) and argue to generalizations (for example, that all pure sodium melts within a specific interval of temperatures). Some scientific arguments take general features of human experience (for example, that the motions of the planets satisfy Kepler's second law) and argue to deeper general laws (such as the law of gravitation). "Transcendental" arguments seem to be distinguished only by the fact that their premises about experience are very, very general (and therefore often banal), and their conclusions are not supposed merely to be the best explanation of the premises or the best scientific conjecture, but demonstrably necessary in order for the premises to be true.

Kant claimed that the way the mind forms experience guarantees that objects of experience must have certain features. Objects are perceived in space, and Kant held that the "empirical intuition" of space—that is, seeing particular things in space—requires that the principles of Euclidean geometry truly describe spatial relations. Events occur in time, and Kant further held that the "intuition" of time requires that events occur in a linear, serial order like the integers. Euclidean geometry and arithmetic are not themselves spatial or temporal intuitions; they are what Kant called the *forms* of sensible intuitions of space and time. Indeed, Kant held that because of these forms of intuition, we can have a priori knowledge that geometry and arithmetic will be true in the world we experience—that is, knowledge not derived from any particular experiences. Kant thought our knowledge of the truth of propositions of arithmetic and geometry was especially puzzling. These propositions appear to be certain and not to be based on experiment of any kind, but their truth did not seem to Kant to be "analytic" either. Although $7 + 5 = 12$, the concept of seven plus five, Kant maintained, does not *contain* the concept of twelve, and hence we could not come to know that $7 + 5 = 12$ merely by analyzing the concept of seven plus five. The doctrine that arithmetic is the pure form of the intuition of time seemed to explain how we could know such "synthetic" claims a priori.

In addition to these "pure forms of intuition," Kant held that certain "pure concepts of the understanding" necessarily apply to any judgements we make about the objects of experience. Kant called these concepts "pure" because he held that they too are not obtained *from* any experience; instead they *form* experience, or they are preconditions of experience of objects. Kant's "pure concepts of the understanding" were taken from the logical categories commonly used in his day (and, indeed, Kant himself wrote a treatise on logic). For our purposes the most important concept of the understanding is that of *causality*. The upshot in psychological terms is that through a process that is not entirely clear, the understanding guarantees that objects of experience are subject to causal regularities. Kant did not hold that the understanding generally reveals a priori *which* possible causal laws are true; that is up to science to determine from experience. The understanding only guarantees that if objects are experienced they and their properties must be subject to some causal regularities or other. But Kant did argue that the understanding also guarantees a priori the truth of many principles of Newtonian physics, for example the law of inertia.

Kant's system seemed to provide an answer to most of the philosophical problems that beset eighteenth-century science. The doubts of metaphysical skepticism were pushed aside as of no relevance to the aims of science or the concerns of humans. The problem of inductive skepticism appeared to be solved. Furthermore, Kant's system justified the feeling—more common in the eighteenth century than today—that geometry, arithmetic, and the fundamental principles of physical kinematics and dynamics are too certain to be based on generalizations from experience and too full of content to be merely disguised definitions.

The influence Kant's system had among scientists and philosophers is hard for us to appreciate today. One example will have to serve by way of illustration. Josiah Willard Gibbs was the first great American-born theoretical physicist, one of the figures most responsible for forming that branch of physics called statistical thermodynamics, which attempts to understand phenomena of heat through the application of probability theory to the mechanics of systems of particles. In the preface to his seminal work, *Elementary Principles of Statistical Mechanics* (1960), Gibbs wrote that his aim was to reduce the phenomena of heat to regularities of mechanics that have an a priori foundation.

3.3 SKEPTICISM AND THE ANTINOMIES OF REASON

Kant's *Critique of Pure Reason* ([1787] 1929) nonetheless contained troublesome seeds of skepticism. Nothing in Kant's picture of the world guaranteed that every meaningful scientific question we might pose about the world of experience can be settled by scientific investigation. Hume's problem may be solved by Kant's picture of the role of the concept of causation, but the logical problem about learning from experience that Plato posed doesn't seem to be solved at all. Kant himself argued that questions arise that cannot be solved empirically or rationally. Kant ([1787] 1929, 384–484) described four "antinomies of pure reason," two of which seem to be clear scientific questions about the world of experience:

1. Does the world have a beginning in time and is space unlimited?
2. Are objects infinitely divisible?

Kant held that these are intelligible questions about the world of experience, and that no experience or a priori reasoning could settle them. He tried to establish as much by giving an a priori proof in each case that the answer is "yes" and in each case another equally good a priori proof that the answer is "no." Kant's "proofs," like most of his attempts at demonstrations, are dubious and equivocal, but a rigorous logical sense exists in which he was correct that these questions cannot be settled by experience.

Recall the scientist and the demon. Suppose the demon may choose from among different possible worlds in which some particular object—a banana, for example— may be divided into exactly two pieces, a world in which it may be divided into three pieces, a world in which it may be divided into four pieces, and so on for every number. Let us also include a world in which the banana can be divided without end. The second question, about infinite divisibility, has a "yes" answer only in the last of these possible worlds. Let us suppose the scientist can do experiments to try to divide the banana or a part of it. One experimental method may not succeed at dividing an object, while another experimental method will work. If the object is divisible, let us assume that an experimental method available to the scientist will divide it, although the scientist can only discover which method, if any, works by trying the method experimentally. Let us further assume that the number of methods of division that might be tried is unlimited. The demon picks one of the possible worlds and in that world the banana has some definite divisibility. The scientist tries experimental methods to divide the banana or its parts, and after each experiment the scientist can conjecture whether the banana is infinitely divisible.

Can the scientist win this game? Not if we use Plato's criterion for winning, namely, that the scientist must not only get the right answer to the question but she must eventually *know when* she had got the right answer. *Indeed, the scientist cannot win this game even if we loosen Plato's conception of knowledge and do not require that the scientist eventually know when she is correct.* Suppose we change things as follows: The scientist wins if no matter what world the demon chooses, the scientist eventually reaches a point at which she guesses the correct answer to the second question in that world, and continues to guess the correct answer in that world ever after. The scientist does not have to be able to *say* when she has reached the point of convergence to the truth. A moment's reflection will show that many games exist that a scientist can win in this way that cannot be won if the scientist is held to Plato's standard. For example, the scientist could now win a game in which the question was whether all samples of water melt at zero degrees celsius. But the scientist cannot win the game on the question of infinite divisibility even with this more charitable criterion of success. In a precise sense, Kant was right that the answer to the question is beyond all possible experiences.

Kant's picture therefore leaves open important practical issues about which scientific questions can and which cannot be settled by which methods. Before and after Kant those issues sparked debates over scientific realism.

3.4 LOGICAL POSITIVISM AND THE LINGUISTIC TURN

Most philosophers and many scientists educated in the late nineteenth and early twentieth centuries were taught Kantian ideas, but by the 1920s the most daring and sophisticated young philosophers of science had come to reject the Kantian picture. The partial disintegration of the Kantian solution was due to several factors. One was the scientific rejection of principles of geometry and physics that Kant had claimed to prove are synthetic a priori truths. The middle and late nineteenth centuries saw the development of non-Euclidean geometries that vied with Kant's Euclidean geometry as descriptions of the structure of space. By the 1920s the general theory of relativity had led a good many scientists to think—contrary to Poincaré—that Euclidean geometry is not true of physical space, and had also led scientists to reject the law of inertia as Newton and Kant understood it. Such results inevitably led scientifically informed philosophers to wonder about Kant's arguments: How good could Kant's demonstrations be if they were refuted by the advance of science?

Another factor in the demise of Kant's system was the development of modern logic at the hands of Gottlob Frege (1972), David Hilbert and others. In 1879 Frege had revolutionized logic, laying the foundations for a theory of mathematical proof that permitted the mathematical study of the structure of mathematics itself. By the 1920s Hilbert, Bertrand Russell and others had developed Frege's ideas into a rich and powerful tool for analyzing language and arguments in mathematics, physics and even philosophy. From the perspective provided by this new tool, Kant's arguments didn't look very sound. Rather than rigorous demonstrations, they seemed to be equivocal, full of gaps, and well short of proving their claims. Kant's response to Hume, for example, seemed inadequate when one moved from talk of justifying a vague principle of causality to questions about how universal laws of nature could be known on the basis of any finite amount of evidence. Kant was left with a sort of picture but without convincing arguments for its details. The Kantian system was replaced by a new philosophical enterprise: unfolding the *logic* of science.

Frege's logical theory focused on formal languages, which were presented as abstract but precise mathematical objects, as definite as the system of natural numbers. Frege's greatest achievement was to show how a variety of mathematical theories could be represented in such formal languages and how the notion of a *proof* could be characterized as a precise mathematical property of certain sets of sequences of symbols in such a language. Early twentieth-century logicians clarified Frege's accomplishment, formalized still more powerful languages and other mathematical theories, and introduced a mathematical representation of how language describes various kinds of facts.

The new subject of *model theory* that emerged concerned the representation of how language describes the actual world or any possible world. Associated with any formal language were an infinity of set-theoretic structures called models. Each model consisted of a set (the *domain* of the model) and special subsets of the domain, or special subsets of pairs or triples or *n*-tuples of members of the domain. The domain of a model was a mathematical representation of the objects in some possible world, and the subsets, and sets of pairs, triples, and so on, of members of the domain were mathematical representations of properties or relations among things in that possible world. The idea, roughly, is that a property corresponds to the set of things that have that property in each possible world. ''Brown'' corresponds in each possible world to the set of brown things in that world.

Parts of the formal language were associated with parts of any model for the language. Certain symbols in the language named members of the domain, while other symbols denoted properties or relations—that is, subsets or sets of pairs or triples and so on of members of the domain. Other symbols in the formal language were used as variables that range over members of the domain. Under such an association between symbols of an abstract formal language, on the one hand, and features of a model, on the other hand, it was possible to define in explicit mathematical terms what it means for a sentence in the formal language to be *true* (or false) in the model.

With this apparatus, one could study the expressive power of various formalized theories entirely mathematically; one could study the conditions under which theories could (or could not) be described by a finite collection of sentences; one could study which properties of collections of models could (or could not) be characterized by any possible theory in a formal language; one could study whether formalized theories were complete or incomplete in the sense that they left no, or some, claims about their models indeterminate; and we could study which properties of formal theories could be determined by algorithms—the way the long division algorithm of elementary arithmetic determines the result of dividing one number by another. By 1930 Kurt Gödel, then a young logician, had proved that a reformulation of Frege's logic by Hilbert and Ackermann is *complete,* meaning that every sentence of a kind of formal language that is true in every model associated with the language is also a sentence that can be proved by their rules of proof.

The logical revolution of the late nineteenth and early twentieth centuries led to revolutionary developments in the foundations of mathematics and eventually to the creation of the contemporary subject of computer science. Along the way it also changed philosophy of science, and that is what interests us here.

Two leaders of this new logical perspective were Bertrand Russell in England and Rudolf Carnap in Austria. Russell had originally been a Kantian—in fact his doctoral thesis was a defense of the Kantian view of geometry in the face of the development of non-Euclidean geometries. Russell, however, subsequently *almost* fully abandoned the Kantian perspective; he corresponded with Frege and made important contributions to the development of logical theory. Carnap, while he made no original contributions to logic, studied with Frege and fully absorbed the new logical methods, and like Russell *almost* entirely abandoned the Kantian perspective. Russell and especially Carnap saw in Frege's mathematical reconstruction of the idea of proof a model for all of philosophy. Philosophy would use the tools of modern logic to reconstruct all of the notions of scientific methodology—the notion of a scientific theory, of testing, confirmation, explanation, prediction, and more. Rather than talking in Kantian terms about "synthesis" or in Lockean terms about "ideas" philosophy would talk about language, about *symbols,* their mathematical relations and their meanings.

This new logical perspective led to a program to replace Kantian epistemology with other solutions to the problems of skepticism, especially metaphysical skepticism. One of the first efforts of the new perspective was to attempt to replace the Kantian synthesis of the world of experience by a *logical construction* of the world of experience. The idea was something like Locke's except that rather than talking vaguely about combining ideas, certain parts of a formal language would be assumed to describe the deliverances of experience, and then new terms would be introduced into the language through formal, logical definitions that reduced the new terms to the terms denoting immediate experiences (see Chapter 4, Sections 4.2 and 4.3). On this view, the external world is whatever satisfies these logical constructions, and the structure of definitions guarantees that what satisfies the constructions, if anything, must be sets of immediate experiences, or sets of sets of immediate experiences, or sets of sets of sets of immediate experiences, and so on. In this way any intelligible talk of the objects of science, the external world, or anything else is reduced to talk about possible experiences. Metaphysical skepticism was avoided in something of the same way it was avoided in Kant, and the possibility of empirical knowledge was explained. Russell sketched such an account and around 1925 Carnap attempted to carry it out in detail in a book he entitled *The Logical Structure of the World* (1969). A similar viewpoint was developed near the same time by the American philosopher and logician C. I. Lewis in his *Mind and the World Order* ([1929] 1956).

Kant's system provided a kind of explanation of mathematical knowledge. How did the new logical turn of the 1920s explain our knowledge of arithmetic and geometry? Not by any attempt to establish that some mathematical knowledge is synthetic a priori. The new philosophers of science unanimously rejected Kant's claims about the source and character of mathematical knowledge; instead, several other alternatives were pursued. One idea was *logicism,* the proposal that just as claims about the world can be reduced by appropriate definitions to claims entirely about experience, so claims in mathematics can be reduced by appropriate definitions to claims about nothing and everything. By appropriate definitions of mathematical objects and operations, mathematical claims were to be reduced to complex *logical truths.* Logical truths are true of everything, true in every possible world, and they can be established by purely logical demonstrations. In this way the certainty and a priori character of mathematics would be accounted for, but only by making mathematics in a sense empty of content. This proposal was carried out in considerable detail by Bertrand Russell and Alfred North Whitehead (1925). A second idea is that mathematical knowledge is tacitly conditional and axiomatic. Many mathematicians had shown how to give rigorous sets of axioms for various fundamental mathematical theories. In the nineteenth century Giuseppe Peano had given a set of axioms for arithmetic, and later Hilbert ([1909] 1971) gave a rigorous set of axioms for geometry. The axiomatic idea about mathematical knowledge was that we don't in fact know a priori that the axioms of arithmetic are true of anything; what we know is only that *if* the axioms of arithmetic are true of a system then certain other claims must necessarily also be true. We don't know a priori, for example, that $1 + 1 = 2$ is true of any particular system, and sometimes it isn't: 1 volume of water mixed with 1 volume of alcohol does not make two volumes of anything because the two liquids are mutually soluble. What we know a priori is that *if* Peano's axioms describe any system of quantities, *then* $1 + 1 = 2$ is true of them. The same is true of geometry. As Einstein put the point, insofar as geometry is about experience it is not certain, and insofar as geometry is certain, it is not about experience.

The philosophical movement Carnap represented came to be known as "logical positivism." That term was applied mostly by German and Austrian philosophers to their point of view, but as we have already noted the perspective was shared by others in England and the United States. The logical-linguistic approach to scientific knowledge underwent a rapid evolution, caused largely by various technical difficulties encountered by attempts at logical reconstruction of human knowledge. One of the very first difficulties was the failure of Carnap's attempt to show how the external world could be constructed logically from descriptions of experiences. Carnap found that he could construct classes reducible to a simple relation between terms denoting experiences, and that some of these classes seem reasonable reconstructions of sensory modalities such as color and sound. But he did not find a satisfactory way to interpret the notion of a physical object or of places and times as such reducible classes. Metaphysical skepticism seemed to threaten once more; fortunately Carnap and the other logical positivists had a stick with which to beat it down.

3.5 THE VERIFIABILITY PRINCIPLE OF MEANING

Locke had claimed that we can only think about ideas, and he held that the source of ideas must be either sensation or internal observation of the operations of our own minds. In a fashion not entirely welcome to Locke, this point of view contains a solution to metaphysical skepticism, since on Locke's view we cannot really think of objects in themselves, only of ideas. What we cannot think of we cannot intelligibly talk of. So when we ask whether the world in itself could be made of nonentities without primary qualities, we are not, on this view, really saying anything intelligible. For metaphysical skepticism and other perplexities, the logical positivists proposed a still more radical solution: the *verifiability principle*.

The verifiability principle holds that a claim is meaningful if and only if it could be verified, that is, if and only if some possible set of observations exists that, were they to be made, would establish the truth of the claim (see Chapter 4, Section 4.4). A claim that is actually false can be verifiable in this sense. It is false that I weigh less than 200 pounds, but a possible sequence of observations exists that were they to be made would establish that I do weigh less than 200 pounds. Because I don't weigh less than 200 pounds, no such observations will in fact be made, but they are possible.

By embracing the verifiability principle, the logical positivists were able to dismiss metaphysical skepticism and metaphysical speculation as sheer nonsense, as literally talk with no meaning, for metaphysical skepticism only gets started by attempting to describe some circumstance whose truth or falsity allegedly is underdetermined by all possible evidence. According to the verifiability principle, no such circumstance can ever be described; words do not have the power to express what is meaningless. (For an example of the application of the principle in a scientific context, see Chapter 5, Section 5.1).

With the verifiability principle available, the failure of Carnap's attempt to show how to construct descriptions of the world from descriptions of experience did not herald the rebirth of metaphysical skepticism, and the principle is still held today by a few philosophers. But the verifiability principle seemed to create as many problems as it solved. The problem, once more, was Plato's.

Consider whether the claim that the melting point of water is zero degrees celsius can be verified. If "verify" means to establish the truth in such a way that we need not reserve the right to change our minds, then Plato's criterion for success in the knowledge game seems to apply: To verify the proposition there must exist a finite sequence of observations such that upon making those observations we can be certain that the proposition is true. So the observations must *necessitate* the truth of the proposition, or in other words, the demon must not have available a possible world in which the observations are true and the proposition in question is false. However, we cannot verify in this way *any* general scientific law. We cannot, for example, use observations of the melting point of water to verify that water melts at zero degrees celsius because it is logically possible (even if we do not believe it) that in the future water will melt at 10 degrees celsius. Thus, according to the verifiability principle, understood in this way, *every scientific law is meaningless*. The verifiability principle saves the world from metaphysics, but at the price of losing the world!

Some, but not many, philosophers have been willing to accept this conclusion. A number of philosophers educated at Oxford endorse it (Dummett 1978, Wright 1980.) Stephen Toulmin, a prominent philosopher of science, proposed for example that scientific theories are not really claims about the world, they are "inference tickets" that lead us to predict new observations from old observations. Toulmin's idea is a form of instrumentalism, much like the view proposed by Osiander, the author of the Copernican preface. In recent decades, however, most people concluded that the verifiability principle is not acceptable. That conclusion is easy to draw when we reflect that no independent arguments for the principle derived from any empirical analysis of how language actually functions. The verifiability principle seems simply to have been a convenient dogma for discrediting metaphysical perplexities. The perplexities won.

3.6 CONFIRMATION, MEANING AND THE "STANDARD CONCEPTION" OF THEORIES

With the failure of attempts to show how the world could be viewed as a logical construction from simple descriptions of experience, and the rejection of the verifiability principle, the problem of separating science from metaphysics began to seem all the more difficult. Kant had solved the problem of metaphysical skepticism by confining the claims of empirical science to the world of experience. The first attempts of the new logical methods to replace the Kantian system with a clearer, more rigorous, more "scientific" understanding of knowledge resulted either in a failure to establish how the world of experience could be known at all, or else a failure to separate the claims of science from those of metaphysical speculation, with all of their underdetermination. To this dilemma philosophers of science in the 1930s responded by using the notions of testing and confirmation to limit meaning. In the end that proved not to help matters much.

One clever solution to the problem of metaphysical skepticism was proposed by Hans Reichenbach (1938): Even if they are expressed differently, two claims mean the same thing if every possible piece of evidence for (or against) one of the claims is equally strong evidence for (or against) the other claim, and vice versa. The usual examples of metaphysical skepticism arise when there appear to be two or more claims that are indistinguishable in this way by any possible evidence. On Reichenbach's proposal, making a case for metaphysical skepticism by articulating two claims that are indistinguishable by any possible evidence is actually making a case that there is no difference in the claims: There is nothing between them to be in doubt about. Reichenbach viewed confirmation as a matter of probability relations: Evidence confirms a hypothesis by raising its probability in accordance with Bayes' rule (see Chapter 2). In his view, two theories say the same thing if they would receive the same probability on every possible observation.

As with the verifiability principle no particular argument arose for this proposal founded on how language is used, but important technical difficulties did appear. Reichenbach had proposed an outline of a characterization of synonymy: A is synonymous with B just if A and B receive the same confirmation on every piece of evidence. Since this synonymy relation depends on an initial distribution of probabilities, there is no way to be sure just what synonymy relation Reichenbach had in mind. Suppose we separate from a formal language a special part or sub-language in which observations are to be reported, and suppose in Reichenbach's spirit we say that two theories are synonymous if and only if they entail exactly the same collection of sentences in this *observation language*. Frege's theory of proof, then, no longer applies, and indeed no effective theory of proof is possible within such a language. So a comparatively clear variant of Reichenbach's proposal would destroy one of the very pillars of the logical movement.

Reichenbach, as well as Hermann Weyl ([1949] 1963), the great mathematician and mathematical physicist of the early part of this century, Carnap and Reichenbach's (and Hilbert's) student, Carl Hempel, developed another perspective. The perspective presented in this chapter follows Carnap's, but all of these philosophers developed very closely related ideas.

In Carnap's picture we can observe ordinary objects and their ordinary properties; we are not separated from the world by a veil of ideas. However, a great deal exists that we cannot observe: events inaccessible in space-time, things too small, and differences in properties too slight for our senses. Science is concerned with both the observed and the unobserved, and indeed even the unobserv*able*. We describe what we observe in language and we state our scientific theories and hypotheses in language as well. Both aspects of language can be formalized. Scientific theories and their claims about the world can be represented in a formal language of any of the kinds Frege, Russell, Hilbert and other logicians developed. The difference between that part of the formal language which represents claims about observations and the more theoretical parts of the formal language are chiefly differences in vocabulary. Many of the symbols that represent the language of science—"photon," "molecular bond," "gene," and so on—are not symbols that denote observed things or properties or events.

By making observations and conducting experiments we can obtain descriptions of particular facts; these descriptions are represented in a formal language as singular sentences in the vocabulary of observation terms. How such facts might be used as evidence for or against *generalizations* that are expressed purely in observational terms seems clear enough. If the generalization is, for example, that all water melts at zero degrees celsius, then observations of water melting at zero degrees celsius count for the generalization, at least until we find a sample of water that does not. Carnap (1936) proposed that singular sentences describing observations stand in a relation of *confirmation* to general claims. The sentence, "This sample of water melted at zero degrees celsius," confirms "All water melts at zero degrees celsius." But how are we to confirm generalizations that are not limited to observation terms? How, for instance, are we to confirm that the atomic weight of oxygen is greater than the atomic weight of hydrogen? Carnap's answer to this fundamental question was a form of conventionalism in which the conventions specify *meanings*. Theoretical terms, Carnap suggested, are literally meaningless until some conditions or rules are given that specify what observable circumstance would count as determining instances of the terms. Carnap called such rules *meaning postulates*. This principle is weaker than the verifiability principle; Carnap did not require that meaningful sentences be verifiable, he required only that they be composed of predicates for which meaning postulates exist (see Chapter 4, Sections 4.3 and 4.4).

Meaning postulates look very much like other generalizations. Although Carnap proposed some restricted logical forms for meaning postulates, what is special about them is their role and their justification. Their role is both to establish the meaning of theoretical terms and to permit the confirmation of theoretical claims. They permit confirmation of theoretical claims by enabling us to infer instances of theoretical generalizations from singular observation statements. If, for instance, the law of Dulong and Petit were taken as a meaning postulate, then we could use it to infer values of atomic weights from measurements of heat capacities. In this way we could confirm the claim that the atomic weight of oxygen is greater than the atomic weight of hydrogen. The justification of meaning postulates is that they are simply *stipulations,* conventions about how we will use theoretical terms. They are therefore, in Carnap's terminology, *analytic* truths, while the claims of science that are not either truths of logic or consequences of meaning postulates are *synthetic* claims.

Contemporary commentators sometimes refer to variants of this conception of the structure of scientific theories as the "standard conception." Its elements are the assumption that scientific language can be formalized, a division between observation terms and theoretical terms, a set of meaning postulates or stipulations relating the two sorts of terms, and an analysis of confirmation in terms of relations among sentences in the same vocabulary. The standard conception is the closest modern philosophy of science ever came to a logical successor to the Kantian picture. The problems of metaphysical skepticism are solved in the standard conception by noting that the terms in metaphysical disputes lack adequate meaning postulates to permit the confirmation of disputed metaphysical claims. A picture of how scientific knowledge is possible emerges: Meaning postulates constrain scientific language, and observations generate observational reports that confirm or disconfirm scientific hypotheses.

3.7 REALISM AND THE LIMITS OF SCIENTIFIC KNOWLEDGE

The standard conception set the framework for logicians to investigate one of the fundamental Kantian questions: What are the limits of possible scientific knowledge? The questions at issue are exactly those that concerned Kant, but now the logical conception permitted logicians to find answers that have real demonstrations rather than Kant's intuitive but inconclusive arguments. This work was begun around 1960 by Hilary Putnam (1965), and has been continued since by a number of logicians and philosophers.

One way to understand such questions is to consider again the knowledge games an imaginary scientist might win against a demon, assuming the standard conception. The demon and the scientist agree on a division between observation and nonobservation terms, and on a set of meaning postulates, and the demon has a set of possible worlds from which to choose. In each of these possible worlds the meaning postulates must be true, but everything else may vary between worlds. Whatever world the demon chooses, the demon must give the scientist the observable facts of that world in some order, and every observable fact must eventually be given. The scientist must conjecture whether some claim is true or false. We can adopt Plato's criterion of success for the scientist or the weaker criterion suggested by Kant's antinomies of reason. We know that the scientist cannot win in any interesting case if we adopt Plato's criterion of success, in which no matter what the demon's choice the scientist's conjectures must eventually converge to the truth and the scientist must know when she has converged to the truth. But what if we grant that the scientist wins even if the scientist may not know *when* she had converged to the truth? Then the question of knowledge is more interesting. In that case, assuming a formal language of the kind developed by Hilbert and Ackermann, whether the scientist can win depends entirely on the logical properties of the claim in question and the logical properties of the meaning postulates. The scientist can win just if the meaning postulates entail that the claim to be decided is true if and only if purely observational claims of a special logical form are true. With the aid of the meaning postulates the claim under investigation must be reducible to a claim of the form, "For every $x, y, \ldots, z$, there exists a $u, w, \ldots, v$, such that so and so," where the "so and so" contains only observational terms and does not contain any expressions (called *quantifiers*) such as "for every" or "there exists." Furthermore, for the scientist to win, the meaning postulates must also entail that the claim under investigation is reducible to another purely observational claim with the quantifiers reversed, that is, to a claim of the form, "There exists an $x, y, \ldots, z$ such that for all $u, w, \ldots, v$, so and so."[1]

Using the standard conception logicians have been able to obtain similar results for a wide range of alternative formal languages and for problems in which the scientist must discover an entire theory rather than settle a preestablished question. Such mathematical analyses can be obtained even when the characterization of possible observation sentences is not by vocabulary, when all the evidence the scientist will see is incomplete, and even when the observation sentences are not limited to statements of particular facts. If we want to go to the trouble, such logical results enable us to say precisely what questions are or are not beyond human knowledge given a definite logical system, a set of meaning postulates, and a characterization of which sentences can be decided directly by observation.

Given the elements of the standard conception, logical analyses of the limits of knowledge give rather pessimistic results: Relative to the logical complexity of the observation sentences, any claim of science that could be known even in the limit must be equivalent to sentences with specially restricted quantifier structure. In addition, of course, what cannot be known in the long run cannot be known in the short run either. On plausible accounts of what we can actually observe, much of what we think we know we could not possibly know. Kant's problem of the antinomies of reason finds a modern form.

Even without detailed logical analyses, by looking at particular scientific theories and enterprises many philosophers sensed that much of scientific belief would prove on close examination to be underdetermined by observation, not just in the short run but even in the long run. Arguments of these sorts were developed in different ways for features of geometry and physics by Reichenbach ([1928] 1957), Salmon (1975), Grünbaum (1973), Glymour (1977), Malament (1977b) and others, and for sciences that depend on deciphering meanings by Quine (1960) and Davidson (1984). These difficulties have been met with at least three different responses.

One response to the apparently strict limitations on the possibility of scientific

---

[1] The following procedure will determine in the limit that "There exists an x such that for all y $F(x,y)$" is true (if it is) and (if it is not) will otherwise not converge to any hypothesis: Let $x_0, x_1 \ldots x_n \ldots$ be an enumeration of all of the objects that could be a value v of x for which "for all y $F(v,y)$" is true. Start by conjecturing *yes* to the claim that "There exists an x such that for all y $F(x,y)$" so long as the data are consistent with "for all y $F(x_0,y)$." If a datum is received that contradicts "for all y $F(x_0, y)$" conjecture *no* to the question. Thereafter, conjecture *yes* provided all evidence so far received is consistent with "for all y $F(x_1,y)$" until this is contradicted, at which point *no* is conjectured and the procedure moves to "for all y $F(x_2,y)$" and so on. If "There exists an x such that for all y $F(x,y)$" is true, then the procedure eventually finds an x_n for which "for all y $F(x_n,y)$ is never contradicted by the data and so the procedure says *yes* ever after. If "There exists an x such that for all y $F(x,y)$" is false, then for each x_n "for all y $F(x_n,y)$" is eventually contradicted by the data, and the procedure changes conjectures infinitely often.

If the denial of "There exists an x such that for all y $F(x,y)$" is logically equivalent to a "There exists an x, for all y such and such" formula, then a similar procedure can be applied. The two procedures, one for the original sentence and one for its denial, can be dovetailed into a single procedure that converges to *yes* if "There exists an x such that for all y $F(x,y)$" is true and converge to *no* if the denial of it is true. It can be shown that a sentence not equivalent to a sentence of the form "there exists an x, for all y, such and such" cannot have its truth verified in the limit.

Realism and the Nature of Theories

knowledge is an argument from the implausibility of miracles. The argument is roughly this: It would be miraculous if the observed features of the world behaved exactly *as if* our best scientific theories were true, even though those theories were in fact not true or not very close to the truth. Since it is irrational to believe in miracles, we should believe our best theories.

The argument has great intuitive appeal because it reflects scientific common sense. Even when some theories provide logically possible explanations of phenomena, the scientific community may reject the theories if the explanations require "miracles." A common sort of objectionable miracle requires contingent parameter values to have infinitely precise values in order for some observed phenomenon to be accounted for. Richard Feynman (1985), for example, recounts discovering theories of fundamental particles and rejecting them because they required that arbitrary parameters have very precise values. The argument based on miracles is in effect a judgement that we are willing to reject many theories because they require circumstances that we regard as having zero probability.

However, we experience a difficulty. Usually we have a pretty good understanding of the properties of theories that scientists at one time or another have developed, but we often have little idea of the properties of theories that scientists have *not* developed. For the argument from miracles to work we need some guarantee that among the vast array of logically possible theories of the world no alternatives exist that generate our observations, or that whatever alternatives do exist require some sort of scientific miracle (for example, that the worlds they describe are very unstable and in them, generating the phenomena we have observed depends on infinitely precise values of parameters). Simply because we do not know of such theories does not mean they do not exist.

Another response to the view that scientific claims reach beyond any possibility of knowledge is to deny that they are really claims at all, or at least not the claims they appear to be. We have already seen one form of that view, instrumentalism, but another form has gained influence in recent decades. Rather than viewing a theory as a body of claims that are true or false, we could view a theory as a (complicated) predicate that is *true of* some systems and not true of others. Rather than a body of claims about all systems, Newtonian dynamics, for example, is viewed as a complex description that is true of some systems and false of others. Instead of having just one intended model—the world—a theory is viewed as a way of specifying a collection of alternative models with which we try to represent, explain and predict aspects of observed phenomena. So Newtonian dynamics forms a complex predicate that applies to many different systems and has many different well-studied models, including, for example, a harmonic oscillator, a damped harmonic oscillator, a 2-body system in which one point mass moves in a closed orbit, a 3-body system, and so on. In explaining phenomena we try to show how observed regularities may be embedded within a model of some known theory, so that any system exhibiting the phenomena may be treated as satisfying the theoretical predicate. The practice of science is to try to show how observed regularities may be embedded within a model of a theory so that the system exhibiting the regularity may be treated as a system to which the theoretical predicate applies.

This conception, sometimes called the *semantic conception of theories,* leads away from asking questions about the limits or foundations of scientific knowledge, for the product of science in this view is not so much knowledge of general propositions but as an understanding of systems of models and how to embed various classes of phenomena within those models. On this conception we may accept and use a scientific theory, but see no sense or point to *believing* the theory. Differing versions of this approach have been developed by a number of philosophers of science including Suppes, van Fraassen (1980), Suppe (1989), Sneed (1971), Giere (1988) and others.

The semantic conception certainly captures one important aspect of how theoretical conceptions are deployed in science. It is not, however, clear that the semantic conception solves any of the fundamental traditional problems facing philosophy of science. In some cases, such as cosmology, there may be no difference between saying that a theory is true and saying that it is true of some system. Again, it is plausible enough to treat Newtonian dynamics as a predicate of systems, or as a class of models that vary in details, but it seems somehow less plausible to treat the atomic theory in chemistry that way. More importantly, even on the semantic conception, one of the aims of science is to *predict* the course of phenomena, and corresponding to the limitations of knowledge there are limitations of reliable prediction. In various settings precise theorems can be demonstrated about such limitations, even in the limit of arbitrarily increasing evidence.

A third response to underdetermination, advocated by Russell (1948), Quine (1969b) and several other philosophers is the conception of *naturalized epistemology*. Granted that starting from nothing but our observations, without any "background knowledge" about how the world works, we could come to know only a very restricted body of claims about the world. Granted that in games with the demon, if the scientist does not know beforehand that the collection of possible worlds the demon might choose is very restricted, then the scientist can only win (even in the more generous sense that does not require that the scientist know when he is right) for a very limited set of questions. Nonetheless, because of how we are in fact constructed biologically and socially, we do not start inquiry utterly ignorant. We have evolved to favor certain behaviors and to organize our sensations in particular ways. Unless hindered in some serious way, infants rapidly learn to identify and reidentify objects, and they learn that objects continue to exist when unobserved; infants have available almost from birth some simple facts about size, distance and perspective. And so on. Society provides us with a great deal of belief about the world, about other people and about social relations. (Of course some of what society leads us to believe is erroneous.) *Assuming* as background the beliefs thus bestowed on us, we can study our own perceptual and cognitive apparatus and our own social structures to discover how beliefs are formed and to determine the reliability (and unreliability) of the processes of human cognition. Such inquiries lead us to devices and procedures to improve the way we acquire new beliefs, and they may even lead us to modify or abandon some of the beliefs common to our culture. Assuming the beliefs—or most of the beliefs—to which we are disposed by nature and culture, we proceed to expand scientific knowledge. Our faith is that in the game we play against nature's demon, the demon has available only those possible worlds in which our background beliefs are true.

Naturalized epistemology is a program for snatching victory from defeat. Meta-physical skepticism wins, true enough, but in practice we are not and could not be in the circumstance of trying to learn from pure ignorance. Assuming the circumstances we cannot help but believe we are in, the powers of science to expand our knowledge are increased—nature's demon has fewer ways to deceive us. Logical and statistical investigations of the possibilities and limitations of scientific inquiry remain valuable critical tools for determining, relative to our background knowledge, which scientific projects and programs can hope to solve which questions. When epistemology is naturalized, normative philosophy of science remains a most important enterprise.

3.8 MEANING, OBSERVATION AND HOLISM

In the last twenty years the standard conception and its solutions to the problems of metaphysical skepticism and justification of scientific belief have come under heavy criticism. One criticism of the standard conception is that there appears to be no objective feature to tell us which scientific claims are meaning postulates. No mark of scientific hypotheses tells us that they are meaning postulates that are not subject to empirical disconfirmation. Even a principle that is explicitly introduced as a stip-ulation about how a new term will be used may, as observations accumulate, come to be abandoned in favor of other claims that historically were confirmed by assuming and using the putative meaning postulate. As Quine put it, any claim can be held true come what may.

Curiously, both Reichenbach and Carnap had anticipated part of this criticism. They had viewed the selection of particular claims as meaning stipulations as a somewhat arbitrary part of the logical reconstruction of science. Their view seems to have been that in order to understand how scientific inference and argument works at any moment, *some* claims must be understood to be functioning as meaning postu-lates at that time. The meaning postulates may change over time, but these changes are themselves alterations in the meanings of scientific terms, not alterations in the content of scientific claims. The picture for which Quine argued was subtly but importantly different. At any moment any accepted scientific claim might be used to justify any other accepted scientific claim; except for rare usages, such as abbrevia-tions, there is not even at a given time any distinction between the two.

Quine's criticisms generated two lines of thought, both premised on the as-sumption that he was correct that no sentences function simply as meaning postulates. One line of thought tried to save the achievements of the standard conception by explaining how scientific claims could be tested and confirmed even without meaning postulates. The other focused on meaning rather than evidence. Without meaning postulates, how do theoretical terms get their meaning? What determines the condi-tions under which theoretical terms correctly apply from the conditions under which they do not correctly apply? One answer went roughly as follows: A theoretical term is generally introduced in order to describe the cause of some aspect of a body of experimental or observational phenomena. The theoretical term then denotes what-ever it is (if it is anything at all) that plays that causal role for that aspect of the phenomena. It may be that the aspects of several phenomena for which a term is introduced have several causes, in which case the theoretical term may "partially denote" each of these features rather than one of them (Field 1973). A more radical view was championed by Quine himself: A theoretical term gets its meaning from the entire body of scientific beliefs in which it is embedded. This view, often called *meaning holism,* further denies any distinct basis for meaning as against belief. It has the consequence that if beliefs about a collection of sentences change, so do the meanings of the terms in the sentences. This doctrine, or rather the phenomenon it alleges—that whenever scientific opinion changes the meanings of all scientific terms change—is sometimes called *meaning variance.*

Just as no clear mark seemed to separate meaning postulates from other scientific claims, so on reflection no clear mark seemed to separate what is observable from what is not observable, and certainly no distinguishing mark seemed suitable for founding a theory of meaning. Depending on details of context and scientific belief, almost any sentence in the language of our theories *might* serve as a report of observations. For example, a scientist who believes that the law of Dulong and Petit is approximately correct might make measurements with a calorimeter and report, "It was observed that the atomic weight of hydrogen in our samples is less than the atomic weight of oxygen in our samples." A physicist checking photographs of particles from an accelerator might say, "A Z particle was observed." In practice we judge the correctness of reports of observations by whether the circumstance claimed to be observed really did occur, and by whether the observers were situated so that they could have distinguished the circumstance from others, for example, from the absence of a Z particle in the picture. Neither of these requirements is met only by sentences in some special "observational" vocabulary. Moreover, how observations are reported is—as Locke noted—very sensitive to the beliefs of the observer. A scientist who holds one theory may honestly describe experimental observations differently from the way the outcomes of the same experiment are honestly described by a scientist who holds a competing theory. However, because of the doctrine of meaning holism, no neutral formulation exists of what has been observed. Observations are unavoidably *theory laden.*

Sometimes the freshest lines of thought in a discipline come from those outside it who sense the larger issues without being entangled in the detailed arguments of specialists. The attack on the standard conception was led by Thomas Kuhn (1970), who at the time had studied physics and the history of science, but not philosophy. Kuhn argued through historical examples that occasionally radical breaks arise in scientific tradition when practitioners follow different lines of research using different standards of argument and making different inferences from observations. Kuhn claimed that the practitioners separated by such a break literally do not understand each other—their scientific languages are *incommensurable—and* they share no methods for resolving their disputes. Between sides in scientific revolutions, success is determined by rhetoric and politics, not by evidence. Moreover, science does not accumulate truths as time passes. When a scientific break occurs, all of the results accumulated by the previous tradition are rejected; either they are dismissed altogether by the new tradition (or *paradigm* as Kuhn called it) or else they are given a new meaning.

Much of the philosophy of science in the last twenty years has centered around arguments over whether to abandon the standard conception entirely, or to save its essential elements by arguing, for example, that the doctrines of meaning holism and meaning variance are false exaggerations, that characterizations of observations exist that are neutral between competing theories, and that Kuhn's and others' historical accounts are incorrect. Rather than entering into the details of these debates (see Chapter 4), let us consider some responses to abandoning the standard conception.

3.9 HOLISM, RELATIVISM AND SOCIETY

One of the first consequences of rejecting the standard conception is a problem about the relevance of evidence. Abandoning meaning postulates meant that confirmation relations involving theoretical terms could no longer be specified for an entire language of theoretical and observational terms. Instead, philosophers of science attempted to characterize confirmation relations that are *relative to a theory*. Different theories will generate different confirmations of hypotheses from the same set of observation statements. This line of thought was undercut by its fundamental inability to resolve the issues about the limits of knowledge, the very questions that had motivated the entire philosophical development leading to the standard conception. One could, for example, analyze logically which claims a scientist could discover in the limit *assuming a theory,* and the answer was essentially the same as in the standard conception but with the entire theory serving in place of a system of meaning postulates. Of course the truth of entire theories is not guaranteed by stipulations about meaning, so how can we know which theories to use? If no theories are assumed then scientists could only win games in which the claim to be decided had the appropriate logical form and no theoretical vocabulary.

Once we abandon the view of the standard conception that the relevance of evidence can be localized, that an experimental outcome bears for or against particular hypotheses but not others, scientific claims appear to become ever more underdetermined. This was the conclusion of Quine (1961), and of Pierre Duhem ([1906] 1954), a distinguished physical chemist who wrote before the standard conception had been articulated. Duhem claimed that in any modern physical experiment virtually the whole of physics is involved; the design of the apparatus and interpretation of its output may involve not only the hypothesis supposedly to be tested, but also principles of optics, thermodynamics, electronics and hence electromagnetism, and so on. If the experiment does not give the expected result, the hypothesis to be tested might be blamed, but from a logical point of view any of these other physical principles might be faulted instead. The much disputed claim that no principled way exists to localize the bearing of evidence is often called the *Duhem–Quine thesis.*

One use of the Duhem–Quine thesis is for what philosophers of science call the *pessimistic induction,* an argument to the effect that the only reasonable conclusion we can have about the present claims of science is that they are false. The argument (Laudan, 1984) is as follows: Every theory we can name in the history of science is, in retrospect, erroneous in some respect. The Newtonian theory of gravitation is incorrect, as is the classical theory of electromagnetism, Dalton's atomic theory, classical physical optics, the special theory of relativity, the Bohr theory of the atom, and so on. The errors of these theories may not matter for most practical purposes, but from a contemporary point of view they are all, strictly, false theories. Since all theories in history have been false, the argument continues, we should conclude that the methods of science do not generate true theories; hence our present scientific theories, which were obtained by the same methods, are false as well.

It is important to see that whatever plausibility the pessimistic induction has depends on rejecting the standard conception of theories and invoking the Duhem–Quine thesis or something very much like it. When we look back at historical cases of false theories, we note that parts of the theory are false, but other parts seem less so. Dalton's atomic theory, for example, contained claims that heat is a fluid, "caloric," and that all atoms have the same volume. In retrospect, these claims seem false. Dalton also held that all atoms of the same element have the same weight, which also seems false, but less so, and in fact *nearly* true. Dalton also claimed that all molecules of a compound contain the same number of atoms of each element, and that seems even today entirely correct. Just as we informally separate parts of Dalton's theory as more and less correct, we also informally weigh how sufficient his evidence was for these various parts of the theory. In retrospect Dalton's evidence for the caloric theory and for the hypothesis about atoms seems insubstantial, and the evidence for the constitution of molecules and the sameness of weights seems better. These separations make some sort of sense in the standard conception, but they are nonsensical on the holistic view of evidence.

Another response to the decline of the standard conception sentiments is represented by the sociology of knowledge movement. In its strongest form (indeed, sometimes called the *strong program*) the view advocated by some sociologists and cultural anthropologists is that the content of science is entirely an artifact of elaborate social customs whose function is often hidden from scientific practitioners themselves. Scientists, on this view, do not really discover anything, although they may think they do. Instead science is an elaborate and expensive social system for deciding what to say, how to talk about the world, and for making social decisions about technical matters (whether to build a nuclear power plant, for example). The scientific community makes such decisions essentially on political grounds; some people have more authority, more influence and more power than others, and so long as they retain these advantages the scientific conceptions they hold are deferred to. There is no normative difference whatsoever between the claims of science and the claims of any religion or religious group; scientific claims have no more warrant, no more justification, no greater rationale, than the claims of Islamic or Christian sects, or than flat-Earth theories or astrology.

Few philosophers of science have much sympathy with the strong program, but its viewpoint should be at least be correctly understood. The position is not (or at least need not be) that the only *facts* are social facts. Undoubtedly a great many facts exist that are not social, but exactly because science is a social enterprise it is claimed to be incapable of giving any warrant that its claims are correct.

The doctrines of meaning variance and meaning holism combined with the rejection of the linguistic observational-theoretical distinction result in a view in which the success of science becomes very mysterious: Meaning changes with belief, and the doctrine of meaning variance applies as much to reports of observations as to other claims in science. Scientists who hold different opinions do not share the same meanings, even if their differences appear to be remote from the subject under discussion. Competing scientific theories, it seems, cannot be tested against a common standard of observations since what the observations mean depends on which theory is believed. No means seem to demonstrate when scientific inquiry can and cannot reliably lead to the truth. When both meaning and the world of experience vary with the beliefs or conceptual scheme of the observer, the great philosophical issues about the possibility of knowledge cannot be answered because they cannot sensibly be formulated. Philosophy of science (and some would say, philosophy generally) comes to an impasse in which there is nothing that can be done; there is nothing for philosophy or for philosophy of science to discover about how to reliably acquire knowledge of the world, or about the limits of such knowledge. This is exactly the conclusion reached by some prominent philosophers, Rorty (1979) for example, who have followed and accepted the attacks on the standard conception, and who recommend giving up the pursuit of philosophical questions about knowledge.

3.10 RELATIVISM AND QUANTUM MECHANICS

Separately from the arguments over the standard conception of theories, modern quantum mechanics has given some philosophers of science reason to think that we cannot acquire any knowledge of the world independently of variable features of ourselves. We cannot because there is no such world.

The quantum theory includes both dynamical variables such as position, momentum, time and energy, and states of a physical system. The theory does not, however, include any physical states in which all dynamical variables have precise values. Any state in which position is perfectly precise, for example, is a state in which momentum is completely indeterminate. The quantum theory restricts how precisely any allowed state can specify both position and momentum and similarly both time and energy. These restrictions are the famous Heisenberg Uncertainty relations. (See Chapter 6, Sections 6.11–6.16 for in-depth discussion.)

A natural response to the quantum theory is to think that microscopic systems have precise values for all of their dynamical variables but the theory simply is unable to determine these quantities precisely. While this view has its supporters, a great deal of empirical and mathematical research has led many physicists and philosophers of science to think it is false. Instead, systems literally do not have a simultaneous precise value for position and momentum; however, if we conduct an experiment to measure position we find one, and if we conduct a measurement to measure momentum we find one. What properties the world exhibits depends on what we ask of it.

Niels Bohr called this phenomenon "complementarity," and the conclusion he drew was a revision of Kant's perspective. In the psychological version of the Kantian picture is a fixed but unknowable world in itself and a fixed "us," and the world we experience is determined by the two together. In Bohr's picture, an unknowable and undescribable world in itself exists, as well as "us," but the "us" is not fixed. We can ask one set of questions about the world and get a coherent set of answers, or we can ask any of many other sets of questions about the world and in each case get coherent answers, but the coherent answers to the several sets of questions cannot be fitted together into a single coherent picture of the world. Changing the experiments we conduct is like changing conceptual schemes or paradigms: we experience a different world. Just as no world of experience combines different conceptual schemes, no reality we can experience (even indirectly through our experiments) combines precise position and precise momentum.

3.11 CONCLUSION: REALISM, RELATIVISM AND PHILOSOPHY OF SCIENCE

The most immediate connection between philosophy of science and the rest of philosophy is through issues about the limits to knowledge and the character and possibilities for justified belief. For a brief while a consensus among the best philosophers of science pursuing the logical program provided a clear framework for exploring these issues and obtaining interesting answers. That consensus has disappeared in the closing years of the century. The community of philosophers of science is fragmented among those who regard the general perspective of the standard conception of theories as correct—although most would modify it in some important way—and those who reject it for various alternatives. The most radical, and at least in one sense the most interesting, fragment argues that because meaning varies with belief and conceptual scheme, and because no linguistic characterization of the observable exists, the epistemological questions that have motivated the enterprise of philosophy of science are unanswerable. Locke's picture, Kant's and Carnap's all ask what can be known by science assuming some things—such as meanings—are fixed. The radicals, however, claim that those things are not fixed, and so the questions have no answer.

Many, perhaps most, philosophers of science reject this relativist view, and the debates over it and how best to mend or replace the standard conception remain central to philosophy of science today. Suppose, however, that we accept the relativist view entirely: meaning and truth vary in some unknown way with aspects of belief or with the experiments we choose to conduct; we observe a world but the world we observe depends on features of our belief, culture, or experiments. Surprisingly, even in this radical picture questions about scientific realism and the limits of scientific knowledge still make sense and have answers; the puzzles of Kant's antinomies still survive. There is information about how best to conduct inquiry that philosophy can still seek after and find.

Even if the relativist picture were correct, philosophers (and others) can investigate *how* meaning depends on belief, and what aspects of sense, reference and experience are altered by what changes in belief, custom or culture. Perhaps more importantly, relativism is not subjectivism; that the world of experience depends in part on variable features of us does not mean that we can have any world of experience we wish. What we experience depends on us and on something not us—Kant's things in themselves for lack of a better term. The world in itself and our conceptual scheme together determine the world we experience, and thus what questions we can and cannot answer. If the world we experience depends on some feature of us and our community—call that feature our conceptual scheme—then there are logical facts about what questions can and cannot be answered in a given conceptual scheme, and about which questions can be answered by changing conceptual schemes. Formal models of inquiry exist in which certain questions can be settled in the limit—scientists can beat the demon—but only if scientists are free to alter conceptual schemes as need be. In these models for some issues successful discovery *requires* scientific revolutions. Even if relativism were true, the limits of knowledge could be investigated and better and worse ways to conduct inquiry would arise. Philosophy of science endures through relativism.[2]

DISCUSSION QUESTIONS

1. Suppose someone knew that two incompatible theories, call them *A* and *B*, are underdetermined by all possible evidence of a certain kind, *E*: Any possible evidence of kind *E* is known to be consistent with *A* and with *B*. Is there any interesting *point* to claiming that, nonetheless, evidence of kind *E confirms A* rather than *B*? If so, what point, and why is it interesting? Is there any interesting sense or point to claiming that, nonetheless, evidence of kind *E* would *justify* such a person in believing *A* rather than *B*? What goals might be served by having a shared confirmation relation that prefers one underdetermined hypothesis to another?

2. One theory about the notion of truth is that a claim that a sentence is true does nothing more than reassert the sentence. According to this view, sometimes called the *redundancy* theory of truth, the sentence, " 'The sky is blue' is true" simply asserts that the sky is blue. The claim that "What Sam believes is true" simply asserts the otherwise unknown set of propositions believed by Sam.

   One advocate of the redundancy theory asks the question: *Why is it practically useful to believe the truth?* Does this question even *make sense* according to the redundancy theory? Why or why not? What expressions using the word "truth" are difficult to account for with a redundancy conception?

3. Suppose after careful logical and psychological study it is found that the array of ordinary beliefs (about ordinary objects, spatial and temporal relations, causal connections, and such) needed as background knowledge for reliable scientific inquiry cannot themselves be reliably acquired from any possible array of facts about elementary experiences such as those available to an infant. What, in that case, should we say about our scientific claims and our ordinary beliefs about the world? Would they be knowledge, dogma, what?

4. Suppose two methods of inquiry are exactly alike so far as their convenience and reliability are concerned, except that method 1 will find out that *A* is true if in fact *A* is true, but method 2 will not. Suppose you are convinced that *A* is not true. Is there any reason for you to prefer method 1 to method 2?

5. What could be meant by the claim that two scientists who speak the same natural language and work in the same discipline literally do not understand one another's claims? What does understanding another person's claims require? Is it sufficient, for example, to be able to describe what the other would say about any case or circumstance? Could there be good evidence that historical figures in science who ascribe to different theories or "paradigms" literally *could not* understand one another? What would such evidence be like?

6. Social approaches to epistemology emphasize the value of procedures for obtaining consensus rather than the value of reliable procedures of inquiry. What ways of organizing rewards, communicating opinions, and making resources available would further the goal of obtaining a

---

[2] The general perspective of this chapter—and the banana—owe a great deal to Kevin Kelly.

consensus of *true* opinions? Is science so organized? What evidence is there that the enterprise of science has goals other than consensus regardless of truth value?

7. If you were in fact a brain in a vat all of whose experiences are illusions produced by computer-controlled stimulations of your nerve endings, what would your words "brain in a vat" signify?

8. Could an ideal scientific theory, that gave the best possible explanation of all possible evidence, nonetheless be false?

SUGGESTED READINGS

Some important sources from the philosophical tradition can be found in

CAHN, STEVEN M. (ed.) (1977), *Classics of Western Philosophy*. Indianapolis: Hackett.

HAMILTON, EDITH and HUNTINGTON CAIRNS (eds.) (1961), *The Collected Dialogues of Plato, Including the Letters*. Translated by Lane Cooper and others. New York: Pantheon Books.

KANT, IMMANUEL ([1787] 1865), *Critique of Pure Reason*. Translated by Norman Kemp Smith. New York: St. Martin's Press.

Two recent, very different, interpretations of Kant's theory of knowledge are

GUYER, PAUL (1987), *Kant and the Claims of Knowledge*. Cambridge, England: Cambridge University Press.

KITCHER, PATRICIA (1990), *Kant's Transcendental Psychology*. New York: Oxford University Press.

Many important papers on the logical revolution and its applications to philosophy of science will be found in

AYER, A. J. (ed.), *Logical Positivism*. New York: The Free Press.

VAN HEIJENOORT, JEAN (1967), *From Frege to Gödel: A Source Book in Mathematical Logic*. Cambridge, MA: Harvard University Press.

Work on the logical investigation of discovery begins with Putnam's seminal paper and continues to this day:

GLYMOUR, CLARK and KEVIN KELLY (forthcoming), *Logic, Computation and Discovery*. Cambridge, England: Cambridge University Press.

KELLY, KEVIN (forthcoming), *The Logic of Reliable Inquiry*.

OSHERSON, DANIEL N., MICHAEL STOB, and SCOTT WEINSTEIN (1986), *Systems That Learn: An Introduction to Learning Theory for Cognitive and Computer Scientists*. Cambridge, MA: Bradford/MIT Press.

PUTNAM, HILARY (1965), "Trial and Error Predicates and the Solution to a Problem of Mostowski, *The Journal of Symbolic Logic* 30: 49–57.

Influential antirealist statements are given in

DUMMETT, MICHAEL (1978), *Truth and Other Enigmas*. Cambridge, MA: Harvard University Press.

FIELD, HARTRY (1980), *Science Without Numbers: A Defence of Nominalism*. Princeton: Princeton University Press.

FINE, ARTHUR (1986), *The Shaky Game: Einstein, Realism, and the Quantum Theory*. Chicago: University of Chicago Press.

GOODMAN, NELSON (1978), *Ways of Worldmaking*. Indianapolis, Hackett.

PUTNAM, HILARY (1981), *Reason, Truth and History*. Cambridge, England: Cambridge University Press.

VAN FRAASSEN, BAS (1980), *The Scientific Image*. Oxford: Clarendon Press.

Systematic realist responses include

DEVITT, MICHAEL (1984), *Realism and Truth*. Princeton: Princeton University Press.

MILLER, RICHARD W. (1987), *Fact and Method: Explanation, Confirmation and Reality in the Natural and the Social Sciences*. Princeton: Princeton University Press.

VISION, GERALD (1988), *Modern Anti-Realism and Manufactured Truth*. London: Routledge.

Four

SCIENTIFIC CHANGE: PERSPECTIVES AND PROPOSALS

J.E. McGuire

Scientific cultures develop and change. The worldviews of the ancient cultures were modified significantly during the Medieval and Renaissance periods. In their turn, these cultures were radically transformed throughout the seventeenth century, the century of the Scientific Revolution and the beginning of Modernity. Likewise, our present scientific and technological cultures bear little comparison with the worlds of the Ancient Greeks, the Medieval metaphysicians, or the worldviews of the nineteenth century. In short, the ways in which we interact with our physical environment, and the ways in which we think about it, have changed and will continue to change.

But how is such change to be understood? What sorts of factors (social, historical, cultural, institutional, and such) determine scientific change? Why do certain views of the physical world become accepted while other and equally plausible views get rejected? Are there forms of continuity through scientific change? If so, what are they and how are they to be characterized? Or are earlier theories and practices radically incommensurate with later theories and practices? Apart from considering scientific development as either continuous or discontinuous, is it at some levels of practice continuous with its past and at others discontinuous? Again, what connection lies between traditional perspectives embedded in the scientific enterprise and innovative changes in theory and practice? Is there only a contingent relationship between tradition and innovation, or is there an intrinsic connection? Are theories accepted because they stand in some logically timeless relation to objective evidence, such as being confirmed, verified, or corroborated? Or are theories accepted because they are human constructs satisfying the cognitive and social norms of the scientific cultures in which they are embedded?

Needless to say, various responses to such questions raise profound issues. Let us begin by considering their scope in a preliminary way. Scientific change can be considered in two chief ways. First, it can be seen as an exercise in logically appraising the evidential consequences of a scientific theory in comparison with those of its rivals. In this framework, justificatory parameters such as degree of confirmation, strength of verification, the management of relevant evidence, or the extent of corroboration are considered, and change is judged in accordance with how well theories pass muster with respect to such criteria. Second, scientific change can be conceived as an historical and social phenomenon. This orientation involves consideration of historical and temporal modalities. How is scientific advance related to its past history and how is it directed to its future? This framework takes seriously the belief that contexts of change and progress dwell in history's unfolding through the present into the future. To judge, then, whether a piece of science is progressive or not involves the retrieval of its past performances and the comparison of these with those of its rivals.

A shift from the first perspective to an historical and social orientation is characteristic of the move to the "new philosophy of science" of the 1960s and 1970s. To sharpen the point, consider the assumptions of two frameworks or scenarios within which issues of scientific change have been addressed.

The first is as follows: At every moment human beings dwell in their immediate, but ever changing, historical and social contexts. They are consequently thoroughly historicized and socialized by contexts that are characteristically different from one another. Given that we dwell in such contexts, their constitutive presuppositions structure our view of history, science, philosophy, and culture. This conception has consequences for truth, the nature of human knowing, and for the objectivity of knowledge and what is knowable. If, as this scenario has it, our norms and standards (both cognitive and social) depend on context, what counts as truth, knowledge, knowing, and objectivity is relative to that perspective. Thus, every cultural reality, science included, carries its own inner logic, its own values and norms, its own ways of carving up experience, its own inner dynamics of change, and must be judged accordingly. On this perspective there are no timeless truths, identical in all spheres of human activity—moral and political, social and economic, scientific and artistic; the cultural worlds in which we dwell are not essentially objective wholes in which all things cohere, but contingent artifacts of our own free making; and finally there is not one true method, based on reason, which yields correct solutions to all genuine problems, and similar in all fields of systematic inquiry. On the contrary, all forms of knowing are historically and socially contingent, and so too are the discourses which express what we claim to know. Accordingly, human knowing, together with what it claims to know, is situated in the temporally bounded worlds in which it dwells. Furthermore, the human self is not a fixed essence that is, but a social construct that evolves and becomes. There are, then, no decontextualized and occurrent objects of knowing to which a self disengaged from context is directed and to which it universally relates, for there is no such self and there are no such objects: Both are illusions of metaphysical reification.

On these assumptions, scientific change becomes a matter not of one perspective on the physical environment metamorphosing into another, but a matter of one scientific culture being substituted for another. This, of course, entails the recontextualizing of theory and practice. So, on this view, there can be no transcultural foundation for objectivity and the universalization of science. Historical, social, sociological, anthropological, constructivist, and rhetorical approaches to scientific change accept to a greater or lesser extent the above scenario.

The second scenario holds fast to a belief in the continuous progress of science, and to the belief that a mind-independent world exists that is objectively knowable. Human beings not only occupy positions in space and in time, they are also knowers. Not only are they knowers, they are knowers who remain self-identical through change. Notice the scope of this assumption. Whatever accounts for the identity through change of the human mind, at all times and places (and thus a fortiori in all cultures, past, present, and future) the human mind itself remains structurally the same. This picture of the situated knower encourages the view that the human mind can predict and retrospect events, occurring along any linear horizon of time and space, from any space-time point within a spatiotemporal matrix.

Accordingly, we have a picture of the human knower disengaged from context and situated in an objective world whose nature differs from the knower's nature. For not only is the space-time matrix independent of human perception, but so too is the physical world that is an object of human knowing. Essential to this picture is also the belief that the mind is occurrently related to the occurrent physical objects which it knows. In its simplest form the belief is this: The mind perceives (say) a tree in the virtue of being able to form an inner mental representation of that outer physical reality (see Chapter 3, Section 3.1).

Within this framework a definite picture of scientific change emerges. There is, so to speak, a God's eye view of the epistemic horizons which we survey. Furthermore, the skeptical challenge posed by the dualism of the mind and the physical universe is met by the construction of semantically coherent theories generated by universally applicable methodologies. Since the time of the seventeenth century, this progressive faith in theory and methodology has gone almost unchallenged, for the belief remains that the methodologies of science give access to a logically coherent account of the real structure of physical reality. Particular theories may come and go; still the well-formed factual basis of our knowledge of the real remains. Accordingly, scientific change is the process of successively incorporating earlier and successful theories into the framework of their successors so that factual and predictive control over nature cumulatively increases over time. In other words, as a result of cumulative scientific change we come to know more and more facts about an objective world that exists beyond. Logical positivism and logical empiricism are compatible in spirit with this second scenario.

Here then are two opposite frameworks of discourse within which problems of scientific change have been addressed. The first perspective is diachronic and takes seriously the temporality and historicality embodied in the making of scientific change. The second is synchronic and conceives the products of scientific practice as detached from context and time. The aim of this chapter is twofold. First, it outlines and critically evaluates the major positions on scientific change now current in the literature. Among perspectives to be discussed are logical positivism and logical empiricism, still considered by many (even by their detractors) to be the "received view'' in the philosophy of science. Secondly, the ''New Philosophy of Science,'' which numbers among its exponents Thomas Kuhn, Imre Lakatos, and Paul Feyerabend, is considered. Next, the major challenges that have been mounted by social theory, sociology, historicism, and the anthropology of science are considered. This chapter provides a critical appraisal of some of the major contributors to the growing and complex literature on scientific change.

4.1 THE DISCOURSE OF THE "RECEIVED VIEW" IN THE PHILOSOPHY OF SCIENCE

''Received view'' means the perspectives on the language and the foundations of science advocated in this century by logical positivism, logical constructionism, logical empiricism, logical atomism and physicalism (included, also, is the ''standard conception of theories,'' Chapter 3, Section 3.7). These labels can be misleading since they evidently cover a wide range of attitudes and positions in regard to the issue of scientific change. Carnap, for instance, shows little or no interest throughout his career in scientific change. Nor, for that matter, does Russell. This is not altogether surprising given the character of their logical, mathematical, philosophical, and linguistic interests. Moreover, both philosophers assume that science is by its very nature, and in virtue of its very methods, cumulative and progressive, and that therefore we can gain a better and better understanding of the world. In contrast, the work of Reichenbach and Popper is relevant to the questions of theory choice and to the comparative merits of competing theories. In fact, questions of theory choice and the nature of scientific change do not become central to the ''received view'' until after the period of the Second World War. However, questions of discontinuity and change in the development of the sciences were central to writers such as Gaston Bachelard (1984) in France throughout the 1930s and beyond.

The purpose of outlining the main commitments and assumptions of the ''received view'' is twofold: (1) To motivate its attitude toward scientific change, and (2) to place this attitude within the larger discourse of positivism. Of course, not all proponents of the ''received view'' accept the same commitments and assumptions, but unless the scope of the discourse of positivism is understood, the issues that motivated the move to the ''New Philosophy of Science'' of the 1950s and 1960s will not be apparent.

4.2 LOGICAL CONSTRUCTIONISM: LOGIC AS THE PHILOSOPHER'S STONE

Let us begin with the program of logical constructionism. This will help to clarify the nature of scientific philosophy, and will serve to introduce the conceptual framework of a more pragmatic mode of analysis still current in the social theory of science, namely, social and linguistic constructionism (see Sections 4.10, 4.11, 4.12, and 4.13).

Logical constructionism is based on two commitments. Western philosophy has attempted traditionally to account for all ways of being and all ways of knowing by appealing to ''privileged'' ontological and epistemological items. Accordingly, the special status of lawful or rule-governed combinations of occurrent elements has repeatedly been proclaimed: Platonic forms, atoms, fields, ordinary physical objects, atomic facts, sense-data, ideas, protocol sentences, bits of information, syntactic structures, and so forth. Furthermore, this view holds that everything that can be experienced, or that is knowable, or which can properly be said to exist, can be translated into, or reduced to, or explained by, privileged and more basic items.

The second commitment holds that the transformation to the privileged basis is to be performed through the medium of a semantically coherent and logically rigorous discourse. For Russell, and following him Carnap, this is the language of Russell and A. N. Whitehead's *Principia Mathematica* (1925). The program of Russell and Whitehead is to show that pure mathematics can be stated in the canonical language of logic using no undefined terms apart from the logical operations of implication, disjunction, class inclusion, and class membership. The motivation is clear: to show that problematic mathematical entities are adequately expressed in this language to the extent they can be constructed out of entities considered to be less problematic. This is the claim that less fundamental entities can be reduced, by the power of the logical language, to a privileged constructive basis without loss of meaning. For example, numbers are defined as classes of classes: Zero is the class of all empty classes, and the number one is the class of all classes each of which is such that any member is identical with any other member. But the use of reductive schema to show that certain items are constructible in terms of other more privileged items leads to problems. Among these is the alleged unproblematic status of classes, and the notorious paradox of the self-membership and nonmembership of classes.

This is logical constructionism in its purest form. It is clear that its motivating feature is the belief that anything that can be said clearly can be said in the discourse of logic, a discourse taken to capture best the fundamental status of the privileged entities and their relations. Thus, anything sayable about something else is not really about that something else, it is about these privileged entities, construed as representing all knowledge in terms of their properties and relations alone. In other words, the techniques of logical constructionism involve a principle of economy acting to reduce epistemic and ontological commitments to a minimum, and to mitigate the possibility of error and confusion.

In the period after the publication of *Principia Mathematica*, Russell applied the techniques of logical constructionism to knowledge of physical objects both in the sciences and in commonsense experience. The basic problem he addressed is this. How can the existence of unperceived entities be warranted, such as fields, forces, subatomic particles, energy quanta, and so forth? One traditional approach is to show that such unperceptible items can be inferred from the content of our immediate experience. But how is that experience to be correlated with these unobservable entities and relations? And what epistemic status is to be accorded to the correlating principles? Alternatively, it can be argued that realism about these unobservables must be presupposed if an adequate explanatory account is to be given of what is in fact observable. This, however, requires a meta-argument to show that adequate scientific explanations cannot be given in purely observable terms, a view that van Fraassen (1980) and Cartwright (1983) have recently opposed.

Russell, in his early work, takes another course which is motivated by his commitment to constructionism: where possible substitute constructions out of known entities for inferences to unknown and unobservable entities. His aim is to show that both unobservable scientific entities and ordinary physical objects, such as flowers and tables, can be constructed out of the incorrigible sense-data of sensory experience ([1912] 1959, Chapter 3 and [1914] 1960, Lecture 3).

In the face of the venerable distinction between the inner world of human experience and an outer world of physical objects, Russell privileges the inner world of sense-data, that is, the colors, shapes and sounds of which we are directly aware in sense perception, and about the nature and existence of which there is thought to be no serious doubt. He still encounters, of course, the ancient skeptical problem of how such private entities relate to outer objects, or how inference to such objects is to be warranted. This he seeks to avoid by showing that all the outer objects of science and of common sense can be construed as complexes of immediate sense-data. Here again is the basic supposition of constructionism, namely, that the problematic can be reduced to, or constructed out of, the more certain and less problematic.

However, in order to avoid the implications of solipsistic reductionism (i.e., the idea that physical objects are constructible out of the actually experienced sense-data of a single observing mind), Russell modified his view. He argued that the basis for construction must include both the actually experienced sense-data of an observer and those that *would be* or *could be* experienced by others *if they had been subjected* to certain experiences under certain conditions. This type of phenomenalism encounters formidable difficulties, many of which Russell recognized in his later work. For example, how do we characterize which actual and possible experiences define a particular physical object under changing conditions of perception without referring directly to that object or to others in our characterization? Again, the position encounters difficulties in specifying the conditionality of the if-then structure of counter factual statements, that is, statements to the effect that if something were the case, or had been the case, then such and such would be, or could have been, the case.

Let us turn now to Carnap's *The Logical Structure of the World* ([1928] 1967). Carnap ([1928] 1967) uses the techniques of construction theory inspired by the methods of Russell and by Wittgenstein's *Tractatus* ([1922] 1955) (see also Chapter 3, Section 3.4). His main conceptual tool is that of reducibility. A concept S is reducible to another set of concepts W if the sentences pertaining to S can be transformed into the sentences of W with the extensional preservation of truth value. The transformation operation is performed by means of a rule, which Carnap calls a constitutional definition. These definitions are arranged into a structure which he calls the constitutional system. This system is a complex of definitions and theorems expressed in the language of *Principia Mathematica*. The known (or knowable) objects that Carnap seeks to place into, or reduce to, the constitutional system are fourfold: cultural objects, other minds, the private experiences of our own minds, and, lastly, physical objects (1967, Part 4).

Carnap ([1928] 1967) uses a principle of tolerance. His purpose is not to provide a description of concept formation, but rather to provide a rational reconstruction of concepts, a notion that later recurs in the work of Imre Lakatos (see Section 4.6). Thus, although he chooses the inner experiences of our own minds as the basis for his solipsistic system, Carnap is at pains to stress that any other basis is in fact possible, and that his choice of the mind's experiences was made for meth-odological and not for metaphysical reasons. In fact, he was later persuaded by Otto Neurath—another member of the Vienna Circle (see Introduction)—that a physicalist basis or language is to be preferred. Carnap ([1928] 1967, Chapter 5) considers philosophy and science and metaphysics in the light of constitution theory, and offers criteria for demarcating scientific questions from metaphysical ones. Thus, the ques-tion "Is there an external world?" is a metaphysical question which must be sharply separated from the scientific question as to whether a physical object can be placed within a set of law-like regularities and a set of orderings in space-time. Accordingly, as with Russell, Carnap is concerned to distinguish metaphysical questions and pseudoquestions from genuine scientific questions, and with Russell to subject sci-entific and commonsense modes of understanding experience to rigorous analysis conducted solely in terms of the language of logic.

4.3 LOGICAL ATOMISM, EMPIRICISM, AND THE UNITY OF SCIENCE

Although they are concerned with the logical syntax of their constructions, Russell and Carnap also manipulate semantics or meanings. But the question becomes: What mean-ings are basic and how are they to be established epistemically? The question of how and where the ultimate source of meaning is established in the analysis of scientific practice remained a central issue in philosophy of science well into the 1970s.

First, let us motivate the doctrine of logical atomism. Thus far we have con-centrated on the methodological side of logical constructionism, namely, the concern to reduce epistemic commitment to a minimum and the concern to display the priv-ileged status of known or knowable objects. Russell and Carnap, however, suppose not only that their constructions have normative force, but also that they clarify the structure of what actually obtains in our experience. Although Russell sometimes proceeds as if his constructions are neutral, at other times he holds (for example) that constructions out of the irreducible data of awareness reveal how the mind actually is, or that the sense-data that make up the appearance of a physical object provide the fundamental basis of our knowledge of what the table really is. This second tendency in his thought is addressed systematically in Russell ([1918] 1956b and [1914] 1960).

Here he endorses explicitly a principle, shared with Carnap, that a metaphysical or world-oriented interpretation of logical constructionism depends on a correspon-dence between an ideal language and the structure of what is real. But how do we choose an ideal language? After all, any number of alternative discursive frameworks are possible. For Russell, the language must be empirical, a desideratum he spells out in terms of the "principle of acquaintance."

Russell tells us, "All our knowledge, both knowledge of things and knowledge of truths, rest upon acquaintance as its foundation" ([1912] 1959, 48). We have acquaintance with anything if we are directly aware of it without the mediation of inference. Thus, we are immediately aware of sense-data—colors, sounds, shapes—possess immediate knowledge through memory, are immediately aware of being aware of something, and can immediately conceive nonparticulars such as universals.

Universals play a crucial rule in extending knowledge beyond particular acts of awareness, for when we know objects, such as a table or an electron, we do not know them by direct acquaintance, but know them by description, that is, through general truths and statements containing universal terms. Such acts of knowing Russell calls "knowledge by description" as opposed to acts of immediate awareness which he calls "knowledge by acquaintance." Furthermore, he persistently affirms that knowledge of what is known by description is ultimately reducible to what is known by acquaintance (ibid., 58).

However, he also tells us, "Every proposition which we can understand must be composed wholly of constituents with which we are acquainted" (ibid.). Russell contends here that in order to speak significantly we have to attach meaning to the language we use; ultimately these meanings must refer to that with which we are directly acquainted. Without meaning there can be no understanding. Consequently, Russell also contends that we understand language only if it refers to what we have experienced by acquaintance, or is defined in terms of expressions that do so refer. His claims are two. First, that unless physical objects are defined in terms of modes of acquaintance, there can be no way of knowing them. And second, and more significant, there can be no way of understanding them without satisfaction of this requirement.

But what is it that Russell proposes to analyze in his ideal language, and what is the structure of this language? Put simply, facts need to be analyzed, and the language is explicitly truth functional in structure. In the world things have various properties and stand in various relations. These properties and relations are facts about those things. For Russell, facts are stated by propositions that are composed of terms and other linguistic items. Some terms are simple; for instance, the term "brown" designates a particular shade of color, and the proper name "Bob" refers to a particular individual. Now our understanding of the term "brown" does not depend on something simpler, for understanding in this case depends on acquaintance with that to which the term refers, that is, a definite shade of color. Thus, the sentence "This is brown" is composed of a simple designating term "this" and the predicate "brown." This sentence expresses the sort of proposition that Russell calls "atomic," and the facts stated by such propositions are atomic facts.

From atomic propositions more complex propositions can be constructed. By joining together atomic propositions with the operators "and" or "or" we can construct propositions which Russell calls molecular. But there are no facts, Russell holds, that corresponds to molecular propositions. What makes the molecular proposition "This is brown and this is green" true, if it is true, is not a molecular fact but two atomic facts, that this is brown and that this is green. Thus, the truth or falsity of molecular propositions depends entirely upon the truth or falsity of the atomic propositions that compose them. In other words, molecular propositions are "truth functions" of atomic propositions.

Russell was aware that this is an ideal picture of language and of its relation to the world. Moreover, he was aware that many propositions that have meaning cannot be captured by truth functional analysis. For example, the meaning of the proposition "All swans are white" is not merely a conjunction of propositions "This swan is white and that swan is white and . . . ," and so on until every swan has been enumerated. For even were it possible to enumerate all swans, it is necessary to stipulate that the swans so enumerated are exhaustively enumerated. This of course reintroduces the generality and as such fails to provide an analysis of it. Russell was therefore forced to abandon truth functional analysis of such propositions and to introduce irreducibly general facts. He also had to abandon a truth functional approach to propositions of the form "x believes that p" since the truth of "x believes that p" is independent of the truth "that p." Despite these difficulties, Russell remained committed to the drive for minimal vocabularies and to the aim of reducing complexity to the logically simple. Although he was haunted by the prospect that what is taken to be logically simple may be subject to further analysis, he never lost faith in the belief that knowledge must ultimately rest on simples if we are to possess meaning and understanding, and that what *is* actually controls the *success* of our objective descriptions.

Like Russell, Carnap also privileges those types of minimal sentences that stand closest to immediate experience. Where Russell speaks of "atomic propositions," Carnap speaks of "protocol sentences." This commitment to observation sentences as the unit which supplies the foundation for meaning and understanding in the sciences is perhaps the leading characteristic of the "received view" in the philosophy of science well into the 1970s. Moreover, one of the central concerns in this empiricist tradition (which also includes Hume, Mach, and Duhem) is the problem not only of specifying how knowledge is based in the simples of experience, but also how theoretical or nonprotocol sentences, which are remote from observational experience, relate to observational sentences.

When Russell and Carnap speak of reducing complex statements and terms to simpler ones, they clearly accept a distinction between "theoretical" and "observational" sentences and terms (see Chapter 2). If, as the empiricist tradition has it, meanings must attach solely to the observational terms of immediate experience, how is meaning accorded to the theoretical terms? Given distinctions between theoretical and observational terms, there are three clear-cut solutions: (1) Show that complex theoretical terms are reducible to, or are constructed out of, observational terms; (2) show that theoretical terms are dispensable in science and are thus eliminable from scientific theories understood as formal systems, and (3) deny the distinction either by showing that observational terms are theory-laden or by arguing that there is really only theory and theoretical terms, and that what we take to be lower-level terms have this status only relative to theory (see Chapter 3, section 3.7).

Carnap ([1928] 1967) places his constructions and his language on a phenomenalist basis—the private data of inner experience—under the influence of Russell. Neurath persuaded Carnap that a physicalist language has certain advantages. The central issue here is not a choice between a phenomenalist as opposed to a physicalist metaphysics, but rather a choice of the fundamental language, in this case, the basic observational sentences which Carnap calls protocol sentences. Thus, Carnap adopts the view (developed also by Russell [1914] 1960, Lecture 4) that the protocol sentences of science are best expressed as quantitative expressions which refer to determinate regions of space-time points (1967, Part 3, C). On this view, all of the theoretical sentences of science can be expressed ideally as equivalent to sentences in the physicalist protocol language. Clearly a physicalist protocol language affords certain advantages over a phenomenalist basis. It is common to different senses, it is intersubjective, that is, directly accessible to all observers, and universal in the strong sense that all sentences in science can be translated ideally into the protocol language. Besides physical protocols, Carnap also advocates the notion of the unity of science. For Carnap the notion is exemplified in the thesis that the language of science can be totally constructed on a physicalist basis, including the languages of psychology and biology.

From first to last, Carnap's intention is to eliminate metaphysics and pseudo-philosophical problems. This drive is no more evident than in the *Logical Syntax of Language* (1937). Here under the influence of the logician Gottlob Frege and the mathematicians David Hilbert and L. E. Brouwer, Carnap distinguishes the "object language" (the language under investigation) and the "metalanguage" (the language that articulates the theoretical account of the object language). His aim is to construct suitable metalanguages in which to practice philosophy, or as he preferred to call it, the logical analysis of language. He generates, in fact, two model languages. Language I is "definite" in the sense that its defining expressions contain no unlimited quantifiers and is constructivist or finitist in character. Language II he advocates as a representation of classical mathematics (1937; Parts 2, 3). Once again he invokes the principle of tolerance, here the notion that linguistic structures are conventional in character and allow the construction of any form of language that is deemed preferable.

For Carnap the logical structure of language is to be replaced by the logical syntax of language. Here he introduces the notion that pseudoproblems in philosophy arise from misunderstandings of how syntax works. For example, in first-level or object-language discourse, language is used referringly to designate its objects. But in a second-level or metalinguistic discourse, linguistic expressions refer to linguistic items at the first level and not to nonlinguistic items. In a time-honored philosophical vein, reminiscent of Plato's *Theatatus*, the thought of the Middle Ages, and the work of Hobbes and Leibniz, Carnap indicates the endemic tendency to confuse these two levels of language. The confusion turns on the simple fact that the "material mode" of speaking is treated as though it were an "object mode"; that is, the metalinguistic functioning of language is conceived as though it were about objects and not about the object language itself. For Carnap it is at the juncture of this confusion that pseudo-philosophical problems begin to emerge.

In a later discussion of ontological commitment which has its roots in Carnap (1937), Carnap again mobilizes resources against pseudoquestions. Carnap (1956a) seeks to show that a thorough empiricism is not incompatible with the use of abstract entities. Two sorts of questions about the existence of entities must be distinguished: internal questions which concern the sort of ontology to which a particular theoretical or linguistic framework is committed; external questions which query the ontological status of frameworks as such (Carnap 1956b, Supplement). For Carnap, acceptance of frameworks raises no issues concerning external questions of ontological commitment. Such external questions raise pseudoproblems of the sort found in the traditional realism-idealism-nominalism debate. The real question is the concrete one of choosing a linguistic framework, and as before, Carnap advocates tolerance.

Logical positivism was always a self-critical movement, and this is no less evident in Carnap's thought than in any other member of the Vienna Circle. Many members of the Circle, Carnap included, began to feel that their philosophical pre-occupation with logically "clean" languages failed to come to grips with the ever-changing realities of scientific practice. Indeed, if empiricism is to deal adequately with the significant advances in science, it must do so without appeal to a priori presuppositions. That is, it must reconstruct scientific knowledge from the bottom up in terms of the data of immediate experience, and not from the top down through recourse to synthetic a priori judgements traditionally thought to be necessary in order to have intelligible experience. But difficulties develop in sustaining a thoroughgoing empiricism. We have already noted the problems Russell encounters with general facts and intentional contexts. Russell (1948) posits a priori postulates which he maintains are necessary preconditions of scientific inference in a manner not unlike the transcendentalism of the Neo-Kantian tradition (ibid., 6). Carnap (1963, 978–979) also transgresses the dictates of a purely empirical account of knowledge. In his later work on degrees of confirmation or inductive probability, he is driven reluctantly to allow that prior probabilities rest on intuition regarding a priori distributions of these probabilities. Such propositions are scarcely anything else but synthetic a priori propositions (see Chapter 3, Sections 3.5, 3.7). Nevertheless, these deep tensions in the empiricist program lay behind Carnap's attempts in the 1930s to loosen its logical and empirical criteria by attempting to reflect more faithfully the open character of actual scientific advance. Here his reevaluation focuses chiefly on the reducibility requirement and on the verification principle of meaning, the scope of which is considerably modified. Indeed, Carnap (1936) agrees with Popper that scientific hypotheses can never be completely verified by observational evidence, and advocates the substitution of the notion of degree of confirmation for that of verification.

Although Carnap never explicitly addresses the notion of scientific change, his work on probability, confirmation, and induction is relevant to the problem(s) of adducing criteria for theory choice. The issue here is how one theory can be shown to be cognitively and empirically better than another theory. In company with most of his contemporaries, Carnap agrees that if a theory is genuinely scientific, it must in some way be responsive to empirically determinable evidence, and that the better theory is comparatively more responsive to the evidence than the lesser theory. But how was this relation to be specified? For Carnap it is in terms of inductive support as spelled out in his probabilistic notion of degree of confirmation (see Chapter 2). Thus, on a given body of evidence, a theory is better grounded than another if its inductive probability is higher.

4.4 VERIFICATION, COGNITIVE MEANINGFULNESS, INDUCTION AND CONJECTURE

Let us turn now to some later developments in the "received view," with reference to the work of Hans Reichenbach and Karl Popper. Reichenbach is the most thoroughgoing and uncompromising empiricist of the group originally centered on Berlin. On the other hand, Popper, although of similar philosophical sensibility, but never a member of either the Vienna Circle or the Berlin group, has been and is a persistent critic of positivism, especially on the issue of how a theory's responsiveness to its evidential basis is to be specified.

The striking thing about Reichenbach's empiricism is that his probabilistic theory of cognitive meaningfulness allows him to escape most of the difficult commitments of earlier positivism. He does not need to consider phenomenalism as did Russell and Carnap; he does not need to regard physical objects as constructions out of sense-data; nor does he need to assume that physical objects are logically equivalent to a finite series of observational reports embodied in protocol sentences. In short, he is able to navigate the central semantic and ontological difficulties that beset classic positivism while still maintaining a robust empiricism.

The key to understanding this "Reichenbachian turn" in the empiricist program lies in the way Reichenbach relates induction and probability to the problem of factual or empirical meaning. Although A. J. Ayer's attempt to explicate a "weak" criterion of verifiability came to grief at the hands of Alonzo Church (see Chapter 3, Sections 3.5, 3.6). Reichenbach never relinquished the notion that a verifiability theory of meaning is possible. For him the question for empiricist philosophy is not how objects are constructed out of experience, but rather how statements about both observable and unobservable objects are verified in present and future experience. Accordingly the context of justification, not that of discovery, is what counts. Moreover, he realized that criteria of cognitive meaningfulness cannot be stated in terms of strict deductive verifiability from a finite set of observational statements, but must invoke the notion of probabilistic verifiability to some specific degree. Thus, Reichenbach avoids the traditional difficulties of reducibility and constructibility by orienting the dynamics of his epistemology toward a probabilistic assessment of the best cognitive outcomes likely to obtain in the long run. Indeed Reichenbach (1938, Chapter 5) is concerned to show the fundamental role that probability plays in accounting for knowledge.

According to Reichenbach, a statement is cognitively meaningful only if in principle it is possible to obtain evidence which will support it to some degree of probability or weight. Depending on whether the weight is high, intermediary, or low, a probabilistic verification will be either supportive or nonsupportive of the statement. Thus, Reichenbach's criterion of cognitive meaningfulness is probabilistic confirmability or refutability.

Closely connected to this epistemic picture is Reichenbach's contention that induction is a rule that can be justified pragmatically in terms of probability frequencies. On this view, induction is a rule-governed activity that directs us to infer or posit that the observed frequency approximates the long-run frequency. We, of course, lack foreknowledge whether a sequence has a limiting frequency or not. Nevertheless, by use of the inductive rule, if the sequence has a limit, induction will capture it; if not, no method is of use. Against this pragmatic approach to induction Reichenbach develops one of his key epistemic notions, the posit. A posit is a statement or hypothesis that is treated as if it were true but which would not be posited if it were known to be false. Posits are epistemic strategies that maximize the information we possess by using the inductive rule (the rule for making posits) in the task of ascertaining the limit of frequencies (see Chapter 2).

Apparently, scientific change is not a problem for Reichenbach. He never doubts that science changes and grows and that there are revolutionary discontinuities in its development. For Reichenbach, the central issues for the philosopher lie elsewhere in the business of giving the most pellucid account of how scientific statements are justifiable. Given his probabilistic understanding of the context of justification and his Baysianism, Reichenbach has much to say that is germane to questions of comparative theory evaluation and choice (see Chapter 2).

Popper's notion of conjecture has much in common epistemically with Reichenbach's posit. For Popper, science is about falsifying basic statements, not about verifying them. Thus, science must construct bold conjectures or severe tests for a theory which, if it passes and continues to survive serious attempts at falsifying it, the theory can be accepted provisionally. Popper is emphatic, however, that a theory or hypothesis can never be established beyond doubt. If a theory survives serious attempts at refutation it is thereby corroborated, corroboration being greater the extent to which the theory is falsifiable (1963, Chapter 10). For Popper, a theory has greater falsifiability to the extent to which it says more about the world and thereby constrains the context of its generalizing power. If this is the case, the theory's basic statements have comparatively greater empirical content. In Popper's view, induction is to be abandoned because it is not justifiably a rational procedure. Thus, strategies for establishing inductive generalizations should be rejected and science should concentrate on conjectural tests and refutations of theories regardless of how they are discovered. In effect, this means that science must forgo the idea of establishing positive support for any theory that goes beyond the immediate evidence for it. Ironically enough, Popper's position is not immune from the procedures of induction. For what is the point in subjecting a scientific theory to severe tests unless we assume that the passing of such tests makes it more likely to do so in the future? This assumption clearly involves the inductive projection of confirming instances into the future. Interestingly, however, Popper's epistemic attitude toward science supports the interpretation of science as a dynamic enterprise which has the potential for changing itself continuously. There are potentially scientific revolutions forever.

Here then, is a powerful discourse. It seeks to reveal the unchanging patterns that lie beneath the surface of scientific practice. For the "received view," science is at once cumulatively progressive, objective, and universal. It is objective in two senses. First, the basic language of science addresses the bar of immediate observational experience, believed to be intersubjectively available to all impartial observers. Moreover, the dream of the "received view" is to capture this experience linguistically in its purest form through a minimal vocabulary which distorts experience to the least extent. Secondly, it is objective in the sense that the basic language is the language of modern logic, a language held to mimic the structures of outer reality. Science is universal according to the "received view" since it holds that the methodological norms of science are invariantly instantiated in various cultures and at various times. Thus, the epistemic attitude of the "received view" embodies an essentializing mode of understanding scientific change. It is essentializing because it attempts synchronically to detemporalize the temporality of science by reducing scientific change to the business of imposing at the logical instant comparative criteria of theory evaluation in the form of either confirmation, verification, falsification, or logical simplicity. Accordingly, the immediate "unit" of change in scientific development is the observational sentence such that a change in empirical content or a rejection at this level leads to an adjustment of the theoretical sentences of the system.

Notice that the "received view" maintains some unquestioned assumptions. No cogent argument is offered for verificationism. Precisely why should meaningfulness consist in what can be perceptually verified? And why should assertive forms of discourse in science be privileged over pragmatic forms? Again, objective truth is assumed to be available, to be communicated through a correct and canonical language, and to be readily distinguished from falsehood. Lastly, there is a reductive reliance on the priority of simples, and on the faith that they can be isolated and decontextualized in the service of theory. The "New Philosophy of Science" is indeed skeptical of these assumptions.

4.5 THE DISCOURSE OF THE GLOBALISTS: THE "NEW PHILOSOPHY OF SCIENCE"

The phrase "The Globalists" refers to philosophers such as Thomas Kuhn, Imre Lakatos, Larry Laudan, and Paul Feyerabend. They are concerned with problems of scientific change, emphatically reject the program of the "received view," and take seriously the historical and temporal dimensions of scientific change and development. The "New Philosophy of Science" means that body of opinion largely influential in the field during the period from roughly 1960 to 1980.

Let us begin by outlining the doctrines and perspectives of the "received view" that the Globalists either reject or put into serious doubt:

1. First and foremost, they reject the theoretical and observational sentence dichotomy. They argue either that no principled distinction is to be made, or that observational sentences are seriously infected by theory. In fact, they privilege theory over observation, whereas the "received view" privileges observation over theory. Recently this whole dialectic has been challenged as misguided by those who see experimentation and experimental techniques as central to scientific practice and change (see Section 4.14).

2. They dismiss the view that the transition from one theory to another is cumulative. They argue that logical and empirical content (even the confirmed consequences of an earlier theory) are not entirely preserved when a theory is replaced by a newer theory. This denies the claim that there is meaning invariance of the observational sentences across theoretical change.

3. They reject the view that theories can be logically assessed at a time through their observational consequences by means either of confirmation, verification, or falsification. There are no such absolute canons of nonrelative knowledge and authority. Theory evaluation is a complex matter involving many factors beyond the idealized logics of justification; and theory change is a diachronic phenomenon involving the career of a piece of science within its changing social and historical contexts.

4. They hold that the distinction between the context of discovery and the context of justification is misguided. Concentration on logical issues of justification skews perspectives on the developmental dynamics of the scientific enterprise and overlooks the wider context in which science grows and changes. For the "New Philosophy of Science" it is important to understand how science comes into being.

5. They cast serious doubts on foundationalism, the view that a disinterested, reflective, and cognitive attitude can disclose privileged epistemic or ontological items to which all else is reducible. "Logical atoms," "protocol sentences," "sense-data," and "increasing empirical content" are cases in point, as indeed is the idea that there is one true theory which can represent the ultimate structure of a mind-independent world.

6. They object to the view that science can be understood solely as an enterprise seeking to establish disinterested knowledge of the decontextualized properties of self-sufficient objects such as electrons, genes, and quarks. To put the point another way, the "New Philosophy of Science" objects to the oversimplified image of science as possessing arcane infallibility, as giving unique epistemic access to what is, as producing unassailable, occurrent, and objective knowledge by the use of absolute and culturally-neutral methods, as giving one true description of the physical world by means of a semantically pure and truth-producing vocabulary.

7. They make explicit the view that scientific theorizing is prior to scientific practice. Indeed, in common with the "received view" the globalists maintain the idea that explanatory knowledge is propositional in content, and thus that all forms of knowing-how are to be transformed into knowing-that. They differ from the "received view" in this respect only to the extent they privilege theory over observation.

4.6 PARADIGMS, LEXICONS, AND INCOMMENSURATION

The main themes of the "New Philosophy of Science" are now presented in the context of the writings of the main contributors. The work of Thomas Kuhn put in doubt the view that science is progressive and cumulative in the increase of its empirical content. For Kuhn, theories are not superseded by their successors because of an accumulation of evidence against them, or because they are either nonverified or refuted, but because they are less good in comparison to the theories that supersede them at choosing new problems and at setting criteria for solving outstanding scientific problems. Consider his view of the emergence of Copernican astronomy. He points to the persistent failure of Polemic astronomy to solve its own problems and puzzles. It failed not only to do this, but in the scientific community there was an increasing sense that the problems of Ptolemaic astronomy were no longer *solvable* within its own framework. To the crisis engendered by this situation the Copernican theory seemed to emerge as a direct response. Here Kuhn stresses the importance of crisis within the scientific community. In many cases the solution to the failure in problem solving is anticipated. For example, in the third century B.C. Aristarchus had in large measure anticipated Copernicus. However, Kuhn points out that, if viewed historically, Artistarchus's anticipation clearly became a viable Copernican possibility only after the crisis generated by the demonstrable failures of the Ptolemaic system itself (Kuhn 1970, Chapter 7).

Kuhn advocates a definite model for understanding scientific change, especially those changes in science that may be called revolutionary. It involves three key notions: paradigm shift, the persistence of outstanding difficulties in the face of a paradigm's declining ability to solve its problems, and incommensurability between the old and the newly emerging paradigm during a period of crisis in science. Kuhn bases his model for scientific change on theories of social change, and sees social and institutional factors as essential parameters for understanding the nature of such transitions in science. In this regard, he rejects the epistemologically motivated distinctions of much of the philosophy of science current in his time: theory versus observation, context of discovery versus context of jurisdiction, continuity versus discontinuity, and verificationism versus conjecture and refutation. These had already been attacked by writers such as Paul Feyerabend and N. R. Hanson. The idea that intellectual advance arises from radical discontinuities had long been an article of faith among the Bachelardians in France (Bachelard 1984). Kuhn, however, produced the first systematic alternative to these entrenched philosophical orthodoxies.

The notions of paradigm and paradigm shift have rightly been criticized as vague and ambiguous. For present purposes this much needs to be said. For Kuhn a paradigm has two distinct connotations. It stands, on the one hand, "for the entire constellation of beliefs, values, techniques, and so on shared by the members of a given community" (Kuhn 1970, 175). Thus a research paradigm is what a scientific community shares, a shared bond of education, acceptance of theories, objectives, values, socialization and professionalization. In this sense, it may be compared to Wittgenstein's (1953) "forms of life" or to Heidegger's (1962) notion of *Dasein*. Although a scientific community may be identified by a shared paradigm, the community itself is a social phenomenon that can be identified and isolated independently of its paradigm. In this sense a paradigm must be understood in sociological and institutional terms as a social complex which expresses the affiliations, techniques, and organization of the scientific research community.

On the other hand, paradigm stands for an important element in the organizational complex, namely, the models that exemplify the explicit rules and criteria that guide the puzzle-solving activities of normal science (ibid.). For Kuhn, normal science does not seek novelties. It seeks, rather, to actualize a paradigm's potential by increasing, for example, the extent of the match between "facts" determinable under the paradigm and the scope of a paradigm's predictions. In this way the paradigm is further articulated by the realization of its puzzle-solving potential. Thus, it is the solved problems within the paradigm's framework which act as exemplars for further puzzle-solving activities under the paradigm. Here Kuhn stresses the cognitive role that successfully solved problems play as standards to be emulated in further research. That is, he rejects the view that science advances simply by applying theories and laws to new experimental and theoretical contexts which may or may not be verified or refuted. Rather, science advances by using the puzzle-solving resources of the paradigm. These Kuhn likens to the use of rules implicit in the solving of crossword and jigsaw puzzles. Just as the rule-governed moves that solve a puzzle insure its solvability by setting discrete boundaries for solution, so it is with the puzzle-solving potential of science. When science performs in this rule-bound way we have normal science. But when its puzzles are no longer solvable by the resources of normal science, they become its problems (ibid., Chapter 4).

Now a close relationship obtains between the proliferation of problems that a paradigm has not yet solved, or solved inadequately, or which are deemed not to be solvable under the paradigm, and the phenomenon of a radical shift to a new paradigm. Thus, the continued loss of puzzle-solving efficacy, and the increase of outstanding problems not amenable to the paradigm, propel the scientific community into crisis which is resolved ultimately by the community coming to share a new paradigm. The new paradigm sets new puzzles and the rules for their solution, and successfully tackles new outstanding problems but not necessarily those of the old paradigm. Here Kuhn insists that the decision to reject one paradigm is always simultaneously the decision to accept another (ibid., Chapters 7, 8).

We must be clear about Kuhn's position at this point. His claim is not that the older paradigm changes gradually into its successor: His claim is that the new paradigm completely replaces its predecessor. This is the doctrine that the old and new paradigms are incommensurable, and that historically the emergence of the one means the destruction of the other. For Kuhn, then, scientific change is reducible to the complete replacement of one set of structures by another (ibid., Chapter 9). It is important to grasp the force of the notion of "complete replacement." Kuhn tells us that the component statements of the rival paradigms are not intertranslatable. If this is so the incommensurability of the paradigms precludes saying that they are logically incompatible since this notion presupposes some measure of intertranslatability, precisely the requirement denied by Kuhn's conception of incommensurability.

In his more recent work, Kuhn still maintains a linguistic approach to incommensurability. In company with the structural linguists he holds that a linguistic item is defined by differences between it and other items in the linguistic field. What is meant by "lion" depends as well on what is meant by "tiger" and on how "lion" and "tiger" differ. If this is so, the reference of a term cannot be secured locally because the relation of a term to its object is based on a network of differences between this term and its object and other terms and objects in the field. Thus, piecemeal translation is not possible, which means that the entire linguistic network indigenous to the language has to be reconstructed in the translator's language. It is of course possible that the language's lexicon is not entirely homologous with the lexicon of the interpreter. If this is so, incommensurability is equated with untranslatability since a complete translation cannot be given. For Kuhn, then, incommensurability results from nonhomologous linguistic networks or lexicons which reflect their cultures as they interpret them. Although a full translation is not possible, incommensurability can be controlled to the extent that we are able to learn the other lexicon or language (Kuhn 1983, 1989). Notice how this notion differs both from paradigm-incommensurability and Kuhn's related claim that no neutral observation language can decide between *theories* since all scientific language is theory-laden.

Before examining the notion of paradigm-incommensurability in its relation to problem solving, let us consider briefly the parallel Kuhn sees between political and scientific development since it throws light on this notion. He points out that both political and scientific revolutions develop in response to increasing dysfunction within the system. This eventuates in the crisis that is a prerequisite to revolution. Furthermore, he notes that revolutions in society "aim to change political institutions in ways that those institutions themselves prohibit" (Kuhn 1970, 93). The resulting tensions lead to resistance, intransigence, and to the partial breakdown of the existing social structures. The deepening crisis at first attenuates the role of the existing structures, but the ensuing divisions lead eventually to competing proposals for the "reconstruction of society in a new institutional framework" (ibid.). Eventually, certain of the reconstructive proposals win the day through persuasion and rhetoric. As in political revolutions, so in scientific paradigm-choice, there is "no supra-institutional framework for the adjudication of revolutionary difference . . ." (ibid.). So, the choice between "competing paradigms proves to be a choice between incompatible modes of community life" (ibid., 94) much in the manner in which one chooses between competing social frameworks.

If we consider the orientation of Kuhn's thought, the parallel he sees between political and social change, on the one hand, and scientific change, on the other, is far from superficial. Indeed, his intuitions about revolutionary change in society seem to inform his model of scientific revolutions. For Kuhn there is no revolutionary change in a society that does not destroy, by its total victory, the social framework that it replaces. Likewise, a scientific revolution, understood as paradigm-change, altogether replaces the paradigm it succeeds and is incompatible with it. Kuhn is not a reductionist. He is not reducing scientific change to social change. His strategy, rather, is to transfer elements from one domain to another, at least by metaphorical extension. The claims of social reductionism are examined in the "strong program" (see Sections 4.9, 4.10, and 4.11 of this chapter).

But if Kuhn's view of the incommensurability that ensues from paradigm-shift is so radical, what of the evident continuity in the development and growth of scientific knowledge, and what of the progressiveness of that growth? Kuhn (1970) addresses this issue in "Progress Through Revolutions," and more squarely in his "Postscript—1969." The key notions in his consideration of scientific progress are the "solved problem" and "problem solvability." For Kuhn they embody "the unit of scientific achievement" (1970, 160). In regard to normal science Kuhn's position is clear enough. Given its allegiance to a shared paradigm a scientific community is an efficient instrument for solving the problems or puzzles that its paradigm defines (ibid., 169). For Kuhn it is axiomatic that the solving of problems is in itself progress. His main difficulty, however, with progress is accounting for the frequent claim that progress is a universal concomitant of scientific revolutions.

For Kuhn, if an accepted paradigm at once defines its problems and posits criteria for their solution, does not its replacement by an incommensurable successor generate a new set of problems (and their criteria of solution) which are definable only within that successor-paradigm? How, then, are the old and new paradigms to be compared with respect to their progressiveness? This can be done only in reference to some element or elements that remain invariant throughout the paradigm-shift, and to which each of the paradigms addresses itself individually. For Kuhn this element is embodied in the efficacy of the new paradigm for solving "some outstanding and generally recognized problem that can be met in no other way" (ibid.). He adds, rather surprisingly, that the new paradigm "must promise to preserve a relatively large part of the concrete problem-solving ability that has accrued to science through its predecessors" (ibid.) as well as generate additional problem solutions.

This is right; successor-theories do solve problems that theories under the predecessor-paradigm could not solve and they do offer solutions to problems that are already solvable. Strictly speaking, however, Kuhn's model of change cannot provide an unproblematic account of these facts. If paradigms are incommensurable, so too are the problems and the criteria of their solutions which they define. What counts as a problem under one paradigm is different from what counts as a problem under another. Because of incommensurability and the consequent absence of any neutral framework beyond the paradigms on which to base an appeal, there is simply no transparadigm criteria of problem-individuation which allows us to say that the paradigms address the same problem.

Kuhn seems to have placed himself into this uncomfortable position for two main reasons. In the first place, he wishes to provide an epistemological substitute for the traditional view that two theories can be said to define their terms differently but still be said to refer to one and the same events, things or processes. He argues vigorously that we have no reason for supposing that (say) the term mass as employed in a Newtonian framework refers to the same "things" as it does in an Einsteinian framework. Both our modes of talking and what we talk *about* change altogether from one framework to the other (1970, Chapter 9). It is one thing to claim, as Kuhn does, that there can be no neutral observation language on which to base an algorithm for evaluating competing theory-choice, but another matter to say that there can be no such language because the proponents "see things differently" or "live in different worlds." Kuhn's proclivity for seeing issues in ontological terms leads him to say repeatedly that "after a revolution scientists are responding to a different world" (1970, 111) without disambiguating clearly enough whether his reference is to conceptual or nonconceptual worlds.

Apart from this tension, in this way he thought (and continues to think) we can best establish the notion that there is no algorithm for theory-choice. Ironically enough, Kuhn's problem-solving approach to the cognitive goodness of the scientific enterprise need not appeal, as he himself recognizes, to the notion that after a paradigm-shift scientists are living in an entirely different world. To dislodge the theory that there is a neutral observation language which can be differently interpreted, ontological extravagances of this sort are hardly necessary and certainly misleading. In subsequent reviews of his position Kuhn recognizes this. Indeed, he has increasingly availed himself of linguistic and pragmatic perspectives to buttress his point that science advances by proliferating new and different rule-bound discourses which set and guide its problem-solving goals.

Secondly, for all his adherence to historical context, Kuhn impaled himself originally on the formal horn of incommensurability because he abstracted scientific activity too far from its appropriate historical and social contexts. He fails, therefore, fully to appreciate the common ground, shared by disparate points of view, which makes any judgement of incommensurability possible. Instead, Kuhn attempts to put everything into question simultaneously. This is impossible as an historicist perspective makes plain. From that perspective, traditions (e.g., paradigms) change and develop and often negotiate internal conflicts by self-criticism and adaptation. If a tradition falls into incoherence either through self-engendered conflict or through challenge from rival traditions, it can critically reconstitute itself by continuous transformation into a new perspective. From this new perspective the tradition can throw fresh light on its elements to reveal the former inadequacy of some and to affirm the present strength of others. Thus the sense of historicality that an historical narrative provides of scientific crises shows that there is no transformation of a scientific tradition into a new perspective that does not preserve some important elements of continuity with its former self. Newton's synthesis of the astronomy, mathematics and mechanics that went before him is indeed just that, a synthesis. This is not to deny that the *Principia* is an impressive achievement. It is a revolution in itself and possessed revolutionary potential for the eighteenth century. But it is also a work that revitalizes the traditions on which it draws by showing their limitations (e.g., Descartes's theory of motion) in understanding the world. At the same time, however, it does not break with Kelper's vision of a true and dynamical astronomy which marries causes and the mathematization of their effects, nor with Galileo's vision of the geometry of motion.

An important aspect of Kuhn's work serves as a natural transition into the philosophy of Imre Lakatos. This is Kuhn's rejection of the idea that knowledge is growing just in case our theories are succeeding in producing better representations of reality. For Kuhn a scientific theory is better than its predecessors only in the sense that it is a better instrument for formulating and solving puzzles, and not because it is a better representation of what the physical world is really like. For Kuhn the idea that there is a match between a theory's power to represent and what is there to be represented is epistemologically dubious. For this reason his work is congenial to current antifoundational perspectives in general philosophy, in rhetoric, and in the humanistic disciplines as a whole. These perspectives hold that what is true, right, and good is justified pragmatically by group consensus arrived at by social practice (on what we agree to accept and reject), and not by privileged access to some extralinguistic reality which our thinking represents.

This is not to claim that Kuhn rejects the existence of a physical world which we seek to explain through science. But it means he has two main commitments. First, scientific explanation is to be accounted for in terms of the successful problem-solving resources of science. On this view, scientific explanation is a matter neither of the theoretical *unification* of phenomena (or of argument patterns) nor of an appeal to theory-independent *entities* as the realist proposes (see Chapters 1 and 2). Secondly, to some extent Kuhn is a social constructivist (see Latour and Woolgar in Section 4.14). However, he does not seem to wish to reduce the "factual" quality of scientific theorizing to social construction. Science, for Kuhn, is still about a theory-independent world. Indeed, science ought to be "concerned to solve problems about the behavior of nature" (1970, 168).

4.7 RESEARCH PROGRAMS AND PROGRESS

Lakatos's work also contains a strong antifoundationalist theme. He takes it for granted not just that we have knowledge, but that knowledge grows. Moreover, the growth of knowledge is not a phenomenon that needs arguing. What is needed, rather, is an account that tells us in what the growth of knowledge consists, and whether it is progressing or not. Furthermore, we can ask these questions of the growth of knowledge by carefully considering the internal features of a body of knowledge over time. This diachronic perspective is an important characteristic of Lakatos's thinking. Thus, in his view, we can talk about scientific advance solely in terms of knowledge and its growth. Lakatos had no need for the notion that knowledge is growing just in case we are approximating more closely to the truth, nor for the idea that knowing is the business of producing more privileged representations of reality. In his view nothing could count as showing that our statements are warranted only if they are compared to an unconceptualized reality. Rather, his position can be compared with Pierce's view that truth is to be replaced by method. Accordingly, truth is just a way of characterizing that which is ultimately acceptable to a community of enquirers who pursue ends by means of certain actions and activities (Hacking 1981a, 131). Thus, instead of asking how well scientific theories represent an extralinguistic reality, Lakatos advocates the development of methodological programs which produce growth of scientific knowledge and allow appraisal of its progressiveness.

For Lakatos the growth of knowledge is always a dynamic phenomena, and the unit by which the nature and direction of that growth is analyzed is the research program. He tells us that in his "methodology the great scientific achievements are research programmes which can be evaluated in terms of progressive and degenerating problemshifts; and scientific revolutions consist of one research programme superseding (overtaking in progress) another" (ibid., 115). In Lakatos's perspective, then, knowledge grows by progressive programs triumphing over degenerating ones. An important feature of his position must be noted at once. He emphatically rejects the view that scientific discovery and advance is to be ascertained by comparatively evaluating two competing *theories*. That is, Lakatos replaces the concept of a *theory* as the basic concept of the logic of discovery by the concept of *series of theories*:

> It is a succession of theories and not one given theory which is appraised as scientific or pseudoscientific. But the members of such series of theories are usually connected by a remarkable *continuity* which welds them into *research programmes*. (Lakatos and Musgrave 1970, 132; italics in the original)

In some respects, Lakatos's concept of a research program resembles Kuhn's notion of normal science. For both of them science is conducted according to rules, and Lakatos tells us explicitly that a research program consists of methodological rules: Some prohibit certain paths of research, the *negative heuristic*, and other rules advocate the paths to pursue, the *positive heuristic* (Lakatos and Musgrave 1970, 132–138).

What are we to understand by a progressing as opposed to a degenerating research program? Lakatos claims to give criteria for progress and nonprogress *within* a program, and also rules for the "elimination" of entire research programs. A program is said to be *progressive* if its theoretical growth anticipates its empirical growth; that is, if it continues to predict novel facts successfully (Hacking 1981a, 117). This means that a theory is progressing so long as its internal ability to produce new effective knowledge continues to outrun the effective knowledge it has already achieved. What matters to a theory is its ability to predict new facts; for the *"building of pigeon holes must proceed faster than the recording of facts which are to be housed in them"* (Lakatos and Musgrave 1970, 188). Effective knowledge, here, means the experimental and theoretical techniques of the program that contribute working control over nature, that is, increase its empirical content. A program is stagnating or nonprogressing if its theoretical growth falls behind its empirical growth, that is, "as long as it gives only *post hoc* explanations either of chance discoveries or of facts anticipated by, and discovered in, a rival programme (*'degenerating problemshift'*)" (Hacking 1981a, 117). The research program that progressively explains more than its rival (in the sense of explain just cited) supersedes it, and the rival can be eliminated from contention and thus retired. Lakatos warns, however, that this procedure of evaluation is far from mechanical. He admits that it is difficult to decide when a program has degenerated hopelessly, or when one of two rival programs has achieved a decisive advantage over the other. After all, a program that is lagging badly behind may stage a comeback, and, in general, there is never anything inevitable about either triumph or defeat.

Lakatos's methodological "ecumenicalism" indicates another feature of his position. He incorporates into the notion of a research program the importance of doing the history of particular episodes in the growth of knowledge. That is, the actual account one constructs of a developing research tradition must include a history of science of that body of growing knowledge. There is no better way in which to judge whether a body of knowledge is a genuine case of growth and progress than to study all the documentary evidence which pertains to it. Accordingly, to test the supposition that knowledge grows by the triumph of progressive programs over degenerating ones, we select an example that illustrates (on the face of it) something that scientists have discovered. Moreover, it ought to be an example about which there is consensus in the field as to its importance. By reading the relevant texts which cover the entire period in the growth of the body of knowledge, and by studying its practitioners, the aim in constructing a research program is to establish what these scientific practitioners were attempting to find out, and how they were trying to find it out.

Let us look now more closely at the notion of a research program and ask whether Lakatos's account of scientific change and progress is more viable than Kuhn's. Lakatos distinguishes the negative heuristic, or the "hard core" of the program from its positive heuristic, or "protective belt." The "hard core" of a program is what its protagonists decide is irrefutable in the sense that it is protected methodologically from refutations. This is a conventionalist strategy, but not one that is maintained at all costs. If the program ceases to anticipate novel facts, its hard core might have to be abandoned. But if the program, while being protected and hardened in this way, continues to predict new phenomena, there is a progressive theoretical shift; if these phenomena are in the end verified, there is also an empirical progressive shift (Lakatos and Musgrave, 1970a, 133–134). The positive heuristic, on the other hand, defines a program's problems, "outlines the construction of a belt of auxiliary hypotheses, foresees anomalies and turns them victoriously into examples, all according to a preconceived plan" (Hacking 1981a, 116). Moreover, the positive heuristic not only dictates the program's choice of problems, it also creates the high degree of autonomy that theoretical science enjoys. A concrete example of these distinctions is Newton's gravitational theory. The irrefutable core is Newton's three laws of dynamics and his law of gravitation. Anomalies produce changes only in the "protective" belt of auxiliary, "observational" hypotheses and initial conditions. The "protective" belt is thus more flexible and resilient than the "hard core" of the program, and is more easily able to turn counterinstances to the program into corroborating instances by the systematic invention of auxiliary hypotheses and models (Lakatos and Musgrave 1970, 135).

Lakatos, like Kuhn, clearly rejects the idea that progress can be evaluated in terms of one theory superseding another because the former is refuted by an experiment that is successfully explained by the latter. That is what Lakatos calls "naive falsificationism" and he calls for a more sophisticated approach (ibid., 93–95). First, progress is not a linear process involving only two competing theories, but a complex process that demands a proliferation of different theories at the same time. Secondly, a counterinstance to a theory can be said to falsify it only *after* it is satisfactorily explained by another theory. And lastly, falsification requires a number of methodological decisions: for example, how to decide on one of many possible interpretations of an experimental situation; whether the experiment falsifies the theory itself or only some auxiliary hypothesis of its "protective" belt; and when and if the theory itself is crumbling and has to be abandoned.

Lakatos clearly defines the relationship between theories in terms of progress, whereas Kuhn compares theories and then attempts to define progress. According to Lakatos's conception, a research program is internally progressive just in case its ability to predict novel facts outruns its established empirical content. A research program is progressive with respect to other programs if it provides an excess of corroborated information in comparison to what they prohibit. Lakatos can, therefore, avoid the entanglements of incommensurability, and at the same time, provide a straightforward account of progress and change in science.

If the basic unit of appraisal is not the isolated theory, or a conjunction of theories, but the "research program," is Lakatos's methodology retroactive? The answer is yes. After all, the thrust of his position is to characterize real cases of growth and to distinguish them from imposters. That a program has enjoyed long-standing progress is ascertainable only after the fact, and this appraisal provides no basis for the claim that it will go on progressing. Furthermore, we can only tell what is progressive and what is degenerating with sufficient hindsight; for at any time a successfully progressing program might become a degenerating one, and vice versa. This means that Lakatos's philosophy provides no forward-looking assessments of presently competing scientific theories (Hacking 1981a, 133–134). At best, the inspection of research programs cautions us to be methodologically lenient. We should be modest about our projects because rival programs may eventually triumph. Moreover, we should be patient if a program is doing badly, since the history of science clearly teaches that prolonged periods are often necessary for one program to supersede another. Thus science is to be judged as an historical development, as an achievement over time, rather than in terms of a situation at a particular time. Lakatos clearly combines history and methodology into a single enterprise, and his insistence on the creative involvement of the history of science with the philosophy of science has been rightly and subsequently influential. However, his equal insistence that history be "reconstructed" synchronically according to preferred rational norms has rightly been rejected by historians.

4.8 PROBLEMS AND PROGRESS

Laudan's position incorporates Kuhn's notion of problem solving and Lakatos's conception of research programs, the latter characterized by Laudan as research traditions. For Laudan any philosophy of science must come to terms with certain persistent features of scientific change (Hacking 1981b, 144–145):

1. The transition from one theory to another is noncumulative. That is, the logical and empirical content (and even the confirmed consequences of an earlier theory) are not entirely preserved when a theory is replaced by a newer theory.
2. Theories are generally not rejected just because they encounter anomalies, nor are they entirely accepted because they are empirically confirmed.
3. Debates about scientific theories are often about conceptual issues rather than about questions of empirical support.
4. Principles and criteria for evaluating scientific theories are not fixed and have altered significantly throughout the course of science.
5. Scientists take many cognitive stances toward theories beyond accepting or rejecting them.
6. Scientific progress cannot plausibly be viewed as evolving or approximating toward the truth.

Each of these features has been challenged. Currently, however, many are accepted by most philosophers of science in one form or another. What is important for our present purpose is that for Laudan they are the touchstone of adequacy for any theory of scientific change. Indeed, he takes his own account of scientific change to be among those that adequately explains these features.

One of Laudan's concerns is to articulate the aim of science in terms other than "truth" or "apodictic certainty." For Laudan the claim that we are progressing toward the truth is empty. No way is possible of establishing that our present theories are more truthlike or on more certain ground than their predecessors. If the notion that one theory is better than another just in case it has more of the truth, or more adequately represents reality, is empty, what is to be substituted as a criterion of progressiveness in science? Laudan claims that the cognitive goals of science may be characterized in many ways. Science can be viewed as aiming at well-tested theories, aiming at theories that predict novel facts, or at theories which have maximum practical applications, and so forth. His own criteria is problem-solving effectiveness. That is, in Laudan's view "*science progresses just in case successive theories solve more problems than their predecessors*" (Laudan 1981, 145).

In Laudan's view this conception of the aim of science sets a goal for science that is epistemically accessible. It also captures the nature of scientific growth. Problem solving divides for Laudan into two broad categories of activity: the solving of empirical and the elimination of conceptual problems. At the empirical level he distinguishes between potential problems, solved problems, and anomalous problems. A potential problem indicates what we take to be the case in the physical world but, as yet, have no explanation for. Solved problems embody cases of actual knowledge achieved by one or more theories. Anomalous problems are such for any theory that has not, or cannot solve them, but which are solved by rival theories. Thus an unsolved or potential problem is never *as such* an anomalous problem. It only becomes so for the theory that cannot solve if it and only if it is solved by rival theories (ibid., 146).

Laudan singles out conceptual problems as especially important. A theory might be in conceptual difficulties if it is internally inconsistent; if it makes claims that are inconsistent with deep, intertheoretical assumptions, such as the conservation of energy; or if it violates the postulates of more general theories to which it is logically subordinate.

In Laudan's view his problem-solving model of scientific advance explicitly recognizes the demands that both conceptual and empirical problems make on a theory. Conceptual difficulties should be minimized, while, at the same time, the theory should solve a maximal number of empirical problems and generate a minimal number of anomalies. Indeed, for Laudan the elimination of conceptual difficulties is as important for scientific progress as increasing empirical support. So much so, that he allows the possibility that a shift from a better supported theory to one less well supported could occur if the less well supported theory had fewer conceptual difficulties.

But what counts as a solution to a problem? Laudan tells us that "a theory solves an *empirical* problem when it entails, along with approximately initial and boundary conditions, a statement of the problem. A theory solves or eliminates a *conceptual* problem when it fails to exhibit a conceptual difficulty of its predecessor" (Laudan 1981, 148). On this view of solvability, many different theories can be said to solve the same problem (empirical or conceptual), and a theory's worth is largely predicated on how many problems it solves. This allows Laudan to assess a theory independently of its confirmed adequacy, or on how well established it is, just in case the theory can be credited with solving an outstanding problem. It also allows him to sever the link between cumulative retention and progress in the context of theory-evaluation. On many pre-Kuhnian accounts of progress, earlier theories were required to be contained in later theories. Other accounts required that the empirical content or confirmed consequences of earlier theories be subsets of the content or consequences of the new theories. This allows the claim that the new theory can do everything that its predecessor can do and more. On Laudan's model, it is possible to assess a theory's progressiveness in terms of its problem-solving efficiency without having to consider the issue of cumulative retention.

But how is problem-solving efficiency to be determined? Laudan recognizes difficulties here. He claims, rightly, that he has only an outline of an account. He tells us to consider a theory by counting the number and weight of the empirical problems it is known to solve; do the same for the number and weight of its empirical anomalies; lastly, assess the number and difficulty of its conceptual problems. In the light of this perspective the theory that comes closest to solving the largest number of important empirical problems, while generating the smallest number of significant anomalies and conceptual problems, is to be preferred (ibid., 149).

The technical problems are immense here. The model is highly qualitative, yet it promises quantitative parameters of evaluation. How a comparative evaluative scale of this sort is to be determined is not clear. Moreover, how are problems to be individuated and counted? Laudan freely admits these difficulties, and notices in mitigation that theories of empirical support also encounter difficulties in identifying and individuating confirming and disconfirming instances (ibid.).

On Laudan's view, science is a rule-governed activity conducted by fundamental commitments and norms which endure over long periods of time. Ultimately our view of scientific change must be seen against these more fundamental commitments. The beliefs that constitute these fundamental views, Laudan calls "research traditions." The elements of their makeup that he stresses differ from Lakatos's characterization of his notion of a "research program." For Laudan, a "research tradition" has two main features: It comprises (1) a set of beliefs about what sorts of entities and processes make up the field of inquiry, and (2) a set of norms of inquiry about how the field is to be investigated, how theories are to be tested, how data is to be collected, and so on. A "research tradition," therefore, binds together a family of theories which are all guided by the same rules or norms, and which all share in the ontology of the research tradition (ibid., 150–152). An example is the Newtonian research tradition which proceeded under the norms of Newton's laws of dynamics and the principle of universal gravitation, together with the ontological claim that all interactions among phenomena (whether chemical, magnetic, biological, or such) had to occur by means of centrally directed forces. This tradition achieved much, although many of the theories that were guided by its norms of research were eventually given up. This makes clear that research traditions are not directly testable like theories. In company with Lakatos, Laudan is insistent that "research traditions" must be illuminated by historical research. Once again, in Laudan's view as in Lakatos's, philosophy of science is blind without history of science. Certainly changes in scientific theories, and the triumph of theories over their rivals, cannot be objects of investigation in abstraction from the social and historical contexts of enquiry in which they were embedded and which propel their specific import. Indeed, in recent work Laudan is engaged in the task of testing empirically and historically various abstract models of scientific change. As yet there is no well-tested theory of scientific change. The matter, however, is worth pursuing. In fact it has recently been taken up by David Hull (1988), who advances a balanced account of the social interplay of reason, argument, evidence, power, prestige and influence in the development of certain of the life sciences.

4.9 ANARCHISM AND "ANYTHING GOES"

Each of the "global" theorists examined thus far is antifoundational in his approach to the cognitive advance of science; each holds that comparative theory-evaluation is not a matter of logically and decontextually assessing theories one at a time, but involves long-term perspectives which can only be established diachronically and historically; and each holds that science is a rule-bound activity. Feyerabend is highly critical of these commitments. His position can be viewed as radical since he also rejects explicitly the claim that science is a rational enterprise.

In Feyerabend's view, successful scientific practice has never proceeded according to rational method at all. Thus, he regards all attempts to characterize the methods of science as totally misguided. For Feyerabend, success in the scientific enterprise depends not on rational argument but on persuasion, rhetoric, propaganda, and practice (Feyerabend 1978; Chapters 3, 4). What any explanation of progress in science must account for is the creativity of the individual theorist or scientific practitioner rather than the methods and authority of science itself. For Feyerabend, the only maxim that "does not inhibit progress is: anything goes" (ibid., 23). He thus advocates anarchism in science, and the proliferation of conflicting and competing theories.

Let us be clear about Feyerabend's position. He makes two central claims: (1) The notion that progress in science is made through the constraints of a paradigm, or through a research program, or tradition is an illusion. These are artificial constructs, unwarranted by the nature of scientific change; (2) also illusory is the notion that science is a problem-solving activity that proceeds by explicit rules and norms. These observations have force. It can be asked in particular whether the problem-solving characterization of science accounts adequately for its explanatory power. Indeed, if there are rules and norms of scientific practice directing its problem-solving activities, what is it that creates explanatory satisfaction through the solving of certain problems? That a problem is solvable says something about the explanatory power of science which is not captured by the problem-solving rubric alone.

How did Feyerabend get into this position? And how can he account for scientific progress in the absence of the machinery he denies? The basis of his position is found in his classic papers, "Problems of Empiricism, I and II" (1965, 1970). Indeed, the views he expresses in his book *Against Method* (1978) are really ideological and rhetorical versions of views already established in the earlier papers. The focus of Feyerabend's attack is what he calls *radical empiricism*, (1965, 154–163). This is the view that at any time only a single set of mutually consistent theories are to be used. Thus, the simultaneous use of mutually inconsistent theories is forbidden. Feyerabend argues for *theoretical pluralism*, and he does so by attacking two articles of faith explicit in radical empiricism: the claim that (1) there must be *consistency* between predecessor and successor theories—that is, new theories must contain or be consistent with the results and the content of the theories they replace (i.e., what Laudan calls cumulative retention); and (2) the claim that there is *meaning invariance* across theory change—that is, he attacks the view that the meanings of scientific terms have to be invariant throughout scientific progress. For Feyerabend both conditions are restrictive. They encourage theoretical *monism*, and discourage theoretical *pluralism* (Feyerabend 1965, 163–168). Only if a theory can be viewed from the perspective of conflicting and alternative theories is there a basis for critically appraising it. Demands for consistency and invariance discourage this development, and encourage adherence to one theory or set of theories long after it is advantageous to do so.

Feyerabend attempts to show that demands for consistency and invariance are not supported by actual science. He argues that Newtonian theory is strictly speaking not consistent with Kepler's law for n-body interactions; that statistical thermodynamics is inconsistent with the second law of the phenomenological theory; that wave optics is inconsistent with geometrical optics (ibid., 168–177). What Feyerabend has in mind here is *logical inconsistency*. Observationally the differences in these perspectives may be too small to detect. Here Feyerabend's view has affinities with Kuhnian incommensurability and with Laudan's rejection of cumulative retention. Feyerabend, of course, does not accept their theoretical accounts of these notions.

His rejection of meaning invariance is a more significant feature of his position. His claim is that both theoretical *and* observational terms change their meanings in the context of theory-shifts. Terms do not mean something in isolation, but only as part of a theoretical system. Thus we do not introduce terms for the phenomena that a theory is to explain, and then bring to bear the theoretical terms of the theory which purportedly explain these observational terms (see Chapter 3, Section 3.7). Everything stands or falls together. If two contexts with basic principles are either contradictory or lead to inconsistent consequences in certain domains, it is reasonable to assume that some of the terms of the first do not occur with the same meaning in the second. Thus, in situations of testing we are dealing with entire sets of partly overlapping, factually adequate, but mutually inconsistent theories (Feyerabend 1965, 174–177). What we have according to Feyerabend, then, is theoretical pluralism as the basis of every test procedure. He argues in some detail that the terms of Newtonian dynamics, such as mass, do not remain invariant when we move from Newton's theory to Einstein's relativistic dynamics. Nevertheless, both theories can simultaneously provide the necessary theoretical background for a test situation (ibid., 169–171).

In Feyerabend's view pluralism is essential to the growth of knowledge. Moreover, the generation of a plurality of theories is not a sign that knowledge is at a primitive stage in its development. On the contrary, it is an essential feature of all knowledge that is not constrained by rules and norms (Feyerabend 1978, Chapter 3). For if there is no meaning invariance across theories, and if theories are logically inconsistent with one another, there is no basis for overarching methodological norms which guide all facets of scientific advance. Kuhn, Lakatos, and Laudan can all agree that the criteria which scientists use in evaluating theories change slowly over time. Moreover, Kuhn and Laudan, but not Lakatos, deny that there are transcultural norms of rationality which guide the practice of science. Feyerabend's position is more radical. In denying the requirements of consistency and meaning invariance he denies that science has an overall methodology which characterizes its activities and aims. Given this conclusion, it is not surprising that Feyerabend thinks it futile to seek for models of scientific change, and pointless to see science as a special sort of enterprise that proceeds by a distinctive set of methodological procedures. On the contrary, we must be willing to suspend adherence to established theories, and to consider the merits of conflicting and alternative theories. In short, we must learn to recognize that science is an anarchistic enterprise that proceeds by the constant proliferation of warring perspectives. What can be said of progress on this perspective? If science is not a rational enterprise it is not a normative one either. But progress is a strong normative notion. What in Feyerabend's position can capture it? Certainly not rhetorical success in making theories plausible to others (ibid., Chapter 4). Again, Feyerabend wants theoretical pluralism to become a *rule* for scientific practice in the future. This, too, is normative and if it is to be successful it is a candidate for *institutionalization* in scientific practice. He says hardly anything of the institutionalization of such a practice nor of the obvious consequence that if "anything goes" pluralism would go.

4.10 SCIENTIFIC CHANGE AS A SOCIAL PHENOMENON: THE IMAGE OF THE COSMOS IN SOCIETY

If the only principle that does not inhibit science is anything goes, little wonder that Feyerabend embraces anarchism. The significance of his work, though, is seen in a different light by those who favor a social approach to science. Kuhn, Lakatos and Laudan all reject the idea of a straightforward logic of verification, refutation, or confirmation whereby theories may be logically assessed in the instant. In place of talking about traditional epistemic notions, such as truth and verisimilitude, they talk of scientific change and progress in terms of problem solving, paradigms and research programs. These notions, as they rightly allow, invite historical and social considerations of scientific advance. Indeed, many now argue that such notions are no more than constructs of social dynamics itself. But Feyerabend rejects these accounts as well, whether they are socially warranted or not. If "anything goes" in the development of scientific research, there can be no privileged account of how and why science advances as it does. Accordingly, it is entirely indifferent whether a sociological, anthropological, historical, psychological, or philosophical account is given of an episode in scientific change.

Sociologists and social historians of science have offered accounts of science which they see implied in the work of Kuhn, Lakatos, and Laudan, and, in the face of Feyerabend's anarchism, have reaffirmed their conviction that science can be construed entirely as an ever-changing social phenomenon. They reject the myth of the "autonomous scientific knower," situated apart from history and social embodiment, and also the vision of scientific knowledge as universalized and decontextualized, perspectives which still linger in the global philosophies of science. In place of these perspectives, social theorists argue that knowledge is present only in its social and linguistic practices by which knowledge is continually adapted to changing social context. They also hold that knowledge is normative: What is the best form of knowledge to have? How can it be augmented? And is it the only form worth having? Unlike the globalists, however, they are emphatic that only the social organization of enquiry can deliver the answers that must cut across traditional disciplinary boundaries.

4.11 MAKING SCIENCE AND SCIENCE MADE: THE STRONG PROGRAM IN THE SOCIOLOGY OF KNOWLEDGE

For two decades the Science Studies Unit at the University of Edinburgh has been associated with an influential approach in the social studies of science. The central tenets of the school are as follows:

1. Social theory adequately describes both the production of science and the product of science; that is, social theory best accounts for the human activity of making science and the human achievement of science made.

2. There is a relationship of dependence between the cognitive order and the social order: A social theory of the mind is the preferable construct insofar as human belief is embedded in social structure. In other words, human belief is socially locatable in the sense that causal laws can be established that show how any belief, scientific or otherwise, is socially generated.

3. The duality of nature versus society is misleading. The notion that physical nature is "out there" can be completely understood in terms of the notion of society "out there." Thus, society and social theory are privileged over nature and the philosophy of science as a basis for analysis.

4. Facts are socially made. Indeed, changes in our beliefs about "facts" and the nature of the "facts" we construct are driven by social interests, by the dynamics of collective needs, and by the power of social forces themselves. Thus, it is not the natural world that constrains our beliefs about nature, but rather the socially generated and opposing interests of competing groups. Indeed, the notions of "need," "interest," and "construct" are central categories in the discourse of the Edinburgh School, and are used with disarming plasticity.

The views of David Bloor are discussed in this section to the extent they are representative of the Edinburgh position. My discussion involves both his earlier post-Mannheimian orientation and his latter post-Wittgensteinian perspectives. Much that is said also characterizes the orientation of Barry Barnes, another well-known member of the school.

Let us begin with Bloor's initial account of the "strong program." He begins with the undoubted conviction that "science is a social phenomenon so we should turn to the sociologist of knowledge" (1976, ix). Moreover, he sees himself as challenging Kuhn, Popper, and Lakatos. In his view it is sociology, not philosophy, that will deliver a meaningful account of the genesis and nature of scientific knowledge. The character of Bloor's claim must be grasped at once. He states it thus:

> Can the sociology of knowledge investigate and explain the very content and nature of scientific knowledge? Many sociologists believe that it cannot. They say that knowledge as such, as distinct from the circumstances surrounding its production, is beyond their grasp. They voluntarily limit the scope of their own enquiries. I shall argue that this is a betrayal of their disciplinary stand point. (Ibid., 1)

For Bloor, sociological accounts of scientific knowledge should not be restricted to its institutional framework or to external factors pertaining to its rate of growth or direction. After all, Kuhn, Lakatos, and Laudan all agree that this perspective is viable. Rather, Bloor's conviction is that sociology can provide a causal account of the creation of scientific knowledge itself, one which uses the resources of social theory alone and is therefore independent of philosophy. It is to this end that he advocated the "strong program."

The program takes knowledge, including scientific knowledge, to be a purely natural phenomenon. Instead of defining knowledge philosophically as justified true belief, Bloor accepts as knowledge whatever informed groups of individuals take it to be. It must, of course, be distinguished from mere opinion. This requirement he proposes to meet by calling knowledge whatever a cognitive community collectively endorses or agrees upon by the pragmatics of social consensus (ibid., 3).

Accordingly, Bloor argues that a sociological account of scientific knowledge should satisfy *four* requirements: It should be causal, impartial with respect to truth and falsity, symmetrical in its modes of explanation, and reflexive in the sense that its explanatory framework should apply to itself as a body of knowledge (ibid., 4–5).

The first requirement seems, on the face of it, reasonable enough. It says that a causal account of the conditions that produce knowledge must be given, not just any causal account, but a causal account which uses *social* parameters. Thus, it requires that the origins of beliefs, and the character of states of knowledge, are to be explained by a theory that utilizes only the principles of social theory. It is evident that Bloor takes social causation to be an unproblematic notion. It is not. Bloor fails to address this issue in his early work. Moreover, he offers no account of what a theory of social causation entails, or a picture of the criteria it should satisfy. At best his position is akin to Humean regularity, for example, if such and such social conditions prevail, then certain types of attitudes tend to prevail. This in itself fails to preclude other modes of explanation.

The second and third requirements are also controversial. They claim that we ought to give a causal account not only of what is taken to be true and rational, but also of what is false and irrational. That is to say, if knowledge is socially determined, false beliefs ought to be produced in the same manner as true beliefs and get accounted for by the same causal framework. This claim is disputed by Lakatos and Laudan and, indeed, earlier by sociologists like Mannheim (1936). Their counterclaim is that sociology can only explain why it is that we are in a state of false belief and error (Laudan 1977, Chapter 7). This happens because "external" factors may intrude, such as the social, the institutional, and the psychological, in the process of establishing knowledge. That is, if the mind goes off the rails, or deviates from the canons of reasonableness, causes apart from the mind must be found. Otherwise, the mind is autonomous in the production of true beliefs, and its methodologies are intrinsically self-justifying in virtue of mental activity itself. The second and third requirements of the "strong program" challenge the view that accounting for knowledge as true belief is independent of causal explanation. Thus, the same style of explanation—social causation—must be able to explain false as well as true beliefs.

The last requirement stipulates that any sociological account of the growth of scientific knowledge should be reflexive, for if a sociological theory is itself an example of a piece of knowledge, to avoid self-refutation it ought to explain itself. The "strong program" must, therefore, satisfy causality, impartiality, symmetry, and reflexivity.

As already indicated, however, Bloor's criteria cannot be reasonably evaluated unless we have some indication of the nature of social causation. Just how do social influences bring about or cause human beliefs? How do beliefs relate causally to institutionalized ways of behaving in society? A step towards answering these questions might consider the ways in which different positions in the social structure seem to correlate with different beliefs. However, correlations are not causes and perhaps not even indicators of underlying causal mechanisms. Moreover, as Bloor himself recognizes, not only must there be an account of how social causes shape beliefs, there must also be an account of human nature, of the believer, the content of whose beliefs gets generated by these same social causes. Unfortunately, Bloor's early work never provides an account of the nature of human believers, or of how social causes bring about the content of their beliefs (1976, 5–10).

Indeed, Bloor never indicates how social causes differ (if at all) from other forms of causation. Furthermore, just how do beliefs derive from the social order? Apparently there is no well-grounded account of the alleged link between the social order and the cognitive order such that the first is seen to be explanatory of the second. Can the second be reduced to the first? Or does a theory of social causation demand the complexities of a multilevel and multivariable model? And how are the notorious difficulties of Marxism regarding the relationship between the economic infrastructure of society and its cultural superstructure to be avoided? Or can that relationship be clarified compellingly? Until these issues are resolved there is no real basis for forsaking explanations of belief formation in terms of reason, intention, motivation, and purpose.

To Bloor's credit, he attempts consistently to locate all knowledge socially, especially in his rejection of Mannheim's category of the "free-floating intellectuals," and in his claim that the ideational content of formal systems, such as mathematics, are not immune to methods of the sociology of knowledge. The notion that intellectuals are classless and socially rootless, and thus the custodians of knowledge not tied to the ideological rationalization of class and status, is dismissed by Bloor as false and itself generated by Mannheim's social and historical position. This raises a further question. If, as Bloor assumes, scientific knowledge is culturally and socially located, how do we allow for ideological distortions of knowledge due to its rationalization through the practices and interests of scientific cultures?

4.12 LANGUAGE GAMES, MENTAL CONTENT, AND A SOCIAL VIEW OF MEANING

In his more recent work, Bloor offers a perspective on human activity which side-steps the issue of social causation. As in his earlier version of the "strong program," he minimizes the role of individual agency by subsuming it under the collectivity of social change and social action, and by stressing the priority of society over the individual. Now instead of relying on an unanalyzed notion of social cause, he argues that language, belief, human reasoning and actions are all natural and sociological phenomena openly intelligible just because they arise from human behavior anchored as it is in material, biological and cultural contexts. Descriptionism, not explanation, is Bloor's objective here, and content is solely a matter of context. The aim is to merge the "what" and the "how" (Bloor 1983).

Bloor's aim is to show that Wittgenstein's later thought "on the natural history of human beings" (1953, 415) can be incorporated into the everyday practices of historical, anthropological and sociological enquiry. Thus, he takes seriously and literally Wittgenstein's claim that, "Commanding, questioning, recounting, chatting, are as much a part of our natural history as walking, eating, drinking, playing" (1953, 25). On this naturalistic perspective, the speaking of a language and the languages we use are "a part of an activity, or a form of life," in just the way that chemical and physiological processes drive our living biological life. These linguistic forms of life, or "language-games," are countless and varied; they come into existence and they become obsolete; they express cultures and subcultures, practices and institutions; they manifest patterns of interlocking usages and activities; they involve streams of interests and needs embodied as they are in the contexts of human behavior; and they are constitutive of everything we can ever claim in the way of knowing.

Bloor's aim is to show that the de facto reality of these "forms of life" is open to empirical investigation, namely, that they come into being through social interests, institutional practices, and human needs. Thus, language has its patterns of usage and meaning only insofar as it expresses itself through the content of ongoing human activities or "forms of life." In this Wittgensteinian vein Bloor breaks decisively with meaning essentialism (i.e., the view that meanings express intentional states or refer uniquely to entities in the world) in favor of the view that language is an interactive tool among communities of speakers, ever changing as the dynamics of social interest change.

Scientific culture, and its changes, is viewed by Bloor through this perspective. It, too, is to be sociologized, naturalized, and treated as a "form of life." Thus, like any other cultural form, science is a natural and social activity which generates special language-games driven by cognitive aims and social interest. Scientists on this view are not autonomous knowers who project intentional meanings onto a compliant reality, but investigators dwelling in special "forms of life" carried by linguistic uses that contextually express the dynamic of interests constitutive of the total scientific culture. For Bloor, as for Wittengenstein, there is no inner Russellian language referring to a privileged mental content which in turn ideally depicts an outer reality. Scientific change is a matter of linguistic redescription and the generation of new discourses compelled by increasing interaction with phenomena and directed by changes in social interests and cognitive needs. Moreover, incommensurability is no problem, since for Bloor no one language-game of the scientific culture can be objectively preferred to any other.

Bloor's position invites an uncompromising relativism, a consequence he eagerly avows. Indeed, he tells us that objectivity and rationality are forged by us as "we construct a form of collective life" (1982, 3); that "Copernicus is undone" (ibid.) and that "human beings are back in the center of the picture" (ibid.) as the measure of all things; that what we take to be universal is "variable and relative" (ibid.); and thus, that the "things we had seen ourselves as answerable *to*, we are now answerable *for*" (ibid.).

Thus, in company with so much that is current in postmodernism he denies the objectivity and rationality of "truthful knowledge," suggesting instead that knowledge, even language itself, is a reflection of power relations within society, and changes when these relations themselves change. Accordingly, there is no standard of knowledge that is transcendent of power and the desire for it, or of authority and the desire for that. Protagoras the relativist has returned to haunt the Platonic realists.

4.13 POLITY AND THE SOCIAL MANAGEMENT OF FACTS

To illustrate further the difficulties of social causation let us now turn to Shapin and Schaffer's interesting and important *Leviathan and the Air-Pump* (1985). These historians are concerned with a concrete case of scientific change, Robert Boyle's researches in pneumatics (with his use of the air pump in that enterprise) and with the controversies this experimental program generated with Thomas Hobbes. In this context they trace the social formation of scientific authority in mid-seventeenth-century England together with the question of legitimizing social space for the experimental way of knowing. They see their work as an exercise in social theory and in the ethnography of science, and their explanatory orientation has much in common with Bloor's position, as well as with the views of micro-sociologists such as T. J. Pinch, Bruno Latour, and Andrew Pickering (Shapin and Schaffer 1985, 14–16). There are also Foucaultian and Wittengensteinian themes woven into their narrative, especially evident in their account of relations of language, society and power. Accordingly, Shapin and Schaffer approach the scientific enterprise as an integrated *pattern of activity*, as a "form of life." Scientific controversies are disputes "over different patterns of doing things and of organizing men to practical ends" (ibid., 15). Moreover, in their view "solutions to the problem of knowledge are embedded within practical solutions to the problem of social order, and that different practical solutions to the problem of social order encapsulate contrasting practical solutions to the problem of knowledge" (ibid.).

The thesis of *Leviathan and the Air-Pump* can be stated simply. How did science together with its social context get co-constructed in the work of Hobbes and Boyle? As Latour rightly indicates. The question Shapin and Schaffer ask is not how Boyle places his ideology of witnessing "facts" experimentally produced in his air pump in the social context of contemporary England, but, rather, how Hobbes and Boyle both construct "a science *and* a context *and* a divide between the two" (Latour 1990, 147). In other words, Shapin and Schaffer attempt to show that Boyle has a science *and* religious polity and that Hobbes has a political theory *and* a science. Moreover, Boyle invents an *artifact*, the laboratory, a special place within which to witness experimental matters of fact produced in the vacuum of his air pump; Hobbes also invents an *artifact*, the State or the Leviathan, by which citizens are represented through social contract. For Shapin and Schaffer these inventions are two sides of the same coin; science and context are one. Indeed, it is the invention of the modern world.

We must be clear about the implications of these claims. Shapin and Schaffer are of course claiming that the business of making science is a social activity, within an institutionalized framework, which is guided by social norms and practices. There is no problem here. In fact, the claim as it stands is unremarkable. However, they are making more serious and controversial claims. These are further evident in the following passage:

> Neither our scientific knowledge, nor the constitution of our society, nor traditional statements about the connections between our society and our knowledge are taken for granted any longer. As we come to recognize the conventional and artifactual status of our forms of knowing, we put ourselves in a position to realize that it is ourselves and not reality that is responsible for what we know. Knowledge, as much as the state, is the product of human actions. (1985, 344)

This passage indicates their belief in a direct relationship between the way knowledge is produced and the sociopolitical context in which it is produced. Indeed, the experimentally produced fact is a social construct, "an artifact of communication and whatever social forms [are] deemed necessary to sustain and enhance communication" (ibid., 25). For all that, however, facts can be viewed as both epistemological and social; epistemological, since we understand best what we construct; social, since they are made by us under controlled conditions. Second, given that scientific knowledge is a product which arises from the experimental way of life, a result cooperatively achieved by designful human activity, what needs explanation is how we act and behave when we engage in the business of making facts.

Now it is one thing to stress the social, the artifactual, and the conventional aspects that characterize the purposive production of experimental knowledge. It is quite another to claim that warranted scientific knowledge is at bottom about us knowers and not about states-of-affairs in nature. That is, the claim that a scientific fact is just a social construct made by specific forms of human activity needs more arguing than Shapin and Schaffer provide. More precisely stated, the distinction between an experimentally made *matter of fact* and the underlying causal reality responsible for the occurrence of that "factual event" in the experimental setting is never clearly made by the authors. Moreover, their analysis seems to rest on a curious shift in reasoning. They start from a perfectly sound methodological question: How does science function *in* society? But they conclude that science is a function *of* society, that scientific knowledge *is* simply about a social construct predicated on social action. From methodological premises alone we are scarcely entitled to draw an ontological conclusion (see Section 4.14). This slide is parallel to moving from a methodological constraint to the effect that only certain properties are to be selected (because they alone can be treated quantitatively and geometrically) to the claim that they alone are the real properties of natural phenomena.

As further evidence of the reductive character of their methodology, Shapin and Schaffer proceed under tacit ontological assumptions. They assume that talk about "social reality" is unproblematic, that explanations advanced in social terms make no unwarranted assumptions about what is the case. In other words, they privilege society over nature: It is the social that is "out there," not the natural. But analysis of persons and their actions does introduce emergent categories. To bring persons under any collectivity is to use identifying criteria not reducible to persons taken individually. Soon we are talking about classes, life forms, communal interaction, and such, each a construct or artifact of the analysis, but none reducible to the basic ontology of persons and actions. The authors are aware that they are dealing with constructs, but they consider them warranted just because they refer to persons, groups, and group interactions. It may be defensible to claim that all knowledge of ourselves as social agents is grounded in the social or political spaces that we ourselves create; it is another matter to claim that scientific knowledge is nothing more than a social construct pertaining to specific forms and patterns of human activity.

Not surprisingly, Shapin and Schaffer also claim that major changes in science are to be explained in social terms alone, that only accounts that trace events back to their sociopolitical context are properly historical. Accordingly, they argue that the rise of early modern science and the rise of the new social order of the seventeenth century are not isolated events. Both are manifestations of one and the same process, namely, a new solution to the problem of order, "Solutions to the problem of knowledge are solutions to the problem of the social order'' (Shapin and Schaffer 1985, 332). The authors take it as undoubted that ''an intimate and important relationship [obtains] between the form of life of experimental natural science and the political forms of liberal and pluralistic societies'' (ibid., 343). Here again they take their cue from Wittengenstein's notion that language is *the* activity which structures all forms of activity. But this conception cannot in itself license the view that the choice of a scientific discourse is at once a solution to what the world is like and to how the scientific community should talk. Shapin and Schaffer, however, strongly suggest this. Certainly, discourses create ''realities.'' It is still an open question, however, whether the practices that produce scientific knowledge are entirely and irreducibly social and linguistic.

Furthermore, linguistic and social affinities are not explanations unless the nature of those affinities—causal or otherwise—is spelled out. Moreover, in explanatory contexts problems of overdetermination should not be underestimated. The authors fail to fully consider these matters. They sometimes write as if changes in the social order and in the scientific order proceed from a common cause, but the nature of that common cause is never isolated and specified. At other times they write as if changes in the social order and those in the scientific order are coproduced. This picture is consistent with the ''strong'' program in social theory and inherits its difficulties. In any event to argue, as they do, that the explanation for both the emergence of modern science and the rise of a new order in Western culture is the formation of a new polity generated by the sociopolitical changes of Restoration England fails to compel immediate assent. Certainly, the formation of the ''experimental way of life'' as a rule-bound act of the ''*social* witnessing'' of facts, performed in the constructed *social* space of the laboratory, presupposes certain sorts of institutions and practices specific to the social forms of mid-seventeenth-century England. In this, Shapin and Schaffer are right. But whether their historiography is universally applicable in accounting for the production of knowledge in the particularities of other social and historical periods is at least questionable.

4.14 TOWARD AN ANTHROPOLOGY OF SCIENCE: THE DISCOURSE OF SOCIAL CONSTRUCTIVISM

Let us turn now to social constructivism as a perspective on scientific change. This orientation has been touched upon in the account of the Edinburgh School and the work of Shapin and Schaffer. It will be well, however, to examine constructivism directly, especially in the anthropological context in which it has been placed by recent literature.

In its strongest formulation constructivism says that scientific facts of a pure and socially unalloyed nature do not exist; but social facts—facts about the existence of the constructions we call ''scientific facts''—do exist. This is open to two interpretations: (1) that scientific theories and their interpretative applications to evidence are sociologically laden in that every application is made within a socially organized context; or (2) that scientific theories and facts are constituted solely by social constructs. On this view, there are no factual referents ''out there'' independent of social constructs to which these constructs refer. The first interpretation is benign; the second is not.

In their *Laboratory Life* (1986), Latour and Woolgar tend to blur these interpretations. They assume methodologically that facts are locally and socially made by the controlled instrumentation of the laboratory. This is sound enough as an anthropological assumption. Unfortunately, they tend to run together the divide between facts understood, produced and maintained under controlled conditions, and the underlying causal realities responsible for these local and laboratory made matters of fact. Recently, however, Latour has affirmed the importance of maintaining this divide (1990, 149).

It is worth stopping to consider the distinction between what is constructed *out of* our laboratory practices and what can be referred to *through* those practices. No doubt science dwells in its practices and no science can transcend the network of those practices. After all, what we understand by truth and reality must arise from our practices and be made intelligible and useful to us through the aims and norms that guide those practices. Nevertheless, what can be referred to in the scientific context does *not* depend on scientific practices although the ability to make such references *does*. That is, we should never infer from the fact that our practices are necessary for access to scientific entities to the conclusion that these entities are *constituted* by our practices. This is a fallacious inference but one often implicitly affirmed in the constructivist literature (see Shapin and Schaffer in Section 4.13). In short, reality and truth are constructs which, like femininity and masculinity, depend on a culture's conventions and norms. Just as that which the construct masculinity can be true of—biological male—is independent of our practices, so the true and real are independent of the constructs of truth and reality, that is, what truth is true of does not depend on our practices. Put otherwise, notions of what counts as truth (e.g., whether it has explanatory uses) should not be confused with our ability to make and often to justify claims about what *exists* both in the context of everyday experience and in that of scientific theorizing (see Horwich 1990). Pragmatists and nonpragmatists alike can agree here.

A sophisticated form of constructivism is found in Latour and Woolgar (1986). Plausibly enough, they propose to treat the laboratory as if it were an institution within an alien scientific culture. More specifically, they see themselves as practicing the methodology of the "anthropology of science" with a view to discovering how it is that the "realities of scientific practice become transformed into statements about how science has been done" (ibid., 29). Unfortunately, in so doing, they commit the original sin of anthropology—cultural imperialism. Indeed, there is a certain hubris in walking into an alien culture, of which they claim to know nothing, taking a single event from the most taxonomical of sciences—biology—and proceeding to dismantle the entire scientific culture by showing that it is in fact a self-deluded copy of their own linguistic culture. This is like describing native magic as bad science, or the clearing between grass huts in terms of the playing fields of Eton.

Latour and Woolgar do claim, of course, that their anthropological perspective on science "entails a degree of reflexivity not normally evident in many studies of "science" (ibid., 30). That is, they hold that the methods they use in studying the practices of the laboratory are similar to those of the practitioners of science themselves. In their more recent work, both Latour and Woolgar have rejected this naive view of the naive anthropological observer who, although he does not know the language and customs of the site being investigated, nevertheless subjects it to an alien metalanguage (Latour 1987, Woolgar 1988).

However, in their assumed role of anthropological observers, Latour and Woolgar (1986) view the Salk laboratory as an extended "text" exhibiting a wide range of linguistic practices and using artifacts as "inscription devices." "Inscription" refers to more than such linguistic acts as writing; it refers to all traces, spots or points on screens or scales, and to histograms, recorded numbers, spectra, and peaks on diagrams, and so on. An "inscription device" is an apparatus used such that it provides some sort of symbolic output. Accordingly, an apparatus that "transforms pieces of matter into written documents" (ibid., 51) is an "inscription device." It is thus "any item of apparatus or particular configuration of such items which can transform a material substance into a figure or diagram which is directly usable by one of the members of the office space" (ibid.). So a scale on an apparatus is an "inscription device" if it provides information about a new compound, a machine if it weighs something, a checking device when it is used to verify an operation. In short, an apparatus is used as an "inscription device" when it is used in an argument such as that involved in the construction of a bioassay profile.

In accordance with this perspective, Latour and Woolgar develop a linguistically sophisticated form of constructivism. Construing the laboratory as a locus of activity, involving communication, persuasion, and the use of apparatuses as "inscription devices," they focus their anthropological investigation on the sociological and the linguistic features of laboratory activities. The laboratory is presented as a "system of literary inscription, an outcome of which is the occasional conviction of others that something is a fact" (1986, 105). But what is the status of the fact to which they have been persuaded? We might think that a fact is something recorded in a scientific article that has "neither been socially constructed nor possesses its own history of construction" (ibid.). Latour and Woolgar think this is a wrong-headed conception of the origin of facts, and they wish to examine how talk of facts appears "to remove the social and historical circumstances on which the construction of a fact depends" (ibid.).

This seems benign enough. But the authors are committed to more than the obvious claim that facts are socially and linguistically constructed in virtue of the directed activities of laboratory research. They claim that facts *are* social constructs. This commitment is unambiguously stated in the following passage:

> Specific to this laboratory is particular configurations of apparatus that we have called inscription devices. The central importance of this material arrangement is that none of the phenomena "about which" participants talk could exist without it. Without a bioassay, for example, a substance could not be said to exist. The bioassay is not merely a means of obtaining some independently given entity; the bioassay constitutes the construction of the substance.—It is not simply that phenomena *depend* on a certain material instrumentation; rather, the phenomena *are thoroughly constituted by* the material setting of the laboratory. The artificial reality, which participants describe in terms of an objective entity, has in fact been constructed by the use of inscription devices. (Ibid., 64)

Accordingly, phenomena and the facts pertaining to them do not obtain in a theory-independent reality, but are constituted and constructed by the social and linguistic processes of the laboratory. Indeed, substances could not be said to exist independently of configurations, such as bioassay, which supervene on items displayed on inscription devices. Thus, facts, far from being objective realities that are discoverable, are themselves constructed by the methodologies of inscription devices.

Latour and Woolgar are nevertheless at pains to deny that facts are just artifacts of social and linguistic practice. They wish to reconcile two facets of "fact" talk:

1. that the term connotes the making or constructing of something; and,
2. that "fact is taken to refer to some objectively independent entity which, by reason of its 'out thereness' cannot be modified at will and is not susceptible to change under any circumstances" (ibid., 175. See also pp. 84 and 87).

Repeatedly the authors claim that they are not antirealists in the sense that substances, and the facts that pertain to them, do not exist; but rather that they do have objective existence, but only *as* constructs. How is this view to be maintained? And how do they combine the claim that facts are not independent of their modes of construction while denying that they are merely artificial?

What the authors claim is that it "*is not just that facts are socially constructed. We also wish to show that the process of construction involves the use of certain devices whereby all traces of production are made extremely difficult to detect*" (ibid., 176; italics in the original). Specifically, they claim that an important inversion in statements and what statements are about takes place in the dialectics of science. In the early stages of the dialectical process there are only statements or linguistic exchanges among scientists, that is, agreements and disagreements. Moreover, the conditions of the construction of these statements are manifestly visible, and seem necessary for purposes of persuasion. However, once widespread communal agreement is reached, the inclusion of these conditions as a means of persuasion is no longer necessary; indeed, they seem to threaten the "fact-like" status of the statements themselves. At this point, what a statement is about takes on a life of its own. It is as if "the statement had projected a virtual image of itself which exists outside the statement" (ibid.). More and more reality is then attributed to what the statement is about, and less and less to the statement itself, "Consequently, an inversion takes place: The object becomes the reason why the statement was formulated in the first place" (ibid.). If the function of literary inscription is the successful persuasion of readers, such that they are most completely convinced when all forms of persuasion disappear, so the "result of the construction of a fact is what appears unconstructed by anyone; the result of rhetorical *persuasion* in the agnostic field is that participants are convinced that they have not been convinced—" (ibid., 240). Indeed, on this view, the various social and linguistic activities that sustain the argument, and which are eventually seen by the participants as irrelevant to the "facts," are the same conditions that constitute and generate the "factual" quality of a given statement itself.

Facts, then, are epiphenomenal projections of the constructive techniques that generate them, and their "objectivity" is a function of the dialectical inversion characteristic of scientific practice. It may seem otherwise to some, especially to practicing scientists. Forgetting the ''constructive'' process that constitutes ''facticity,'' scientists and their readers often revert unconsciously to the notion that facts are ''out there,'' and that their existence is to be revealed by the techniques of scientific discovery. But this is an illusion—Latour and Woolgar insist—an illusion because those statements warranted by scientific consensus as ''fact-statements'' stand alone long after their constructive procedures have become invisible. Their claim is consistent with a common position in the sociology of knowledge that knowledge is both socially constructed and socially located. Indeed, their narrative is an instantiation of the process of objectification (the reification or ''making real'' of the constructed knowledge) and of internalization (the unconscious ''hardwiring'' of the epistemic outlook provided by the objectified knowledge). Thus, the structure of the facts that are ''out there,'' either locally or cosmically, resonates with the structure of our minds precisely because we put them there (Berger and Luckmann 1967). This in itself fails to dislodge the view that the human mind is itself is an integral part and product of the natural order of things.

It is one thing to stress the social, the artifactual, and the conventional aspects that characterize the purposive production of scientific knowledge. It is another to claim that warranted scientific knowledge, and talk about the ''factual quality'' of consensual statements, is at bottom about the activities of scientific knowers and not about states-of-affairs in nature. There would appear to be a fallacious shift in reasoning. Latour and Woolgar begin with a perfectly sound methodological question: How is scientific knowledge about facts generated *in* the cognitive community of the laboratory? But they conclude that that knowledge is no more than a function *of* the special linguistic and social practices of the community, that scientific facts *are* social constructs predicated on social action. From methodological premises about how people behave we can scarcely derive ontological conclusions about what is the case in nature. This is to confound *how* knowledge is produced, with what that knowledge is *about*. That is, it is an instance of the fallacy discussed above: The shift from the dependency of knowledge on practice to the claim that the objects of that knowledge are constituted by those practices.

But for Latour and Woolgar to talk about the ''factual'' quality of scientific statements is to talk *about* connected *meanings* in the social space of scientific discourse. That is, scientific ''factuality'' is a construct arising from hermeneutical activity of linking meanings through theoretical unification. But an important distinction is blurred. We certainly construct connected meanings when we interpretatively unify inscriptional items manifest on laboratory apparatus, but the arrangements of points and spots that we interpret are not themselves meanings; rather they have meaning. Moreover, these arrangements of patterns are not themselves *caused* by the effective linkage of meanings. They are caused by underlying phenomena the properties of which are represented by the arrangements of inscriptional items. Latour and Woolgar wish to ''reconceptualize'' these relationships. In fact, they argue that physical objects, including living systems, can be treated as if they are ''material dictionaries'' (1986, 48). This means that any material thing that can be labeled, encoded or recorded can be treated through its linguistic representation. This includes rats injected with a liquid by a syringe. Here the ''material dictionary'' is the record of their behavior after injection. Given the orientation of Latour and Woolgar, this perspective runs the risk of turning the claim "*as if* things are material dictionaries" into the claim "they *are* material dictionaries." Thus, analogies become identities, and, in general, causes tend to become merged with the behavioral effects that they cause, and things become their properties.

The point is this: Causes, meanings, and effects must not be confounded. This is especially so in the case of effects which are produced by the use of sophisticated laboratory techniques. Here, of course, the scientist encodes relationships linguistically and theoretically between inscriptional items of an apparatus; but through these encoded meanings the scientist is also able to intervene and interfere with nature predictively and creatively. This is not to say that theoretical entities are there to be discovered by just anyone. Many scientific effects do not exist without the intervention of certain kinds of laboratory apparatus. But those effects are there, and they are real because they have to do with the world, if only the local world of the laboratory and not with our linguistic conventions. Accordingly, certain entities are in part theory-independent and scientific theories say something about them. Of course, this is not to claim that every theory-independent entity that is postulated exists—such as caloric. It does mean that some entities remain and are experimentally manipulatable long after the theories that first postulated and managed them are rejected. Accordingly, local realism about scientific entities is inherently plausible.

Talk of causes, things and properties raises the fundamental issue involved in Latour and Woolgar's program. They maintain the strong claim that a change in construction *constitutes* a change in phenomena as such. Recall that their argument can be construed as claiming that scientific facts do not exist, but social facts—facts about the existence of the constructions we call "scientific facts"—do exist. If this is so, they are making an ontological claim in the manner of the social realist. The realist takes certain statements involving reference to social facts, entities and properties as literally true and as involving genuine ontological commitment (see Ruben, 1985, Chapter 2 and Papineau 1978, Chapter 1). For example, the realist holds that the belief that "Scotland is part of the United Kingdom" is literally true, that there are no candidates, social or nonsocial, to which the social particulars "Scotland" and the "United Kingdom" are reducible. Thus, the social realist is concerned to show that certain social factors are irreducible, that is, that they are not candidates for reductive identification with other entities. Examples of reductive identification are attempts to identify mental states with brain states or types of behavior, moral properties with natural properties, numbers with sets of sets, knowledge with justified true belief, and physical objects with sets of sense-data. Since Latour and Woolgar make a strong ontological claim, namely, that scientific facts just are objective social constructs, they need to give an *ontological* justification of this claim to avoid the ontological double-talk implied in their rhetorical packaging of "fact." Thus, they must give a noncircular account to show that in each case in which there is talk of a "scientific fact" there *is* an appropriate "social fact" which is a necessary and sufficient condition for it. The thrust of necessity and sufficiency is that there *cannot be* the relevant "scientific fact" without there being an appropriate "social fact," and vice versa. The modality required is strong. It is metaphysical (in contrast to epistemic or methodological) since it claims that there cannot be any *possible* "scientific fact" that is not reductively identifiable with a "social fact," (Ruben, 1985, 5–7). These requirements must be met; otherwise, social constructivism remains a benign *methodological* claim (while still masquerading in ontological dress) to the effect that "scientific facts" are simply *eliminable* on the grounds that only socially and linguistically constructed entities are epistemically accessible to our cognitive abilities. Thus, in the absence of direct ontological arguments, Latour and Woolgar cannot make good their putative claim that "phenomena are *thoroughly constituted* by the material setting of the laboratory" such that a change in construction entails a change in phenomena and a change in our knowledge concerning them.

4.15 SCIENCE AS PRACTICE

The perspectives discussed all turn on various interrelations of the categories of theory, method and evidence: The received view stresses minimalist observation statements tied to evidence and sensation; the new philosophy of science concentrates on paradigms and on theories and the methods they drive; the social theoretic emphasizes the making of facts and the types of social consensus that justify their acceptance. In the more recent literature, however, other categories have come to the fore: Scientific practice, the audience and context of science, experimental and technological networks, and actants rather than agents (i.e., the view that there is no relevant distinction in the context of making science between people and things). Furthermore, many of the traditional divides that oppose inner states of knowing to outer objects of knowledge are rejected, along with the tidy-looking philosophical dichotomies they engender.

4.16 THE PRACTICES OF SCIENTIFIC CHANGE

The view that science is driven by practice rather than by theory or observation raises perspectives on scientific change. The claim that practice is central to the understanding of science is found in the writings of Heidegger ([1927] 1962, Division 2, Part 4), Bachelard (1984) and Polyani (1974). Also Bourdieu (1977) has written an influential account of practice and practices, and Hacking (1983) has argued that experimentation has a life of its own independent of theory. Only in recent years, however, has practice become a central focus in the writings of both philosophers and historians of science. Thus, there is work on the activities of laboratory technicians, the practices of experimentation, the practices of theoreticians, and the practices involved in the technical applications of science, to mention a few of the areas of investigation. Writers concerned with these topics deny that *all* practice is theory-laden, and also that experimental practice is just theory pursued in a different guise. They therefore deny both empiricist accounts of science, which posit pure observations of data (unmediated by experimental practices) as the basis for theory choice, and theory-dominated accounts which hold that experimental practices are largely structured via the theory-ladeness of observations. Furthermore, they situate the making of science and products of science firmly within the cultural sites within which these phenomena are produced, sustained, and developed. Thus, their orientation differs from microsociological investigations, such as Latour and Woolgar's, to the extent it goes beyond individual laboratory sites to include the various cultural networks of changing scientific practice. Indeed, in paying attention to the actual processes by which scientific knowledge is produced (and not relying on the various *images* of science in the professional and popular literature of science) they attempt to reconstruct the complex networks of skills, competences, negotiations, persuasions, and intellectual and material resources from which stable patterns of scientific practice emerge.

For those who emphasize the interactive dynamics of practice, the collective not the individual is therefore at center stage. Since in their view science is largely skill-based, network-based and laboratory-based, it can be located somewhere between the activities of individuals who pursue it and the material, cultural and cognitive frameworks which they inhabit. Moreover, Pickering (1984), Galison (1987), Holmes (1985) and Hull (1988) stress those nonverbal techniques and transparent coping mechanisms employed by experimenters which play a central role in the process of theory construction. In these contexts, scientific understanding is a ''knowing how,'' a matter of having a ''feel'' for the go of things, of reasoning practically to desired outcomes, dispositions not captured by the notion that understanding is a ''knowing that'' abstractly expressed through verbal, representational and propositional relationships. These writers also point to the construction of common research cultures and networks that routinize important patterns of skills which travel beyond the confines of the individual laboratories which give them birth.

Discussing three types of scientific practitioner further illustrates this orientation: the technicians, the experimentalists, and the theoreticians. It is now increasingly realized that we can no longer take technicians' work as transparent, and their activities and roles as invisible, if we are to establish a fuller understanding of the nature of scientific knowing and the character of scientific practice. The transparency of technicians in the business of making science results largely from the established habit of distinguishing those in the laboratory who are custodians of technical knowledge (and hence deemed authorities) and those who are merely skilled in manipulations—the technicians. This divide tends to obscure the symbiotic relationship between scientist and technician, and also the different ways in which the organizational economies of individual laboratories evaluate and focus the significance of the two groups in the making of laboratory knowledge. Certainly, if viewing the significance of technicians' work solely in terms of the activities of those involved explicitly with theory and with theory-driven practice persists, the importance of that work in the ongoing business of constructing and testing scientific knowledge will go largely undetected. To this extent an important element of continuity through scientific change will be missed. Also missed will be the full significance of the displacement, replacement, redistribution, and reorganization of skills, and the ongoing process whereby enskilled technology is substituted for human skills. Appreciation of these features throws scientific change into a new light.

Experimental practices also provide insights into ongoing science and scientific change. In good part, most accounts of experimental practice are contextualist and emphasize the symbiosis of practices, techniques, discourses and concepts. Moreover, these accounts range from microsociological studies, such as *Laboratory Life*, to full-fledged attempts to delineate scientific cultures, their mutual interactions, and their relationships to the larger cultures within which they are sited. Furthermore, the literature is concerned to clarify the ongoing interaction between theory and practice, while at the same time reflecting the fact that each has relative autonomy in different levels of discourse.

The autonomy of practice is well documented in Frederic Holmes's (1985) study of Lavoisier. He shows that ongoing experimental practice need not be driven by the need to test explicit theory as an outcome. Much that Lavoisier did in the laboratory over time, far from being driven theoretically, proceeded as an unfolding narrative guided by regulative rather than logical principles. Only when the problems set by the investigative goals had been solved did theory enter as the verifying presence necessary to link the various experimental narratives into a coherent theoretical structure. In this context, we have the appropriate ambiance for the dialogue between theory and practice.

Notice how Holmes's position differs from the laboratory dynamics portrayed by Latour and Woolgar. Theirs is the model of competition: personality, institutional affiliation, rank of researcher, the nature of the research, and its capital for future investigations, and such, count as much as strength of argument, cogency of evidence, and style of reasoning. In other words, the radical contingency of the power game, and the contingent thrust of persuasion, count as much for scientific change as anything else that is happening in the laboratory (see also Latour 1987).

Andrew Pickering (1984) argues that the symbiosis of practices and theories is driven by interests structured in terms of perceived opportunities. In some respects his emphasis on interest is reminiscent of the Edinburgh school. However, Pickering's concern is with the symbiotic relationships between theoretical and experimental practitioners. In his view their practices intentionally reinforce one another. The work of one group justifies the work of the other, and vice versa. Experimentalists routinely refine experimentally produced phenomena with the interests of the theorists in mind who use their work. Conversely, the theorists adapt their practices to the phenomena produced by the experimentalists. Both groups mutually contribute to the confluence of traditions which generate self-contained and autoreferential contexts of developing practices. In Pickering's view, it is not an objectively conceived natural world that propels the dictates of scientific practice, it is the opportunities that the scientific actors perceive for furthering their theoretical and experimental interests. Hull (1988) has illuminated many of the aspects of scientific practice discussed by Pickering, Galison and Holmes. He develops an evolutionary model of the selection processes of scientific choice and of scientific development. The model is sufficiently general to capture both social and conceptual change while remaining sensitive to the data of his inquiry and empirically nonvacuous. Hull in no way opposes the scientific belief that the practices of science can and do causally interact with the causal forces of the nonhuman world.

Of importance is the notion of practicing a theory. This is not just a question of understanding a theory's formal expression in its textbook setting, but is rather the business of adopting and transmitting through practice a set of mental technologies suggested by contextualized applications of the theory to problem solving. These mental technologies are embodied in patterns of skills present in the very conceiving and solving of problems. Such techniques are not inherent in the published literature on the theory, but are learned, exercised, and transmitted in the manner of a guild of masters and apprentices.

The reference to activities of the guild raises the question of the relationship between tradition and innovation, between exercising practices and developing them. Clearly, exercising scientific practices is a goal-oriented activity pursued in order both to extend the reach of the practices and to enact further their inherent values. Successful practices have histories and are bearers of the traditions that inform them. Thus, practices embody both the traditions that drive them and the creative modification these traditions permit as more of the values inherent in the practices are realized. Clearly, there is not merely a contingent relationship between tradition and innovation, but one that intrinsically connects traditions with their practices and the innovations that arise through their exercise.

From this perspective the careers of theories are inseparably connected with experimental practices and technologies. This image suggests that theories function not so much as Platonic paradigms, but as devices that guide experimental reasoning and enact scientific experience. Whenever theories are exercised in an experimental setting they are expressed through human practices and by the technologies that serve them. Just as the norms which guide practical deliberations are modified through application to changing situations, so theories, as expressed through experimental practices, are modified through adaptation to events produced and localized in the laboratory. Thus scientific change in its laboratory setting is a constant interplay between the place where experimental events are produced, the means by which they are produced, and the significance of what is produced. On this view, the significance of experimental practices is partly underwritten by the theory that helps to guide those actions. It is also imparted through the narrative account of what was involved in the performing of an experiment, for example, the place of experimentation, the design and reliability of the instrumentation, and the manner in which the experiment was performed. The narrative, however, should not be mistaken for the experiment itself. In the context of experimental practice, scientific change and the production of scientific knowledge present a perspective different from that current in the standard literature. Here contingent forces are at work to humble any remaining illusion that human inquirers necessarily have cognitively reliable access to reality.

The modern research complex makes unique demands on our understanding of scientific change. In the process of making laboratory science in its current highly technical and instrumental contexts the human role is with difficulty demarcated within the professional, technical and institutional structure in which it is embedded. Latour (1987) has coined the term "actant" to refer to this feature of modern science in the making. It is a notion that cuts across the divide between the collective and the individual and thus appears to avoid the sociological biases built into the perspectives of the social study of science. However, it also affects the very category of "scientific change," for if there is no principled distinction between the human and the nonhuman in the production of scientific resources, the evaluative criteria for judging change, epistemic and otherwise, are systemically open to reconceptualization.

4.17 CONCLUSION

The problem of how to account for scientific change is a ramified one. The accounts examined here harbor difficulties. The "global theorists" advance categories that are overdetermined relative to the detail of actual cases of scientific change. The social theorists, for their part, are wedded to methodologies which tend to reduce scientific change to a preferred basis. They, too, want change to lie in a procrustean bed. All have something in common: They essentialize; they privilege one perspective at the expense of others. Moreover, as the discourse of positivism reminds us, the intersection of the diachronic and the synchronic is there to challenge us.

Two observations are in order. What is the nature of scientific knowledge itself? Is it constitutively bound up with human interests, the "will to power," such that the notions of truth and reason are constructable on that basis alone? Or is scientific knowledge "about" something other than text, language and socially warranted action? Second, what role does history play in contextualized accounts of scientific development? It can be argued: an important one, especially in recounting the origin of those crises which often herald scientific progress through change. The history of a science, theory, practice or argument provides essential background for understanding and evaluating its credentials. Thus the case for the superiority of a theory is made effectively by showing narratively that it has thus far outdistanced its competitors by avoiding their defects while incorporating their strengths. No norms are invariantly instantiated; as Feyerabend and others point out, what is now best may later fail. Certainly we may "construct," "deconstruct," "rhetorize," "historicize," "sociologize," and "analyze." But in evaluating and understanding scientific change we ignore the diachronic of history at our peril. For it is from history's background perspectives that the past of change and its directions for the future emerge.

DISCUSSION QUESTIONS

1. Choose from among the challenges to the "received view" regarding scientific development the one you regard as most persuasive. How best can the "received view" be defended against it?

2. Discuss the concept of a scientific "revolution" in the light of the various accounts of scientific change that have been presented.

3. The scientific community believes that its practices continue to interact with a nonhuman world. Realists hold that only their perspective on this belief can account fully for scientific progress. Outline and discuss a constructivist response to the realists' view.

4. Can scientific innovation be understood apart from diachronic accounts of how science has developed?

5. In his *Science and Values* (1984), Laudan argues that we must relativize the assessments of scientific progress to the aims and criteria we *now* believe in. This seems at odds with his earlier notion that science is progressing if it has problem-solving efficacy against the aims and standards of its time and context. Discuss and assess these two perspectives. Are they incompatible or are they reconcilable?

6. Choose a piece of science (e.g., a law, a theory) with which you are familiar and show whether it improves over its competitors in the field or over what was believed of the phenomena prior

to the advent of this piece of science. In your discussion consider the notions of control and predictive power over nature.

7. Carefully distinguish scientific change from scientific progress. How are they related and how do they differ? Can an account of one also be an account of the other?

8. Consider the differences between an account of scientific change which reduces change over time (the diachronic) to the notion of the total and successive *replacement* of one framework by another framework and one which sees scientific change as the gradual transformation of earlier perspectives into later perspectives.

SUGGESTED READINGS

FEYERABEND, PAUL (1978), *Against Method*. London: Verso Press. A provocative attack on the pretensions of scientific rationality and an antidote to the notion that scientific change is methodologically driven.

HACKING, IAN (ed.) (1981), *Scientific Revolutions*. Oxford: Oxford University Press. This collection contains a number of fine articles on various aspects of scientific change. Hacking's introductory essay offers a comprehensive discussion of central issues.

LATOUR, BRUNO and STEVE WOOLGAR (1986), *Laboratory Life: The Construction of Scientific Facts*. Princeton: Princeton University Press. An early attempt to produce an anthropology of science which treats scientific practice and change as a linguistic phenomenon.

LAUDAN, LARRY (1977), *Progress and Its Problems: Toward a Theory of Scientific Growth*. Berkeley and Los Angeles: University of California Press. A good source for the transformation from the old to the new approaches to scientific change.

Five

PHILOSOPHY
OF SPACE AND TIME

John D. Norton

Philosophy of space and time holds a special place in twentieth-century philosophy of science. Einstein advanced his two theories of relativity, the special in 1905 and the general in 1915, at about the same time as the new movement in philosophy associated with the Vienna Circle, logical positivism and logical empiricism was forming. The founders of this movement—especially Carnap, Reichenbach and Schlick—were fascinated by Einstein's work and sought to bring out its philosophical implications. As a result, some of the clearest applications of their general ideas can be found in the philosophy of space and time. Some examples of these applications are the subject of Part 1 of this chapter and are:

1. An application of the verifiability criterion in Einstein's special theory of relativity. (See "verifiability principle," Chapter 3.) Its target is the absolute state of rest of Newton's theory of space and time, (section 5.1).

2. Antirealist claims about the geometry of space and about the simultaneity of distant events. (See "realism," Chapter 3). The geometry of space is claimed to be chosen by convention and not a factual discovery about the real world. Within special relativity, it is claimed that whether two spatially separated events are simultaneous cannot be determined factually but must be stipulated conventionally, as long as the events cannot causally affect one another, (sections 5.2 and 5.3).

3. A reduction of spatiotemporal relations to causal relations in the causal theory of time. (See "reduction," Chapter 8). All talk of "before" and "after" is claimed to be reducible to talk of what can causally affect what, (section 5.3).

Part 2 contains more challenging material. It provides an introduction to the spacetime methods now commonly used in philosophy of space and time. It has been written for nontechnical readers insofar as very little prior mathematical knowledge is presumed. The major elements of spacetime theories are presented via geometrical pictures and physical metaphors so that the major task of the reader is reduced to visualizing these pictures and metaphors. As we will see, these modern methods are themselves of considerable philosophical interest for they embody an automatic method for determining which elements of a spacetime theory can be chosen by convention and which are fixed by reality.

Part 3 contains applications of these methods. They are used to develop two results. The first is due to David Malament. It supports the claim that if one believes the causal theory of time one cannot also consistently believe the thesis of the conventionality of simultaneity in special relativity. The second is the "hole argument," which originated in Einstein's work. It seeks to establish that a strict form of realism about space and time—a "substantivalist" position—leads to indeterminism in very many spacetime theories.

$E = mc^2$. Boxed text, such as this, contains explanatory mathematical material that can be skipped by a reader less interested in technicalities.

Part I: Basic Questions

5.1 THE PRINCIPLE OF RELATIVITY

5.1.1 Newton's Absolute Space

Prior to Einstein, the dominant view of space and time was embedded within the mechanics of Newton, the most successful of all scientific theories. In the exposition of his mechanics, his *Mathematical Principles of Natural Philosophy* ([1687] 1962), Newton had distinguished absolute from relative spaces. Relative spaces are the spaces of common experience and a relative space is associated with each observer. The relative space of the reader is the space of the room in which the reader sits and its extension outside the room. If the reader is travelling in an airplane, the relative space will be the space of the cabin and its extension outside the airplane. Now the one process can be described in many different relative spaces. Consider a child riding a carousel at a fairground and a proud parent watching outside from the security of a park bench. In the relative space of the parent and bench, the child is orbiting about the central axis of the carousel. However in the relative space of the carousel, the child is at rest and the world spins about him.

The discussion of relative spaces was not merely an entertaining flourish by Newton but an essential preliminary to the development of his mechanics. He was seeking to lay down the laws that would govern the motions of all bodies. The laws were general descriptions of the behavior of bodies and these descriptions were to be given in one or other relative space. His first law, for example, asserted that in the absence of a net-impressed force a body would remain at rest or in uniform motion in a straight line, that is, in inertial motion. However a body that was at rest in one relative space may be moving nonuniformly in another. This is precisely what we visualized in the case of the parent and the child riding the carousel. To which relative space should Newton refer his laws? If they held in the relative space of the observing parent, then they could not hold in the relative space of the carousel and vice versa. Ideally we would choose a motionless space. Which space could this be? It is not the space of the parent, fixed with respect to the earth. If we believe Copernicus, the earth rotates on its axis and orbits the sun. Again the relative space of the sun is not a motionless space. The sun is a star, one of many in our galaxy, and it orbits the galactic center. As long as we try to base the mechanics on the relative space of some definite object such as the earth or sun, then we risk that object moving along with its relative space with respect to some other object. Again, nothing guarantees that this next body might not itself be in motion or come to be in motion with respect to yet another body. An infinite regress threatens.

Newton solved his problem by denying that the relative space of any body was motionless other than by accident. He announced in a Scholium in Book 1 of his *Mathematical Principles* ([1687] 1962) that of all possible spaces there was one that was eternally motionless, independently of any body that might be associated with it, "Absolute space, in its own nature, without relation to anything external, remains always similar and immovable." It was to this absolute space, the ultimate arbiter of motion and rest, that his laws were to be referred.

5.1.2 The Problem of Verification

Newton's solution raised an immediate problem. How was this absolute space to be distinguished from the many other relative spaces? The obvious answer was to turn to the laws of Newton's mechanics themselves. Absolute space could be identified as that space in which Newton's laws held. Consider a region of the universe remote from massive bodies and a test body in it free of impressed forces. If the body's motion was referred to absolute space, according to Newton's first law it would be at rest or moving uniformly in a straight line. Unfortunately this obvious condition fails to pick out a unique space as absolute space. One relative space which satisfies the condition is the one in which the test body is at rest. But there are others. Consider the relative space of an observer moving uniformly in a straight line with respect to the test body. That observer will see the test body in uniform motion in a straight line, so that the observer's relative space will be one in which Newton's first law holds. In fact there are infinitely many such spaces since there are infinitely many such observers possible, each moving at a different velocity with respect to the original test body. Even allowing for the remaining laws, it turns out that the condition that absolute space be that space in which Newton's laws hold fails to specify a unique space as absolute space. The condition picks out an infinite set of spaces, the inertial spaces, which are the relative spaces of a set of observers moving uniformly with respect to one another in inertial motion.

We should not think that this failure was an oversight of Newton that could be remedied by a small addition to his laws. In fact it was crucial to his theory that his laws work equally well in every inertial space. In formulating this theory, he had to reconcile two apparently contradictory assertions. The first was the Copernican hypothesis that the earth was not at rest but it moved at great speed, spinning on its axis and orbiting the sun. The second was the common fact of everyday experience that we earthbound observers could notice no mechanical effect on the earth's surface attributable to this supposed motion. In Newton's theory, these two assertions were reconciled by noting that, even though the earth spins and orbits, the motion of an observer fixed on the earth's surface is very nearly inertial. (The situation is not so different from that of an athlete running around the circumference of a very large circular stadium. Because of the large size of the stadium, a small segment of the track is almost straight, so that for any brief period the athlete is running in almost exactly a straight line. This would not be the case were the athlete to run in a circle of much smaller radius.) Thus at any instant Newton's laws hold to very good approximation in the relative space of an earthbound observer and to this approximation the observer will not notice any motion of the earth. The approximation is a good one. It requires very sensitive measurements to detect deviations of an earthbound observer from inertial motion. The best known example is the Foucault pendulum experiment, which can be found operating in many science museums. Were Newton's laws to hold in just one inertial space, then this reconciliation would collapse, for the motion of the earth is still only approximately inertial. Over time, the earthbound observer slowly migrates from one inertial space to another as the earth completes its daily rotation and annual orbit of the sun. If significantly differing laws of motion were to hold in each inertial space, then these differences would be revealed to an earthbound observer in the course of the migration and indicate prominently the observer's motion.

5.1.3 The Elimination of Absolute Rest

The precarious compromise of absolute space prevailed for over two centuries in spite of the discomfort of many of Newton's critics. What finally brought the issue to a head were the developments in the theories of light, electricity and magnetism of the nineteenth century. That century saw the revival of the theory that a light ray was a wave and that wave turned out to be the oscillation of electric and magnetic fields. In particular, through the work of such physicists as Maxwell, Hertz and Lorentz, light and its fields were pictured as waves in a medium known as the luminiferous ("light bearing") ether. This ether was a medium that was assumed to pervade all space and it provided physics with another preferred state of rest akin to Newton's immobile absolute space.

By the time Einstein advanced his special theory of relativity in 1905, he had come to see that the status of this luminiferous ether was very similar to that of Newton's absolute space. Newton's laws entailed that no mechanical experiment could distinguish inertial motion with respect to absolute space from rest. Correspondingly, a long series of actual experiments in the nineteenth century had failed to detect the earth's motion relative to the ether. The most accurate and best known of these was the celebrated Michelson–Morely experiment of 1887. Moreover Einstein peered into the innermost heart of the Maxwell–Hertz–Lorentz theory and concluded that as far as *observable* magnitudes were concerned, the theory entailed that inertial motion was not distinguishable from rest. His example concerned a magnet and an electrical conductor such as a wire loop. If the magnet moves through the conducting loop, a current will be induced in the conductor. Alternately, if the conductor moves over the magnet, a current will again be induced in the conductor. It does not matter which of the magnet or conductor is at rest. The same *observable* thing happens, the induction of an electric current, whenever there is *relative* motion between the magnet and the conductor and the magnitude of that current is determined solely by the magnitude of the relative motion. Absolute rest plays no role as far as observables are concerned.

This example launched Einstein's famous paper "On the Electrodynamics of Moving Bodies" ([1905] 1952b) in which he unveiled his special theory of relativity. Referring to this example, he continued:

> Examples of this sort, together with the unsuccessful attempts to discover any motion of the earth relatively to the "light medium," suggest that the phenomena of electrodynamics as well as of the mechanics possess no properties corresponding to the idea of absolute rest. They suggest rather that, as has already been shown to the first order of small quantities, the same laws of electrodynamics and optics will be valid for all frames of reference for which the equations of mechanics hold good [in all inertial spaces]. We will raise this conjecture (the purport of which will hereafter be called the "Principle of Relativity") to the status of a postulate . . . The introduction of a "luminiferous ether" will prove to be superfluous inasmuch as the view here to be developed will not require an "absolutely stationary space" provided with special properties. . . . (Ibid., 37–38)

With these words, Einstein introduced the principle of relativity to this theory. The principle asserts, in effect, that the laws of mechanics, electrodynamics and optics are to hold equally well in all inertial spaces. (Einstein picks out the inertial spaces indirectly as those "for which the equations of mechanics hold good.") Newton's quest for the one truly immobile space among them is to be abandoned.

Of course Einstein was not consciously applying the verifiability principle (Chapter 3) when he wrote these words. The criterion was not formulated until over twenty years later. However if we are to identify any justifiable, scientific applications of the criterion then this case surely would be included. The absolute states of rest to which he referred—those of mechanics and electrodynamics—had defied all attempts at experimental identification. Moreover the physical theories involved in each actually predicted that the states of rest admitted no observational consequences which would enable their identification. This is the canonical circumstance for which we seek when we apply the verifiability criterion: the postulation of an entity or state of affairs devoid of observational consequences that would admit its verification. We are thereby enjoined to despise the entity or state as an idle metaphysical conception and to banish it from our discourse.

5.2 CONVENTIONALITY OF GEOMETRY

5.2.1 The Rise and Fall of Euclidean Geometry

The geometry developed by the ancient Greeks is one of the great scientific success stories. Here we see a theory of space which reached its mature form in antiquity and survived so well in that form that even within the last 100 years its standard exposition, Euclid's *Elements* (1956), was still usable as a practical text. In the 1910s, an influential school of thought in philosophy of space and time adhered to an ingenious explanation for this success. Followers of the eighteenth-century German philosopher Immanuel Kant held that the geometry of Euclid *must* hold for all our experience. They urged that the spatial organization of our experiences did not come from external reality, the "things in themselves," but were introduced by our minds in the process of organizing our sensations into something intelligible. This circumstance can be clarified with a parable. Imagine a man compelled always to wear rose-colored glasses. Everything he sees will have a rose tint. He can be sure that every object he will ever encounter will have this tint. However this certainty does not reflect anything about external reality. It results from an unavoidable component of his apparatus of perception. So for the Kantians the truth of Euclidean geometry was guaranteed because that geometry was necessarily imposed onto experience by our minds.

An awkward development of the nineteenth century made the necessity of this imposition less than obvious. It was discovered that it was possible to have consistent geometries based on postulates disagreeing with those of Euclid. Thus even if geometry is imposed by our minds onto experience, it ceased to be clear that the geometry imposed necessarily had to be Euclidean and not one of those "non-Euclidean" geometries. The awkwardness became a serious embarrassment when Einstein completed his general theory of relativity in 1915. That theory entailed that the actual geometry of our own space was non-Euclidean, although the deviations observable in our vicinity from Euclidean geometry were very small.

The new philosophies of science emerging at that time retained the basic idea that the Kantians applied widely: some parts of a theory are reflections of reality and some parts are provided by the organizing mind. However, in the new view, that part contributed by the mind was no longer of fixed or necessary character. It was allowed to vary and it could be chosen at whim or as a convention. To continue the above parable, the man still has to wear glasses, but now he can choose freely the color himself. Thus a new theme entered philosophical analyses of scientific theories, the division of the conventional components from those reflecting reality. The debate over the placement of this division survives in part today in debates over realism and antirealism.

Geometry attracted special attention. It was urged by a number of thinkers such as the French physicist-mathematician Henri Poincaré and more insistently by the German philosopher Hans Reichenbach that the choice of a geometry for space is a matter of convention, for there is no independently true geometry in nature. Thus the geometer could choose whether he would work with a Euclidean or a non-Euclidean geometry. Needless to say one would expect such choices to favor Euclidean geometry wherever practical, since Euclidean geometry is both simple and familiar. But this choice, so the conventionalist thesis asserts, is not to be confused with a discovery of the true geometry, for there is no such thing.

5.2.2 An Argument for Conventionality

How is the conventionality of geometry to be established? Einstein ([1921] 1954a, 235–236) develops the simplest and still best known version of the argument for conventionality of Euclidean geometry and attributes the argument to Poincaré. By itself, the argument runs, a geometry G tells us nothing observable about space. Rather it tells us something about certain idealized structures such as rigid rods which do not actually exist. In order to derive observational consequences about real bodies, we need to resort to physical theories P dealing with such topics as elasticity and thermal expansion to correct for the deviations in the real bodies' behavior from the ideal behavior.

For example, imagine that we wish to check Pythagoras's theorem for the case of a large right angle triangle with sides of length 30 and 40 feet enclosing the right angle. According to the theorem, we expect the hypotenuse of the triangle to be 50 feet and, to check this expectation, we construct the triangle using steel tapes stretched between three points. The catch is that we cannot just pin together three tapes of length 30, 40 and 50 feet. Such steel tapes will always sag a small amount no matter how tautly they are pulled, so that we will need tapes slightly longer than 30, 40 and 50 feet to span the true distances. We can always reduce the amount of sagging by pulling harder on the tapes, but this will in turn stretch the tapes by a small amount since steel is elastic. Thus a *very* accurate check of Pythagoras's theorem with steel tapes can only be carried out if one makes a series of complicated corrections to the behavior of the steel tapes, allowing for their deviations from the behavior of ideally rigid measuring rods. These corrections exploit physical theories of the gravitational and elastic deformation of bodies.

In general terms, observational consequences follow only from $G + P$, the conjunction of the geometry G with the physical theories P. This means that we are free to make conventional modifications to the geometry G as long as we modify our other physical theories P accordingly so that the observational consequences remain unchanged. Thus the one set of observational consequences can be accounted for equally by a large number of conventionally chosen geometries.

5.2.3 Universal Forces, Coordinative Definitions and the Heated Metal Plate

It is at first a little hard to visualize how these conventional elements will appear or, for that matter, just what a non-Euclidean geometry might look like. The following illustration will help. It has been used in various forms by Poincaré, Reichenbach and others and has been modified here for brevity. We imagine a large circular metal plate 10 feet in diameter in a Euclidean space. Since a theorem for circles in Euclidean geometry is that

$$\frac{\text{Circumference}}{\text{Diameter}} = \pi$$

the circumference of the plate will be approximately 31.4 feet (see Figure 5.1). These figures translate into the following operations: A rigid one-foot ruler can be laid 10 times across the diameter and roughly 31 times around the circumference. Were the geometry of the space not Euclidean, then in general we would not obtain π as the ratio of the circumference of the circle to its diameter, but some other ratio larger or smaller than π, when we performed these operations.

Now imagine we discover that the disk is heated in the center but remains cool at its periphery. Imagine also that we carry out the measuring operations with a rod that is not rigid but which will expand when heated as real rods do. (For simplicity, we allow the rod to warm to the temperature of the disk each time we lay it down so that it expands accordingly to the temperature at that part of the plate.) It now follows that if the ruler can be laid 31 times around the circumference, then it will not be possible to lay it the full 10 times across the diameter due to the thermal expansion of the rod towards the center of the plate.

These measurements can be used to justify two different geometries for the plate. First we arrive at Euclidean geometry if we correct the measurements for the thermal expansion of the measuring rod by means of an appropriate physical law for thermal expansion. Second we arrive at a different geometry if we make the alternate physical assumption that no thermal correction is needed and that the rod always measures true distance no matter how its temperature varies. This latter geometry will not be Euclidean. We have, in place of the Euclidean result, the new result:

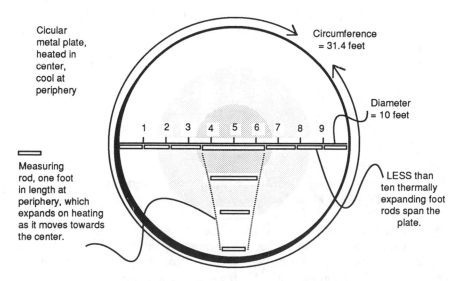

Cicular metal plate, heated in center, cool at periphery

Circumference = 31.4 feet

Diameter = 10 feet

1 2 3 4 5 6 7 8 9

Measuring rod, one foot in length at periphery, which expands on heating as it moves towards the center.

LESS than ten thermally expanding foot rods span the plate.

Figure 5.1 Geometry of a heated metal plate.

$$\frac{\text{Circumference}}{\text{Diameter}} > \pi$$

since fewer than ten rod lengths will be needed to span the diameter of the disk. Our choice of geometry then amounts to our choice of which are the rigid rods, the corrected or uncorrected rods. Notice that we *cannot* select the rigid rod by checking for the one that remains constant in length under transport. For what remains constant in length is a matter that can only be known if we already know what is the geometry. We are in a vicious circle. We break it, Reichenbach urges in his influential *Philosophy of Space and Time* ([1928] 1957), by realizing that our choice of rigid rod is a matter of definition. It is a definition that coordinates aspects of physical reality with idealized components of an abstract theory, so Reichenbach calls the definition a "coordinative definition." Now precisely because our choice of rigid rod is a matter of definition, there is no factually correct choice, only conventional preferences. Therefore, since each choice of a rigid rod produces its own geometry, there can be no correct geometry, only our conventional choices.

We would probably be less likely to choose the uncorrected rod as our standard of rigidity (or, better said, congruence of intervals) because we know that various materials expand differently when heated by the same amounts. We might not like to have a material-dependent definition of congruence, especially since the differential thermal expansion of different materials would provide a way of mapping out the temperature changes across the plate, which we might then want to interpret as a cause of length distortion. In the place of thermal deforming forces, Reichenbach considers what he calls "universal forces," which he defines as deforming all materials by the same amount and for which no insulating walls are possible. The situation remains essentially the same as that of the heated metal plate. A universal force will deform our measuring rods, so that the decision to correct or not to correct for such a force amounts to a conventional choice of geometry. In particular the usual assumption that there are no universal forces is a definition leading to a conventional choice of geometry, just as is the decision not to correct for thermal deformation in the case of the heated metal plate.

The notion of universal forces allows Reichenbach to mimic Einstein's figurative "$G+P$" in a concrete way. If we use some physical rod to measure a distance, then its observed length is related to its true length by the equation:

$$\begin{pmatrix} \text{Observed} \\ \text{length} \end{pmatrix} = \begin{pmatrix} \text{True length} \\ \text{according to} \\ \text{geometry } G \end{pmatrix} + \begin{pmatrix} \text{Correction required by} \\ \text{universal force governed by} \\ \text{physical theory } P \end{pmatrix} \qquad (1)$$

This equation tells us how to accommodate virtually any conventionally chosen geometry G to a fixed set of observed lengths. We simply choose a universal force whose corrections will make the above equation hold.

The essential point, stressed by the conventionalists, is that some such choice of congruence condition must be made as a definition to break a vicious circle. We cannot know factually what universal forces prevail until we know the geometry; but we cannot know the geometry until we know what universal forces prevail.

> Reichenbach actually supplies a more precise version of equation (1), which uses the mathematical apparatus introduced below in Section 5.5. If γ is the metric of a space when universal forces are assumed to be absent, then we can choose any other metric γ' to be the metric of the space as long as there is an object F satisfying
>
> $$\gamma = \gamma' + F \qquad (1')$$
>
> We treat F as the potentials of a universal force field and choose our criterion of congruence to be one which corrects for the deformation due to F. This choice leads us to conclude that the metric of the space is γ'.

5.2.4 Is Everything Conventional?

Can the arguments that were used to establish the conventionality of geometry be used to establish the conventionality of other laws or structures? It seems that they can. To use them to reveal any law as conventionally chosen, all we need is that no observational consequences follow from the chosen law alone. The derivation of these consequences must involve another physical law. We can then make conventional changes to the first law as long as we adjust the second in a way that preserves the observational consequences.

As an example, take Coulomb's law in electrostatics, which describes how one electric charge exerts a force on another. In its field form, it asserts that every electric charge generates an electrostatic field potential φ in the space around it and that this field diminishes in intensity inversely with distance r from the charge, that is, as $1/r$. Now in its field form Coulomb's law *by itself* entails no observable consequences. To recover conclusions in terms of the observable motions of charges, we need the force law, which describes how the field φ acts on other charges. That law asserts that the force exerted on a unit test charge is directly proportional to the quantity "grad φ," which measures the rate at which the field φ changes as we move from point to point in space. But how do we know that the force law is correct? We can only check it independently if we already know what is the field φ. But this we do not know until we have in hand a law, such as Coulomb's, which tells us the fields produced by given charges. We are in a familiar vicious circle. We break the vicious circle with a coordinative definition. We *define* that the field φ is such that the force on a unit test charge is $-$grad φ. Compatibility with observational consequences now leads to Coulomb's law: The field φ diminishes as $1/r$ as we move away from its source. The crucial point is that we are free to choose an alternative coordinative definition.

Examples such as these raise the following questions. Is the conventionality of geometry a position peculiar to geometry and a few other similar fields? Or is it part

For example, we can conventionally replace Coulomb's law with another in which φ weakens as $1/r^3$ with distance r from its source. To preserve the original observational consequences we adjust the force law by defining the force on a unit test charge to be $-\text{grad}\ \sqrt[3]{\varphi}$. Alternatively, we could mimic the introduction of universal forces in geometry in (1) by defining a second field f which mediates the interaction of charges as does the electrostatic field φ. If φ is the electrostatic potential in the case of vanishing f, we can conventionally choose the electrostatic potential to have any other form φ' as long as f is such that

$$\varphi = \varphi' + f$$

is satisfied. The observational consequences remain unchanged provided we adopt a new force law in which the force on a unit test charge is given by $-\text{grad}(\varphi' + f)$. Just as in the case of universal forces in geometry, we have no independent access to the field f so that it is a matter of convention as to whether we set it to zero or not.

of an indiscriminate antirealism in which any law is judged conventional if the law fails to entail observational consequences without the assistance of other supplementary laws? The latter view appears to be a version of the Duhem–Quine thesis (see Chapter 3). This thesis states that it is impossible to test the individual laws of a theory against experience. We can only test the entire theory. Any attempt to test individual laws will fail since we can always preserve any nominated law from falsification by modifying the other laws with which it is conjoined when we derive observational consequences from it.

One conventionalist response is to insist that this general antirealism differs from the true cases of the conventionality of geometry and other related cases such as the conventionality of simultaneity to be discussed in the next section. What distinguishes these latter cases is that the conventionality depends on a very small vicious circle that must be broken by a definition since no independent factual test is possible for the individual components of the circle. The Duhem–Quine thesis does not restrict the manner in which we might protect a law from falsification. We might have to do so by a complicated and contrived set of modifications spread throughout the theory. Some of the components modified may be subject to independent test and thus not properly susceptible to conventional stipulation.

Grünbaum (1973, Part 1) provides an alternate escape from the construal of the conventionalist position as a form of indiscriminate antirealism. He bases the conventionalist thesis on a claim peculiar to space. He urges that space has no "intrinsic" metrical properties, the properties that determine the distances between points, so that these metrical properties must be provided conventionally by us as a definition of congruence.

5.3 THE CAUSAL THEORY OF TIME AND THE CONVENTIONALITY OF SIMULTANEITY IN SPECIAL RELATIVITY

5.3.1 Reichenbach's Constructive Axiomatization of Relativity Theory

Spatial geometry is one of a number of aspects of theories of space and time which, it has been urged, can be chosen conventionally. One might well ask what a theory of space and time might look like if it is built in a way that tries to take full account of these various conventions. One of the early and most significant answers comes in the form of Reichenbach's ([1924] 1969) axiomatization of special and general relativity. Axiomatizations usually seek to condense the physical content of a theory into the fewest axioms of the widest possible scope. Since such axioms inevitably employ the highest-level theoretical structures which mix factual and conventional elements, such axiomatizations do not help the conventionalist sort the conventional from factual content of the theory. To this last end, Reichenbach devised the "constructive axiomatization." Its axioms are statements which are as close as possible to immediate sense experience and thus the basic, factual content of the theory. Of course unfiltered sense experience would lead to an entirely unmanageable axiomatization. Thus Reichenbach employed axioms which rely only on lower-level, prerelativistic theories and which are as close as practical to experiential statements. The axioms are couched in terms of such primitives as "events," "real points," "signals," "rods" and "clocks." The theory is built from axioms such as (from Reichenbach 1924, 29)

> Axiom 1,1. There is no signal chain such that its departure and its return coincide at [a real point] P.

In the course of developing the full theoretical structure of relativity theory, Reichenbach found that he had to supplement his axioms, which contained the theory's experiential content, with definitions of the form of the "coordinate definitions" discussed above in 5.2.3. These definitions comprise the conventional content of the theory and present an integrated picture of the interplay of factual and conventional elements in relativity theory.

5.3.2 The Causal Theory of Time: The Reduction of Time to Causation

An idea implicit in Reichenbach's axiomatization of relativity was what Reichenbach (1956) later called the "causal theory of time." The theory asserts that the temporal order of events is reducible to causal relations between the events. In other words, when we make some assertion about the temporal order of two events, we are really only making assertions about the possibility of one causally affecting the other. Thus Reichenbach ([1928] 1957) offered as a "topological coordinative definition of time order: (p. 136). "If E_2 is the effect of E_1, then E_2 is called later than E_1" (ibid.). More generally, E_2 is later than E_1 if E_1 can causally affect E_2.

Reichenbach assumed that if E_1 and E_2 are causally related we can distinguish the cause from the effect and developed his so-called "mark theory" as one way to support this view. Later workers, such as Grünbaum (1973, Chapter 7), have shown that Reichenbach's attempts here failed and have reconstructed the causal theory on the basis of a symmetric relation of causal connectibility which does not distinguish cause from effect.

The showpiece of the causal theory of time is the possibility of providing an axiomatization of special relativity solely in terms of causal relations so that in some sense the entire space and time of special relativity is reducible to causal relations. The earliest such axiomatization is due to Robb (1914, 1921) in the 1910s and a recent one is due to Winnie (1977).

> A major problem for the theory is the extension to general relativity. As we see in Section 5.8, causal relations in relativity theory depend only on the light cone structure. It has been known since the late 1910s from the work of Kretschmann and Weyl that the specification of the light cone structure in general relativity does not fully determine the complete metrical structure of the spacetime. It would seem that the remaining underdetermined part of the metrical structure cannot be reduced to causal relations. One escape for the conventionalists is to represent the causal structure of spacetime by both light cone and affine structure together.

5.3.3 Conventionality of Simultaneity

One of the conventional definitions contained in Reichenbach's axiomatization of special relativity has become the subject of special attention in philosophy of space and time. It presumes that there is considerable conventional freedom in our determination of which events are simultaneous in special relativity with respect to a given inertial space. (This conventionality should not be confused with the "relativity of simultaneity" in special relativity, which is discussed in Section 5.8 and which involves a change of simultaneity relations with change of inertial space.)

The basic problem is the following. We select two points in an inertial space of special relativity. How are we to judge which events at the first point are simultaneous with which events at the second? This problem is not entirely unfamiliar to us. Imagine that we are celebrating a birthday in two cities A and B separated by a large distance and that we would like to start lighting the candles on both cakes at the same instant. We could synchronize the two events by means of a long-distance telephone call between the two parties at the time of the lighting. However we might prefer an indirect method. We ensure that identical clocks are located at each party and, prior to the festivities, we synchronize the clocks by means of the long-distance phone call. Once synchronized we can use the clocks to determine which later events are simultaneous at the two places.

If we pare the foregoing procedure down to its barest elements, we have exactly the classic procedure Einstein introduced in his 1905 special relativity paper, "On the

Electrodynamics of Moving Bodies,'' for establishing the simultaneity of distant events. We have identical clocks located at points A and B of an inertial space. We synchronize them by transmission of a signal. We send a signal from the A-clock which is reflected instantly at the B-clock and returns to the A-clock (see Figure 5.2). Were infinitely fast signals possible, we could synchronize the clocks easily. The emission, reflection and return of the signal would all happen instantaneously. So if the emission is arranged for when the A-clock reads 12 noon, we would set the B-clock to 12 noon when the signal arrives. However in special relativity, it is usually assumed that no signal can travel faster than light. Indeed conversations are not transmitted infinitely fast by a telephone system, but at the speed of light. Thus, even if we use the fastest signals available—light signals—there will be a delay at the A-clock between the emission of the signal at A and its return to A. In setting the clocks, we have to decide which event at the A-clock between the emission and the return is simultaneous with the event of reflection of the signal at the B-clock. The situation is shown in Figure 5.3, where the various events at each clock have been spread out in the vertical direction.

If $T_{A\text{-emission}}$ is the time of emission of the signal of the A-clock and $T_{A\text{-return}}$ the time of its return, then Einstein chose *as a definition* that the event exactly half way between the two was the one that was simultaneous with the signal's reflection. This event has the A-clock time coordinate

$$T_{1/2} = T_{A\text{-emission}} + \tfrac{1}{2}(T_{A\text{-return}} - T_{A\text{-emission}}) \tag{2}$$

What other choices did Einstein have? The causal theory of time rules out any event at A prior to $T_{A\text{-emission}}$. Such an event could send a signal travelling slower than light to arrive at the reflection event at B, so that the event would be judged as before the reflection. Similarly any event at A after $T_{A\text{-return}}$ would be judged as after the reflection at B. Since the upper limit to the speed of signals is that of light, no event at A between $T_{A\text{-return}}$ and $T_{A\text{-emission}}$ can interact causally with the reflection at B. Therefore any of these events could be chosen as simultaneous with the reflection at B and the clocks synchronized accordingly. The A-clock time of this event is given by

$$T_{\epsilon} = T_{A\text{-emission}} + \epsilon(T_{A\text{-return}} - T_{A\text{-emission}}) \tag{2'}$$

where ϵ must be chosen so that $0 < \epsilon < 1$. The conventionality of simultaneity resides in the conventional freedom to set the value of ϵ anywhere in this interval.

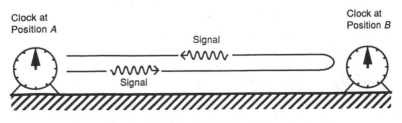

Figure 5.2 Synchronizing clocks by reflection of a signal.

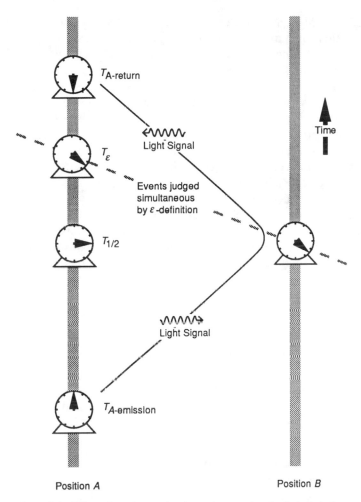

Figure 5.3 Effect of maximum signal speed on synchronization of clocks.

The argument can be strengthened and divorced to some degree from the causal theory of time if it is worked into the familiar form of a vicious circle argument. We could know that the A-event at $T_{1/2}$ is factually simultaneous with the reflection at B as long as we knew that light propagates isotropically, that is, at the same speed in all directions, for then light would take the same time to travel from A to B as from B to A. But to know how fast light travels from one place to another a known distance away, we need to have synchronized clocks at both source and destination so that we can measure the transit time of a light signal and then compute its velocity. Now we can only have *synchronized* clocks if we already know which events at spatially separated places are simultaneous—and the circle is closed. We break this vicious circle by stipulating a value for ϵ. The choice of $\epsilon = 1/2$, "standard" simultaneity, amounts to the stipulation that light travels at the same speed in both directions of the signalling set-up.

If the light signalling method fails to determine a factual simultaneity relation in special relativity, might not some other method? Much of the literature on the subject of conventionality of simultaneity has been devoted to investigating such alternative methods of synchronizing spatially separated clocks and seeking to reveal definitions equivalent to the setting of a value for ϵ in them. See for example Salmon (1977) to get a clear sense that no such convention-free, alternative method is likely to be found. Note that this literature urges the conventionality of the ''one-way'' velocity light, that is, the velocity between two spatially separated points. The round trip velocity is not taken to be conventional since only one clock at the common source and destination is needed for its measurement.

We return to the conventionality of simultaneity in Section 5.11 to see one of the most dramatic reversals in debates in the philosophy of space and time. David Malament has recently derived a theorem in special relativity which, he urges, shows that the causal relations of special relativity do *not* leave the simultaneity relation underdetermined and thus the relation cannot be set conventionally within the causal theory of time. He shows that the only nontrivial simultaneity relation definable in terms of the causal relations of special relativity is the familiar standard simultaneity relation of $\epsilon = 1/2$.

Part II: Theories and Methods

The purpose of this part is to introduce the methods now used almost exclusively in recent work in philosophy of space and time. These methods differ from those used in Part I in several important ways.

1. There is less emphasis on theories of a space and time as a set of law-like sentences. Rather the theories are approached semantically (see Chapter 3). Thus the activity of the theorist becomes akin to that of the hobbyist model builder, who seeks to represent a real sailboat by constructing a model that captures as many of its properties as possible. The space and time theorist builds models which are intended to reflect the spatial and temporal properties of reality. However the theorist's models are not constructed out of balsa, glue and string, but out of abstract mathematical entities such as numbers.

2. Theories of space and time—including Newton's theory of space and time—are worked into a spacetime formulation. Thus when Newton's theory is compared with its relativistic rivals, all the theories are formulated in the same manner, ensuring that the differences observed are true differences and not accidents of differing formulations.

3. A major theme of Part I was the separation of the conventional or arbitrary elements of a theory from the factual or, as we now say, ''physically significant'' elements. A means of effecting automatically this separation is built into the notions of ''covariance'' and ''invariance'' to be explained here in Part II.

5.4 A SIMPLE THEORY OF LINEAR TIME

Let us begin by developing a very simple theory of time whose main purpose is to illustrate the use of models and the notions of covariance and invariance in a setting far simpler than the spacetime theories to which we will soon turn. The basic temporal facts of some physically possible world are that it has infinitely many instants, extending indefinitely into the past and future. The set of instants is homogeneous: Every instant is exactly like every other. The set is also assumed to be isotropic: The future and past directions are exactly alike. To capture and make precise these loosely stated facts, let us develop the following sequence of time theories.

5.4.1 The One Coordinate System Formulation

Let us select as the model for our theory the manifold of all real numbers **R**. Each real number in **R** represents a particular instant (see Figure 5.4). This representation relation is a *coordination* of the instants of the physically possible world with the mathematical structure **R** so that the relation is commonly called a *coordinate system*. We can infer many of the temporal properties of the physically possible world from the coordination. For example, the fact that there is no greatest real number represents the fact that there is no last instant, so that the world persists through indefinitely many instants into the future. Similarly, the denseness of **R**—the fact there is always another real number between any two given real numbers— represents the denseness of time. It models the fact that every temporal interval can be divided so that indivisible time "atoms" are disallowed.

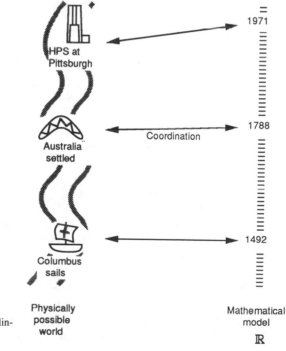

Figure 5.4 Coordinate system for the linear theory of time.

Physically possible world

Mathematical model

R

5.4.2 The Standard Formulation

Unfortunately we cannot construe every property of **R** as representing a property of the physically possible world. For example **R** is anisotropic; the direction of increasing real numbers is distinct from that of decreasing real numbers. However, we posited that the physical instants form an isotropic continuum. Similarly, **R** is inhomogeneous; the real number 0 is distinct, for example, from every other number. However, we posited that the physical instants form a homogeneous continuum.

A simple device enables us to designate systematically which are the physically significant properties of the models. To deny physical significance to the anisotropy of **R**, we expand the coordinations of the physically possible world with **R** allowed by the theory. We now allow a new coordination reflected about 0 (see Figure 5.5). Those instants coordinated with 0, 500, 1000, 1500 and so forth in the original coordinate system are now coordinated with 0, -500, -1000, -1500 and so forth in the new system. We call the transformation connecting the two coordinate systems a reflection about 0. If we allow that both the original and reflected coordinate systems are equally good representations of the continuum of physical instants, then the anistropy of **R** no longer enables us to pick out a preferred direction in the continuum of physical instants. The direction picked out by increasing real numbers in one coordinate system is the opposite direction to the one picked out by increasing real numbers in the reflected coordinate system.

Similarly we deny physical significance to the inhomogeneity of **R** by allowing all the coordinate systems produced from the original by a translation of the original coordinate system. For example, in the original coordinate system the instant to which 0 is assigned is singled out as special when compared to the one to which 500 is assigned. We can remove this special status by allowing a second coordinate system in which the latter event is now assigned the value 0. This new coordinate system is produced by translating the original by 500. Figuratively this amounts to "sliding" down by 500 each of the real values coordinated to each instant by the original coordinate system to form the new coordinate system. See Figure 5.5. We ensure that the inhomogeneity of **R** accords no special status to any physical instant by allowing into the theory all coordinate systems produced by a translation from the original by *any* real value. Thus, given any physical instant at all, we can always find a coordinate system in which that instant is assigned the value 0 or, for that matter, any other real value you care to name.

5.4.3 Covariance and Invariance

In sum, the standard formulation of the theory has the original coordinate system as well as all those produced by the coordinate transformations of reflection and translation. Let us call these the *standard* coordinate systems of the theory. The set of reflections and translations form a group of transformations (see the following box) which essentially only means that we never leave the set of transformations if we invert or combine them. It is called the *covariance group* of the theory. Alternatively, we say that the theory is covariant under reflections and translations.

The advantage of the standard formulation over the one coordinate system

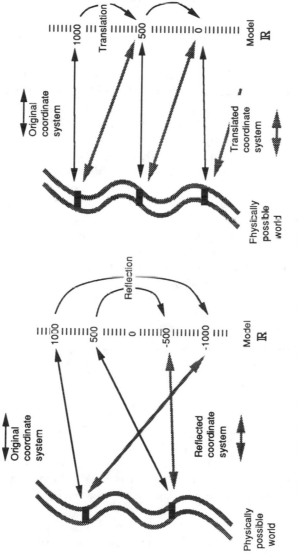

Figure 5.5 Coordinate transformations for the standard formulation of the linear theory to time.

formulation is that it enables us to distinguish the physically significant or factual properties of the theory's model from the arbitrary ones. Those properties are exactly the ones that remain the same in all the coordinate systems of the theory. We can state this important conclusion in another way. By an *invariant* of a transformation, we mean something that remains unchanged under the transformation. Thus we arrive at a principle of paramount importance to all theories of space and time:

> The factual or physically significant quantities of a theory of space and time are the invariants of its covariance group.

All other quantities can be chosen arbitrarily or conventionally. For example, the fact that one coordinate system assigns the real value 27 to some instant is not invariant and thus not physically significant. A different coordinate system will in general assign a different value to the instant. Thus the choice of coordinate system is an arbitrary or conventional stipulation. However if the difference of coordinate values of two instants is 100, then it will be ± 100 in all standard coordinate systems. Thus we conclude that the absolute value of coordinate differences (i.e., the difference as a positive number) in standard coordinate systems is invariant and therefore physically significant. These coordinate differences are interpreted as duration or physical time elapsed, such as might be read by a physical clock.

The strategy of characterizing geometric structure as the invariants of groups has a venerable history. It dates back to Felix Klein's "Erlangen program" of the 1870s in which Klein set out to use the strategy to unify the treatment of the diverse geometries discovered in the nineteenth century.

If the original coordinate system assigns the real value T to some physical instant i, then a new coordinate system produced by a reflection about 0 assigns the new value T' to i where

$$T' = -T$$

and a translation by K assigns the new value T'' to i where

$$T'' = T - K.$$

Combining we can now represent the covariance group of the standard formulation as the set of all transformations given by

$$T^* = At - K$$

where A is $+1$ or -1 and K has any real value. Formally this set of transformations is a group since it satisfies the three conditions:

1. The set contains the identity transformation.
2. Every transformation's inverse is in the set.
3. The composition of two transformations of the set is in the set.

5.4.4 The Generally Covariant Formulation

The adoption of a generally covariant formulation of the theory provides a way of making more explicit just what are the physically significant quantities of the theory. To arrive at the formulation, we expand the allowed coordinate systems to include all those which can be transformed to the original system by smooth invertible transformations on **R**. Figuratively this means that the allowed transformations of the theory include not just reflections and "slidings" (translations) of the coordinate system but just about any arbitrary "stretching and squeezing" which preserves the smoothness of the coordinate system and the uniqueness of the identification of all instants. However, we cannot leave the theory in this state for we can no longer represent duration by coordinate differences. Coordinate differences are certainly no longer invariant under the arbitrary transformations now allowed. To recover the ability to represent duration invariantly, we must explicitly introduce a new mathematical structure into the theory.

Consider some very small duration between two instants which have coordinate values 1000 and 1001 in a standard coordinate system (see Figure 5.6). The coordinate difference—call it "ΔT"—equals 1 and it is the duration between the two instants. Now introduce a new coordinate system which has been stretched linearly to double the size of the original system, so that to instants originally assigned values 0, 500, 1000, 1001 and so forth are now assigned values 0, 1000, 2000, 2002 and so forth. The coordinate difference in the new system between the same two instants— call it "Δt"—is now equal to 2. To recover the original duration we must multiply the new coordinate difference by a scale factor of 1/2. This scale factor is the extra geometrical structure which we need. Every coordinate system of the generally covariant formulation must be supplied with this scale factor to enable assertions about duration to be made. In general for a *small* duration between two instants whose coordinate values differ by Δt in some coordinate system we have the invariant result:

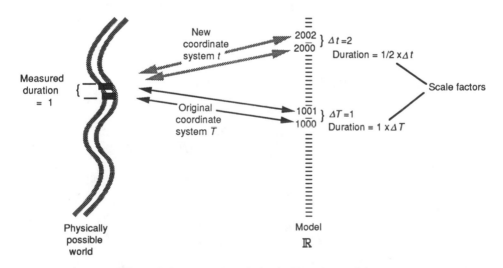

Figure 5.6 Temporal metric for the linear theory of time.

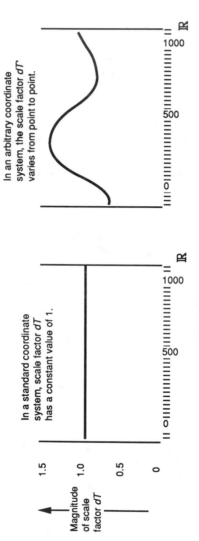

In a standard coordinate system, scale factor dT has a constant value of 1.

In an arbitrary coordinate system, the scale factor dT varies from point to point.

Figure 5.7 Scale factor dT in different coordinate systems.

$$\text{Invariant duration} = \text{Scale factor } X \, |\Delta t|$$

The scale factor of a standard coordinate system is unity. It is 1/2 for the linearly stretched system described here. In some arbitrary coordinate system, the scale factor will have a value that varies from instant to instant according to how much the coordinate system has been stretched or squeezed in the transformation from a standard coordinate system (see Figure 5.7). There is a simple rule—see equation (3) in the following box—for computing how the scale factor will change under an arbitrary coordinate transformation. The existence of such a rule means that the scale factor is a *covariant quantity*: Once we know its value in one coordinate system we invoke its characteristic transformation law to find its value in any other coordinate system. Alternately, such quantities are known as *geometric objects*. See Figure 5.8 for a pictorial representation of this transformation law.

In order to comply with the standard notation, let us represent the scale factor by "*dT.*" The scale factor *dT* (together with all its transforms) is known as a "covector" or "one-form" and, with regard to its function in the theory, might also be called a "temporal metric" since it is responsible for assigning measurable time

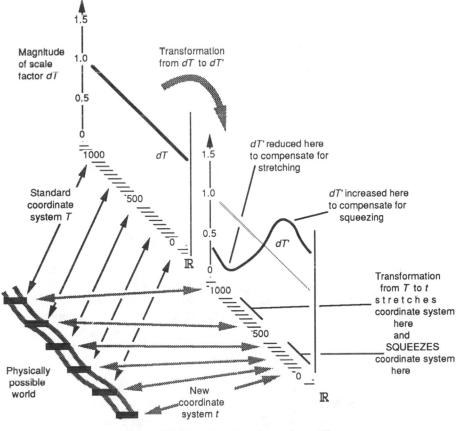

Figure 5.8 Transformation of scale factor *dT*.

durations to the intervals between instants. (For completeness we note a technical complication. A covector dT assigns positive or negative measures to intervals between instants according to their directions. Because we want no anisotropy in the continuum of instants, only the absolute values of the measures assigned have physical significance.)

The generally covariant formulation of the theory has models of the form $<\mathbf{R},dT>$, where the angle brackets ''$<,>$'' denote an ordered pair. Every time we change coordinate systems we generate a new scale factor dT. Thus the model set of the theory contains infinitely many models

$$<\mathbf{R},dT>,<\mathbf{R},dT'>,<\mathbf{R},dT''>, \ldots$$

where dT, dT', dT'', ... can all be transformed into one another and thus represent the one covariant quantity or geometric object. There is a natural division of labor between the two structures of the pair that form the models. The fact that the physical world can be coordinated with $\mathbf{R}$ gives us its topological properties: Briefly, its instants form a linear continuum with no end points in either direction. Unlike the standard formulation, the coordination with $\mathbf{R}$ gives us no information on the physically measurable duration between instants. Such information is provided by the temporal metric dT, the second member of the pair.

The model $<\mathbf{R},dT>$ is typical of those used in theories of time, space and spacetime. The models of the theories we now turn to all have the general form

$$<\text{manifold, geometric object, geometric object, } \ldots >$$

The first member of the model, the manifold, represents the topology of the time, space or spacetime in question. Thus it tells us how many dimensions a space has and

---

Let T be a standard coordinate system and ΔT the coordinate difference between two very close instants so that ΔT is also the duration between the instants. We now transform to a new coordinate system t, which need not be a standard coordinate system. We have immediately

$$\text{Duration of interval} = \Delta T = \frac{\delta T}{dt} \Delta t$$

and we can identify dT/dt as the scale factor dT in the coordinate system t. For example, if $t = T^3$, then the scale factor is given by $dT/dt = 1/(3T^2) = 1/(3t^{2/3})$. If we now consider another coordinate system t' with dT' equal to dT/dt', then the chain rule for differentiation, $\dfrac{dT}{dt'} = \dfrac{dt}{dt'} \dfrac{dT}{dt}$, gives us the general transformation law for dT:

$$dT' = \frac{dt}{dt'} \, dT \tag{3}$$

(3) is the characteristic transformation law for covectors or one-forms.

Philosophy of Space and Time

gives us information on its global topology. In the simple linear time theory, time was globally like a line, extending indefinitely into past and future. However, we might want to model a time that is cyclical so that the past and future join. We would then not use **R** as the manifold, but another one-dimensional manifold that is closed like a circle. There are many manifolds more complicated than **R** that the theorist can choose in building models. The remaining members of the model are the geometric objects such as dT that are "painted" onto the canvas of the manifold. They provide the nontopological properties of the space. Thus if we want to know the time elapsed between instants, we look to a temporal metric. In a theory of space, we look to a spatial metric to tell us the distance between two points along some curve. Such a theory is the subject of the next section.

5.5 EUCLIDEAN SPACE

The theory of a Euclidean space is very similar in structure to the linear time theory. Let us consider the case of a two-dimensional Euclidean space. The generalization to the three-dimensional case is entirely straightforward.

The theory's models are built with two-dimensional manifold $\mathbf{R}^2$, where $\mathbf{R}^2$ is

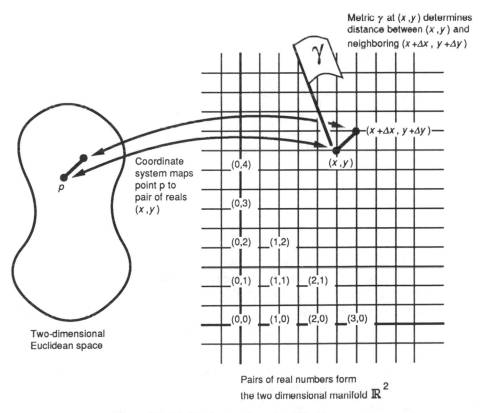

Figure 5.9 Model of a two-dimensional Euclidean space.

the manifold whose points are all the pairs of real numbers. Informally we picture the manifold $\mathbf{R}^2$ as the set of all pairs of real numbers laid out in a two-dimensional table that is without holes and that extends indefinitely. That this manifold can be coordinated with a physical Euclidean space reflects the fact that the space has all of its topological properties (see Figure 5.9). The theory is to be generally covariant. Therefore we allow any coordination between the physical space that is produced by a smooth transformation from the original. These transformations include all manner of translations, rotations, reflections, "stretchings" and "squeezings" that preserve smoothness of the coordinate system and the uniqueness of the labelling of the points.

Our theory cannot yet determine the distances between the points of the space. This information is provided by the geometric object γ which is the metric tensor of the space. This object is defined at every point of $\mathbf{R}^2$ and encodes the distances from that point to the points neighboring it. The metric tensor can be used to determine the length of curves in a Euclidean space by breaking up the curves into a sequence of small segments, determining the length of each segment and adding.

In sum, the models of the theory are pairs of the form

$$<\mathbf{R}^2, \gamma>$$

Since the theory is generally covariant, infinitely many coordinations will be allowed between the physical space and the manifold $\mathbf{R}^2$. Just as in the case of of the linear time theory, as we transform from one coordinate system to another, we may have to modify the scale factors forming γ to retain the invariance of the judgements of length which it hands down. Thus the model set of the theory will be infinitely large:

The distance $\Delta 1$ between a point with coordinates (x,y) and a neighboring point $(x + \Delta x, y + \Delta y)$ is given by the quadratic form

$$\Delta 1^2 = \gamma_{11}\Delta x^2 + \gamma_{12}\Delta x \Delta y + \gamma_{21}\Delta x \Delta y + \gamma_{22}\Delta y^2 \qquad (4)$$

where the coefficients γ_{12} and γ_{21} are equal. The matrix of the four values of these coefficients

$$\begin{bmatrix} \gamma_{11} & \gamma_{12} \\ \gamma_{21} & \gamma_{22} \end{bmatrix}$$

represents the quantity γ in the relevant coordinate system. In certain special coordinate systems—the Cartesian coordinate systems—the coefficients reduce to an especially simple form ($\gamma_{11} = \gamma_{22} = 1$, $\gamma_{12} = \gamma_{21} = 0$) and (4) becomes

$$\Delta 1^2 = \Delta x^2 + \Delta y^2 \qquad (4')$$

which is a version of Pythagoras's theorem. A formulation of the theory of Euclidean space which uses only Cartesian coordinate systems is a standard formulation of the theory.

$$<\mathbf{R}^2,\gamma>,<\mathbf{R}^2,\gamma'>,<\mathbf{R}^2,\gamma''>,\ldots$$

The quantities γ, γ', γ'' . . . transform into one another under transformation between different coordinations and jointly represent the one geometric object.

5.6 SYMMETRY PRINCIPLES

Symmetry principles provide a precise way of giving mathematical expression to important physical properties of space and time. In the theories of linear time and Euclidean space in Sections 5.4–5.5, symmetry principles express the homogeneity and isotropy of time and space. In the spacetime theories to follow, symmetry principles will also express the relativity principles of the theories.

The idea of symmetry used in analyzing these theories is no different in essence from the common notion of symmetry applied to everyday objects. One familiar type of symmetry is the bilateral symmetry exhibited (approximately) by the human form. To see the symmetry, imagine a transformation that switches the left- and right-hand sides of the body so that the left hand changes place with the right, the left foot with the right and so on. This transformation, a reflection about the central plane, is a symmetry of the human form since it leaves the form unchanged. Another type of symmetry is rotational symmetry exhibited, for example, by a cylinder. If we rotate the cylinder any number of degrees about its central axis, the rotated shape will coincide exactly with the unrotated shape (see Figure 5.10).

These examples illustrate the two essential elements of symmetry. First, one has a transformation, such as a reflection or rotation. Second, the transformation leaves something unchanged. The transformation is known as a *symmetry transformation* or, more briefly, a *symmetry* of that thing.

These same ideas can be applied to a Euclidean space as well. As a stepping-stone to this application, consider a pattern, such as we find on wallpaper. These patterns can exhibit symmetries. The pattern shown in Figure 5.11 exhibits a reflec-

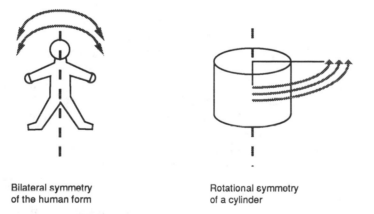

Bilateral symmetry
of the human form

Rotational symmetry
of a cylinder

Figure 5.10 Symmetries of common objects.

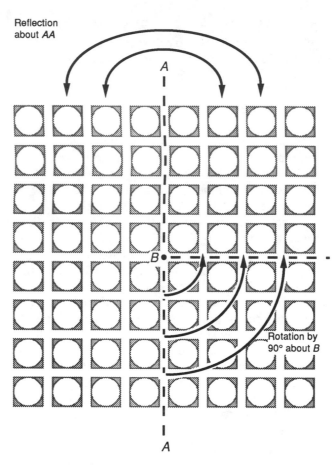

Figure 5.11 Symmetries of a decorative pattern.

tion symmetry since a reflection of the pattern about axis AA leaves the pattern unchanged. Similarly, the pattern exhibits a rotational symmetry. If the pattern is rotated by 90 degrees about the point B, then the pattern remains unchanged. In a Euclidean space $<\mathbf{R}^2, \gamma>$, the manifold $\mathbf{R}^2$ behaves like the paper and the metric γ is like the pattern painted on it. A transformation on this space that leaves the space unchanged is a symmetry transformation (or just symmetry) of the space. Three types of symmetry transformations are exhibited by this space as shown in Figure 5.12: a reflection about any axis, a rotation by any angle about any point, and a translation by any distance in any direction.

These symmetries of a Euclidean space express the space's homogeneity and isotropy. To say the space is homogeneous just means that every point and its geometry is exactly like every other point and its geometry. Thus if observers examine the geometry in the vicinity of one point of the space and then translate their viewpoint to any other point, then the geometry observed should remain unchanged.

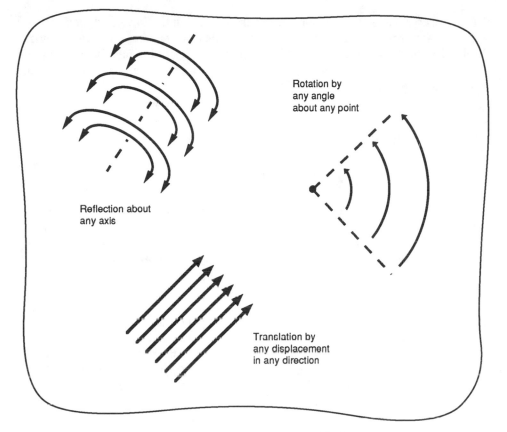

Rotation by
any angle
about any point

Reflection about
any axis

Translation by
any displacement
in any direction

Figure 5.12 Symmetries of a two-dimensional Euclidean space.

But this merely says that any translation on the space leaves the space unchanged. That is, any translation is a symmetry of the space. Similarly, to say that the space is isotropic just means that every direction in the space is exactly like every other.

If the covariance group of a formulation of a theory of time, space or spacetime coincides with the group of its symmetry transformations, then that formulation is a *standard formulation* of the theory. A formulation of the theory of Euclidean space restricted to Cartesian coordinate systems is a standard formulation. Standard formulations tend to be simpler mathematically. However, they can be misleading since explicit mention of the geometric structures present tends to be simplified out of the formulation's equations. Thus the Euclidean metric γ is rarely mentioned in a standard formulation of the theory of a Euclidean space.

Thus if observers examine the geometry of the space as it lies in some direction at any point and then rotate their viewpoint by any number of degrees, then the geometry observed in the new direction should be the same. Again this merely says that any rotation about any point on the space leaves the space unchanged so that all such rotations are symmetries of the space.

5.7 NEWTONIAN SPACETIME

5.7.1 Transition to a Spacetime Formulation

In this section, let us develop a generally covariant, spacetime formulation of Newton's theory of space and time, modified to be compatible with the principle of relativity. To have such a formulation of the Newtonian theory for work in philosophy of space and time is very important, even though the new formulation is more complicated than the traditional one. Much philosophical interest exists in comparing the Newtonian theory with the theories of special and general relativity. The relativistic theories are presented most clearly in their generally covariant, spacetime formulations—general relativity necessarily so since no other formulation is known. For our comparisons to be reliable, we must carry them out on theories formulated in the same way. Otherwise our conclusions may well pertain not to true differences between the theories but only to differences between their methods of formulation. Section 5.10 discusses some of the damage that has been done by failing to use uniform formulations in such theory comparisons.

5.7.2 Formation

The Newtonian spacetime theory is produced by combining the theory of linear time with that of Euclidean geometry and just a little further structure. We begin with a Newtonian universe and take "snapshots" of its contents at all instants. These snapshots are simply three-dimensional Euclidean spaces (although for the figures we continue to suppress the third dimension and represent the space as a two-dimensional Euclidean space). Since each snapshot is taken at a different time, each of them can comprise an instant in the linear time theory. We construct the Newtonian four-dimensional spacetime by taking each of the three-dimensional Euclidean spaces and "stacking them up" in a linear continuum (see Figure 5.13). If we picture the spacetime as a deck of cards, then the geometry on each card (instantaneous snapshot) is given by a Euclidean metric γ. The temporal structure, as we proceed through the deck from card to card (instant to instant), is given by the temporal metric dT.

The deck of cards pictured shows us exactly where the theory as described so far is incomplete. Many ways are possible to stack up cards, as shown in Figure 5.13. Which is the right one? If we have points $A, B, C \ldots$ at rest in the space, then an acceptable stacking is one that places the points $A, B, C \ldots$ in each instant exactly on top of one another so that points at rest can be pictured as straight lines penetrating vertically through the stack. Moving points can also be represented as lines that penetrate obliquely through the stack. (To see this, imagine a point which moves from A to B to C as time proceeds from 0 to 1 to 2. It will be represented by a line that intersects A on the snapshot at time 0, B at 1 and C at 2.) In particular, we will represent points that move uniformly in a given direction—that is, move inertially—as *straight* lines penetrating the stack obliquely.

The stack of instants forms a four-dimensional manifold, each of whose points is an event, a point in space at a given time. Each instant is a three-dimensional surface in that manifold, technically a "hypersurface." These hypersurfaces are sets of simultaneous events, so they are called "*hypersurface of simultaneity*." The lines representing moving and motionless points are their *worldlines*. The encode the entire history of each point's motion.

5.7.3 Principle of Relativity

The spacetime theory as described so far incorporates absolute rest. In assuming that there is only one correct way to stack the instantaneous snapshots, we have singled out the points A, B, C of Figure 5.13 as absolutely at rest. In section 5.1, we discussed the principle of relativity in terms of interpenetrating absolute and relative spaces. In the spacetime context, such spaces are represented by *frames of reference*. Consider the points of a relative space. Each point will be a worldline penetrating the stack of instants. The totality of points of the space will thus be represented by a dense bundle of worldlines penetrating the stack. If the space is an inertial space, then the corresponding bundle will be a bundle of straight lines as shown in Figure 5.14 and will be called an *inertial frame of reference*.

In effect Newton supposed that one of these inertial frames of reference was special and represented an absolute state of rest. Thus for him the only correct stacking of the surfaces of simultaneity would be one that aligned the points of this absolute frame. The principle of relativity requires that all inertial frames are to be equivalent so that all inertial states of motion are equivalent. Geometrically this amounts to saying that all directions in spacetime picked out by inertial frames are equivalent. Thus if we consider two inertial frames, such as in Figure 5.14, we should not think of either as having properties different from the other. Unfortunately, because of the limitations of drawing pictures of inertial frames, one frame is drawn as penetrating the stack vertically and the other obliquely. This difference is not reflected in the actual geometric structure of a Newtonian spacetime.

The situation is closely analogous to the isotropy of a Euclidean space. All directions in such a space are equivalent. However if we draw a picture of these directions, such as in Figure 5.15, one points up the page in the 0-degree direction and another across it in the 90-degree direction. Since a Euclidean space admits rotations as a symmetry, we can erase any suggestion that a given direction in the space is preferred by rotating the space so that the given direction is at the 0-degree position and noting that the space is unchanged.

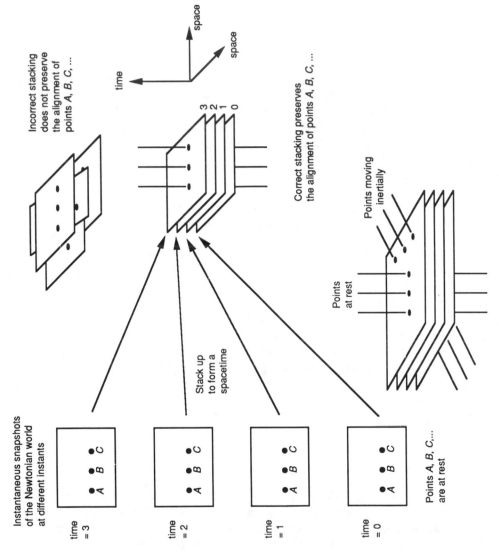

Figure 5.13 The formation of a Newtonian spacetime.

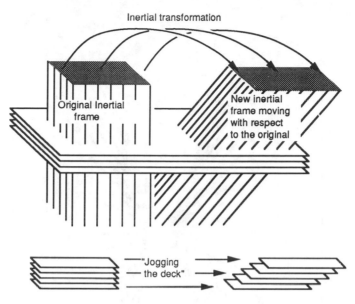

Figure 5.14 Inertial transformation in a Newtonian spacetime.

Similarly, the rules for stacking the deck of hypersurfaces of simultaneity must allow us to restack the deck so that it is aligned according to any inertial frame of reference. This means that any frame can be transformed to the zero velocity state. Let us call the transformation that maps inertial frames into inertial frames, shown in Figure 5.14, an "*inertial transformation*." Figuratively it corresponds to realigning the hypersurfaces of simultaneity in a manner akin to jogging a deck of cards. What we have concluded is that an inertial transformation cannot change the spacetime in the same way that a rotation does not change a Euclidean space, so that the picking out of any inertial frame as uniquely at rest is a purely arbitrary stipulation. That is, the principle of relativity is a symmetry principle:

> *Principle of relativity in a Newtonian Spacetime:* An inertial transformation is a symmetry of a Newtonian spacetime; it leaves the spacetime unchanged.

5.7.4 Models of a Newtonian Spacetime

To summarize, the models of a Newtonian spacetime are quadruples

$$<M, dT, h, \nabla>,$$

where M is a four-dimensional manifold each of whose points represents an event. This manifold is sliced into instants, that is, hypersurfaces of simultaneous events. The measurable time elapsed as we move from instant to instants is given by the temporal metric dT. Each of the hypersurfaces of simultaneity is a Euclidean space with its own Euclidean metric γ; the structure h combines all of them into a single

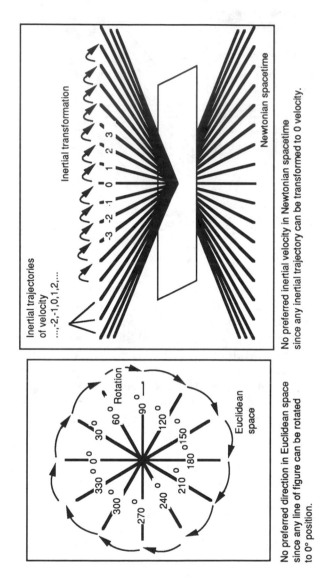

Figure 5.15 Symmetries of Euclidean space and Newtonian spacetime.

geometric object. Finally we need a structure which will dictate the allowed ways of stacking instants. That structure is the *affine structure* ∇ of the spacetime. The affine structure specifies which of all the curves in the four-dimensional manifold *M* are the straight lines. (Notice that neither structure introduced so far—neither the Euclidean metric of each hypersurface of simultaneity nor the temporal metric—gives us any way of determining which are the straight lines that penetrate through the hypersurfaces.) We require that the instants be stacked in such a way that the trajectories of inertially moving points coincide with the straight lines of the manifold's affine structure. This rule will be compatible with the principle of relativity, if we require that the affine structure ∇, as well as temporal and spatial metrics *dT* and *h*, admit inertial transformations as symmetry transformations.

> We recover a standard formulation of Newtonian spacetime theory by adopting standard coordinates *T* from the linear time theory and *X*, *Y* and *Z* from Euclidean century and combine them to form a coordination between the Newtonian spacetime and $\mathbf{R}^4$. The straight lines of the affine structure ∇ are now just what you would expect: the set of all lines given by the linear relations between the coordinates including $T = aX = bY = cZ$, for all real values *a*, *b* and *c*. A typical inertial frame is given by the set of all such straight lines parallel to the *T* axis. An inertial transformation that transforms this frame to a frame moving at velocity *V* in the *X* direction is given by
>
> $$T' = T, \qquad X' = X - VT, \qquad Y' = Y, \qquad Z' = Z.$$

5.8 SPECIAL RELATIVITY

5.8.1 Relativity of Simultaneity

Einstein developed his special theory of relativity in 1905 axiomatically as the consequences of two postulates: the principle of relativity and what we now call the *light postulate*. The latter postulate asserts that the velocity of light has the same constant value ($c = 300,000$ km/sec) in all inertial spaces. On first acquaintance, it seems that no theory free of logical contradiction could be based on these postulates. How could the velocity of light remain the same in all inertial spaces? Surely if we transform to inertial spaces moving successively faster in the direction of a light ray, the light ray's velocity must be diminished as we catch up with it until it is finally brought to a standstill. The light postulate asserts that we can never catch the light ray. No matter how fast we go in chasing it, it always moves away from us at the same speed, 300,000 km/sec. What Einstein realized was that this state of affairs was possible if we were prepared to forgo some commonly assumed properties of space and time. One of the most important concerned simultaneity. In the Newtonian theory it had been assumed that two events either were or were not simultaneous. In special relativity, things ceased to be so simple.

Consider Einstein's standard simultaneity relation defined in Section 5.3. Assume that clocks A and B of Figures 5.2 and 5.3 have been synchronized by Einstein's light signaling procedure so that they are in standard synchrony (at least according to an observer at rest with respect to them). If we now transform our viewpoint to an inertial space in which clocks A and B are moving together in the direction from A to B, we no longer agree that the two clocks are in standard synchrony. In the new inertial space, the light signal will have to traverse a greater distance on its outward journey than on its return journey. For on the outward journey it must catch a B-clock that flees from it, whereas on the return journey the A-clock rushes forward to meet it. If the light postulate is correct and the speed of light has the same constant value in the new inertial space in both directions, then the outward journey must take more time than the return journey, so that the event of the reflection of the signal at B cannot happen midway between its emission and return at A—at least according to an observer in the new inertial space. That is, the clocks cannot be in standard synchrony in the new inertial space.

Thus in special relativity judgements of whether two clocks are in standard synchrony and, therefore, whether two events are simultaneous depend on the choice of inertial space to which the judgements are referred. This result is known as the *relativity of simultaneity*. It should not be confused with the *conventionality* of simultaneity discussed in Section 5.3. The relativity of simultaneity applies even after a particular definition of simultaneity has been chosen, such as standard $\epsilon = 1/2$ simultaneity above, and arises when we change inertial spaces. The conventionality of simultaneity arises within a single inertial space.

5.8.2 Minkowski Spacetimes and the Lorentz Transformation

The four-dimensional spacetime formulation of special relativity was discovered by Hermann Minkowski in 1907. Its spacetimes are called Minkowski spacetimes in his honor. A Minkowski spacetime is much like a Newtonian spacetime. Both are based on four-dimensional manifolds of events. Moving points in each are curves, and points moving inertially are straight lines, so that inertial frames of reference are still bundles of parallel straight lines. However, the most prominent landmark of a Newtonian spacetime, its unique divisibility into hypersurfaces of simultaneity, is not present in a Minkowski spacetime. For the relativity of simultaneity entails that each inertial frame defines a different slicing of the spacetime into hypersurfaces of simultaneous events. A hypersurface of simultaneity of a given inertial frame of reference is said to be *orthogonal* to the curves of the frame.

The transformation between inertial frames of reference in special relativity is called a *Lorentz transformation*. The relativity of simultaneity makes it more complicated than in the Newtonian case shown in Figure 5.14. For in the Lorentz transformation, the slicing of the spacetime into hypersurfaces of simultaneity must be

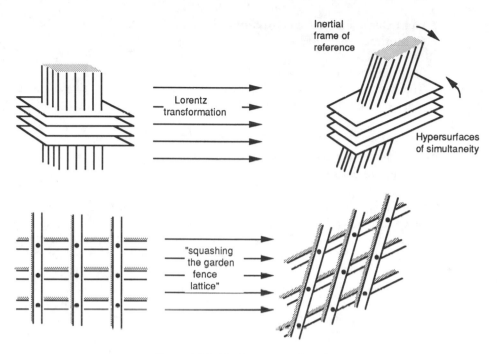

Figure 5.16 The Lorentz transformation

adjusted to the new frame so that the hypersurfaces of the slicing remain orthogonal to it. The transformation is pictured in Figure 5.16. Where the inertial transformation of a Newtonian spacetime is mechanically akin to "jogging a deck of cards," the Lorentz transformation is mechanically akin to "squashing the garden fence lattice."

Finally we note that the Lorentz transformation is a symmetry of the Minkowski spacetime so that the principle of relativity holds just as it does in the Newtonian spacetime of the previous section.

In a standard formulation of special relativity, the standard coordinates X, Y, Z and T correspond to space and time measurements made by instruments at rest in the frame whose worldlines are the T curves. The Lorentz transformation, which transforms this frame to one moving at velocity V in the X direction, is given by

$$T' = \beta(T - VX/c^2) \qquad X' = \beta(X - VT) \qquad Y' = Y \qquad Z' = Z$$

where

$$\beta = 1/\sqrt{1 - V^2/c^2}.$$

5.8.3 Light Cone and Causal Structures of a Minkowski Spacetime

An infinite number of curves pass through any given event of a Minkowski spacetime. The light cone structure of the Minkowski spacetime at that event is simply the division of the curves *at that event* into three classes: those that represent

1. points moving at velocity c, the velocity of light ("light-like");
2. points moving at velocity less than c ("time-like");
3. points moving at velocity greater than c ("space-like").

The name "light cone" arises from the fact that the light-like curves form a cone through the event as shown in Figure 5.17. The time-like curves all fall within the cone and the space-like curves outside the cone. The light cone structure of the entire Minkowski spacetime is the specification *at every event* of the above three-way division.

Time-like curves can be the worldlines of massive particles. Light-like curves can be worldlines of light signals. The usual assumption in special relativity is that no causal process such as a particle or signal can travel faster than light so that space-like curves *cannot* be the worldlines of any particle or signal. Under this assumption, the light cone structure takes on special significance for the philosophy of space and time, for it is equivalent to the causal structure of the spacetime. More precisely, if we know the light cone structure of the spacetime then we can construct an exhaustive catalog of which pairs of events can causally interact with one another in the spacetime. We do this by finding all pairs of events which could be connected by the trajectory of a particle or signal, that is, by a curve that is everywhere time-like or light-like (see Figure 5.18). The resulting catalog is the causal structure of the spacetime. Conversely, if we know this catalog, then we can reconstruct the light cone structure.

5.8.4 The Minkowski Metric

As a *spacetime* theory, the Newtonian theory is rather complicated. It requires three distinct structures to be specified: dT, h and ∇. As a spacetime theory, special relativity is far simpler. The functions of the three Newtonian structures is performed by just one, the Minkowski metric η. Thus models of special relativity are of the form

$$<M,\eta>$$

where M is a four-dimensional manifold and η a Minkowski metric. The properties of η are very similar to those of a Euclidean metric γ (see the following box) since η also assigns lengths—called "intervals"—to curves. The metric η picks out which are the time-like, space-like and light-like curves by the intervals it assigns to them. It assigns a zero interval to light-like curves. It assigns a positive interval to time-like

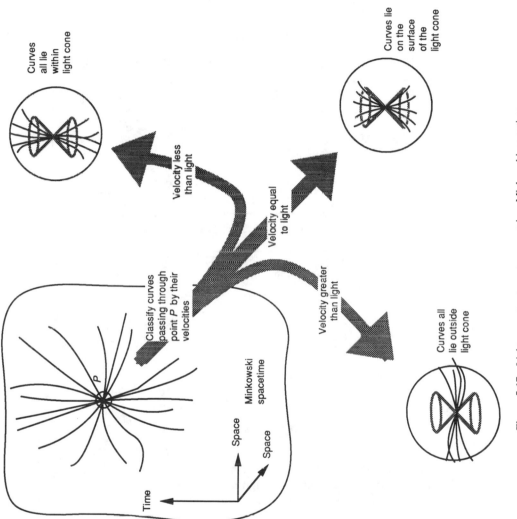

Figure 5.17 Light cone structure at an event in a Minkowski spacetime.

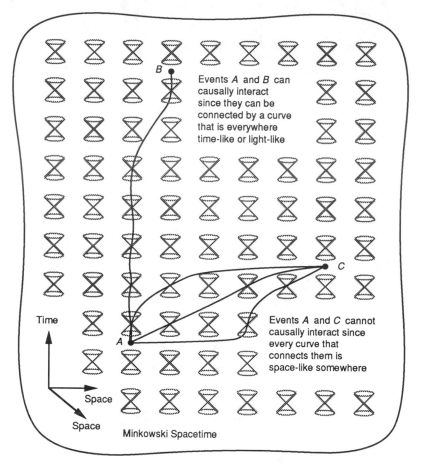

Events *A* and *B* can causally interact since they can be connected by a curve that is everywhere time-like or light-like

Events *A* and *C* cannot causally interact since every curve that connects them is space-like somewhere

Time

Space

Space

Minkowski Spacetime

Figure 5.18 Light cone structure determines causal structure.

In a standard coordinate system *X, Y, Z, T* of special relativity, the Minkowski metric is associated with the differential form

$$\Delta s^2 = \Delta T^2 - \Delta X^2 - \Delta Y^2 - \Delta Z^2 \qquad (5')$$

which is fully analogous to the differential form (4′) of the Euclidean theory. The differences between the metrics of the two theories derive entirely from the differences in sign between forms (4′) and (5′). The transition to the generally covariant formulation introduces four arbitrary spacetime coordinates, x^0, x^1, x^2, x^3 and the equation (5′) generalizes to

$$\Delta s^2 = \eta_{00}(\Delta x^0)^2 + \eta_{01}\Delta x^0 \Delta x^1 + \ldots + \eta_{23}\Delta x^2 \Delta x^3 + \eta_{33}(\Delta x^3)^2 \quad (5)$$

which is analogous to (4). The explicit representation of the metric η is the symmetric matrix of coefficients η_{ik}, where $i,k = 0, 1, 2, 3$.

Philosophy of Space and Time

curves. This interval is the time elapsed as measured by a clock moving with the particle represented by the curve. The Minkowski metric assigns an imaginary interval to space-like curves. The absolute value of this interval is the spatial length of the curve should the curve lie fully in a hypersurface of simultaneity.

5.9 GENERAL RELATIVITY

5.9.1 Physical Foundations

General relativity is Einstein's relativistic gravitation theory and is a modification of special relativity that incorporates gravitation. It was completed by him in 1915 and is probably his greatest contribution to physics. The novelty of the theory is the way that gravitation is treated. Prior to general relativity, it was customary to think of a gravitational field as a distinct entity that could be added to a spacetime. Thus gravitation-free spaces were possible. In general relativity, the gravitational field is combined with the same structure that determines lengths and times so that a gravitation-free space is no longer possible.

The chain of ideas that led Einstein to general relativity began in 1907 when he was struck by a remarkable property of gravitation known since the time of Galileo. When a gravitational field deflects the motion of a body, the amount of deflection is independent of the nature of the body and, in particular, the mass of the body. This property is a very special property of gravitational fields and is not shared, for example, by electric fields. If the motion of a charged body is deflected by an electric field, then the greater the charge on the body, the larger the deflection. It was as though the trajectories of bodies falling in a gravitational field were already laid out in spacetime and any falling body would have to follow them, whatever its mass. Now a Minkowski spacetime just happens to have trajectories with exactly this property. These are the trajectories of inertially moving points, the straight time-like worldlines defined by the Minkowski metric. Any body moving inertially in a Minkowski spacetime follows these trajectories in a way that is independent of the mass of the body. Since these trajectories have exactly the unique, characteristic property of gravitation, Einstein was drawn to conjecture that a Minkowski spacetime was actually already a special case of a spacetime with a gravitational field and that spacetimes with more general gravitational fields could be constructed not by adding further structures to the spacetime but by modifying what was already there.

5.9.2 Principle of Equivalence

This conjecture was formulated and justified in a vivid manner in a thought experiment. Einstein imagined a physicist enclosed in a box in the supposedly gravitation-free space of special relativity. He then imagined that the box was accelerated uniformly in some direction. All free objects in the box would fall to one side with the same acceleration. The observing physicist, Einstein argued, could explain this phenomenon in two equally good ways. He could say that the box was accelerated. Alternately, because of the special property of gravity, he could say that the box was unaccelerated but that a homogeneous gravitational field was acting on the box. Einstein's "principle of equivalence" asserts that the two states of affairs—uniform acceleration in a gravitation-free space and a homogeneous gravitational field—are fully equivalent or, in our words, exactly the same state of affairs. Reduced to its briefest form, the thought experiment shows us that supposedly gravitation-free special relativity already incorporates gravitation—to see that gravitation is already there, transform to a uniformly accelerated space to make a homogeneous gravitation field manifest.

5.9.3 Generalizing Special Relativity

What characterizes the gravitational fields of special relativity is the following property: If two test bodies have initial velocities identical in magnitude and direction, they will continue to move so that the distance between them remains the same (see Figure 5.19). We are interested in more general gravitational fields such as those produced by the earth. In these more general cases, the distance between the above two bodies would not remain constant but would converge or diverge as the bodies fell. To construct general relativity, Einstein replaced the Minkowski metric η of special relativity with a more general metric g which would allow this convergence or divergence. In the new theory, unrestrained particles still follow the straight time-like curves of the spacetime, just as they did in a Minkowski spacetime. However the "straight" lines defined by the new more general metric g no longer behave in the way that we expect straight lines to behave. For example, two "straight" lines that are initially parallel need not remain at a constant distance from one another as

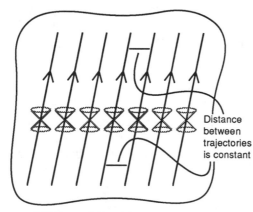
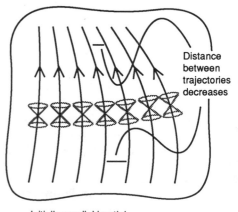

Initially parallel inertial trajectories in a Minkowski spacetime of special relativity

Distance between trajectories is constant

Distance between trajectories decreases

Initially parallel inertial trajectories in a spacetime of general relativity

Figure 5.19 Inertial trajectories in special and general relativity.

Philosophy of Space and Time

we proceed along them (see Figure 5.19). Such results are typical in the geometry of curved surfaces, such as the surface of a sphere, and the mathematical techniques used in general relativity were originally developed in the context of problems of curved surfaces. As a result, talk of "curvature" is common and we routinely distinguish the "flat" spacetime of special relativity from the "curved" spacetimes of general relativity.

In sum, the models of general relativity have the form

$$<M,g>$$

where M is a four-dimensional manifold and g is a generalization of the Minkowski metric η. Since every distinct distribution of masses in the universe produces a distinct gravitational field, there will be very many different models in the theory. In particular, a nonuniform matter distribution will produce a nonuniform gravitational field. As a result, the models of general relativity will, in general, have no nontrivial symmetries, so that we cannot formulate relativity principles of the type seen in the flat Newtonian spacetime theory and special relativity.

Part III: Applications

5.10 CONFUSIONS OVER COVARIANCE

Misunderstandings of the significance of the covariance group of a theory have been responsible for more than their fair share of confusions in philosophy of space and time. Let us review two important examples.

5.10.1 The Generalization of the Principle of Relativity

One of Einstein's best known claims for his general theory of relativity is that it extends the principle of special relativity to accelerated motion. We noted in the previous section that the spacetimes of general relativity admit no nontrivial symmetries in general, so that we cannot formulate a relativity principle of the type formulated in Newtonian theory or special relativity. Thus Einstein's claim has proved increasingly difficult to defend and its defense has required stratagems of increasing complexity. (Friedman 1983 makes the case against the claim especially clear.) The simplest and most common argument for the claim is not a good one. It merely notes that general relativity is a generally covariant theory. However, general covariance by itself cannot sustain the claimed generalization of the principle of relativity since every spacetime theory we have examined in this chapter has been given generally covariant formulation. They cannot all satisfy a generalized principle of relativity! The illusion that general covariance and an extension of the principle of relativity are synonymous depends most commonly on the simple mistake of incautiously comparing two theories formulated in different manners: general relativity in its generally covariant formulation with special relativity in a standard (i.e., nongenerally covariant) formulation. In its standard formulation, the Lorentz group is both the theory's covariance group and its symmetry group, the group of its symmetry transformations. As we have seen, the principle of relativity is associated with the symmetry group so that a theory that extends the principle would need to expand that symmetry group. In the transition to general relativity, we do expand the covariance of the theory from Lorentz covariance to general covariance, but since the geometric structure of general relativity in general admits no nontrivial symmetries, we actually reduce the symmetries admitted by the theory. Those who have failed to keep the symmetry and covariance groups of special relativity conceptually distinct easily fail to see the significance of this reduction and fall into the trap of thinking that they have also somehow automatically extended the principle of relativity. Had the two theories been compared from the start with both in their generally covariant formulations, this problem might never have arisen.

5.10.2 Conventionality of Simultaneity

Winnie (1970) showed that we can generalize a standard coordinate system of special relativity to a new coordinate system with time coordinate t_ϵ in such a way that events with equal t_ϵ are judged simultaneous by some ϵ-criterion. It is sometimes thought that this fact *by itself* is sufficient to vindicate the conventionalist claim. This is obviously false since all that has been shown is that we can extend the covariance of the theory so that it can use t_ϵ coordinate systems. We have seen that it is possible to extend the covariance of the theory even further to general covariance, which allows arbitrary coordinate systems. Indeed we have seen that we can give generally covariant formulations of every spacetime theory considered so far. If we can automatically read the t coordinate of any of these formulations as giving a criterion of simultaneity, then we could vindicate the strangest of simultaneity relations, including nonstandard simultaneity relations even in Newtonian spacetimes. What is needed is some independent means of arguing that the t coordinate of a given formulation does represent a possible simultaneity relation, such as the causal theory of time seeks to provide for t_ϵ.

5.11 MALAMENT'S RESULT

One of the most dramatic turns in the debate over the conventionality of simultaneity was provided by Malament (1977a). Contrary to most expectations, he was able to prove that the central claim about simultaneity of the causal theorists of time was false. He showed that the standard simultaneity relation was the only nontrivial simultaneity relation definable in terms of the causal structure of a Minkowski spacetime of special relativity.

Let us give a more precise version of Malament's result and outline the ingenious method he used to establish it. To begin, recall that we saw in Section 5.8 that the causal structure of a Minkowski spacetime is equivalent to its light cone structure. Recall also that the standard simultaneity relation is inertial frame dependent so that unless we specify an inertial frame in some way we should expect no interesting results at all. Malament picks out an inertial frame by specifying one of its worldlines O as the worldline of the Observer for whom the simultaneity relation is to be defined. Thus the basic question becomes:

What simultaneity relations are definable in terms of the light cone structure of a Minkowski spacetime and the worldline O of an inertially moving observer?

Malament first shows that

The relation of standard simultaneity is definable in terms of O and the light cone structure.

The proof involves the construction shown in Figure 5.20. We pick any event e on O and seek the hypersurface of events s simultaneous to e in the inertial frame of O according to the standard criterion. We have found that hypersurface if the following condition is satisfied. Let a be any event on O prior to e. The set of all possible light signals emitted from a must intersect s and, when they are reflected back to O upon intersection, they must all arrive at O at the same event b. The hypersurface s is, of course, orthogonal to O.

Malament's central result is that

The relation of standard simultaneity is the only binary relation definable in terms of the light cone structure and the worldline O provided

(i) the relation cannot be trivial insofar as it relates every event to every other event, or fails to relate events on O to events not on O;
(ii) the relation is an equivalence relation.

Condition (ii) is required if the relation is to partition the events of the manifold into disjoint sets of mutually simultaneous events such as, for example, the hypersurfaces of simultaneity of the standard case.

The proof of the result depends on the fact that the worldline O and the light cone structure admit certain symmetries. For example, in the rest frame of O these structures single out no preferred spatial direction and thus remain invariant under spatial rotation about O. Thus any relation defined exclusively in terms of O and the light cone structure will be unable to pick out a preferred direction and, therefore, must admit the same rotational symmetry. So if p and q are related by the simultaneity relation and f is any rotation about O, then the rotated events $f(p)$ and $f(q)$ must also be simultaneous (see Figure 5.21). We can now repeat this argument for all the

224

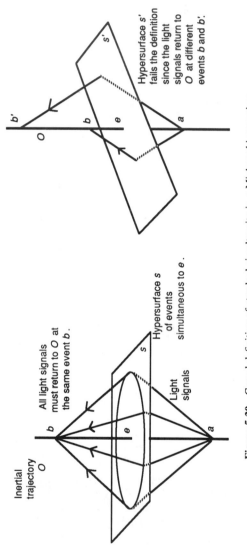

Figure 5.20 Causal definition of standard simultaneity in a Minkowski spacetime.

remaining symmetries of O and the light cone structure. The basic idea is that every symmetry of O and the light cone structure must also be symmetries of any simultaneity relation it defines. These symmetries are the translations, scale expansions (uniform magnifications and reductions) and reflections about a hypersurface orthogonal to O, all of which must map O back into itself. They are shown in Figure 5.22. Malament then showed that the standard simultaneity relation is the only binary relation satisfying (i) and (ii) and remaining invariant under these symmetries. Without going through the proof, we can easily satisfy ourselves of the plausibility of the result. Assume that the simultaneity relation is such that it will slice up the spacetime into hypersurfaces of mutually simultaneous events, as in Figure 5.22. Then it is intuitively evident that only a slicing by orthogonal hypersurfaces will remain invariant under the symmetries listed.

The major weakness of Malament's analysis lies in the sensitivity of his basic result even to small changes in the conditions assumed. The analysis depends on the assumption that the simultaneity relation be definable by the following list of structures:

light cone structure, the inertial worldline O.

It is crucial that this list be preserved since the slightest change in it seems to be sufficient to defeat Malament's basic result. For example, we could ask what simultaneity relation is definable if we add to the list another inertial worldline O' with a velocity differing from O. Through the construction of Figure 5.20, we can define at least the standard simultaneity relation of the O frame and the standard simultaneity relation of the O' frame; but the latter is a nonstandard relation with respect to the O frame. More generally, Peter Spirtes (1981, Chapter 6) has shown

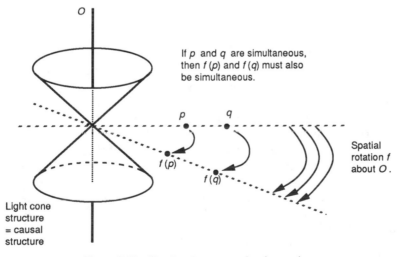

If p and q are simultaneous, then $f(p)$ and $f(q)$ must also be simultaneous.

O

p q

$f(p)$

$f(q)$

Spatial rotation f about O.

Light cone structure = causal structure

Figure 5.21 Simultaneity preserved under rotation.

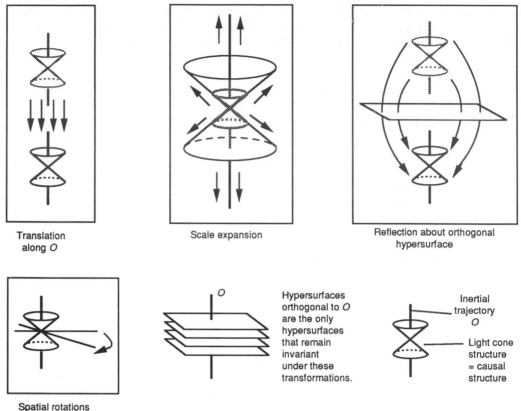

Figure 5.22 Symmetries of the light cone structure of a Minkowski spacetime and an inertial trajectory O.

that merely adding a temporal orientation to the list—that is, the ability to distinguish past from future—is sufficient to enable definition of infinitely many nonstandard simultaneity relations. However, before modifying the construal of causal definability by adding or subtracting from the list, we would need to find very good reasons for doing so.

Adolf Grünbaum (private communication) has pointed out that the need to postulate (ii), that the relation is an equivalence relation, rather than derive it, is another weakness of Malament's challenge to the causal theorists. It eliminates by decree any simultaneity relation that does not partition the spacetime into disjoint sets of mutually simultaneous events. An example of such a relation is the relation "is not causally connectible" which has been called the relation of "topological simultaneity" in the literature (Grünbaum 1973, 203). The latter relation fails to be transitive: Events A and B may each not be causally connectible to a third event, while being causally connectible to each other on a time-like worldline. Therefore this relation cannot partition the spacetime into disjoint sets of mutually simultaneous events.

5.12 REALISM ABOUT SPACETIME STRUCTURES

5.12.1 Spacetime Substantivalism and the "Hole" Argument

Isaac Newton is usually singled out as the canonical realist in the context of theories of space and time and most especially so for his treatment of the absoluteness of his absolute space and absolute time. Their absoluteness arises in a number of senses which have been dissected admirably in Earman (1989). The sense we are concerned with here is that of independence. Absolute space, as we saw in Section 5.1, and absolute time are asserted to have existences entirely independent of the things they contain. This doctrine is the "substance" view or the "substantivalist" view. It owes its somewhat unfortunate name to the view that substance is that which can exist independently. A better name, with fewer distracting connotations, might have been simply the "independence" view. Clearly, the substantivalist position can be formulated analogously for spacetime theorists.

The view is an extreme form of realism concerning spacetime. It arises fairly naturally for realists who seek to construe theories of spacetime as literally as possible. Such a construal automatically sees the divisions between the different structures of a theory as reflecting natural divisions between the actual structures of the physical world. The substantivalist position gives expression to the reality of one of the most important divisions in physical theories, that between spacetime and the matter it contains. The position has become increasingly attractive with the revival of realism in philosophy of science and the problems facing the nonrealist programs of conventionalism and relationalism in spacetime theories.

The "hole" argument (Earman and Norton 1987) is based on ideas advanced by Einstein in 1914, 1915 and 1916 and seeks to establish that acceptance of spacetime substantivalism in a very broad class of spacetime theories forces acceptance of an odious form of indeterminism. (See Chapter 6 for a discussion of determinism.) In informal terms the argument establishes that the substantivalist is forced to insist that there are differences between certain physically possible worlds, even though not just observation but the laws of the theories themselves cannot pick between them.

5.12.2 Presuppositions of the Argument

To make the argument more precise, we must settle several questions left vague. The term "spacetime" is ambiguous insofar as it is unclear as to what specific entity it refers. Let us assume that "spacetime" means the manifold M of our models so that the substantivalist attributes the substantival properties to M or to what M represents in the physically possible worlds. (Other choices are possible here, and in many such cases the hole argument can still be made to apply, as shown in Norton 1989.) The "very broad class of spacetime theories" mentioned is what we call "local spacetime theories." These are generally covariant, spacetime theories of the type considered in this chapter, including versions of Newtonian spacetime theory, special and general relativity. The most important instance of a theory to which the argument applies is general relativity, our current best spacetime theory, which is available only in local formulation. We will develop the hole argument as it applies to general relativity.

Finally we need some more precise theoretical statement of the substantivalist doctrine. The phrase "independent existence" conjures pictures to our intuitions, but without restatement it cannot be analyzed by the machinery of this chapter. Unfortunately there seems to be no precise and satisfactory construal of the doctrine. The claim that spacetime, represented by a bare manifold M, can exist independently of its contents translates naturally to the claim that there is a possible world modeled by the bare manifold M. However, this claim is routinely denied by every spacetime theory we have seen so far. They invariably require that the manifold M be supplemented by further structures in order to produce models of physically possible worlds. Fortunately we do not need a precise construal of spacetime substantivalism to complete the argument. We need only a necessary commitment of spacetime substantivalists.

That commitment arose in Leibniz's famous debate with Newton's representative Samuel Clarke. In their correspondence, Leibniz asked if the world would be changed if God had placed its bodies into space in such a way that East and West were exchanged but all other relations between the bodies were preserved. Leibniz noted that there would be no discernible difference and he urged that no change had actually been effected. However, he realized that the Newtonian substantivalist must nonetheless insist that the world would be different for its bodies would now be located in different spatial locations.

In spacetime theories, the analogue of Leibniz's spatial rearrangement of bodies retaining all other relations between them is a transformation on the manifold M and associated transformation of geometric structures defined on M. If $<M,g>$ is a model of general relativity and h a transformation on M then h transforms g into the new metric $g' = h(g)$. From the general covariance of general relativity, we know that $<M,g'>$ will also be a model of the theory. The two metrics g and g' will in general assign different metrical properties to the same event in M. Spacetime substantivalists must take seriously this rearrangement of properties on the spacetime manifold M and they must hold that the two models $<M,g>$ and $<M,g'>$ represent different possible worlds. That is, spacetime substantivalists must *deny*:

> *Leibniz equivalence*: Two intertransformable models of a spacetime theory, such as $<M,g>$ and $<M,g'>$, represent the same possible world.

This denial is immediately awkward for the substantivalists. Since the metrics g and g' are intertransformable, they are clones of one another. For every property of g we can find a corresponding property of g' by consulting the transformation h. These corresponding properties include all observable properties, so that both g and g' agree on all observables. (The metrics g and g' disagree only on how their properties are to be spread over the manifold M and these differences of spreading cannot be translated into observable differences.) Thus substantivalists must insist that $<M,g>$ and $<M,g'>$ represent two distinct worlds, even though they are worlds whose differences could not be discerned by any observation. In the heyday of logical positivism and the verifiability criterion, this conclusion alone would have been sufficient

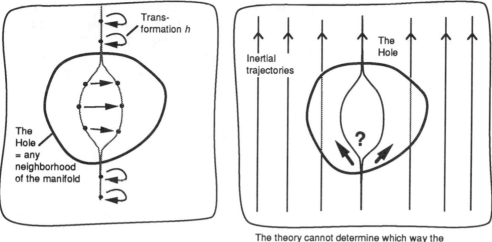

The theory cannot determine which way the freely falling particle will traverse The Hole.

Figure 5.23 The hole argument.

grounds for rejection of the substantivalist position. The hole argument, however, leads the substantivalist to an even worse result.

5.12.3 The Argument

To arrive at the odious form of indeterminism promised, we select any neighborhood of the manifold M. We call it the "hole" for historical reasons associated with Einstein's first use of an early form of the argument. We select any transformation h on M which is the identity outside the hole but comes smoothly to differ from it inside the hole (see Figure 5.23). Then g and $g' = h(g)$ will be the same everywhere outside the hole but will come smoothly to differ within the hole. It now follows that even with a full specification of the spacetime everywhere outside the hole, the theory will be unable to tell us how the spacetime will develop into the hole. For if the model of the spacetime assigns the metric g to the manifold outside the hole, then the theory will allow the metric to develop as either g or g' into the hole and cannot determine which is the correct development.

If we recall that the metric determines the inertial trajectories of the spacetime, then we can see just how disastrous is this result. Given the fullest specification of the spacetime outside the hole, the theory will be unable to determine the trajectory along which a particle in free fall will traverse the hole, even though its trajectory before and after the hole is known exactly. As is explained in Chapter 6, this is an extremely awkward form of indeterminism, for the hole might be both of very small spatial size and temporal duration. Even given a full specification of the fields in its future, past and everywhere else in space, the theory is still unable to specify what happens inside the hole.

The substantivalist is driven to this indeterminism by the need to deny Leibniz equivalence. If the substantivalism were to be given up, Leibniz equivalence could be accepted. Then both the original model and its diffeomorphic copy could be said to represent the same physically possible world, and the indeterminate nature of the development of the fields into the hole would be a mathematical curiosity of no physical significance. Otherwise the substantivalist must adhere to the physical distinctness of two states of affairs whose distinctness is opaque to both observation and the laws of the theory in question.

DISCUSSION QUESTIONS

1. Compare the application of the verifiability criterion as described in Section 5.1 in the context of the principle of relativity with some of its other applications.
2. How are we to approach two theories of space and/or time which have identical observational consequences? Consider whether we are free to choose conventionally between them. (You may find it helpful to consider the examples of Newton's theory of space and time with and without absolute rest and Euclidean geometry with vanishing and nonvanishing universal forces.)
3. Outline some of the virtues and vices of the reduction of temporal or spatiotemporal structure to causal structure offered by the causal theory of time.
4. Adjudicate in the debate between a conventionalist and a realist over the geometry of space or the simultaneity relation in special relativity.
5. Compare the axiomatic way of formulating theories of space and time (such as used by Euclid and many others) with the model theoretic or "semantic" method used in this chapter.
6. Einstein often acknowledged that his discovery of the theories of relativity owed a debt to the reading of various philosophers, notably Hume and Mach. Read the introductory sections of Einstein ([1905] 1952b) and Einstein ([1916] 1952a) (they are not at all hard to follow!) and try to identify those parts of his development dependent on overt philosophical considerations and, if you can, pin down their source.

SUGGESTED READINGS

EARMAN, JOHN (1989), *World Enough and Space-Time: Absolute versus Relational Theories of Space and Time*. Cambridge, MA: Bradford/MIT Press. More advanced work.

EINSTEIN, ALBERT ([1917] 1954), *Relativity: The Special and the General Theory*. 15th ed. Translated by Robert W. Lawson. New York: Crown. Einstein's popular exposition of relativity.

———. ([1922] 1956), *The Meaning of Relativity*. 5th ed. Translated by Edwin Plimpton Adams, Ernst G. Straus and Sonja Bergmann. Princeton: Princeton University Press. Einstein's own text on relativity theory.

FRIEDMAN, MICHAEL (1983), *Foundations of Space-Time Theories: Relativistic Physics and Philosophy of Science*. Princeton: Princeton University Press. More advanced work. Self-contained exposition of spacetime theories.

GEROCH, ROBERT (1978), *General Relativity from A to B*. Chicago: University of Chicago Press. Modern popularization of general relativity.

GRÜNBAUM, ADOLF (1973), *Philosophical Problems of Space and Time*. 2d ed. Dordrecht: Reidel. Classic, wide-ranging source.

HORWICH, PAUL (1987), *Asymmetries in Time: Problems in the Philosophy of Science*. Cambridge, MA: Bradford/MIT Press. Good introduction to the problem of the direction of time.

PAULI, WOLFGANG ([1921] 1958), *Theory of Relativity*. Translated by G. Field. New York: Pergamon Press. Classic text on relativity theory in the older style.

REICHENBACH, HANS (1956), *The Direction of Time*. Edited by Maria Reichenbach. Berkeley and Los Angeles: University of California Press. Classic source for the problem of the direction of time.

————. ([1928] 1957), *The Philosophy of Space and Time*. Translated by Maria Reichenbach and John Freund. New York: Dover. Classic source, especially for conventionalist viewpoint.

SALMON, WESLEY C. (1980), *Space, Time and Motion: A Philosophical Introduction*. 2d ed. Minneapolis: University of Minnesota Press. Good general introduction.

SKLAR, LAWRENCE (1974), *Space, Time, and Spacetime*. Berkeley and Los Angeles: University of California Press. Good general introduction.

TAYLOR, EDWIN F. AND JOHN ARCHIBALD WHEELER (1963), *Spacetime Physics*. San Francisco: Freeman. Modern standard text on special relativity.

WALD, ROBERT M. (1984), *General Relativity*. Chicago: University of Chicago Press. Modern standard text on general relativity.

Six

DETERMINISM IN THE PHYSICAL SCIENCES

John Earman

In this chapter we examine the doctrine of determinism, both because it is of intrinsic interest and because it serves as a vehicle for introducing a number of foundations problems in classical, relativistic, and quantum physics. Those whose primary interest lies in the determinism–free will problem may wish to read Sections 6.1–6.5 and then skip to Section 6.17 where free will is discussed.

6.1 A THOUGHT EXPERIMENT

In order to get a sense of what the doctrine of determinism involves and of why it has exerted such a strong appeal, you are invited to perform the following thought experiment. Describe the state s_1 of the room in which you are now reading this book—give the location of the furniture, the lighting conditions, the temperature distribution, and so forth. Wait twenty-four hours and record your description of the new state s_2. Now imagine that sometime in the future you find that your description of the state s_3 matches the previous description for s_1. Again wait twenty-four hours and record your description of the new state s_4. Would you then expect your descriptions of s_2 and s_4 to match?

Not necessarily. But normally one would expect that any difference between the descriptions of s_2 and s_4 can be traced to one of three sources. First, your descriptions of s_1 and s_3 may not be fine grained enough. For example, the macroscopic temperature distribution may be the same for the two days in question, but the velocities of some of the air molecules may be different and these differences may eventuate in noticeable macroscopic differences on the following days. Since we are conducting a thought experiment, we are free to imagine that the descriptions of the states are so detailed that the sameness of the descriptions guarantees the sameness of the physical states. Second, the system contains a "free agent"—you—and you might exercise your free will and decide to get up and walk around the room in the second instance while in the first instance you remained slumped in your chair. This supposition already begs a central question of the determinism–free will problem, for one major philosophical tradition holds that determinism is incompatible with free will. We will return to this problem in Section 6.17. Since we are now engaged in an attempt to gain an understanding of what determinism means, it is fair to ignore the free will problem and to imagine that the system contains only inanimate objects. Third, the difference between s_2 and s_4 might be traced to the fact that the system is subject to external influences. Locking the doors, shuttering the windows, and insulating the walls serve to screen out some but not all of these influences. No matter how the room is constructed, its envelope will not shield against gravitational forces, and according to Newton's theory of gravity, every massive bit of matter in the universe exerts a gravitational tug on every other bit. The only way to meet this difficulty is to exercise the license of thought experiments and to imagine either that there are no outside influences or that the room has been expanded until it has swallowed up all those influences.

With these three loopholes plugged, the normal expectation is that the sameness of s_1 and s_3 will be matched by the sameness of s_2 and s_4. Such an expectation has been elevated to the status of a metaphysical principle, as in G. W. Leibniz's Principle of Sufficient Reason, according to which nothing happens without a sufficient reason why it should be so rather than otherwise. Although Leibniz sometimes emphasized the theological interpretation of Sufficient Reason, according to which God's choice of which possible world to actualize is motivated by the desire to bring into being the best of all worlds, he also intended it to have a causal meaning, as when he formulated it as the principle that *"there is nothing without a reason, or no effect without a cause."* (Leibniz 1970, 268) With the three loopholes closed, a difference between s_2 and s_4 without a difference between s_1 and s_3 would be an effect without a cause. Consequently, Immanuel Kant would have seen a violation of his Law of Universal Causation according to which everything that happens presupposes something from which it follows according to a rule.

It might be objected that we have neglected a potential fourth loophole in that the differences in the times at which s_1 and s_3 occur can make a difference for the subsequent developments of the system. Leibniz would have had no sympathy for this point of view. For Leibniz, time itself cannot be a cause because on his relational conception of time, time is nothing over and above the order of successive states of the world. A twentieth-century version of these sentiments is to be found in Herbert Feigl's remark that

> "Same causes, same effects" makes sense only if there is such a neutral medium as space-time which thus is no more than a *principium individuationis*. Differences in effects must always be accounted for in terms of differences in the *conditions*, not in terms of purely spatio-temporal location. (1953, 412)

This is an example of how determinism is linked to issues in the philosophy of space and time; we will encounter other examples below.

Setting aside metaphysics for methodology, it has been held that fruitful scientific inquiry is promoted by the assumption that determinism is true. Thus, Max Planck wrote that the law of causality is a "heuristic principle, a signpost and in my opinion the most valuable signpost we possess, to guide us through the motley disorder of events and to indicate the direction in which scientific inquiry must proceed in order to attain fruitful results." (1932, p. 26; translation mine) One can easily appreciate the scientific horse sense of Planck's remark. To begin inquiry under the opposite presupposition—that differences in subsequent development of system need not be due to differences in the earlier history—seems to amount to a counsel of scientific despair, an admission that the world is a motley confusion of events rather than a collection of orderly and predictable processes. The scientific discernment of the order—the "fruitful results" of which Planck spoke—consists in large part in the discovery of the laws of nature that govern the temporal evolution of the systems under study. If the laws so far discovered fail to guarantee a deterministic evolution, then, Planck is saying, that is a sign that more laws remain to be discovered.

The problem of laws of nature was raised in Chapter 1. Nothing said in the present chapter will help to resolve the problem except in the sense that an interesting concept of determinism militates in favor of two features of the solution. First, determinism as understood here presupposes that there is a distinction between laws and initial conditions, and to the extent that the correct account of laws implies that this distinction is vague, then so too is the doctrine of determinism. Second, if there is no distinction between lawful and accidental generalizations, and any true empirical generalization counts as a natural law, then the claim that the world is deterministic degenerates to a near triviality.

Supposing that we knew what the laws of nature are, it remains to say in more precise terms what it means for these laws to be or fail to be deterministic. This task will be tackled in the following section in terms of a concrete example.

6.2 NEWTONIAN GRAVITATIONAL THEORY

Imagine a world whose physical contents consist entirely of material particles idealized as point masses. The mass of each particle is a positive number which, by the laws of nature, remains constant through time. Newton's second law of motion asserts that the product of the mass of the particle by its acceleration at any moment is equal to the net impressed force acting on the particle at that moment. (Velocity is defined as rate of change of position, and acceleration is defined as rate of change of velocity.) This assertion is empty until supplemented by another law specifying the nature of the forces. For the case of gravitational interactions, Newton's law of gravitation states that the net force acting on particle i is the sum of the forces exerted by all the other particles and that the gravitational force exerted upon particle i by particle j is an attractive force acting along the line joining i and j and is proportional to the product of their masses m_i and m_j and inversely proportional to the square of the distance between them. This exhausts the physics of Newtonian gravitational theory for point masses. The demonstration of the deterministic character of the theory belongs to pure mathematics. By the *state* at time t of a (finite) system of particles let us understand the specification of the position and momentum (= mass × velocity) of each particle at t. A *history* of the system for some interval of time is a specification of the state at each instant in the interval. A history is said to be *physically possible* relative to some set of laws just in case it satisfies those laws. In the case in point, we check for physical possibility by computing the acceleration $a_i(t)$ of each particle i at each instant in the interval and then verify that the product $m_i a_i(t)$ equals the Newtonian gravitational force on i at t. The system is said to be *Laplacian deterministic* (for some time interval) just in case for any pair of physically possible histories (for that interval) sameness of state at any instant in the interval implies sameness at any earlier or later time in the interval. The mathematical theory of ordinary differential equations guarantees, first, a richness of physically possible histories: for any given state (with noncoincident particle positions), there exists a physically possible Newtonian history which extends the state into the past and future. And, second, it guarantees the uniqueness of the extension: For any given state, there exists only one extension. (Actually, these theorems hold only locally in time and guarantee existence and uniqueness only for a finite interval. The implications of this point will be discussed below in Section 6.7.)

One can naturally wonder why "state" was defined in the above way. The answer, of course, is that this definition lends itself to the deterministic character of Newtonian mechanics. This self-serving answer does not beg the question of determinism since we can raise the question with respect to different sets of possible state variables. The only caution is that it is essential that the definition of state be confined to genuinely instantaneous values, for obviously if bogus variables that code up information about the past and future were allowed, determinism would trivialize.

The label "Laplacian" was attached to the above definition both to credit Pierre-Simon Laplace, an early champion of this form of determinism (see Section 6.5 below), and also to indicate that there are other varieties of determinism. For example, the laws might be such that while the instantaneous state of a system does not suffice to determine the history, a finite segment may. For present purposes, let us concentrate on the Laplacian brand of determinism, but the reader is invited to speculate on other forms. Note also that the above definition requires a unique prolongation into both the past and the future. To be more detailed and pedantic, we could distinguish between *future* and *past Laplacian determinism*, which require that sameness of state at any time implies sameness at any later (respectively) earlier time. The distinction is purely pedantic if the laws of nature display a symmetry property called *time reversal invariance*, for then future and past Laplacian determinism stand or fall together. As far as is presently known, the only fundamental laws of physics that fail to have this symmetry property have to do with exotic weak interactions of elementary particles.

6.3 DETERMINISM, MATERIALISM, AND MECHANISM

As illustrated by Newtonian gravitational theory, the modern doctrine of determinism gained currency from examples of systems that are both materialistic—in the sense that they are composed of chunks of matter—and mechanistic—in the sense that the laws of their operation are force laws forming part of classical mechanics. But deterministic systems need not be materialistic, and while the epithet "mechanistic" may be applicable purely in virtue of their being deterministic, the systems need not be mechanistic in any deeper sense.

As an example, consider the source-free electromagnetic field. On the modern view of fields, there is no material "stuff" here, only the pure electric **E** and magnetic **B** fields construed as entities in their own right and not as states of an underlying medium. The immaterial nature of these fields is emphasized by the fact that unlike material particles, different **E** (or **B**) fields do not exclude one another but can be freely superposed. The field laws of electromagnetism, as codified by James Clerk Maxwell, state that

$$\nabla \times \mathbf{E} = \frac{\partial \mathbf{B}}{\partial t} \qquad \nabla \times \mathbf{E} = 0$$

$$\nabla \times \mathbf{B} = -\frac{\partial \mathbf{E}}{\partial t} \qquad \nabla \times \mathbf{B} = 0$$

In contrast to the case of Newtonian mechanics, the initial data for the electromagnetic field are more circumscribed. Consistent with Newton's laws of motion the velocity or rate of change of position of a particle can be freely specified independently of position; but the time rates of change of **E** and **B** are not free to choose since the left-hand pair of Maxwell equations fixes these quantities once **E** and **B** have been specified. Further, while Newton's laws do not constrain the possible initial positions and velocities of the particles, the right-hand pair of Maxwell's equations requires that the initially specified **E** and **B** fields be divergenceless. Nevertheless, an analogous form of determinism holds for the electromagnetic field. Let **E** and **B** be specified over all space at some initial instant, subject only to the constraints that $\nabla \cdot \mathbf{E} = \nabla \cdot \mathbf{B} = 0$ at that instant. Then there exists a unique solution to Maxwell's equations that satisfies the given initial data, and for any earlier or later time the constraints are also satisfied. Again this result is purely mathematical and does not depend upon the physical interpretation of **E** and **B**. They might just as well be the intensities of activities of an immaterialistic Cartesian mental substance. In this way the notion of determinism is just as applicable to the nonmaterial as the material, to the mental as well as the physical.

6.4 DETERMINISM, CHANCE, AND CHAOS

The theory of probability arose in part as an attempt to quantify the risks associated with games of chance. It is therefore a testimony to the power of the vision of determinism that one of the most eloquent expressions of the vision occurs in Laplace's pioneering work on probability, *A Philosophical Essay of Probabilities* ([1814] 1951):

> All events, even those which on account of their insignificance do not seem to follow the great laws of nature, are a result of it just as necessarily as the revolutions of the sun. In

ignorance of the ties which unite such events to the entire system of the universe, they have been made to depend upon final causes or upon hazard, according as they occur and are repeated with regularity, or appear without regard to order; but these imaginary causes have gradually receded with the widening bounds of knowledge and disappear entirely before sound philosophy, which sees in them only the expression of our ignorance of the true causes. (p. 3)

Even assuming that Laplace is correct and that the seeming hazards connected with games of chance—dice, roulette, and the like—are due to our ignorance of the true causes, there remains the task of characterizing the origins of the intuitive difference between systems that behave in an orderly fashion and those which seem to behave in a haphazard way.

A strong clue is found in J. C. Maxwell's insistence on a distinction between two maxims: (1) same causes always produce same effects, and (2) like causes produce like effects. The first maxim is an expression of determinism. The second implies the first but it implies something more: that, as Maxwell puts it, "small variations in initial circumstances produce only small variations in the final state of the system." He continues:

In a great many physical phenomena this condition [i.e., (2)] is satisfied; but there are other cases in which a small initial variation may produce a very great change in the final state of the system, as when the displacement of the "points" causes a railway train to run into another instead of keeping its proper course. (1952, 13–14)

The suggestion then is that seemingly chancy systems are ones which obey maxim (1) but not (2); that is, they are deterministic but their temporal evolution is unstable in that they display sensitive dependence on the initial conditions. The point can be illustrated with the help of the following crude example of two ways of dropping cannon balls (see Figure 6.1). In case (a) a ball is released from a certain height and is allowed to fall to the ground under the action of the Earth's gravity (air friction neglected). Suppose that we wish to know with some finite accuracy $\epsilon > 0$ the final resting place of the ball in the mud. Then for any initial state x (position), v (velocity) of the (center of mass of the) ball, there will be a finite Δx, and a finite Δv such that any other initial state within the range $x + \Delta x$ and $v + \Delta v$ will eventuate in a final position which is the same to within ϵ as the final resting place that eventuated from x, v. In that sense, like initial conditions lead to like final positions. In case (b) a number of fixed scattering centers are added to simulate a pinball game. In this case for some choices of initial state x, v and desired accuracy ϵ there will be no finite latitude Δx and Δv, no matter how small, such that all initial states within that latitude of the original will result in final positions differing no more than ϵ from one another. In that sense, like causes do not produce like effects, and any ignorance of the exact initial state will make the final result seem ϵ-chancy.

In recent years the study of deterministic but "chaotic" systems has been fashionable, and as with every fashion there is an accompanying media hype, with some of the more popular presentations suggesting that the study of chaos has led to a revolution in scientific thinking. (A good but somewhat breathless popular presen-

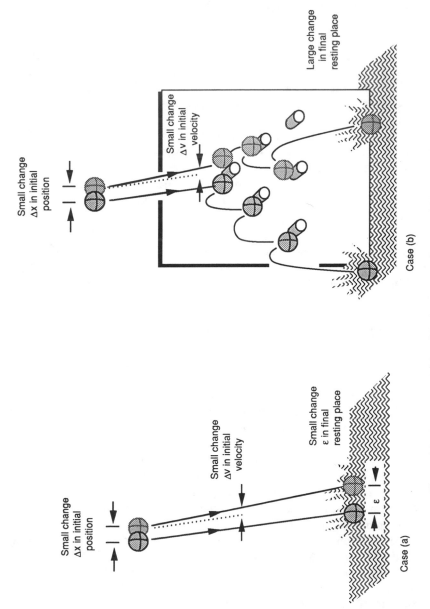

Figure 6.1 Extreme sensitivity of final state to small changes in initial conditions.

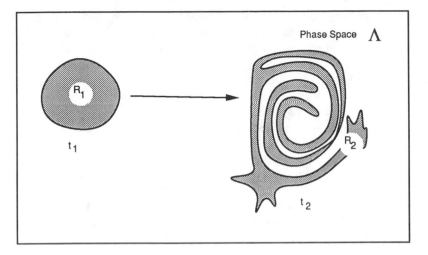

Figure 6.2 Deterministic evolution in phase space.

tation is Gleick 1987.) In actuality, the basic ideas are not new, many having been around for a century or more. What is new is the progress in proving hard mathematical results about chaotic systems and, perhaps just as significant, the availability of computers powerful enough to simulate models of very complex systems. The developments in this field are too technical to review here, but some of the ideas can be introduced by means of a mathematical device called the *phase space* Λ of a system. This is an imaginary space constructed in such a way that the points of the space correspond one-to-one to the allowed states of the system of interest. In the case of a system consisting of N Newtonian point particles, Λ would need to have $6N$ dimensions, $3N$ to code the spatial positions and $3N$ to code the velocities. A history of the system is given by a curve $\gamma(t)$ in Λ parameterized by time. Determinism means that for each $\gamma \in \Lambda$ exactly one such history through γ will satisfy the laws governing the system.

This apparatus can be used to model the behavior of two kinds of systems, *conservative* and *dissipative*. To explain the difference, consider some nice region R_1 of Λ (see Figure 6.2). Think of each point in R_1 as a possible initial state of the system at t_1. Then under the action of determinism, R_1 will be transformed at the later time t_2 into some determinate region, say, R_2. In general, the shape of R_2 will be different from that of R_1. But for conservative systems, the volume of R_2 will be the same as that of R_1 whereas for dissipative systems there is generally a loss of volume.[1]

In familiar examples of dissipative systems, the dissipative mechanism—for example, the friction between a tea cup and a marble rolling around in the cup—forces a wide range of initial states to tend towards a final equilibrium state—the marble at the bottom of the tea cup. But in chaotic systems the simple equilibrium states are replaced by *strange* attractors. In the Lorenz system, which models a

---

[1] The proof of the conservation of volume for classical Hamiltonian systems is known as Liouville's theorem. Of course, all of this presupposes a precise definition of measure for regions of phase space, a topic whose details are too technical for the present presentation.

problem in weather prediction, the phase space is ordinary Euclidean three-space. A trajectory starting near the origin moves to the left and makes some loops around the left attractor A, then veers to the right to loop around the right attractor A', and then to the left, and so on in a seemingly random pattern that depends sensitively on the initial state (see Figure 6.3, which shows the projection onto the $Y=Z$ plane). If this is the way the weather does work, then the beating of a butterfly's wings can have a major impact on the long-range forecast.

For conservative systems instability is manifested in a rapid spreading out of phase points. In the most extreme case, the points starting in a small volume element will become uniformly spread over all of Λ in a time τ that is short compared with macroscopic time scales. If Λ is partitioned up into cells $C_1, C_2, \ldots, C_N$ that correspond to conditions which are macroscopically ascertainable by means of an experiment that operates on a time scale τ, then the rapid spreading property means that a knowledge of which of the C_i the phase point occupied in the past offers no clue as to which cell the phase point will occupy when the experiment is next performed. From the macroscopic point of view the system seems to behave like a random mechanism in that it hops unpredictably from macrostate to macrostate.

The upshot is that some classical systems exhibit a peaceful coexistence between determinism and chaos—determinism at the microlevel is not only compatible with chaos at the macrolevel but actually gives rise to it. This detente naturally colors the way those with classically trained intuitions perceive the indeterminism of quantum physics (see Sections 6.12–6.16).

6.5 DETERMINISM AND PREDICTION

The most frequently cited formulation of determinism is not the one quoted in the preceding section but another from Laplace's *Analytical Theory of Probabilities:*

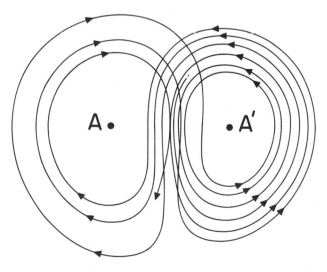

Figure 6.3 The Lorenz attractors.

Determinism in the Physical Sciences

Given for one instant an intelligence which could comprehend all the forces by which nature is animated and the respective situation of the beings who compose it—an intelligence sufficiently vast to submit these data to analysis—it would embrace in the same formula the movements of the greatest bodies of the universe and those of the lightest atom; for it, nothing would be uncertain and the future, as the past, would be present to its eyes. ([1814] 1951, 4)

Laplace's intelligence, or demon as it is often called, is one of the most vivid images to emerge from the entire literature on determinism. It is also one of the most pernicious, for some philosophers are not content to use Laplace's demon merely as an illustrative device to add some color to an already defined doctrine of determinism but want rather to understand the meaning of determinism in terms of the prediction tasks that can or cannot be carried out by such a demon. The most pressing need of such an approach is to specify the powers the demon is allowed to have. This much is clear: Determinism is concerned with scientific predictability, not with soothsaying, precognition, divine foreknowledge, and so forth, and so the demon must not be endowed with any such powers. But exactly what is covered by the "and so forth"? We come close to the mark if we respond that the demon is to have only the requisite mathematical ability, say the ability to solve the differential equations that constitute the relevant laws. But this still misses the mark; for determinism is purely and simply a claim about the existence and uniqueness of solutions, and whatever knowledge that a demon endowed with whatever abilities has about the truth of this statement is entirely ancillary.

The demonology becomes really pernicious with attempts to give the demon a human face. Thus, Karl Popper suggests that Laplace's demon should be construed as "a super-scientist" (1982, 34). This means, among other things, that "The demon, like a human scientist, must *not* be assumed to *ascertain initial conditions with absolute mathematical precision;* like a human scientist he will have to be content with a finite degree of precision" (ibid.). It follows that for unstable systems Popper's demon will not be able to carry out various prediction tasks and so, on Popper's understanding of "scientific determinism," that these systems will be nondeterministic. The preferable conclusion is that the failure here is a failure on the part of the humanized demon and not a failure of determinism per se. We started with a vocabulary containing "determinism," "instability," and "predictability," each with a relatively clear and distinct meaning, and it can only do mischief to form a mishmash of all three.

Here Popper might appeal to the authority of Joseph Larmor who added a note to Maxwell's two maxims (discussed in the preceding section), stating that "it is only in so far as stability subsists that principles of natural law can be formulated: it thus perhaps puts a limitation on any postulate of universal physical determinancy such as Laplace was credited with (Maxwell 1952, 13)." But it is hard to believe that Maxwell himself would have subscribed to such a sentiment since he was surely aware that precise principles of natural law have been formulated although they do not conform to Larmor's demand. In any case, the recent work on chaos theory is a direct refutation of Larmor's sentiment.

In the following sections we encounter other ways in which determinism and predictability come apart.

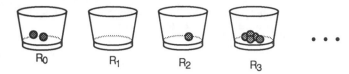

Figure 6.4 A register machine.

6.6 DETERMINISM AND COMPUTABILITY

The modern digital computer is just a very special and very limited kind of deterministic system with a discrete set of states. The schematic structure of digital computing is usually explained in terms of a device called a Turing machine, in honor of Alan Turing who was the father of the theory of effective computability. But here we will study another device, called a *register machine,* which is equivalent to a Turing machine in terms of computational ability and which is more directly adapted to explicating the notion of an effectively computable function of the natural numbers N.[2]

The hardware of a register machine consists of an unlimited number of registers $R_0, R_1, R_2, \ldots$. For sake of concreteness we will think of the registers as urns, each of which holds a finite number of balls (see Figure 6.4). Register R_0 holds any integer from 1 to $M > 0$ while the other registers hold any nonnegative integer. (In computerese, R_0 is used to code the internal state of the machine while the other registers are used as memory storage.) A state $\{n_i\}$, $i = 1, 2, \ldots$, of the machine is just a list giving the number of balls in each of the registers. The change of state is governed by a deterministic transition law which is specified by a "program" consisting of a finite list of instructions $I_1, \ldots I_M$, subject to the covering rule that if R_0 holds the number n, then instruction I_n is carried out. Instruction I_M is the halting instruction which says that the state remains the same. The remaining instructions $I_1, \ldots I_{M-1}$ can be of one of four forms. A zero instruction says to take all of the balls out of some specified urn R_n, $n > 0$, and then to add one ball to R_0, leaving the other registers undisturbed. A successor instruction says to add one ball to some specified R_n, $n > 0$, and one to R_0, leaving the other registers undisturbed. A transfer instruction says to transfer the contents of some specified R_n and R_m, $n,m > 0$, and to add one ball to R_0. Finally, a jump instruction says to compare the contents of some specified R_n and R_m, $n,m > 0$, and to either add one or $p > 1$ balls to R_0 according as R_n and R_m do or do not contain the same number. (If the result of adding p balls to R_0 is greater than M, the understanding is that M balls are to be placed in R_0.) Such a machine is said to compute a function $f\, N \to N$ just in case for any $x \in N$ in the domain of f, if the machine started in state $<1, x, 0, \ldots, 0>$ it eventually halts in state $<M, f(x), \ldots>$, and for any $x \in N$ not in the domain of f, the machine does not halt if started in the same state. To get a feeling for how these machines work, the reader is invited to write a program for manipulating the balls according to the above rules in such a manner that the machine computes $f(x) = x^2$.

---

[2] For more details, see Cutland (1980). The usual definition of register machine does not include the register R_0 used here to encode the internal state of the machine. Register machines, Turing machines, and other types of computers are discussed in Chapter 11.

A function of the integers will be said to be *effectively computable* just in case there is a register machine that computes it. It is intuitively evident that half of this definition is justified: If a function can be computed on a register machine, then it certainly deserves to be called effectively or mechanically computable. The other half, that any function of the natural numbers that is effectively, mechanically computable must be computable by a register machine is called Church's thesis. We will not discuss the evidence for this thesis save to say that it is considerable; in particular, many other plausible alternatives to the register or Turing machine characterizations turn out to yield extensionally equivalent definitions of effectively computable functions.

It is natural to wonder what functions can be computed on a machine which is similar to a register machine but which operates nondeterministically, say, because of the introduction of a random element. The answer is that relaxing the assumption of determinism does not allow any additional functions to be computed. But the nondeterministic machines may reduce the number of steps needed to reach an answer.

One would also like to know how to bring together the notion of computability explained here with the kind of determinism just discussed, say, in Newtonian particle mechanics. Suppose, for example, that $x(t)$ is the position function of a particle in some solution to Newton's laws of gravitation. Is $x(t)$ an effectively computable function of t? Thus far we have no means of answering such questions since the concept of computability has been defined only for functions of the natural numbers. But there is a standard way to pass from the natural numbers to the rationals and thence to the real numbers, and concepts needed in this passage can be effectivized so as to extend the register machine notion of computability to functions of the reals. Then just by cardinality considerations alone $x(t)$ may not be effectively computable since there are more solutions to Newton's laws than there are effectively computable functions. (There are uncountably many solution functions and only a countable number of effectively computable functions.) If the initial positions and velocities are effectively computable real numbers, then $x(t)$ will be an effectively computable function. When the laws of time evolution are partial differential equations, as in field theory, then the situation is interestingly different in that effective computability for initial data is not necessarily preserved by deterministic evolution.

Finally, one can wonder about the concept of analogue as opposed to digital computability. In particular, can any deterministic system be regarded as an analogue computer which "computes" its solution functions? Such questions cannot be answered without the help of a general theory of analogue computability, which at present does not exist.

6.7 DETERMINISM AND OPEN SYSTEMS

The natural conclusion to draw from the above review is that determinism is part of the marrow of both the methodology and the content of classical physics: The fundamental laws discovered in classical physics are deterministic; the process of scientific inquiry is most fruitfully conducted under the assumption that determinism is true; and in general the concept of determinism is pervasive, finding its way into such diverse areas as the analysis of chaos and effective computability. It is now time to add a serious caveat to this first impression, for there is a sense in which determinism stands on unfirm ground in classical physics.

Recall from Section 6.2 that the existence and uniqueness theorems for the initial value problem in Newtonian gravitational theory were modest in refusing to make claims about all future and past times. One reason for that modesty is that when collisions of point particles occur, Newton's gravitational force law blows up. We can deal with this difficulty either by ignoring it on the grounds that collisions are relatively rare in the class of all solutions allowed by Newton's laws (there are some precise results to this effect) or else by finding a method of extending the solution through the singularity (e.g., for collisions in one spatial dimension, the model of the elastic bounce can be used to regularize the solution). However, there is a second and more mind-boggling reason for the modesty, namely, the possibility of noncollision singularities in which the solution fails to exist after some finite time because all of the particles have disappeared, not because they dissolve and fade away, but because they accelerate so hard that they escape to spatial infinity. Before worrying about the physical possibility of such a feat we first have to understand why it is conceptually possible.

The laws of Newtonian physics all obey an important symmetry or invariance principle: They are the same in every inertial frame. Different inertial frames move uniformly and rectilinearly with respect to one another, the spatial transformation between two such frames having the form $\mathbf{x}' = \mathbf{x} + \mathbf{v}t$ where v is the relative velocity between the frames. This leads to the Galilean velocity addition law: if $\mathbf{u}$ and $\mathbf{u}'$ are respectively the velocities of a particle as measured with respect to the unprimed and the primed frames, then $\mathbf{u}' = \mathbf{u} + \mathbf{v}$. Since there is no upper bound on $\mathbf{v}$, it follows that $\mathbf{u}'$ can be made as great as we like and consequently that Newtonian laws of motion cannot impose a finite upper bound on the speed with which a particle can move. It only remains to note that in the absence of such a bound, a body can escape to spatial infinity in a finite time even though it never attains an infinite velocity. (The reader should draw the spacetime world line of such a particle.)

For a long time physicists wondered whether this conceptual possibility could be implemented by a system of point mass particles moving under the action of their mutual Newtonian gravitational forces. Within the last few years they have convinced themselves that a noncollision singularity is possible with five particles—though they never collide, their mutual interactions can be arranged so that all disappear to spatial infinity in a finite time, (see Gerver 1984 for details). Since the Newtonian laws are invariant under time reversal, the temporal mirror image of this process, with five particles appearing from spatial infinity, is equally physically possible.

To appreciate the relevance for determinism, return to the thought experiment of Section 6.1. That construction was designed as an intuition pump to work up the expectation of a deterministic outcome. But the pump works only if various loopholes are closed, one of which was that the system might be open to outside influences. With particles acting at a distance, the only safe way to assure the system is closed is to extend the system to include all the particles that there are. But we now see that safe is not sure since even if the system consists of the entire physical universe, it may not be closed to outside influences in the relevant sense.

Determinism in the Physical Sciences

Determinists may be disconcerted that what was supposed to be the paradigm of classical determinism at work turns out to undermine confidence in determinism, but they might try to shrug off their disappointment with the remark that the example involves the idealization of point particles and, in particular, it draws on the infinite potential well of these particles as the energy source for powering the escape to infinity. The embarrassment returns, however, with the realization that the openness of the universe is not a peculiarity of this example but is a widespread feature of classical physics. Moving from particle mechanics to the classical field laws governing the evolution of continuous field quantities acting by contract, we find just such an openness rampant. The paradigm case of such a field law is the Fourier heat diffusion equation, which implies that heat waves propagate with an actually infinite velocity, making it no trick at all for heat effects to appear from or disappear to spatial infinity. In mathematical terms there is a solution ϕ_s (s for surprise) such that $\phi_s(x,t) = 0$ for all $t \leq 0$ but $\phi_s(x,t) > 0$ for $t > 0$. Like Maxwell's equations, the Fourier equation is linear so that if ϕ_1 and ϕ_2 are solutions then so is $\phi_1 + \phi_2$. By adding ϕ_s onto an arbitrary solution ϕ determinism is totally wrecked. Determinism can be restored by imposing various boundary conditions at infinity, but such a move serves to underscore the moral that in the classical setting determinism does not stand on its own feet but needs the help of various crutches.

6.8 DETERMINISM AND THE NATURE OF SPACE AND TIME

The discussion in the preceding section helps to reveal how the fortunes of determinism are linked to questions about the structure of space and time, a point that will be brought into sharper focus in this section and also in the following sections where relativistic spacetimes are contrasted with classical spacetimes. The present section also aims to demonstrate the linkage between determinism and the ontological status of space and time.

On a relational conception of motion, any meaningful statement about motions must be translated into statements about relative motions of physical bodies. In particular, to say that body A accelerates is, on the relational view, elliptical for saying that A accelerates relative to some other system of bodies. This view is clearly inconsistent with the spacetime framework assumed in the preceding section for in the presence of inertial frames, acceleration is an absolute quantity (under a Galilean transformation connecting two inertial frames, acceleration is an invariant, $\mathbf{a}' = \mathbf{a}$) making it meaningful to speak of the spatial acceleration of A independently of A's relation to other bodies. If inertial frames are banished so as to create a spatiotemporal environment congenial to the relational conception of motion, then the Galilean transformations are replaced by a much wider set of transformations, which we can dub the relational transformations. Just as Newtonian laws of motion are invariant under the Galilean transformations, so whatever laws of motion the relationist may propose must be invariant under the relational transformations. But the latter are so broad as to seemingly preclude any interesting sense of determinism of the Laplacian variety. For among such transformations are those which coincide with the identity map for all $t \leq 0$ but which are nontrivial for $t > 0$, leading to the result that there are solutions of the relationist equations of motion that coincide for all past times but diverge in the future. (For a more detailed discussion of this and other problematic features of classical determinism, see Earman 1986.)

Those who are relationists about motion also tend to be relationists about space and time, and those who are relationist about space and time will respond that the splitting of physically possible histories that is supposed to be violation of Laplacian determinism is but an illusion. For to be a relationist about space and time is to reject the idea that space is literally a container for bodies and spacetime is a container for physical events and to hold that the absolutist's space and spacetime are merely artifices for representing the mutual relations among physical bodies and physical events. For the relationist the splitting histories spoken of are, therefore, seen as merely different ways presenting the same physical situation, and no more indeterminism is involved than in, say, describing the fall of the Roman Empire first in Latin and then in English.

No adjudication of this dispute between the absolute and relational conceptions will be attempted here, (see Earman 1989 for details on the absolute-relational controversy). But note that to create a safe environment for classical determinism, either the structure of spacetime must be sufficiently rich or else a relational construal of spacetime must be adopted.

6.9 DETERMINISM AND THE SPECIAL THEORY OF RELATIVITY

The source-free Maxwell equations introduced in Section 6.3 are Laplacian deterministic without any help from boundary conditions at infinity to ward off electromagnetic space invaders. At least this is so if the equations are placed in their proper setting—Minkowski spacetime (see Chapter 5); for in this setting it can be shown that Maxwell's equations imply that electromagnetic effects propagate at exactly the speed of light c.

More generally, the light cone structure of Minkowski spacetime seems to function as an invisible barrier to space invaders, making the relativistic setting safe for Laplacian determinism. Alas, life is not so simple. We have to contend with the possibility of *tachyons,* swift particles which do not do the impossible of crossing the light barrier but which nevertheless revive the specter of space invaders because their world lines are represented by space-like curves. This specter can be controlled if there are no free tachyons and if all tachyon world lines terminate in tardyon world lines which stay confined within the light cones. Frank Arntzenius (1990) has shown how to construct a Lorenz invariant theory in which two tardyons interact by exchanging tachyons. The theory is globally Laplacian deterministic in the sense that the state on some appropriate global time slice of Minkowski space suffices to fix the past and future. The theory, however, is not locally deterministic since, for example, the state on the local time slice S picture in Figure 6.5 does not suffice to fix the state within the future cone subtended by S, for a tachyon that does not register on S can unpredictably enter the future cone. In any case, this toy theory is highly artificial, and controversy continues about the extent to which it is possible to construct physically plausible theories of tachyons that are compatible with the demands of relativity

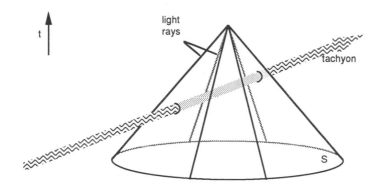

t

light
rays

tachyon

S

Figure 6.5   Tachyons defeat local determinism.

theory. Until these controversies are settled, the fate of special relativistic determinism remains unsettled.

If relativistic determinism is spared the challenge of faster than light particles then another breach is opened between determinism and prediction since the same light cone structure that serves to shield against space invaders also serves to prevent finite embodied observers from receiving enough information about initial conditions relevant to the determination of a future event before the event occurs (see the Discussion Questions for further discussion of this point).

6.10 DETERMINISM IN THE GENERAL THEORY OF RELATIVITY

Einstein's general theory of relativity (GTR) raises two types of problems for determinism, the first having to do with the causal structure of spacetime and the second having to do with the ontological status of spacetime.

To pose the first problem, let us understand by a *causal curve* a smooth spacetime curve whose tangent is everywhere nonspace-like. Such a curve represents the world line of an idealized causal process propagating with a speed $\leq c$. (We are here ignoring the possibility of tachyons.) A space-like hypersurface S is said to be a *Cauchy surface* just in case S meets every causal curve without end point exactly once. Such a surface is an appropriate initial value surface for launching global Laplacian determinism since every causal process must leave its imprint on S and, thus, one can hope to determine the past and future behavior of a process from its behavior on or near S. Conversely, the hope of using initial data on a time slice S to determine the future and past behavior of a process is vain if that process does not register on S. The problem for general relativistic determinism is that various cosmological models satisfying the basic laws of GTR, Einstein's field equations, do not possess a Cauchy surface. Figure 6.6 shows three ways in which the Cauchy property can fail. In Figure 6.6(a) the surface S_a lacks the Cauchy property because the causal curve α closes on itself and does not meet S_a. In Figure 6.6(b), S_b is not a Cauchy surface because the curve β, an analogue of a classical space invader, runs off to spatial infinity without meeting S_b. In Figure 6.6(c), collapsing matter produces a naked singularity. The time slice S_c fails to be a Cauchy surface because the causal

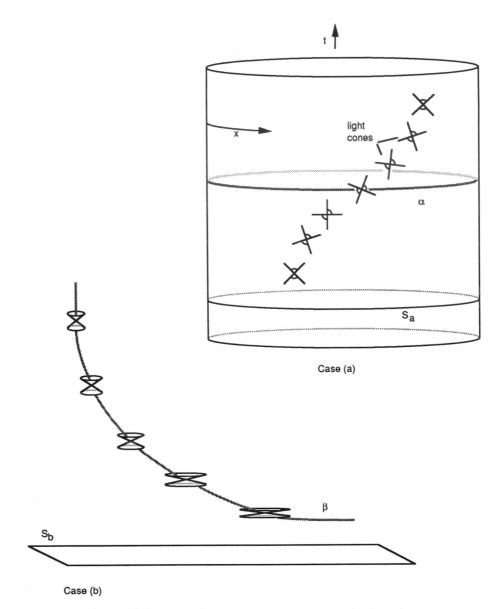

Figure 6.6a,b General relativistic spacetimes lacking a Cauchy surface.

curve γ emerges unpredictably from the singularity (which strictly speaking is not part of the spacetime).

The determinist can respond that Einstein's field equations are not sufficient to pick out the genuine physical possibilities and that additional strictures are needed, in particular, strictures that will rule out the situations illustrated in Figure 6.6. Such a response need not be question begging since there are independent reasons to doubt that these situations are physically possible. Closed causal curves such as that in Figure 6.6(a) give rise to paradoxes of time travel (e.g., it would seem that observers

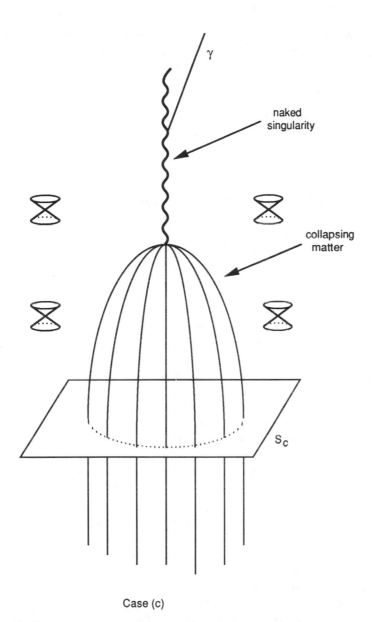

γ

naked
singularity

collapsing
matter

S_c

Case (c)

Figure 6.6c   General relativistic spacetime lacking a Cauchy surface.

could travel into their own past and shoot themselves-at-an-earlier-time). The behavior pictured in Figure 6.6(b) is often accompanied by an unphysical instability, for example, the introduction of a small amount of matter on S_b leads to singular behavior at a later time. And the singularity pictured in Figure 6.6(c) violates the cosmic censorship hypothesis which asserts that the singularities that develop in physically reasonable cases of gravitational collapse are hidden in the interiors of "black holes," the exteriors of which admit a Cauchy surface. This hypothesis is currently the focus of a lively debate among theoretical physicists. Whatever the ultimate decision on the issues raised by the models in Figure 6.6, it is evident that the fate of general relativistic determinism is intertwined with some of the thorniest foundations problems in GTR.

The second threat to general relativistic determinism arises even on the presumption that the first has been swept aside and attention is focused on cosmological models with Cauchy surfaces. This threat has already been discussed in Chapter 5 under the heading of the "hole argument" and will not be rehearsed again here.

6.11 DETERMINISM AND QUANTUM MECHANICS

When first told that nonrelativistic quantum mechanics (QM) is a nondeterministic, stochastic theory, we might naturally expect that the deterministic laws of temporal evolution of classical mechanics are replaced by a nondeterministic law by which the present state does not fix the future state but, say, only a probability distribution over future states. But in the first instance this is not what we find in QM; indeed, the basic equation of temporal evolution—the Schrödinger equation—is in some respects more deterministic than its classical analogue. Consider again the case of point mass particles interacting via a $1/r^2$ force. In the classical mechanical treatment of this problem we had to worry about both collisional and noncollisional singularities. In the quantum mechanical treatment, however, neither of these worries materializes since the solution probably exists for all time. (Technically, the point is that the Hamiltonian operator for this process is self-adjoint so that the evolution operator is unitary and, thus, is defined for all time; see below.) Furthermore, the quantum evolution does not show the kind of sensitive dependence on initial conditions that translates into classical chaos.[3] From whence then comes quantum indeterminism? To answer this question it is necessary to delve into some technical aspects of QM.

6.12 QUANTUM FORMALISM AND QUANTUM INDETERMINISM

QM replaces the classical notion of state with a vector[4] Ψ in a special kind of vector space, called *Hilbert space*. Before launching into details it may be helpful to recall some of the facts about the more familiar case of a Euclidean vector space. For example, the velocity of a particle in classical mechanics can be represented by a vector in a three-dimensional Euclidean space. Any such velocity vector v can be decomposed into a linear sum of three arbitrarily chosen but linearly independent

---

[3]At least this is so if the nearness of states is measured in the usual norm, that is $\|\psi\| = (\psi, \chi)^{1/2}$, where $(,)$ is the inner product of the Hilbert space; see Section 12. Since Schrödinger evolution preserves the norm two states that are initially close remain close. It is currently a matter of lively debate as to how to define the notion of quantum chaos.

[4] More precisely, the quantum state is given by the one-dimensional subspace spanned by the vector; for $\psi' = \exp(ir)\,\psi$, r a real number, and ψ both represent the same state in that they give exactly the same probabilities for measurement outcomes, as seen further in Section 6.12.

basis vectors $\mathbf{u}_1, \mathbf{u}_2, \mathbf{u}_3$: $\mathbf{v} = \alpha_1\mathbf{u}_1 + \alpha_2\mathbf{u}_2 + \alpha_3\mathbf{u}_3$, where the expansion coefficients α_i are real numbers. Because of its Euclidean structure the vector space has a natural notion of inner product which associates a real number $(\mathbf{u}, \mathbf{v})$ with each pair of vectors and which defines a notion of length $\| \mathbf{v} \| \equiv (\mathbf{v}, \mathbf{v})^{1/2}$ of vectors and a notion of orthogonality $\mathbf{u} \perp \mathbf{v} \equiv (\mathbf{u}, \mathbf{v}) = 0$. The basis vectors $\mathbf{u}_i$ can be chosen so that they are pairwise orthogonal and of unit length, in which case the expansion coefficients α_i give the length of $\mathbf{v}$ in direction i. Finally, we can define linear operators or gadgets that eat vectors and spit out vectors in such a way as to respect the linear structure of the space; that is, the operator $\mathbf{O}$ obeys the rule $\mathbf{O}(\alpha\mathbf{u} + \beta \mathbf{v}) = \alpha\mathbf{O}(\mathbf{u}) + \beta\mathbf{O}(\mathbf{v})$ for any real numbers α and β. A simple example of such a linear operator is a projection that takes a vector to its component in a certain direction.

All of this familiar material carries over to Hilbert spaces, but with some subtle but major changes. In the first place, the complex numbers replace the real numbers: Euclidean vector spaces are closed under linear combinations obtained by multiplying vectors by real numbers and taking their sum; Hilbert spaces are closed under linear combinations obtained by multiplying vectors by complex numbers and taking their sum. Second, in some physical applications the Hilbert spaces have to be infinite-dimensional. This second feature will be downplayed here since many of the basic foundations problems can be raised for the finite case.

Physical quantities in QM (or "observables" as they are often called) are represented by a special class of linear operators called *self-adjoint* operators.[5] As in the case of Euclidean vector spaces, an operator $\mathbf{A}$ on Hilbert space can be thought of as an animal that eats vectors and spits out other vectors. A vector φ is said to be an *eigenvector* of $\mathbf{A}$ just in case when $\mathbf{A}$ eats φ it spits out $a\varphi$, where the number a is called an *eigenvalue* of $\mathbf{A}$. If $\mathbf{A}$ is self-adjoint, then the eigenvalue a must be a real number, an important result since one of the basic postulates of QM is that the possible results of measuring an observable A are given by the eigenvalues of the corresponding operator $\mathbf{A}$.

One of the ways in which indeterminism emerges in QM is in the prediction of measurement outcomes. Suppose that the observable A is being measured. Express the state vector Ψ as a linear combination of the eigenvectors φ_i of $\mathbf{A}$:
$\Psi = \Sigma \; \alpha_i \; \varphi_i$, $\mathbf{A}_i = a_i\varphi_i$. (This can always be done in a unique way for a self-adjoint operator A.) Then another basic postulate of QM asserts that upon measurement of A the probability of obtaining the result a_k is $|\alpha_k|^2$. In general the numbers $|\alpha_k|^2$ lie strictly between the extreme values of 0 and 1 required in a deterministic treatment.

A second way in which QM is indeterministic concerns the temporal evolution of the state Ψ. In Section 6.11 it was stated that this evolution is governed by the deterministic Schrödinger equation. That statement is only half correct, at least according to the von Neumann formulation of QM. On von Neumann's account the quantum state Ψ changes continuously and deterministically until a measurement is made, at which point it jumps discontinuously and nondeterministically into the

---

[5] For bounded operators, self-adjointness of $\mathbf{A}$ means that $\mathbf{A}^+ = \mathbf{A}$ where the adjoint $\mathbf{A}^+$ of $\mathbf{A}$ is defined by $(\mathbf{A}^+ \varphi, \chi) = (\varphi, A \chi)$ for all φ and χ.

eigenstate φ_i corresponding to the eigenvalue a_i obtained for the observable A being measured—the so-called collapse of the superposition.

What is disturbing about the second indeterministic feature is not so much the indeterminism per se as the conundrum it seems to involve. For a measurement is presumably a physical interaction between an object system and a measurement apparatus. As such it should be describable in the same terms that apply to other interactions and, thus, the evolution of the joint object-plus-apparatus system should obey the Schrödinger equation. Alas, there are convincing demonstrations that Schrödinger evolution is incompatible with the collapse of the superposition. Leaving aside the technicalities, the gist of the problem is that because of the linearity of the Schrödinger equation, it predicts that after interaction the object-plus-apparatus system is left not in a definite state but in a superposition of coupled object-apparatus states. This conundrum, commonly referred to as the quantum measurement problem, has sparked a number of desperate attempts at a resolution, only two of which will be mentioned here. First, it has been suggested that the measurement interaction is not purely physical and that the collapse is brought about by the action of a conscious observer. Such a revival of Cartesian dualistic-interactionism is hard to swallow. Second, it has been suggested that no collapse takes place and that instead the world splits into many copies, in one of which the value a_1 is realized, in another of which the value a_2 is realized, and so forth. Such an ontological inflation is as hard to swallow as Cartesian dualism. (See Wheeler and Zurek 1983 for a collection of papers on the quantum measurement problem. For the "many worlds" interpretation, see DeWitt and Graham 1973.)

The first form of quantum indeterminism suggests to the classically trained intuition that QM is incomplete and that quantum probabilities are expressions of our ignorance of the exact state of nature. The second form of indeterminism and the measurement problem to which it leads suggests that something is rotten at the core of QM. Perhaps the conjectured incompleteness of QM is also responsible for this defect. Before exploring this notion further, we turn to an argument which is designed to prove the suspected incompleteness of QM.

6.13 THE EPR PARADOX

One of the most frequently cited papers on the foundations of QM appeared in 1935 in the *Physical Review* under the somewhat awkward title of "Can Quantum Mechanical Description of Reality Be Considered Complete?" It carried the names of three authors, Albert Einstein, Boris Podolsky, and Nathan Rosen, and consequently is referred to as EPR. (The paper was in fact written mainly by Podolsky; see Fine 1981.) The influence of this paper is belied by its length of less than four journal pages. Although it contains some technical terms which may be unfamiliar to you, you ought to be able to reconstruct the structure of the argument of EPR. So before reading further in the text, turn to the Appendix of this chapter where EPR is reprinted and try to identify the conclusion of their argument, the premises, and the steps leading from the premises to the conclusion.

In paragraph 3, EPR state what they regard as a necessary condition for *com-*

pleteness of a physical theory; namely, "*every element of the physical reality must have a counterpart in the physical theory.*" While declining to define "physical reality," they provide in paragraph 4 a sufficient condition or Criterion of Reality:

> If, without in any way disturbing a system, we can predict with certainty (i.e., with probability equal to unity) the value of a physical quantity, then there exists an element of physical reality corresponding to this physical quantity.

The purpose of their argument appears at first to be to establish that QM is an incomplete theory (I). To understand their rationale for this conclusion, recall that two operators **A** and **B** are said to *commute* just in case $[\mathbf{A}, \mathbf{B}] = \mathbf{A}\,\mathbf{B} - \mathbf{B}\,\mathbf{A} = 0$, that is, for every vector, the result of first operating by **A** and then by **B** is the same as first operating by **B** and then by **A**. (In the case of unbounded operators, which are defined on at most a dense set of Hilbert space, we would have to be more careful about domains of definition.) Introducing the abbreviation (NSV) to stand for the assertion that when the operators corresponding to two physical quantities do not commute, the two quantities cannot have simultaneously sharp values (or simultaneous reality, as EPR say), the EPR argument can be reconstructed as follows:

either (I) or (NSV)
If not-(*I*) then not-(NSV)
∴ (*I*)

(This follows the reconstruction of Fine 1981.) The argument form is evidently valid, so the evaluation of the argument reduces to the acceptability of the premises.

EPR try to justify the first premise by arguing that if (NSV) fails, then QM is incomplete. If (NSV) fails the physical quantities in question would have simultaneous reality (simultaneously definite values), and by the completeness condition these values must be part of a complete description. EPR conclude that

> either (1) *the quantum-mechanical description of reality given by the wave function is not complete* or (2) *when the operators corresponding to two physical quantities do not commute the two quantities cannot have simultaneous reality.*

The first sentence indicates that they interpret the completeness condition to mean that the counterpart in the theory of the physical quantity must allow prediction of the value of the quantity with probability 1. Under this reading the second sentence is true since for no state does QM assign probability 1 to definite values of observables corresponding to noncommuting operators.

To establish the second premise EPR employ a thought experiment involving a pair of correlated particles. The version used here is similar in spirit to the original but has the dual advantages of lending itself to actual laboratory trials and of linking to Bell's theorem to be discussed. In the set up pictured in Figure 6.7, a pair of particles leave the source and travel in opposite directions in space until they encounter analyzers, each of which contains settings that correspond to the measurement of various

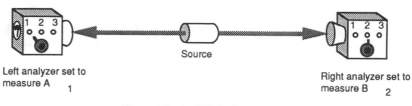

Source

Left analyzer set to
measure A $_1$

Right analyzer set to
measure B $_2$

Figure 6.7 An EPR-Bell experiment.

observables A_1, A_2, . . . on the left particle and various observables B_1, B_2, . . . on the right particle. It is arranged that $[\mathbf{A}_i, \mathbf{B}_j] = 0$ for any i and j even though the $\mathbf{A}_i$ and the $\mathbf{B}_j$ do not pairwise commute among themselves. Though not needed until later, it is also convenient to assume that all the observables in question are bivalent, taking on the values (say) ± 1. Finally, we appeal to the principles of QM to establish the existence of states in which the left and right observables are perfectly anticorrelated in the sense that with probability 1 the outcome of measuring A_i on the left is ± 1 just in case the result of measuring B_i on the right is ∓ 1.

For the time being it suffices to concentrate on two pairs of observables A_1, B_1 and A_2 and B_2. Now suppose that we make a choice of the setting on the left analyzer. Depending upon whether we choose to measure A_1 or A_2 we can, from the measurement result, predict with certainty the value for B_1 or B_2. Applying the Criterion of Reality we conclude that the values of the quantities B_1 and B_2 are simultaneous elements of reality and, therefore, that QM is incomplete.

On further reflection this seems a little too quick. For how does it follow from the fact that measurements made at different times on the left particle allow us to assign definite values to B_1 and B_2 in such a way that these quantities have simultaneously definite values? This concern is addressed in the penultimate paragraph of EPR:

> One would not arrive at our conclusion if one insisted that two or more physical quantities can be regarded as simultaneous elements of reality *only when they can be simultaneously measured or predicted*. On this point of view, since either one or the other, but not both simultaneously, of the quantities P [B_1] and Q [B_2] can be predicted, they are not simultaneously real. This makes the reality of P[B_1] and Q[B_2] depend upon the process of measurement carried out on the first [left] system, which does not disturb the second [right] system in any way. No reasonable definition of reality could be expected to permit this.

What should now be clear is that the real conclusion of EPR is not that QM is incomplete (I) but rather that it is incomplete under the assumption of locality (Loc) in the sense that what is done on one wing of the experiment does not influence what happens on the other. (That this is what Einstein intended is made clear by his later paper Einstein [1948] 1971.) Exactly what this locality assumption amounts to will be discussed, but before turning to that matter it is worth noting a few other puzzling features of the EPR argument.

The first noteworthy feature is that the second premise has been incorrectly advertised since the antecedent not-(I) plays no role in the analysis of the thought experiment. The premise is more correctly stated as: if (Loc) then not-(NSV), and the ultimate conclusion of the argument is not that QM is incomplete but that it is either incomplete or nonlocal.

The second feature is the puzzle of why it is necessary to consider measurements of observables corresponding to noncommuting operators. Measure, say, A_1 on the left and obtain a value (say) $+1$. Infer with certainty that B_1 has a value of -1. So by the Criterion of Reality the right particle has a value of -1 of B_1 after the measurement. But by (Loc) the measurement on the left cannot create a value of B_1 on the right. So that value must have been possessed at the time of the measurement, and since QM does not predict that value with certainty, the theory is incomplete. The obvious answer to the puzzle would cite the strong psychological pressure to invoke what was taken to be one of the most characteristic features of QM—the existence of noncommuting observables. The EPR invocation takes the following form. Measure A_1 on the left and obtain a value (say) $+1$, and infer with certainty that the particle on the right is an eigen state χ_1 of B_1 with eigenvalue -1. Measure A_2 on the left and obtain (say) -1, and infer that the particle on the right is an eigen state χ_2 of B_2 with an eigenvalue of $+1$. But by (Loc) no real change can take place on the right wing as a consequence of what was done on the right wing. Thus, EPR conclude, it is possible to assign two different state vectors to the same reality. Since this is an inconsistency if those vectors represent different real states of affairs, QM is incomplete. This pattern of argument, which is repeated in Einstein ([1948], 1971) defies the structure set out above for the EPR argument.

Finally, we can wonder why, questions of completeness aside, (Loc) is not violated in the thought experiment. Whether or not the QM state description is complete, it does describe objective physical features of the system, for example the long-run relative frequencies of outcomes, and so objective features on the right wing are influenced by what is done on the left wing. Einstein's reply would presumably have been that no violation of (Loc) is involved if probability is given an ensemble interpretation and measurement is the process of selecting a subensemble of systems characterized by the values obtained in the measurement. Ironically, EPR experiments, under a more detailed analysis, bring into question this interpretation.

6.14 INCOMPLETENESS AND BELL'S INEQUALITIES

Einstein's conclusion that QM is either incomplete or nonlocal was supposed to have a pejorative force since he took it as a goal of scientific theorizing to produce a theory that was at once both complete and local. The EPR experiments, however, can be used to challenge the presumption that reality itself is local and complete at the quantum level. For EPR, QM is incomplete because it does not contain counterparts of every element of reality: the A_i and B_j are supposed to have simultaneous definite values even though there are no counterparts of these values in the theory. However, the assumption of ontological completeness—the value definiteness of the observables—turns out to be in conflict with the testable statistical predictions of the theory.

Let us suppose that the description of QM has been "improved" in line with the sentiments of EPR so that the new theory gives a complete description in that for any point λ in the new state space Λ, the theory assigns definite values to A_1, A_2, A_3 and also to $B_1, B_2,$ and B_3. The notation $(\pm, \pm, \pm; \pm, \pm, \pm)$ will be used to denote the region of Λ where the observables have the designated values; for example $(+, -, +; -, -, +)$ is the region where $A_1 = +1, A = -1, A_3 = +1$ and $B_1 = -1, B_2 = -1,$ and $B_3 = +1$. Also in keeping with the aims of EPR, the QM probabilities assigned to compatible pairs of observables are interpreted as measures of our ignorance of the location in Λ of the actual state λ. By the anticorrelation condition, $Pr((A_1 = +1)\&(B_2 = +1))$ is the sum of the measures x and y assigned respectively to the regions $(+, -, +; -, +, -)$ and $(+, -, -; -, +, +)$ (see Figure 6.8). Likewise, $Pr((A_1 = +1)\&(B_3 = +1))$ is the sum of the measures

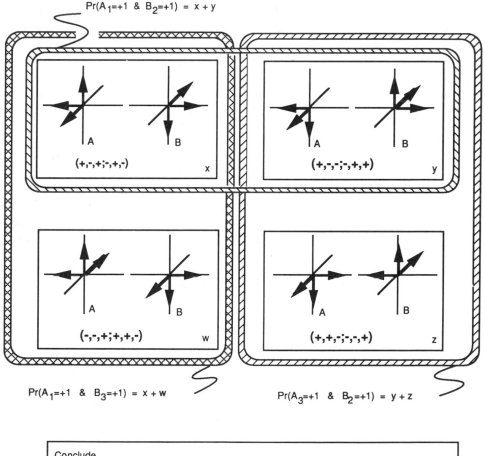

Pr(A_1=+1 & B_2=+1) = x + y

Pr(A_1=+1 & B_3=+1) = x + w   Pr(A_3=+1 & B_2=+1) = y + z

Conclude		
Pr(A_1=+1 & B_2=+1)	$\leq$ Pr(A_1=+1 & B_3=+1) +	Pr(A_3=+1 & B_2=+1)
x + y	x + w	y + z

Figure 6.8 A form of Bell's inequalities.

y and z assigned respectively to $(+, -, -; -, +, +)$ and $(+, +, -; -, -, +)$. And finally, $Pr((A_3 = +1)\&(B_2 = +1))$ is the sum of the measures x and w assigned respectively to $(+, -, +; -, +, -)$ and $(-, -, +; +, +, -)$. Thus, since $x + y$ must be less than or equal to $y + z + x + w$,

$$(B) \; Pr((A_1 = +1)\&(B_2 = +1)) \le Pr((A_1 = +1)\&(B_3 = +1))$$
$$+ \; Pr((A_3 = +1)\&(B_2 = +1))$$

The inequality (B) is a member of a family of inequalities known as *Bell's inequalities*. (The derivation given here is due to Wigner 1970).

To see the relevance of (B) to the issues at hand, we need only add the punch line which consists of two parts. First, there are QM states that are provably in violation of (B). Thus, insofar as the statistical predictions of QM are respected, QM does not lend itself to an extension to a "complete" theory of the kind EPR envisioned. Of course, this might be taken as further proof that something is rotten at the core of QM. But now we add the second half of the punch line: When the relevant experiments are performed, the statistics bear out the predictions of QM and violate (B). The incompletability of QM is therefore not a defect of the theory but, apparently, a reflection of the way Nature operates on the quantum level.

6.15 QM AND LOCALITY

Let us now turn from the incompleteness prong of the EPR dilemma to the non-locality prong. Here the news is both good and bad. To understand the tidings we need to separate two senses of locality, namely, *setting independence* and *outcome*

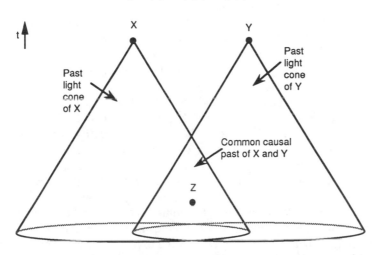

Figure 6.9 Reichenbach's principle of common cause.

independence (see Jarrett 1984, whose terminology is somewhat different from that used here). Setting independence requires that the probability of an outcome on one wing of the experiment is independent of the choice of what is measured on the other wing or, equivalently, of the setting of the measuring device on the other wing. QM is fully in accord with this demand, and as a consequence the QM correlations between the two particles do not allow messages to be sent from one wing to the other by means of a manipulation of the setting on the apparatus on one side and an observation of a change in the outcome statistics on the other side. (There can be no Bell telephone.)

Outcome independence demands that the particular measurement outcome on one wing does not affect the probability of the outcome on the other wing, that is, $Pr(A_i = a/B_j = b) = Pr(A_i = a)$, or, equivalently, $Pr((A_i = a)\&(B_j = b)) = Pr(A_i = a) \times Pr(B_j = b)$. This condition can be violated by QM probabilities since, for example, in the anticorrelated state QM predicts that $Pr(A_1 = +1/B_1 = +1) = 0 \neq Pr(A_1 = +1)$. This by itself is not the bad news. After all, correlations among distant events are daily facts of life. The bad news is rather that the correlations cannot be explained in accord with what Reichenbach called *the principle of common cause* (PCC). Consider events X and Y that occur at relatively space-like regions. If X and Y are not probabilistically independent, that is, $Pr(X\&Y) \neq Pr(X) \times Pr(Y)$, then PCC demands that there is an event Z in the common causal past of X and Y (see Figure 6.9) which induces conditional probabilistic independence, that is, $Pr(X\&Y/Z) = Pr(X/Z) \times Pr(Y/Z)$. Equivalently, Z probabilistically screens X and Y off from one another: $Pr(X/Y\&Z) = Pr(X/Z)$ and $Pr(Y/X\&Z) = Pr(Y/Z)$. Applying this to the present case, X and Y would be the outcomes of measurements performed on the two wings of the EPR experiment in such a manner that the measurement events are relatively space-like and Z would be the obtaining of some condition λ at the sources of the pairs of anticorrelated particles. But in the case of perfect anticorrelation, the requirement that $Pr((A_i = a)\&(B_j = b)/\lambda) = Pr(A_i = a/\lambda) \times Pr(B_j = b/\lambda)$ implies that $Pr(A_i = a/\lambda)$ and $Pr(B_j = b/\lambda)$ are both 0 or 1. Thus, the PCC essentially forces us into the situation described in the preceding section. Even when the requirement of perfect anticorrelation is dropped, the PCC can be shown to lead to a second family of Bell inequalities each of which is more complicated than (B) but which collectively are also violated by QM statistics. Thus, just as Nature speaks against the extendibility of QM to a "complete" theory it also speaks against the extension to a theory satisfying the PCC.

Just how bad is this bad news? Perhaps not so bad after all. Consider again the deterministic case. In the setting of field theory it is plausible that determinism be implemented locally by what has been called *Einstein locality*; that is, the state in a region R should be determined by the state on a slice S contained in the past light cone of R. In generalizing from the deterministic to the stochastic case the most natural generalization of Einstein causality is not the requirement that probabilities for outcomes in region R be screened off from the outcomes in the relatively space-like region R' (see Figure 6.10) by events in the common causal part of R and R' but rather the requirement that the probabilities of outcomes in R be determined by the state on S. This latter requirement is one which we would expect to be satisfied in a relativistic quantum field theory (see Hellman 1982a).

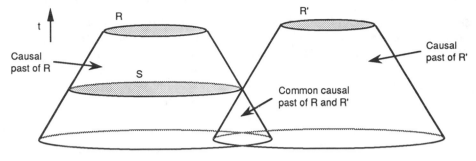

Figure 6.10 Einstein locality.

6.16 QM AND DETERMINISM

It might seem that what has already been said above suffices to undermine determinism in the quantum domain. For the attempt to foist the assumption of value determinateness onto QM and to understand QM probabilities as expressions of our ignorance of preexisting values was shown via Bell's inequalities to be incompatible with verifiable quantum statistics. So determinism, which implies value determinateness, is also incompatible with the results of the EPR-Bell experiments.

This last step is a little too quick. To see why, recall classical phase space introduced in Section 6.4 and think of quantum observables as analogous to classical observables such as "the roulette ball lands on red 10." Determinism means that for any initial state $\lambda_0 = \lambda(t_0)$, the state $\lambda(t)$ at any later time t is uniquely fixed; and by the completeness of states, if the roulette experiment is performed at t, $\lambda(t)$ fixes a definite value for the observable in question—say, $+1$ if the ball does indeed land on red 10, -1 otherwise. But the kind of value determinateness needed for the proof of the Bell inequalities is of a different kind; it requires that for every $\lambda \in \Lambda$ a definite value is assigned to the observables, and this is not done in any direct way in our classical analogue since for some and perhaps all the points lying on the uniquely determined trajectory through λ_0 the roulette experiment may not be performed. Nevertheless, it might be thought that an indirect assignment can be made since determinism will imply the truth (or falsity as the case may be) of "If the roulette experiment were performed at t, then the result would be red 10." But to imagine that, contrary to the fact, the roulette wheel is spun at t is to imagine that we move off the uniquely determined trajectory through λ_0 to some nearby trajectory where the counterfactual spin is realized. But there may be many such actualizing trajectories, all equally near to the original, and in some of them red 10 many eventuate and in others not. Indeed, the instability of classical systems that display randomness and chaos at the macrolevel would seem to underwrite such a finickiness with respect to red 10. In counterfactual logic there is no guarantee that either "If the roulette wheel were spun, then the result would be red 10" or "If the roulette wheel were spun, then the result would not be red 10" is true. The same may be true of quantum observables even on the supposition that there is an underlying deterministic mechanism.

There is yet another and more plausible way that counterfactual conditionals can be used to defeat determinism in the quantum domain. Suppose that a great many repetitions of the EPR-Bell experiment are performed and that on the nth trial the left appartus is set to measure, say, A_2 while the right appartus is set to measure B_1, and suppose that on this trial $A_2 = +1$. The assumption of *counterfactual definiteness* asserts that if the right apparatus had been set differently (say to measure B_2) the outcome on the left would still have been $+1$. Under this assumption we can derive yet another form of the Bell inequalities which are again provably inconsistent with quantum statistics. Now the assumption of counterfactual definiteness is implausible if the world is at base stochastic, for then if the experiment is repeated, even under identical conditions, there is no reasonable expectation that the outcome will be the same. But if the world is deterministic and, more particularly, if Einstein causality holds so that the outcome on the left is determined by local conditions on the left, then counterfactual definiteness is much more plausible (see Hellman 1982b). In this way the Bell inequalities can be combined with the observation of experimental violations of the inequalities to provide a plausibility argument against determinism in the quantum domain.

6.17 DETERMINISM, INDETERMINISM, AND FREE WILL

The commonsensical position regarding free will was forcibly expressed by Dr. Johnson, "Sir, we *know* that our will is free, and *there's* an end on't." But, with apologies to the good doctor, we know no such thing. What we know is that in normal circumstances our actions are accompanied by the feeling of freedom—we feel in control, we feel that our actions are up to us. However, we have to face the possibility that our feelings are misleading, and worry over this possibility is not a purely academic one since the application of moral and legal sanctions for wrongful acts is widely held to turn on the question of whether the agent acted freely.

The attempt to view moral agents as a stream of events runs into difficulties whether the stream moves deterministically or indeterministically. One longstanding tradition in philosophy is labeled *incompatibilism* (or *hard determinism*). It asserts that if determinism is true, then there is no free will. Another venerable tradition can be labeled *supercompatibilism* (or *hard indeterminism*).[6] It asserts not only that determinism is compatible with free will but also that determinism is required for free will. Or to put it in a form that parallels our formulation of compatibilism, if indeterminism is true, then there is no free will. But since either determinism or indeterminism must be true, the only way both positions can hold is for there to be no free will.

Incompatibilists assume that if agents acted freely they must have had available to them genuine alternatives. The test typically used for the availability is that they "could have done otherwise." Using a tracing back construction the incompatibilist then argues that in a deterministic setting the agents could not have done otherwise. For trace back the causal antecedents of the act in question to a time before which the agents were born. The physical state of the universe at that time together with the relevant laws of physics then uniquely fixes the action (at least insofar as it is physically characterizable); but surely the earlier state is beyond the agents' control. In the Discussion Questions the reader is asked to consider a detailed version of this

---

[6] The more usual terminology speaks of *compatibilism* or *soft determinism*. The view labeled here as *hard incompatibilism* is defended by Hobart (1966); see below.

tracing back construction. For present purposes it suffices to consider a cruder but still intuitively effective means of motivating the incompatibilist viewpoint. It starts from the idea that what counts for freedom is the power to do otherwise *in the same circumstances*. (If I am in a straightjacket and unable to play tennis, it is cold comfort indeed to be told that I have the power to play tennis in other circumstances, namely, when I am not strapped up.) It continues with the idea that if I have the power to do X, then there must be possible cases where I do X. But if the circumstances include the physical state just prior to the action in question, then determinism implies that there is no physically possible case where I do other than I actually did while the same circumstances obtain and, hence, that I could not have done otherwise in the same circumstances.

Turning to the opposite side of the coin, supercompatibilists claim that an undetermined action is a misnomer—it is something that just happens to the agents rather than something they do. R.E. Hobart's (1966) version of this claim was that if the agent's behavior was undetermined, "it [was] just as if his legs should suddenly spring up and carry him off where he did not prefer to go" (p. 70). This may not seem correct since we do not seem to have any trouble distinguishing cases where agents' actions are in accord with their desires from cases where a spontaneous jerk or twitch produces behavior not in accord with their desires. But it does pose the challenge of giving an account of agency in the indeterministic setting. In the first place we cannot maintain that Jane was the author of her action in the strong sense that her choosing or willing, together with the underlying physical state, determined her action A; for whatever exactly this phrase means it is presumably incompatible with the consequence that, because of indeterminism, in the very same circumstances—including her choosing and willing—her behavior might not have included A. It remains open that Jane's choosing or willing indeterministically caused A. Here we can appeal to two ideas: First, that, as in QM, indeterminism does not entail complete lawlessness but may be expressed by means of probabilistic laws; and second, that Jane's choosing or willing is a probabilistic cause of A in that it raises the lawful probability of A's happening. It is doubtful, however, that such a probabilistic link is strong enough to sustain a robust sense of agency. If the probabilistic laws involved are like those of QM, then given all of the particulars of the circumstances, including Jane's choosing and willing, it was still a matter of chance that her behavior included A. Her choosing and willing were not causes of A in the sense that counts in the courts to determine responsibility in tort cases: that is, but for her willing and choosing, A would not have occurred. (This was pointed out by Arthur Fine (forthcoming). See Nozick (1981) for an attempt to describe indeterministic choices.)

In a desperate bid to rescue agency from the incompatibilists, Chisholm (1982) proposes to remove the self from the flow of events in the physical world. The supercompatibilist is also circumvented because, according to Chisholm, the self determines or brings about an action in manner that cannot be translated into talk about events determining or causing other events.[7] The price of this rescue is to remove the free actions of mankind from the ken of science, for science codifies its results in laws, deterministic or indeterministic, stated in terms of connections between happenings in the world of space and time.

[7] Chisholm's (1982) doctrine is suggested by Aristotle's dictum that "A staff moves a stone, and is moved by a hand, which is moved by a man."

It seems then that the attempt to locate human agents in nature either fails in a manner that reflects a limitation on what science can tell us about ourselves, or else it succeeds at the expense of undermining our cherished notion that we are free and autonomous agents.

DISCUSSION QUESTIONS

1. No general definition of determinism, either for theories or for physical systems, has been given in this chapter. Remedy this defect by providing what you take to be the most fruitful definition.

2. Suppose that any true and contingent universal generalization that makes no reference to particular individuals or particular spatiotemporal locations is counted as a law of nature. Then the doctrine of determinism trivializes. Discuss.

3. Consider a case of Newtonian particles that escape to spatial infinity in a finite time. (Draw the spacetime history of such an escape particle. How can it reach spatial infinity without ever going infinitely fast?) It would seem that in such a case mass and particle number are not conserved. Does the violation of these conservation principles show that this scenario is not genuinely physically possible?

4. Is deterministic prediction possible in Minkowski spacetime, the setting for special relativity theory? (a) Define the *domain of deterministic prediction* $DP(R)$ for a region R of Minkowski spacetime to consist of all those points which (i) are *not* in $C^-(R)$ but (ii) *are* in $D^+(C^-(R))$. Here, $C^-(R)$ (the *causal past* of R) consists of all those points in the past light cone of R, and $D^+(X)$ (the *future domain of dependence* of X) consists of all those points p such that any causal curve through p can be extended in the past direction until it meets X. Show that for typical regions R of Minkowski spacetime, $DP(R)$ is empty. (b) Justify the definition given in (a). (c) Do the results of (a) and (b) show that deterministic prediction is not possible in relativity theory?

5. (Alternative derivation of Bell's inequalities) Let

$$\mathcal{A}_i^+ \equiv \{\lambda \in \Lambda : A_i = +1\}$$

etc.

$$\mathcal{B}_i^+ \equiv \{\lambda \in \Lambda : B_i - +1\}$$

In this notation, the anticorrelation condition says that

$$\mathcal{A}_i^- \cap \mathcal{B}_i^- = \varnothing = \mathcal{A}_i^+ \cap \mathcal{B}_i^+$$

Show that

$$\mathcal{A}_i^- \cup \mathcal{B}_i^- = \Lambda = \mathcal{A}_i^+ \cup \mathcal{B}_i^+$$

and that

$$\mathcal{A}_1^+ \cap \mathcal{B}_2^+ = (\mathcal{A}_1^+ \cap \mathcal{B}_2^+ \cap \mathcal{A}_3^+) \cup (\mathcal{A}_1^+ \cap \mathcal{B}_2^+ \cap \mathcal{B}_3^+).$$

Use these results to derive

$$Pr((A_1 = +1)\&(B_2 = +1)) \leq Pr((A_3 = +1)\&(B_2 = +1))$$

$$+ Pr((A_1 + 1)\&(B_3 = +1)).$$

6. Evaluate the following attempt to show that determinism and free will are incompatible: A necessary condition for agents to have acted freely is that they could have done otherwise; but determinism means that no one could have done other than they did in fact do. To establish the second claim, consider the following argument offered by Peter van Inwagen (1982). Let L stand

for the laws of physics and let P_0 and P_1 stand respectively for the propositions that express that physical states of the universe at times t_0 and t_1. Choose t_0 to be a time before which person J was born. If J did not in fact raise his hand at t_1 and determinism is true, then J could not have raised his hand at t_1 because:

a. Determinism means that P_0&L entails P_1.

b. If J had raised his hand at t_1, then P_1 would be false.

c. If (b) is true, then if J could have raised his hand at t_1, J could have rendered P_1 false.

d. If J could have rendered P_1 false, and if P_0&L entails P_1, then J could have rendered P_0&L false.

e. If J could have rendered P_0&L false, then either J could have rendered P_0 false or else J could have rendered L false.

f. But at t_1 J could not have rendered P_0 false because the state of the universe before J was born is beyond J's control.

g. And at t_1, or any other time, J could not have rendered L false because no human agent has the power to determine what is and what is not a law of nature.

h. Therefore, J could not have raised his hand at t_1.

7. Bell's theorem uses probabilistic inequalities to show that the assumptions of EPR are incompatible with QM. A new proof which dispenses with these inequalities is given by D.M. Greengberger et al., "Bell's theorem without inequalities," *American Journal of Physics 58* (1990): 1131–1142. Summarize the main steps of the proof.

SUGGESTED READINGS

EARMAN, JOHN (1986), *A Primer on Determinism*. Dordrecht: Reidel. See this text for further details of determinism in classical and relativistic physics.

HUGHES, R. I. G. (1989), *The Structure and Interpretation of Quantum Mechanics*. Cambridge, MA: Harvard University Press. A readable introduction to quantum mechanics and its foundations problems.

CUSHING, JAMES T. and ERWIN McMILLAN (eds.) (1989), *Philosophical Consequences of Quantum Theory: Reflections on Bell's Theorem*. Notre Dame: University of Notre Dame Press. Discusses Bell's inequalities and their implications.

BEROFSKY, BERNARD (ed.) (1966), *Free Will and Determinism*. New York: Harper & Row.

WATSON, GARY (ed.) (1982), *Free Will*. Oxford: Oxford University Press.

Two excellent collections of articles on the determinism–free will problem.

Appendix: Can Quantum-Mechanical Description of Physical Reality Be Considered Complete?*

Albert Einstein, Boris Podolsky, and Nathan Rosen

In a complete theory there is an element corresponding to each element of reality. A sufficient condition for the reality of a physical quantity is the possibility of predicting it with certainty, without disturbing the system. In quantum mechanics in the case of

* Originally published in *Physical Review*, 47, 777–80 (1935). Copyright, America Physical Society. Permission granted by The Albert Einstein Archives, The Hebrew University of Jerusalem, the editors of *The Physical Review*, and the editors of the Einstein Papers, Boston University.

two physical quantities described by non-commuting operators, the knowledge of one precludes the knowledge of the other. Then either (1) the description of reality given by the wave function in quantum mechanics is not complete or (2) these two quantities cannot have simultaneous reality. Consideration of the problem of making predictions concerning a system on the basis of measurements made on another system that had previously interacted with it leads to the result that if (1) is false then (2) is also false. One is thus led to conclude that the description of reality as given by a wave function is not complete.

1.

Any serious consideration of a physical theory must take into account the distinction between the objective reality, which is independent of any theory, and the physical concepts with which the theory operates. These concepts are intended to correspond with the objective reality, and by means of these concepts we picture this reality to ourselves.

In attempting to judge the success of a physical theory, we may ask ourselves two questions: (1) "Is the theory correct?" and (2) "Is the description given by the theory complete?" It is only in the case in which positive answers may be given to both of these questions, that the concepts of the theory may be said to be satisfactory. The correctness of the theory is judged by the degree of agreement between the conclusions of the theory and human experience. This experience, which alone enables us to make inferences about reality, in physics takes the form of experiment and measurement. It is the second question that we wish to consider here, as applied to quantum mechanics.

Whatever the meaning assigned to the term *complete*, the following requirement for a complete theory seems to be a necessary one: *every element of the physical reality must have a counterpart in the physical theory*. We shall call this the condition of completeness. The second question is thus easily answered, as soon as we are able to decide what are the elements of the physical reality.

The elements of the physical reality cannot be determined by *a priori* philosophical considerations, but must be found by an appeal to results of experiments and measurements. A comprehensive definition of reality is, however, unnecessary for our purpose. We shall be satisfied with the following criterion, which we regard as reasonable. *If, without in any way disturbing a system, we can predict with certainty (i.e., with probability equal to unity) the value of a physical quantity, then there exists an element of physical reality corresponding to this physical quantity*. It seems to us that this criterion, while far from exhausting all possible ways of recognizing a physical reality, at least provides us with one such way, whenever the conditions set down in it occur. Regarded not as a necessary, but merely as a sufficient, condition of reality, this criterion is in agreement with classical as well as quantum-mechanical ideas of reality.

To illustrate the ideas involved let us consider the quantum-mechanical description of the behavior of a particle having a single degree of freedom. The fundamental concept of the theory is the concept of *state*, which is supposed to be completely characterized by the wave function ψ, which is a function of the variables chosen to describe the particle's behavior. Corresponding to each physically observable quantity A there is an operator, which may be designated by the same letter.

Determinism in the Physical Sciences

If ψ is an eigenfunction of the operator A, that is, if

$$\psi' \equiv A\psi = a\psi, \tag{1}$$

where a is a number, then the physical quantity A has with certainty the value a whenever the particle is the state given by ψ. In accordance with our criterion of reality, for a particle in the state given by ψ for which Eq. (1) holds, there is an element of physical reality corresponding to the physical quantity A. Let, for example,

$$\psi = e^{(2\pi i/h)p_0 x}, \tag{2}$$

where h is Planck's constant, p_0 is some constant number and x the independent variable. Since the operator corresponding to the momentum of the particle is

$$p = (h/2\pi i)\partial/\partial x, \tag{3}$$

we obtain

$$\psi' = p\psi = (h/2\pi i)\partial\psi/\partial x = p_0\psi. \tag{4}$$

Thus, in the state given by Eq. (2), the momentum has certainly the value p_0. It thus has meaning to say that the momentum of the particle in the state given by Eq. (2) is real.

On the other hand if Eq. (1) does not hold, we can no longer speak of the physical quantity A having a particular value. This is the case, for example, with the coordinate of the particle. The operator corresponding to it, say q, is the operator of multiplication by the independent variable. Thus,

$$q\psi = x\psi \neq a\psi. \tag{5}$$

In accordance with quantum mechanics we can only say that the relative probability that a measurement of the coordinate will give a result lying between a and b is

$$P(a, b) = \int_a^b \bar{\psi}\psi dx = \int_a^b dx = b - a. \tag{6}$$

Since this probability is independent of a, but depends only upon the difference $b - a$, we see that all values of the coordinate are equally probable.

A definite value of the coordinate, for a particle in the state given by Eq. (2), is thus not predictable, but may be obtained only by a direct measurement. Such a measurement however disturbs the particle and thus alters its state. After the coordinate is determined, the particle will no longer be in the state given by Eq. (2). The usual conclusion from this in quantum mechanics is that *when the momentum of a particle is known, its coordinate has no physical reality.*

More generally, it is shown in quantum mechanics that, if the operators corresponding to two physical quantities, say A and B, do not commute, that is, if $AB \neq BA$, then the precise knowledge of one of them precludes such a knowledge of the other. Furthermore, any attempt to determine the latter experimentally will alter the state of the system in such a way as to destroy the knowledge of the first.

From this follows that either (1) *the quantum-mechanical description of reality given by the wave function is not complete* or (2) *when the operators corresponding to two physical quantities do not commute the two quantities cannot have simulta-*

neous reality. For if both of them had simultaneous reality—and thus definite values—these values would enter into the complete description, according to the condition of completeness. If then the wave function provided such a complete description of reality, it would contain these values; these would then be predictable. This not being the case, we are left with the alternatives stated.

In quantum mechanics it is usually assumed that the wave function *does* contain a complete description of the physical reality of the system in the state to which it corresponds. At first sight this assumption is entirely reasonable, for the information obtainable from a wave function seems to correspond exactly to what can be measured without altering the state of the system. We shall show, however, that this assumption, together with the criterion of reality given above, leads to a contradiction.

2.

For this purpose let us suppose that we have two systems, I and II, which we permit to interact from the time $t = 0$ to $t = T$, after which time we suppose that there is no longer any interaction between the two parts. We suppose further that the states of the two systems before $t = 0$ were known. We can then calculate with the help of Schrödinger's equation the state of the combined system I + II at any subsequent time; in particular, for any $t > T$. Let us designate the corresponding wave function by ψ. We cannot, however, calculate the state in which either one of the two systems is left after the interaction. This, according to quantum mechanics, can be done only with the help of further measurements, by a process known as the *reduction of the wave packet*. Let us consider the essentials of this process.

Let $a_1, a_2, a_3, \ldots$ be the eigenvalues of some physical quantity A pertaining to system I and $u_1(x_1), u_2(x_1), u_3(x_1), \ldots$ the corresponding eigenfunctions, where x_1 stands for the variables used to describe the first system. Then ψ, considered as a function of x_1, can be expressed as

$$\Psi(x_1, x_2) = \sum_{n=1}^{\infty} \psi_n(x_2)u_n(x_1), \tag{7}$$

where x_2 stands for the variables used to describe the second system. Here $\psi_n(x_2)$ are to be regarded merely as the coefficients of the expansion of ψ into a series of orthogonal functions $u_n(x_1)$. Suppose now that the quantity A is measured and it is found that it has the value a_k. It is then concluded that after the measurement the first system is left in the state given by the wave function $u_k(x_1)$, and that the second system is left in the state given by the wave function $\psi_k(x_2)$. This is the process of reduction of the wave packet; the wave packet given by the infinite series (7) is reduced to a single term $\psi_k(x_2)u_k(x_1)$.

The set of functions $u_n(x_1)$ is determined by the choice of the physical quantity A. If, instead of this, we had chosen another quantity, say B, having the eigenvalues $b_1, b_2, b_3, \ldots$ and eigenfunctions $v_1(x_1), v_2(x_1), v_3(x_1), \ldots$ we should have obtained, instead of Eq. (7), the expansion

$$\Psi(x_1, x_2) = \sum_{n=1}^{\infty} \varphi_s(x_2)v_s(x_1), \tag{8}$$

where φ_s's are the new coefficients. If now the quantity B is measured and is found to have the value b_r, we conclude that after the measurement the first system is left in the state given by $v_r(x_1)$ and the second system is left in the state given by $\varphi_r(x_2)$.

We see therefore that, as a consequence of two different measurements performed upon the first system, the second system may be left in states with two different wave functions. On the other hand, since at the time of measurement the two systems no longer interact, no real change can take place in the second system in consequence of anything that may be done to the first system. This is, of course, merely a statement of what is meant by the absence of an interaction between the two systems. Thus, *it is possible to assign two different wave functions* (in our example ψ_k and φ_r) *to the same reality* (the second system after the interaction with the first).

Now, it may happen that the two wave functions, ψ_k and ψ_r, are eigenfunctions of two noncommuting operators corresponding to some physical quantities P and Q, respectively. That this may actually be the case can best be shown by an example. Let us suppose that the two systems are two particles, and that

$$\Psi(x_1, x_2) = \int_{-\infty}^{\infty} e^{(2\pi i/h)\ (x_1 - x_2 + x_0)P} dp, \tag{9}$$

where x_0 is some constant. Let A be the momentum of the first particle; then, as we have seen in Eq. (4), its eigenfunctions will be

$$u_p(x_1) = e^{(2\pi i/h)px_1} \tag{10}$$

corresponding to the eigenvalue p. Since we have here the case of a continuous spectrum, Eq. (7) will now be written

$$\Psi(x_1, x_2) = \int_{-\infty}^{\infty} \psi_p(x_2)u_p(x_1)dp, \tag{11}$$

where

$$\psi_p(x_2) = e^{-(2\pi i/h)(x_2 - x_0)p}. \tag{12}$$

This ψ_p however is the eigenfunction of the operator

$$P = (h/2\pi i)\partial/\partial x_2, \tag{13}$$

corresponding to the eigenvalue $-p$ of the momentum of the second particle. On the other hand, if B is the coordinate of the first particle, it has for eigenfunctions

$$v_x(x_1) = \delta(x_1 - x), \tag{14}$$

corresponding to the eigenvalue x, where $\delta(x_1 - x)$ is the well-known Dirac delta-function. Eq. (8) in this case becomes

$$\Psi(x_1, x_2) = \int_{-\infty}^{\infty} \varphi_x(x_2)v_x(x_1)dx, \tag{15}$$

where

$$\varphi_x(x_2) = \int_{-\infty}^{\infty} e^{(2\pi i/h)(x-x_2+x_0)p} dp = h\delta(x - x_2 + x_0). \tag{16}$$

This φ_x, however, is the eigenfunction of the operator

$$Q = x_2 \tag{17}$$

corresponding to the eigenvalue $x + x_0$ of the coordinate of the second particle. Since

$$PQ - QP = h/2\pi i, \tag{18}$$

we have shown that it is in general possible for ψ_k and φ_r to be eigenfunctions of two noncommuting operators, corresponding to physical quantities.

Returning now to the general case contemplated in Eqs. (7) and (8), we assume that ψ_k and φ_r are indeed eigenfunctions of some noncommuting operators P and Q, corresponding to the eigenvalues p_k and q_r, respectively. Thus, by measuring either A or B we are in a position to predict with certainty, and without in any way disturbing the second system, either the value of the quantity P (that is p_k) or the value of the quantity Q (that is q_r). In accordance with our criterion of reality, in the first case we must consider the quantity P as being an element of reality, in the second case the quantity Q is an element of reality. But, as we have seen, both wave functions ψ_k and φ_r belong to the same reality.

Previously we proved that either (1) the quantum-mechanical description of reality given by the wave function is not complete or (2) when the operators corresponding to two physical quantities do not commute the two quantities cannot have simultaneous reality. Starting then with the assumption that the wave function does give a complete description of the physical reality, we arrived at the conclusion that two physical quantities, with noncommuting operators, can have simultaneous reality. Thus the negation of (1) leads to the negation of the only other alternative (2). We are thus forced to conclude that the quantum-mechanical description of physical reality given by wave functions is not complete.

One could object to this conclusion on the grounds that our criterion of reality is not sufficiently restrictive. Indeed, one would not arrive at our conclusion if one insisted that two or more physical quantities can be regarded as simultaneous elements of reality *only when they can be simultaneously measured or predicted*. On this point of view, since either one or the other, but not both simultaneously, of the quantities P and Q can be predicted, they are not simultaneously real. This makes the reality of P and Q depend upon the process of measurement carried out on the first system, which does not disturb the second system in any way. No reasonable definition of reality could be expected to permit this.

While we have thus shown that the wave function does not provide a complete description of the physical reality, we left open the question of whether or not such a description exists. We believe, however, that such a theory is possible.

Seven

PHILOSOPHY
OF BIOLOGY

James G. Lennox

Does biology deploy concepts, patterns of explanation, theories and research methods which are fundamentally different from those found in the physical sciences? If so, is this in principle the case, or is biology destined to gradually merge with chemistry and atomic physics as a subdiscipline? Or to put the question from the standpoint of the objects of biological research—what, if anything, is special about living things? These are the underlying questions which motivate much of the philosophy of biology today.

In practice, biology consists of a loosely connected set of disciplines ranging from those which study interactions among large groups of organisms spread across space and time—ecology, biogeography, paleontology—to those which focus on biochemical processes at the molecular and submolecular level. Those theoretical achievements in our century which have managed to unify these disciplines to some extent are Darwinian evolutionary theory and genetics, and it is not surprising that these theoretical disciplines have been the focus of most philosophical attention (see the introductory texts by Hull 1973 or Rosenberg 1985). The goal of this chapter is to explore certain philosophical questions regarding a central component of Evolutionary Biology, the theory of natural selection. In the following chapter, Dr. Kenneth Schaffner will focus attention on the biomedical sciences, in which the concepts and methods of the biochemist and molecular geneticist play a central role.

The scientific explanations provided by evolutionary biology are formulated with concepts such as "inclusive fitness," "adaptation," "design," "niche," and "genetic drift," and are often overtly historical and/or teleological in character. In this chapter, then, our goal is to understand these concepts and types of explanation. This will contribute to understanding why it is that evolutionary theory is a central unifying theory within biology. Doing so is all the more important given the recent revival of the Christian fundamentalist claim that evolutionary biology is no more scientific than is the biblical account of life's history in the book of Genesis. By the end of this chapter you will be able to evaluate this claim.

In a manner similar to Chapter 5, this chapter will approach the philosophical questions raised by current evolutionary theory by tracing the historical emergence of those questions as the theory developed. Just as so many of the philosophical issues in contemporary physics can be traced back to Einstein, so in evolutionary biology most, if not all, roads lead back to Charles Darwin.

7.1 THE DARWINIAN ORIGINS

Scientific theories do not spring on stage fully developed, with their range fully established and with their empirical credentials in hand. They may begin as the recommendation of an alternative kind of answer to traditional questions or from the recognition of new questions in need of an answer. They develop historically, often in ways unimagined by their originators. Questions having to do with the proper way to define a theory's basic concepts, the nature of its empirical support, the best way to formulate and interpret its explanations, the relation of its concepts and principles to other theories in other sciences, emerge from this development—and the form answers to these questions take often helps to shape its further development. In order to understand and evaluate the answers to these philosophical questions, it helps to have a sense of a theory's history. It will serve us well, then, to begin with the following summary of the Darwinian theory of evolution, penned by its originator, Charles Darwin:

> If during the long course of ages and under varying conditions of life, organic beings vary at all in the several parts of their organisation, and I think this cannot be disputed; if there be, owing to the high geometrical powers of increase of each species, at some age, season, or year, a severe struggle for life, and this certainly cannot be disputed; then, considering the infinite complexity of the relations of all organic beings to each other and to their conditions of existence, causing an infinite diversity in structure, constitution, and habits, to be advantageous to them, I think it would be a most extraordinary fact if no variation ever had occurred useful to each being's own welfare, in the same way as so many variations have occurred useful to man. But if variations useful to any organic being do occur, assuredly individuals thus characterised will have the best chance of being preserved in the struggle for life; and from the strong principle of inheritance they will tend to produce offspring similarly characterised. This principle of preservation, I have called, for the sake of brevity, Natural Selection. (Darwin [1859] 1964, 126–127)

This model of evolution by natural selection was first outlined by Charles Darwin (1809–1882) in a sketch written nearly twenty years before the above summary, from *On the Origin of Species,* was published. It provides us with an *explanatory pattern,* identifying in abstract form the kinds of processes and conditions that will produce modifications in species by means of natural selection. By referring to this passage as providing an *explanatory pattern* I mean that, while it is not itself an explanation, it *does* tell the reader the kinds of entities and processes to be mentioned in an ideal evolutionary explanation (Kitcher 1985a, 132–139; Brandon 1990, 159–183).

The Darwinian model of evolution by natural selection begins by specifying five fundamental features of organic life:

1. organic populations are parts of an *ancestor-descendant history;*
2. the members of such populations *inherit* traits from their ancestors and *pass them on* to their descendents;
3. they also *vary* with respect to those heritable traits;
4. owing to their tendency to increase their numbers geometrically, the members of such populations *compete* with each other for limited resources;
5. the environment in which they live is infinitely complex and constantly changing.

Given these facts, Darwin tells us that the probabilities are high that (a) some of the variations mentioned in (3) will put their possessors at a competitive advantage relative to others; (b) the organisms with these *advantageous* variations will have the best chance of surviving; and (c) they will thus tend to leave a disproportionate number of offspring with these variations in the next and succeeding generations.

Darwin spent the first four chapters of the *Origin*—the pages leading up to the summary quoted above—providing strong warrant for the truth of (1)–(5). Much effort has been spent in recent years trying to make a conclusion about evolutionary change follow deductively from some reconstruction of that summary. This misses the point. Darwin was presenting a *causal model,* not a deductive proof, and the above passage's primary concern is to specify abstractly a set of processes which, were they to interact in certain ways over long periods of time, would "mechanically" produce certain patterns of change.

At the core of Darwinian evolutionary theory then, is a set of abstract propositions which identify the causal basis of natural selection. While we now know that it is possible for evolutionary change within populations to take place without the operation of natural selection, the principle of natural selection remains the essential core of evolutionary explanation.

Given the central role of the processes of variation, inheritance and selection in this model, Darwin's candid admission of ignorance regarding the mode of operation of all three is remarkable. At the opening of his discussion of the laws of variation he explains that to describe them as due to chance "serves to acknowledge plainly our ignorance of the cause of each particular variation" (Darwin [1859] 1964, 131). Early in the first chapter he notes that "[t]he laws governing inheritance are quite unknown" (ibid., 13). And finally, immediately after the summary of his theory quoted earlier, Darwin seeks to disarm his critics with the following disclaimer:

> Whether natural selection has really thus acted in nature, in modifying and adapting the various forms of life to their several conditions and stations, must be judged of by the general tenor and balance of evidence given in the following chapters. (ibid., 127)

The point of this remark is that, at the time Darwin presented the theory, no one, himself included, had actually observed the operation of selection in wild populations, though Darwin had done a careful review of the ways in which breeders of domestic animals and plants produce distinct varieties. In Darwin's view, the best evidence for the theory's truth was its ability to explain, by reference to the above causal model, a vast number of apparently unconnected biological generalizations—patterns found in the fossil record, in the current distribution of animals and plants around the world, in their development, anatomy and classification. In Darwin's time this sort of support was referred to as a "consilience of inductions,"[1] and it was with considerable frustration that he noted how reviewers of the *Origin* conveniently ignored this part of his argument.

Nonetheless, ignorance regarding the mechanisms of inheritance, variation and natural selection made it possible for critics to voice reasonable doubts about the theory put forward in the *Origin*. For natural selection is *not* an inevitable consequence of combining *any* mechanism of inheritance with *any* degree and kind of variabilty under *any* conditions of competition, as a number of reviewers were quick to observe. It all depends on the nature of inheritance, on the amount, sources and extent of variation and on the dynamics of populations. Furthermore, even had Darwin established that species could originate through natural selection, that would not establish it as the only source of new species. Darwin had left much work to be done.

And much of it *was* done, by a legion of biologists between 1860 and 1950, many of whom gave little thought to the relationship of their work to Darwin's. In the process of trying to work out the laws governing the formation of hybrids in plants, a scientifically trained monk named Gregor Mendel derived certain basic theorems about the process of inheritance. Mendel determined that one "factor" or "element" from the reproductive cells of each parent plant (to be coined a "gene" by W. Johannsen in 1909) combined to determine the character of each trait of the offspring, that these factors continued to exist, independently of each other, in the cells of the offspring, and were distributed in the next generation according to the laws of chance. (Mendel's experiments will be discussed in more detail momentarily.)

Mendel's results were published in 1865, but they were not integrated into the study of inheritance and variation until after Mendel's work was independently cited as precedent by three different researchers in 1900, and its importance championed by the British geneticist William Bateson. Study of the processes involved in the production of the reproductive cells (gametes) and fertilization were becoming experimentally more sophisticated during the same period, and in the early 1900s the first hints emerged that the *chromosomes,* microscopically visible structures within the cell nucleus, found in varying but species-constant numbers throughout the animal and plant kingdoms, might somehow "carry" Mendel's "parental factors" (*genes*).

---

[1] Vaguely assumed to derive from Sir Issac Newton's ([1687] 1962) Rules of Philosophizing, this notion was popularized in the nineteenth century by Whewell (1837) and Herschel ([1830] 1987). In Whewell's words, ". . . the evidence in favour of our induction is of a much higher and more forcible character when it enables us to explain and determine cases of a kind different from those which were contemplated in the formation of our hypothesis." Compare Herschel, Chapter 6, Sections 181–182.

This suggestion was initially based on interesting similarities between the behavior of chromosomes during the production of male and female sex cells (called *meiosis*) and their eventual union (*fertilization*) on the one hand, and the behavior of Mendelian factors on the other. In both cases there was an initial halving of the number of factors, followed by a pair-wise recombination which restored the original number.

Elegant experiments with populations of fruit flies in the laboratory of T. H. Morgan at Columbia University beginning in 1910 established the association of Mendelian factors, or genes, with chromosomes beyond reasonable doubt. Work by Morgan and many others opened up a host of inquiries regarding the physical characteristics of these genes. Elaborate experimental methods were created for determining their positions on the chromosome, and their patterns of transmission, as well as the precise character of the processes of cell division and fertilization.

During the same period numerous researchers were beginning to investigate the causes of *genetic mutation* (apparently random changes in genes) and to measure rates of mutation in experimental populations. Specific questions regarding the role of the environment in gene expression led to the important distinction between the *genotype* and *phenotype* (that is, between the *inherited source* of a trait and its *observable expression*). Initially, research intended to test, modify and extend the Mendelian model of inheritance focused on cases of discontinuous variation originating as "mutations," that is, cases in which the variations in a trait were few and discrete, with no intermediate forms—for example, populations with red and white flowers and no intermediate colors. Some, such as Hugo De Vries and T. H. Morgan, went so far as to suggest that mutation was the crucial process in creating new species, relegating selection to the less "creative" role of eliminating unfit mutants.

Darwin, on the other hand, had insisted that small, continuous variations served as the material for selection, and that selection produced species by a slow process of "adding up" these small differences in certain directions. A number of his followers, involved with the careful measurement of changes in continuous variations, thus saw themselves as the defenders of orthodox Darwinism, upholders of Darwin's motto, "Natura non facit saltum" (Nature makes no leaps). Their careful mathematical analyses revealed patterns in the transmission of characteristics from parent to offspring populations that could not be explained along simple Mendelian lines. Thus in the period from 1900–1915 there appeared to be a conflict between "mutationist" and "selectionist" theories of evolution. Eventually, however, researchers were able to demonstrate that such patterns could result from the interactions of more than one pair of genes (*polygenic* inheritance), or of interactions between the *genotype* and the environment. Gradually then, Darwin's insistence that slight, continuously varying traits were the "raw materials" upon which natural selection operated became compatible with a modified Mendelian mechanism of inheritance.

Mendelism also helped Darwinism resolve one of its longstanding problems. Though admitting his ignorance of the laws of inheritance, Darwin often imagined that the different characteristics of parents would *blend* in their offspring, resulting in characteristics *intermediate* between those of the parents. For example, a mating between a long-legged and a short-legged wolf would produce cubs with legs intermediate in length between those of the parents. An astute critic of the *Origin*, Fleeming Jenkin, pointed out that when such a theory of inheritance is combined with

Darwin's insistence that selection acts only on slight individual differences, even a trait with a very great selective advantage would have very little evolutionary impact (Jenkin [1867] 1983, 312–320). Darwin recognized the power of Jenkin's argument, and systematically changed later editions of the *Origin* to stress a model whereby selection favors not rare individuals with slight differences, but all the organisms with variations tending in the advantageous direction. In the following passage, for example, after noting that even if a strong advantage were conferred on an individual bird with a strongly curved beak, blending would prevent selection changing the species in this direction, he goes on to suggest a solution:

> [B]ut there can hardly be a doubt . . . that this result would follow from the preservation during many generations of a large number of individuals with more or less strongly curved beaks, and from the destruction of a still larger number with the straightest beaks. ([1872] 1962, 101)

Here, Darwin is trying to solve a problem which arises by assuming that the units of inheritance blend during mating. Mendelian genes, however, do not blend when they combine—each generation sees a reshuffling of these stable and independent units of inheritance, influenced by occasional mutations and migrations of new genes into the population. This picture of inheritance laid to rest worries that the blending of slight variations with the population norm would prevent selection from affecting any serious evolutionary change. One of the initial triumphs of neo-Darwinism, in fact, was to show that a single mutation, conferring only a slight reproductive advantage on its possessor, could spread rather rapidly through a large interbreeding population. Initial resistance to a Mendelian mechanism of inheritance had, by the 1930s, given way to whole-hearted acceptance. (For a more detailed and less anachronistic telling of the story, see Provine 1971.)

A crucial development in insuring the integration of Mendelian genetics with a Darwinian evolutionary theory was generalizing Mendel's results from the special case of self-fertilizing hybrid crosses to what one actually finds in nature, namely, randomly interbreeding wild populations. This step in the development was so small as to be recognizable only in retrospect. In response to a casual question put to him during a cricket match by the experimentalist R.C. Punnett, the mathematician G. H. Hardy pointed out that Mendel's laws, derived from the crossing of pure lines followed by repeated self-fertilization of the resulting hybrids, could be generalized to apply to large randomly breeding populations.[2] Neither Punnett nor Hardy had any idea of the significance of this conversation for evolutionary theory—without realizing it, they had created population genetics.

Population genetics is the statistical study of patterns of inheritance and variation in organic populations. Recall that Mendel experimented with populations of pea plants which differed discontinuously from one another with respect to seven different characters. Mendel sought to discover what would result from systematically mating the plants with these discrete traits of the same character. For example, in one case (see

---

[2] The founding document is a brief note written by a mathematician to correct an error made by a biologist about the implications of Mendel's laws for "mixed" populations. See Hardy ([1908] 1959).

Table 7.1) Mendel fertilized ova from pea plants with Tall stocks with pollen from those with Short stocks. The resulting first filial generation (F1) of plants were all Tall. He then took careful steps to insure that all these F1 plants were *self-fertilized* (i.e., pollen from a given plant fertilized that plant's own egg cells). The result of these "selfings" was that the character (Tall) which had appeared in *all* the plants of the F1 generation appeared in approximately ¾ of those of the second filial generation (F2), while the alternative parental character [Short] appeared in approximately ¼ of the total number of F2 plants. This 3:1 ratio of Tall to Short plants that Mendel observed in the F2 generation was to be expected on three assumptions: (1) that each parent's contribution to their offspring remains *distinct* in the F1 generation—they neither blend with nor destroy one another; (2) that these contributions tend to combine, during self-fertilization, according to the laws of chance; and (3) that one parental character is *dominant,* that is, that when a seed possesses two *different* parental characters the resulting plant always appears like one rather than the other parent, in this case like the Tall parent. (Remember that a common assumption at this time, one which Darwin sometimes seemed to share, was that the offspring would have a characteristic intermediate between those of the parents. Mendel's results contradicted this assumption.)

If these assumptions were true, then the F2 Tall plants, while they all *looked* alike, must be a mixture of two different sorts of plants; those which were of the pure Tall type, and some which were hybrids of Tall and Short, with Tall dominating. The Short plants, on the other hand, must be purely Short, for if they were hybrids the Tall character would dominate and they wouldn't appear short.

Allowing these F2 plants to self-fertilize, therefore, constituted an immediate *test* of these assumptions. If these assumptions were correct, then all the Short plants should produce nothing but Short offspring; but while approximately one third of the Tall plants should produce nothing but Tall offspring, the other two thirds should produce a mixture of Tall and Short plants, in a ratio of 3:1. The results of Mendel's experiments are represented in the following tables. As you can see, the F3 results strongly confirmed the three hypothesized assumptions.

The form of these experiments, and a number of Mendel's experimental decisions, suggest that they were designed with the above hypothesis in mind. Mendel

TABLE 7.1

Pure Parental Types	T (Tall)		t (Short)
Cross fertilization	T	x	t
First Filial Generation (F1)		All T	
Self-fertilization			
Second Filial Generation (F2)		3T:1t	
Self-fertilization			
Third Filial Generation (F3)			

	Parents	Offspring
	⅓ of T	All T
	⅔ of T	3T:1t
	All of t	All t

begins by noting that the character which apparently disappeared in the F1 hybrids reappears in a statistically regular way in the F2 generation, and reappears in precisely the parental form. (The actual ratios were never quite those noted above. His results *approximated* those; the approximations became closer the larger the sample, and those are the ratios to be expected given random segregation, chance recombination and the dominance of one character over the other, the three assumptions mentioned earlier. In the experiment just reported, for example, the actual numbers were 787 Tall/277 Short, yielding a ratio of approximately 2.84:1 (See Mendel [1865] 1966, 13.)

Mendel went on to perform a series of complicated *back cross experiments* (crossing both pure parental plants of both sexes with both types of hybrids) in order to confirm a *causal theory* he had constructed which would explain the above changes in the distributions of parental characters in subsequent generations. The theory was that the pollen cells and egg cells of plants possess independent "factors"—genes as we now say—for each character trait. Each act of fertilization thus combines two factors, the combinations being governed simply by the laws of chance. The factors responsible for different characteristics assort independently of each other, as Mendel discovered by experimenting with combinations of characters simultaneously. When two different genes for a particular trait combine in one zygote, one somehow "dominates" the other (a condition which is now called *heterozygous*). Such unobservable factors in the zygote "determine" the development of the observable characters of the organisms in each generation. This theory can be represented as the production of different gene combinations, now called *genotypes,* in each generation as in Table 7.2.

TABLE 7.2

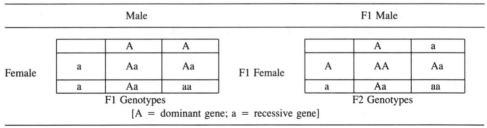

	Male				F1 Male		
		A	A			A	a
Female	a	Aa	Aa	F1 Female	A	AA	Aa
	a	Aa	aa		a	Aa	aa
	F1 Genotypes				F2 Genotypes		

[A = dominant gene; a = recessive gene]

(This manner of displaying the results of matings is called the *Punnett Square,* named for R. C. Punnett whom you met earlier in this section.)

Mendel's insights into "the laws governing the distributions of characters in hybrids" can be transformed into a formula representing the *ratio* of different genetic makeups (termed *genotypes*) in a population formed by the random mating of individuals with different forms of the same gene (*heterozygotes*). If we represent the different forms of the gene at the same locus (known as different *alleles*) by **A** and **a** respectively, that formula will look like this:

AA:2Aa:aa.

The *frequencies* of the different genotypes can then be represented as follows:

$$p^2 + 2pq + q^2 = 1$$

where p = the frequency of **A**, q = the frequency of **a**, and p + q = 1. (You may have noticed that the Hardy-Weinberg Principle is derived from the frequency of each of the alleles **A** and **a** by an application of two of the axioms of probability you learned in Chapter 2, namely, Rule 3 [special addition] and Rule 4 [special multiplication]. See Chapter 2, Section 2.7.)

This formula (which can be expanded for traits with more than two forms and for applications to characters produced by multiple genes) expresses the idea that, under conditions of random mating, the genotypic frequencies of an indefinitely large population should remain constant from one generation to the next. Such a population is in a state biologists refer to as *Hardy-Weinberg equilibrium* (HWE henceforth). (It turns out the formula worked out by Hardy during a cricket match had been presented by W. Weinberg during a lecture in Stuttgart in January of 1908.) The recombination of genes during mating in a population in HWE is "fair"—all one needs to know are the initial frequencies of alleles, and the laws of chance will determine their frequencies in subsequent generations.

An *informal "proof"* of HWE: Assume the frequencies of parental genotypes are as given by HWE, i.e., p^2 **AA** + $2pq$ **Aa** + q^2 **aa**. Then the gametes produced by the parent population will be p^2 **A** + pq **A** + pq **a** + q^2 **a** (that is, all the gametes of the **AA** parents will be **A**, ½ those of the **Aa** parents will be A and ½ **a**, and all the gametes of the **aa** parents will be **a**). So the frequency of the A allele is p and of the **a** allele q, as we can easily see.

Frequency of **A** = $p^2 + pq = p(p+q)$; but $p + q = 1$; so frequency of **A** = p.
Frequency of **a** = $q^2 + pq = q(p+q)$; but $p + q = 1$; so frequency of **a** = q.
Assuming no disruptions to random mating, these gamete frequencies will produce genotype frequencies exactly the same as those found in the parent population.

The Hardy-Weinberg Law thus gives us a "base line" with which we can compare actual *changes* in frequencies of alleles across generations of reproductive communities. Deviations from this base line indicate a disruption of this equilibrium of genotypic frequencies across generations. A number of factors may lead to such disruptions: a variety of forms of alteration of the genetic material (*mutation*), the *migration* of new genes into the population (which will change the initial frequencies), random changes in frequencies arising from sampling error (known as *genetic drift*), and *selection* favoring one genotype over another. Assuming other disruptive forces have been corrected for or ruled out for the moment, population genetics seeks to build into its models *the notion that a change in the frequency of a particular genotype is a measure of its relative fitness.*

Let us take an imaginary example to illustrate how this works. Suppose we are studying a population of land snails with shells of two different colors (Brown and Yellow) which in the laboratory we have determined are due to a single pair of alleles (**B** incompletely dominant over **Y**). Taking a census we find the initial genotypic frequencies are as follows:

$$P1 \quad \mathbf{BB} = 2000 \quad \mathbf{BY} = 4000 \quad \mathbf{YY} = 1600.$$

We wait one generation, and take a new census and find

$$P2 \quad \mathbf{BB} = 1600 \quad \mathbf{BY} = 2800 \quad \mathbf{YY} = 800.$$

Now from this we can easily calculate the survival rates for each of the three genotypes.

Survival rate (λ)

$$\mathbf{BB} = \frac{1600}{2000} = 0.8 \quad \mathbf{BY} = \frac{2800}{4000} = 0.7 \quad \mathbf{YY} = \frac{800}{1600} = 0.5$$

Next we calculate the relative fitness of each by assigning the genotype with the highest survival rate the fitness value 1, and setting the others relative to it.

Fitness (W)

$$\mathbf{BB} = 0.8/0.8 = 1 \quad \mathbf{BY} = 0.7/0.8 = 0.875 \quad \mathbf{YY} = 0.5/0.8 = 0.625$$

Next we can calculate the selection coefficient (s), which represents the *reduction* in fitness of a genotype, by subtracting the fitness value from unity $(1 - W)$.

Selection Coefficient (s) $\quad \mathbf{BB} = 0 \quad \mathbf{BY} = 0.125 \quad \mathbf{YY} = 0.375.$

To summarize: This population of snails is not in HWE. From $P1$ to $P2$ there was a significant shift in the frequencies of the genotypes in question, the homozygous Brown snails increasing in frequency from roughly 0.357 of the total in $P1$ to roughly 0.444 in $P2$, the heterozygotes increasing much more slowly, and the homozygous Yellow population decreasing in frequency significantly.

In the above example, the values for W and s were determined solely on the basis of relative survival frequencies. On the face of it, you might find this rather odd. After all, it was Darwin's original insight that differences in fitness or selective advantage were, under certain conditions, the *causes* of differences in survival frequencies. Does not population genetics treat measures of the *results* of differences in fitness as if these represented the *property* of fitness itself? Hold on to that question, for it is a good one. The concepts of fitness and selection embedded in the theoretical machinery of population genetics are problematic, and will be among our chief concerns in Section 7.3.

Putting that question aside for the moment, we calculate the frequencies for each genotype in subsequent generations by multiplying the fitness value determined for each genotype by its initial frequency. Thus if we are able to determine the frequencies of alternative genotypes in a population, and fitness values for these alternatives, we should be able to establish with precision the change in frequency of a particular allele across generations due to differences in fitness.

Conceiving of fitness in this manner had the effect of changing the way biologists thought about the evolutionary process. Evolutionary change, among population geneticists at least, came to be conceived as gradual changes in the frequencies of alleles at a given locus—changes in the gene pool as they were called. To the extent that evolutionary change is change in the genetic makeup of a population across time, this seemed to give us a mathematically precise way of characterizing evolutionary change on the basis of the three central components of Darwin's model—existing variation, inheritance and fitness. At the same time, however, the concept of fitness became more and more closely identified with the measurement used within these models (actual reproductive rate), and less and less closely identified with the complex adaptive interactions with the environment which Charles Darwin viewed as the essence of natural selection.

Thus, as elegant as this Genetic Model of Natural Selection seems to be, those variables W and s, representing relative fitness and the coefficient of selection, hide more than a little mischief. One of the principal achievements of the so-called neo-Darwinian theory of evolution, the integration of the theory of natural selection with population genetics, has also produced confusion about the concepts of fitness, adaptation and natural selection, and therefore about the nature of a Darwinian explanation. After we look at some simple yet realistic examples of explanation in evolutionary biology, we must try to sort out some of these confusions.

7.2 SOME EVOLUTIONARY EXPLANATIONS

What sorts of questions do biologists look to the theory of evolution to answer, and what sorts of answers are given? There are too many for us to look at even a representative sample here. We can, however, begin to appreciate its scope by briefly looking at four examples of evolutionary explanation in which the facts to be explained are of different sorts, the evidential support for the premises of the explanations varies considerably, but in which the concepts of natural selection, fitness and adaptation play an important role. This will allow us to keep philosophical discussion of these concepts anchored in reality.

7.2.1 Case 1. The Evolution of the Horse

While more the exception than the rule, a growing number of evolutionary histories are supported by a rich collection of fossils. The lineage leading up to the modern genus *Equus,* of which the horse is a species, is one such sequence. Especially after the combination of rich data coming from North American deposits with earlier data from Europe, a number of patterns in these fossils could be discerned. Once these patterns were established, they gave rise to a number of questions.

One such pattern emerged from a study of the teeth in this lineage, represented in Figure 7.1. It will be noticed both that the ratio of molar tooth height to length increases through time, and that there is an *acceleration* in the rate of change through time. Why did the molars in this lineage get taller relative to their (horizontal) length, and why did this change accelerate across time? Here is the sort of explanation evolutionists offer:

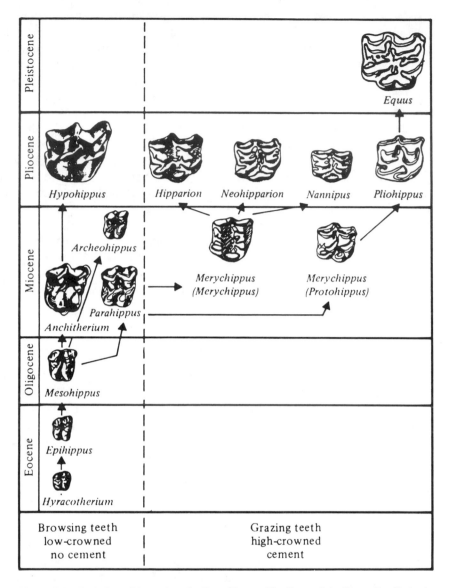

Figure 7.1 Evolution of horses' teeth. From *Horses: The Story of the Horse Family in the Modern World and Through Sixty Million Years of History* by G. G. Simpson. Copyright © 1951 by Oxford University Press, Inc. Reprinted by permission.

There is little doubt that it arose because Merychippus and its descendants abandoned the habit common to all earlier horses of browsing on leaves, and took the newly evolved grasses as their main food. Other horse lineages, which although now extinct, survived alongside the grazing horses for many millions of years, continued to browse on leaves, and in these there was no increase in the rate of evolution of tooth shape. (Maynard Smith 1972, 254)

We should note the evidence, not presented here but assumed, upon which one of the premises of this explanation rests: There was a newly evolved food source, the grasses, at the time this evolutionary change began; and there is paleoecological evidence associating the change of molar height with a grazing environment, while the more stable tooth height is associated with a browsing environment. This association encourages the evolutionary biologist to make what has been termed *the adaptationist assumption:* The change in relative tooth height occurred as an adaptation to a change in the environment, in this case a change in available nutritional resources.

But a different sort of assumption is crucial to this argument which is not mentioned here, though it often is in textbook presentations of this example. You can identify it if you ask, "Why should the shift to a grassy environment be correlated with the increase in relative height of teeth?" The answer lies in the high silicate content of the grasses, which makes for increased wear per unit of vegetation consumed. There is an *engineering assumption* here; relatively taller teeth facilitate the use of this newly evolved energy source, for such teeth will last longer.

Finally, an assumption is made about the change in question being due to heritable factors, a reasonable assumption given what we know about the genetics of current members of the family in question. Thus there is an assumption, not directly testable, made about the *heritability* of the trait in question. This example suggests, in outline, the following explanation:

1. Like any animal population, populations of *Mesohippus* will possess considerable heritable variation, including a mutation rate of around 10^{-6} for any given trait in each generation.
2. As the environment developed a new source of nutrition, those members of these populations which possessed variations allowing them to exploit this underexploited source of energy were at a slight advantage relative to other members of these populations.
3. A relatively taller molar was one such advantageous variation.
4. Thus parents with this heritable variation (other things being equal) left a greater number of offspring than their cohorts.
5. Thus the frequency of this trait—and of the gene(s) coding for it—increased, relative to the alternatives, in subsequent generations.
6. This trait thus spread through the population, and the overall character of the population diverged in this respect from its ancestors.

From this we may abstract the *neo-Darwinian pattern of explanation:*

1. There is *heritable variation* for trait **t** along Mendelian lines.
2. The environment provides an *adaptive opportunity*.
3. There exists *differential adaptation* among variants.
4. This produces *differential transmission* of traits.
5. This produces *adaptive character change* in the population over time.

7.2.2 Case 2. Sickle Cell Anemia and Malaria

Hemoglobin, the human oxygen transport vehicle in red blood cells, is found in three distinct forms in certain East African populations. These forms differ in virtue of one amino acid substitution controlled by a single gene. We may refer to the different types of hemoglobin by the letters **A**, **AS**, and **S**, and the different genotypes **aa**, **as**, and **ss**. The **a** allele shows incomplete dominance over **s**, so that there are three possible phenotypic expressions. The **aa** genotype produces hemoglobin with the optimal oxygen transport capabilities (**A**), **as** leads to red blood cells with a combination of hemoglobins (**AS**) which are thus somewhat less efficient in this respect, while **ss** leads to hemoglobin (*S*) which leads to the deformation of the cells known as "sickling." This produces a number of severe pathophysiological effects which are often fatal.

The puzzling fact, requiring explanation, is that ecological genetical studies have established that the **s** allele is found in these populations at a frequency as high as 23%, whereas it ought (especially given the lack of dominance) to be, on the theory of natural selection, virtually absent.

The explanation in this case appeals to the concept of *heterozygote advantage* (also known as *heterosis* or *overdominance*). (For a general discussion of this example, including the general mathematical population genetic model for such cases, see Ayala 1982, 106–111. For a nice discussion of current molecular knowledge of the hemoglobins, including a discussion of the molecular mechanisms involved in sickle-cell anemia, see Rosenberg 1985, 73–83.) A correlation was discovered between populations in which the **s** allele was present at high frequencies in the presence of mosquitoes carrying the malaria-producing parasite *Plasmodium falciparum.* Epidemiological studies then revealed that having **AS**-hemoglobin provided resistance to malaria. Thus in environments where the probability of contracting malaria is high, persons with the **as** genotype are at an advantage when compared both to the **aa** and to the **ss** homozygotes. Compare the two maps in Figure 7.2 showing the distribution of **AS**-hemoglobin and of *Plasmodium falciparum,* respectively.

A number of features relevant to later discussion distinguish this example from Case 1. First, it is concerned with a different type of question. The question to be answered concerns the genetic and phenotypic makeup of a present population rather than a change that has taken place over geologic time in a population. Second, the question at issue provided a *challenge* to evolutionary theory. How can natural selection, which favors advantageous traits, explain the presence of a gene for a lethal trait at such high levels? Population genetic models had, in fact, shown that such a situation was theoretically possible under conditions known as heterozygote superiority. This case provided an opportunity to deploy the model, to see if the theory could in fact rise to the challenge.

This explanation also differs from Case 1 in that it relies far less on untested assumptions regarding the genetics of the population in question. The genetics— indeed the molecular genetics—responsible for the phenotypic differences in question (differences in hemoglobin types) is well understood. Also, the population data required to deploy the machinery of population genetics is readily available. It is, in every sense of the word, a textbook case of a population genetic success.

Equally important to note, however, is that the success of this explanation, as

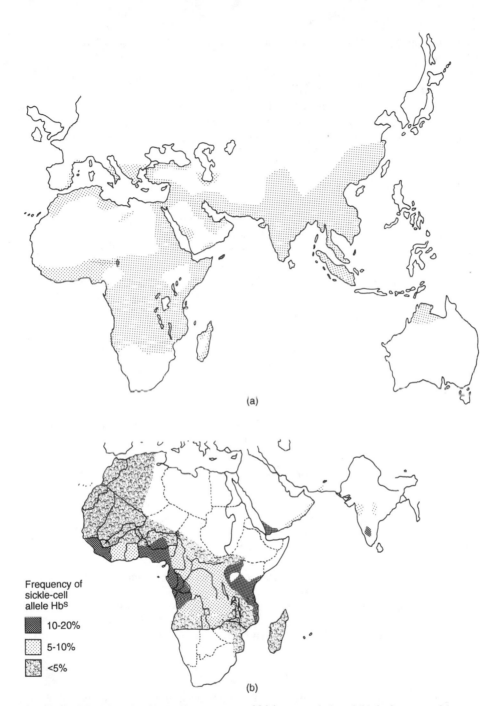

(a)

Frequency of sickle-cell allele Hb^S

10-20%

5-10%

<5%

(b)

Figure 7.2 Maps showing (a) the occurrence of falciparum malaria and (b) the frequency of the Hb allele, which causes sickle-cell anemia, both in the Old World. From *Modern Genetics*, Second Edition, by Ayala and Kiger (Menlo Park, CA: Benjamin/Cummings Publishing Company, 1984). Reprinted by permission.

in Case 1, depends crucially on evidence supporting the adaptive advantage of the heterozygote *phenotype* **AS** over the other phenotypes in a specific, shared environment. The relative adaptive advantage of this phenotype in a malarial environment is a crucial feature of the explanation of the unexpected genotypic frequencies. It is also important to see that, whereas Case 1 dealt with a long-term evolutionary trend involving the origins and extinctions of a number of species, and thus drew much evidential support from paleontology and paleoecology, Case 2 provides us with little or no understanding of this sort.

7.2.3 Case 3. Horse toes

Let us return to the horses for a third example, in this case an example of the use of so-called "optimality models" in evolutionary explanation. In such explanations models borrowed from engineering or economics predict the best design for a structure, or the best behavioral strategy, given a certain adaptive problem to be solved. These predictions are then compared with the actual structure or behavior in question. Reasonable fit is taken as evidence that the structure or behavior was "designed" (in the biological case, by natural selection) as a solution to the adaptive problem.

The fossil record records a number of correlated changes in the evolution of the modern horse. One of these, correlated with the changes in dentition discussed in Case 1, is the development of hoofed species from an original ancestral species with toes (see Figure 7.3). Here is a brief summary, from one of its originators, of an explanation of the development of the *Equus* hoof. The question for which the following explanation is an answer is this: Why did *Equus* evolve a single, elongated toe (i.e., a hoof) while the muscles controlling its motion are located in the hip and linked to this "toe" by means of elongated tendons?

> [In] galloping a horse must accelerate and decelerate its legs with each stride, and this uses up a lot of energy. The energy expended can be reduced by lightening as far as possible the lower part of the leg, since this is the part of the leg which must be moved fastest. By concentrating the muscles in the upper part of the leg, the lower part is lightened, and the energy used up in galloping reduced.
>
> The reason for having a single toe is less obvious. The cross-sectional area of the bones in the foot must be sufficient to withstand the compression and bending stresses imposed while galloping. A single cannon bone has a greater resistance to bending than would four or five bones of the same total cross-sectional area. Hence a five-toed horse would require bones in its feet of greater total weight than a single-toed horse. Thus the single toe, like the concentration of muscles near the hip, reduces the weight of the foot, and consequently the energy needed for running. (Maynard Smith 1972, 16)

These structural "design" features of the horse's leg are all viewed as adaptations for galloping. Such explanations assume that such complex characteristics are the product of a long history of selection for those heritable variations with the most advantageous consequences for their possessors—the adaptation assumption. Thus does "nature optimize" among the available alternatives to produce the wonderful adaptations around us. Such explanations, as both their critics and defenders agree, contain assumptions about three things:

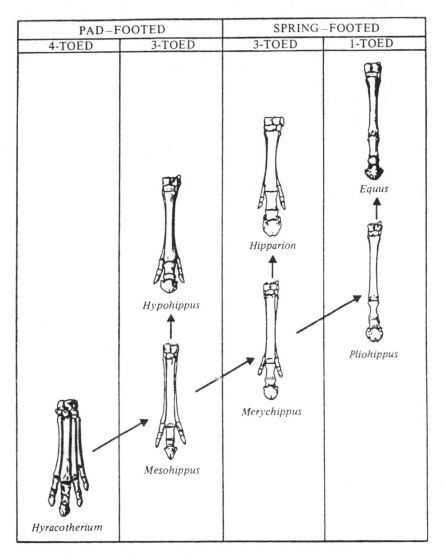

PAD—FOOTED		SPRING—FOOTED	
4-TOED	3-TOED	3-TOED	1-TOED

Equus

Hipparion

Hypohippus

Pliohippus

Merychippus

Mesohippus

Hyracotherium

Figure 7.3 Evolution of horses' feet. From *Horses: The Story of the Horse Family in the Modern World and Through Sixty Million Years of History* by G. G. Simpson. Copyright ©1951 by Oxford University Press, Inc. Reprinted by permission.

1. the range of possible phenotypes;
2. the feature being optimized;
3. the heritability of the traits in question.

For example, in the analysis of animal gates from which the above example is taken, it is assumed (and in the case of the horse we have fossil evidence in support of the assumption) that the alternative structures discussed were among the variations available for selection; and it is assumed that the feature being optimized was efficiency of energy expenditure at a given speed. This led the authors to predict that the time an animal was completely off the ground would increase as a function of speed but decrease as a function of size, a prediction borne out in the case above.

Explanations of this sort are often based initially on the use of mathematical models developed in engineering, economics and the theory of games. Such explanations are sometimes offered in the absence of empirical support for assumptions (1) and (3) and with little more than common sense behind the choice of (2). More extreme critics of such explanations have suggested that the whole program of framing them is misguided, arguing in effect that ingenious practitioners of "the adaptationist program" will always be able to tinker with their assumptions sufficiently to match apparently rigorous tests, while employing assumptions which are not independently testable. These criticisms will be discussed at the end of the chapter.

7.2.4 Case 4. Darwin's Finches

The finches of the Galapagos Islands, off the Pacific coastline of South America, were among the species that first set Darwin thinking along evolutionary lines. They have continued as a fertile testing ground for evolutionary biology up to the present. In 1981, for example, the journal *Science* published an elegant report (Boag and Grant [1981] 1982) of research on *Geospiza fortis,* the ground finch of the island Daphne Major. This study aimed to explain changes in this population of birds as a result of intense selection favoring birds with certain traits. As so often with such studies, this specific explanation was intended to help establish the validity of a general pattern of evolutionary explanation.

The facts to be accounted for are these. Over a two-year period, during which a major drought occurred, the size of the population of ground finches under investigation shrunk from approximately 640 to 85. A number of characters related to overall body size and beak dimension were under study. The population means for these characters shifted toward larger body and beak size among the surviving birds. Two questions were of interest to the investigators:

1. Were the changes in the population due to natural selection?
2. What were the precise mechanics of the selection process?

During the period of rapid population decrease, the drought also led to a general decline in abundance of seeds upon which the finches fed, with the smaller seeds declining more rapidly than the larger (seeds were compared on a "size in millimeters/hardness in newtons" index). Careful study indicated that between 1976 and 1977 medium to large seeds went from being 17 percent to 49 percent of the sample population's diet. Furthermore, "large birds ate larger seeds than smaller birds, suggesting that small birds disappeared *because* they could not find enough food" (Boag and Grant [1981] 1982, 176, emphasis added). In addition to this problem, small birds began to feed on a smaller seed that they normally ignored because it was the only small seed produced during the height of the drought. This seed was high in latex, and led to matted and lost plumage in the smaller birds. Many smaller birds in this condition were found dead during the dry season in that year. Larger birds turned to a food source normally ignored as well, an extremely large seed not available to the smaller birds. Boag and Grant concluded:

> It is reasonable to infer natural selection from the greater survival of large birds because about 76 percent of the variation in the seven morphological measurements and in principal component 1 scores [scores based on a combination of measurements] is heritable. (Ibid., 178)

The explanation that emerges from this study is clear:

1. a severe change in the abiotic environment (drought) had a differential effect on the normal food supply of the population sample under study;
2. certain variations within the population, especially in beak dimension, allowed certain members of the population to survive during this change better than others—in particular, the change in food supply differentially affected the mortality rates of smaller birds;
3. incidental to this, but with potentially significant evolutionary consequences, because the mean size of females is considerably smaller than males, the sex ratio altered from 1 male:1 female to 6 males:1 female during the period of the study.

The study takes on added significance because it is a paradigm of an ''evolutionary bottleneck,'' a situation where a population is put under intense selection during which only a small, biased sample of the population survives to contribute genetically to subsequent generations. It has been postulated by many evolutionary theorists that such events may play a central role in evolutionary processes. Were it the case that most evolutionary change were of this character, it would explain patterns in the fossil record in which long periods of relative stability are ''punctuated'' by rapid periods of speciation. Boag and Grant conclude:

> [G]iven the many small, isolated, relatively sedentary, and morphologically variable populations of Darwin's finches and the high spatial and temporal variability of the Galápagos, this type of event provides a mechanism for rapid morphological evolution. Occasional strong selection of heritable characters in a variable environment may be one of the keys to explaining the apparently rapid adaptive radiation of the Geospizinae in the Galápagos. (Ibid., 180)

These four explanations differ in many respects: the types of questions they address, the nature of the empirical support for their premises, the extent to which they are historical in character, and so on. Nevertheless they all involve appeal to random, heritable variations, natural selection, fitness, and adaptation. With these cases as background, it is time to explore certain philosophical problems connected with the place of these concepts at the core of evolutionary biology.

7.3 EXPLANATION AND NATURAL SELECTION

The Darwinian theory of evolution rests on two fundamental principles:

1. the principal cause of evolutionary change is natural selection;
2. the variations upon which selection operates arise at random with respect to the adaptive requirements of organisms.

The remainder of this discussion focuses on philosophical difficulties which arise in trying to understand these two principles; and in particular, on the meaning of the concepts "fitness," "adaptation," "selection," and "chance" within evolutionary theory. This section considers the first three; Section 7.4 considers whether a theory based on adaptation and selection is a teleological theory; and Section 7.5 considers the concept of chance.

7.3.1 The Meaning and Definition of Fitness

We began with Darwin. We saw in his characterization of his theory that three things are kept distinct: (1) the differential adaptive values of variations; (2) the differential preservation of the organisms with those variations; and (3) the differential reproductive rates of those organisms. As so often, Darwin was wise to make these distinctions, and many conceptual confusions in recent evolutionary controversies can be traced to failures to make them, or at least to make them clearly. Historically, we have traced these confusions to the rush to "operationalize" the concept of fitness, that is, to define the concept as if it referred to values arrived at when measuring it. For example, it is not uncommon to see the general theory of natural selection stated in the following way:

> Most people are familiar with the basic theory of natural selection. Organisms vary in a heritable fashion. Some variants leave more offspring than others; their characteristics, therefore, are represented at a greater frequency in the next generation. (Wilson 1984, 273)

Notice that in this description of the theory of natural selection, the only explanation offered for the greater frequency of certain characteristics in the offspring population than in the parent population is that the parents with those characteristics leave more offspring. Darwinian Fitness often receives a similar treatment. Take, for example, the following glossary entry for "Fitness" in a highly regarded primer in population genetics:

> **Fitness** The reproductive contribution of an organism or genotype to the following generations. (Ayala 1982, 240)

Such formulations have given rise to a family of objections to evolutionary theory which, if valid, are devastating. Yet as we saw in Section 7.1, these formulations arise naturally from the fact that the values supplied for the variable W, which is termed ''relative fitness,'' are simply measures of the relative reproductive contributions of different genotypes.

7.3.2 Fitness as Tautology

Biologists and philosophers have repeatedly alleged that the theory of evolution by natural selection ''rests on a tautology,'' or that the principle of fitness or natural selection is a tautology. Ever on the lookout for a priori grounds for dismissing evolutionary biology, the ''scientific creationists'' parrot this objection. In order to make initial sense of it, we first need to know what it means to refer to a statement as a tautology. (For a good discussion, see Sober 1984b, Chapter 2.)

The meaning of ''tautology'' which coincides most closely with the use one finds in such criticisms of evolutionary theory is: *a proposition in which what is predicated of a subject term is already implicit in the meaning of that term.* In the case at hand, to say

''The fitter of the three genotypes increases its representation in the next generation'' (T)

appears to be making a claim about a genotype that was not already implied by referring to it as the fitter one. But, as we have seen, the way population geneticists sometimes talk (as in the definition of ''Fitness'' quoted above) all ''fitter'' *means* is ''increases representation in the next generation.'' Thus, for those already committed to the above definition of ''Fitness,'' T is equivalent to

''The one of the three genotypes which increases its representation in the next generation increases its representation in the next generation'' (T')

Given the above explication of tautology, definitions are tautologies. Definitions play an important role within scientific theories, making explicit something implicit in the meanings of the theory's key terms. The mere fact that T is a tautology is not cause for dismissing it. Rather, the problem is that T is reasonably taken to embody an *explanation* of why a certain genotype increases in frequency. But if T is really simply equivalent to T', then the claim that one organism is fitter than another doesn't embody an *explanation* of evolutionary change at all, but simply a *description* of it.

7.3.3 Fitness as Triviality

Thus what is really troubling about these formulations of the theory of natural selection is not that T is a tautology, but rather that it *poses* as an explanation. Given a standard definition of fitness, the statement T, which we expect to be either a predictive or explanatory claim, turns out to be neither. On the contrary T is simply a trivial corollary of the definition of fitness (see Brandon 1990, 11–12 for a fuller discussion).

The theory originally formulated by Charles Darwin, however, is a *causal* theory, according to which differences in the fitnesses of different parent organisms *produce* the eventual changes in gene frequencies in subsequent generations. These changes of frequency, not to mention changes in the observable characteristics of organisms, should be *explained by* the theory of natural selection, not merely described by it. The characterization of the theory of natural selection and of the concept of fitness provided in the above quotations suggest that evolutionary biology is not *real science* at all, but just bookkeeping. It appears a poor and distant relative of Darwinism.

7.3.4 Fitness as Propensity

Can we revive a robust Darwinism without giving up the gains of population genetics? As a first step in the right direction, imagine the following situation. Two genetically identical zebras in the same environment are standing next to one another. One is in a dry riverbed, the other is standing next to it. A flash flood occurs, carrying the one zebra off, leaving the other behind. The one left behind reproduces successfully on ten occasions before it dies. Now according to the definitions offered above fitness is determined simply by actual reproductive success, which implies that the zebra which avoids the flood and has ten offspring is clearly the more fit. On the other hand, the same definition implies that fitness values be ascribed to genotypes, which suggests these two zebras should have the same fitness value. Defining fitness in this manner is doomed to such paradoxes because it cannot distinguish changes in gene frequency due to differential adaptation from other causes of such changes.

To avoid this dilemma, the concept of fitness must refer to something besides *actual* reproductive success. It has thus been suggested by a number of philosophers that "fitness" within the theory of evolution refers to a *propensity* possessed by given genotypes or alleles to increase in frequency in subsequent generations. (Brandon 1990, 14–24, now prefers to refer to this as the propensity interpretation of adaptedness. This interpretation of fitness is defended in Brandon 1978, Mills and Beatty 1979, Burian 1983, and Sober 1984.) The zebra example can then be dealt with by arguing that both zebras have the same *expected* fitness, the same *propensity* to reproduce relative to different genotypes, though fate intervened to prevent one of them from *realizing* its reproductive potential. Differential reproductive rates of different genotypes is then to be viewed as potentially, but not necessarily, due to differences in fitness.

This is clearly a step in the right direction, but problems arise with this suggestion. This interpretation encourages us to think of fitness as a "dispositional property" of the organisms in question—just as different materials display differences in elasticity, so different members of a population differ in their capacities to leave offspring. Following this model to its logical conclusion, we would expect to find underlying physical differences which would explain these different reproductive capacities, just as different structural properties of various materials explain differences in elasticity. Fitness differences (i.e., differences in reproductive capacity), however, are not based on physical differences in this manner. Differences in fitness, as Darwin constantly stressed, depend on a complex and variable constellation of *functional relationships* between the organism and its environment. This interpretation of fitness is, then, faced with the difficulty of making the interpretation abstract enough to cover all these various adaptive relations which underlie reproductive potential.

Defenders of the "propensity interpretation" of fitness sometimes respond to the difficulty by, in effect, embracing it. For example, two defenders of the propensity interpretation respond to the above problem as follows:

> As long as there is evidence of fitness differences independent of the differences in reproductive success they are invoked to explain, the explanations are potentially acceptable. There is no reason why the extra evidence has to be the same kind in every case. (Brandon and Beatty 1984, 345)

We can use two of our example explanations to see their point. In Case 4 above, larger beak size *in relation to* a shift in the size of available seeds made one finch more fit than another, while in Case 1 molars with higher crowns and more enamel *in an emerging grazing niche* made one ancestral horse more fit than another. In each case evidence exists for the assignment of certain values which is independent of actual reproductive success, though the physical basis of the different reproductive propensities may differ from species to species or from one environment to another.

But this response also has difficulties, which can be seen by thinking about our usual way of giving empirical content to dispositional terms. "Rubber is elastic" may initially be understood as based on a simple counterfactual claim—"if a rubber object were subjected to certain forces and then released, it would return to its initial size and shape." But typically, within a scientific theory of the material in question, this counterfactual is supported by an understanding of the structural properties of that material. In accounting for rubber's tendency to "rebound," we look for a molecular-mechanical account of its microstructure which will account for its being flexible rather than rigid. Further, we expect that a suitably *abstract* theory will account for the property of elasticity *as such,* whatever material we may be talking about. To the extent that the disposition is the same, the physical basis for it should be. We must be careful, then, not to demand an account of the fitness of *organisms* which is as concrete as an account of the elasticity of *rubber.* An ideal physical theory should provide a *general* (underlying physical) account of elasticity, true for *all* elastic solids. Such a theory must be abstract enough so that differences of chemical constitution of both material and surrounding physical conditions—such as temperature, pressure, gravitational field—will not require a different theory.

Similarly, an ideal evolutionary theory should provide a general account of fitness. If fitness is a dispositional property of organisms, there ought to be an abstract account of the underlying basis of fitness as such, whatever type of organism we are discussing. The fact that no such account seems to be part of current evolutionary theory has led philosophers to describe fitness as a *supervenient* property of the organisms to which it belongs (Sober 1984b, 48–50; Rosenberg 1985, 154–168).

7.3.5 Fitness as Supervenient

Supervenience, as used in this context, can be defined as follows:

A relationship between a nonphysical property and an object in which (a) the objects with that property may differ in the physical characteristics underlying the property and yet (b) any object with those physical characteristics will have the nonphysical property in question.

Take the property of being a clock, for example (Rosenberg 1985, 73). There is no one set of physical mechanisms which characterizes all objects which function as clocks. (Think of how different the inner workings of digital and analogue watches are, for example.) However, anything with any of the appropriate but different physical mechanisms will so function. Being a clock is, therefore, supervenient on each of the appropriate physical structures.

This concept has played a central role in attempting to understand the relationship between states of mind and their physical basis in functional terms. In that context, part of the motive for developing the concept of supervenience was to describe our cognitive activity in a way which avoided attributing it to a nonphysical entity (the mind) and also avoided *identifying* types of mental and physical states. There is a precisely analogous motivation for claiming that fitness is a supervenient property: By referring to fitness as supervenient, philosophers are saying that there is always *some* underlying physical basis for fitness differences, but that one can make no immediate inference from the knowledge that one organism is fitter than another to the *actual* physical differences underlying that difference. In fact, given the crucial role of environment-organism interactions in generating differences in fitness, "underlying physical differences" has to be taken very generously to include such interactions. Take our Case 2, for example: A person who is a heterozygote regarding the sickle-cell gene has one fitness value in Eastern Africa, but a very different one in Western Pennsylvania. This is not because the person's physical makeup has changed, but because a relevant feature of the environment has.

Describing fitness as a supervenient property is valuable for two reasons. First, it is a commitment to naturalism, to the idea that however difficult fitness may be to grasp abstractly, it refers to a property of physical objects. Second, it has focused our attention on the problem of identifying fitness too closely with an actual physical characteristic of the organism, however abstractly described. Nonetheless it seems not to take us very far in our search for a *positive* account of fitness. By focusing on two of our examples we can at least point the search in a fruitful direction.

7.3.6 Fitness and Adaptation: A Proposal

In both our "sickle-cell anemia" and our "horse hoof" examples, the evolutionary explanations provided depended on evidence of a causal relationship between relative competitive success or optimal functioning in an environment on the one hand and relative reproductive contribution on the other. Thus, in order for (i) competitive success or (ii) optimal functioning to translate into components of an evolutionary explanation, the characteristics at the basis of these attributions must (a) be heritable and (b) lead to reproductive advantages. Neither (a) nor (b) follows necessarily from (i) or (ii). That is, in order to make use of adaptive differences between organisms in an evolutionary explanation we must do one of two things. We must make plausible assumptions that these adaptive differences are inherited in the population under study (which, if possible, should be evaluated at some point) or, when possible, include the results of ecological genetical studies in our explanation. (For the most commonly used methods, see Endler 1986, Chapter 6; Sheppard 1975, Chapter 12.) But our four case studies confirm that these are *precisely* the kinds of premises that evolutionary biologists *do* appeal to when offering selectionist explanations of evolutionary changes. An elegant example of such an appeal can be found in Case 2, in the studies which demonstrated a correlation of high levels of a lethal recessive gene with malarial environments, determined the genetic basis of sickle-cell anemia and used population genetic models of heterosis to explain frequencies of the s-allele in malarial environments.

In fact, the contingent nature of the relationship between adaptational differences and reproductive differences is crucial to a plausible account of contemporary evolutionary theory. Major steps in the evolutionary process may result from processes other than natural selection. These processes may give rise to changes in the genetic makeup of a population which are *not* due to natural selection or adaptation. Thus any account of the theory of natural selection, such as Wilson's discussed earlier, which simply equates changes in populations due to differential reproductive success with changes due to selection, will be unable to distinguish two very different sources of evolutionary change.

By the same token, there are ways of discovering that natural selection is operating to *maintain* gene frequencies at given levels so that the lack of changes in gene frequency does not necessarily imply absence of natural selection. But once again on Wilson's description of the theory, absence of gene-frequency changes would *imply* a lack of natural selection. Ultimately, then, the concept of fitness is dependent on the more fundamental concept of *adaptation*.

Let us take just one more step in this direction. The above arguments suggest that *differential adaptedness* is the physical property which underlies some differences in expected reproductive success (see Brandon 1978, Chapter 6; Burian 1983, Chapter 11; Brandon 1990). This leads to a further suggestion: As Darwin's own formulation of the theory suggested in 1859, there needs to be a *principle of differential adaptation* at the center of the theory of evolution by natural selection.

Thus, the analogue to an underlying physical theory in evolutionary biology is an *abstract theory of adaptation*. And in specific cases, it is only when differences in adaptive value of phenotypes produce differences in reproductive propensity that actual reproductive differences can be treated as measures of fitness. The problem of providing such an abstract theory is a serious one, but it avoids one seemingly

> *Principle of Differential Adaptation:* When differences in adaptation underlie and explain differences in reproductive success, it is appropriate to refer to such differences in reproductive potentials as differences in fitness.

intractable problem from the start. We do not need to suppose that there is some hidden physical property common to all organisms underlying their fitness propensity.

Evolutionary theory requires, then, a concept which identifies abstractly those concrete properties which biologists identify as adaptive differences. What sorts of properties are these likely to be? It is worthwhile again thinking back to our four cases. In those cases, the concrete properties of the organisms relevant to *selection* are not traits such as tooth length, blood type, bone length or beak size. Rather (keeping the same four examples in mind) the *selectively relevant properties* are *functional* properties such as grass-mastication ability, optimal malarial-resistance–oxygen transport capacity, ability to maximize speed per unity of energy, ability to grasp available seeds. Evolutionary theory is still in its exploratory stages in this area, but contributions to such an abstract account of adaptation are to be found in the engineering and game theoretic models used by students of adaptation, in population ecology and in bioenergetics, the study of biological activity from the standpoint of energy efficiency. Attempts to incorporate such models of organisms-environment interaction into a general "nonequilibrium thermodynamic" interpretation of evolutionary biology, while premature, are extremely suggestive. (For contributions from engineering and game theory, see Williams 1966, Wainwright et. al. 1976, Maynard Smith 1978; from bioenergetics, Lehninger 1971; from population ecology, Levins 1966, Pianka 1974; and from nonequilibrium thermodynamics, Brooks and Wiley 1986, with the papers in Weber, Depew and Smith [eds.] 1988.)

The property of being adaptive is a *relational* property—in referring to a trait as an adaptation, one identifies constitutional differences that *may* confer functional advantages in given environments. Because such functional differences must always be specified relative to a common environment, no account which refers simply to a physical feature, for example, "hemoglobin of type A," is sufficient. Only in environments where this trait confers some advantage in organic function on its possessor will it be an adaptation.

The argument presented, then, has been for distinguishing *adaptiveness* from *fitness,* which for our purposes we may agree to think of as a dispositional property related to reproductive success. Calculations of changes in genotypic frequencies, on the view of evolutionary biology which has emerged from our discussion, constitute measures of changes in fitness only under certain conditions. In particular, such changes indicate differences in fitness only if we have independent evidence that those changes are due to adaptive differences, that is, to functional differences relative to a common environment.

7.3.7 Fitness of Replicators and Adaptedness of Interactors

Organic systems do two things which are fundamental to the evolutionary process—they replicate themselves and they *interact* with their environment. The conclusion of this section may be restated by saying that differences in adaptation are to be understood as differences among interactors in a common environment. When these differences are heritable and are relevant to reproductive success, they will lead to different rates of replication among replicators (see Brandon 1985, Dawkins, 1982, Hull 1981b, Mitchell 1987). Only with such differences will evolutionary change take place. However, different rates of replication can occur for reasons other than differential adaptation.

7.3.8 Some Real Virtues of Population Genetics

Strong objections have been made in this chapter to the apparent explanatory sterility of the population geneticist's account of fitness. But let us conclude by indicating the sorts of things for which population genetics is especially valuable. First, at the theoretical level, the models of the various theoretically possible outcomes of different levels of selection on different phenotypes provide a powerful engine for extending neo-Darwinism. Take the following question. "Suppose we have a large randomly mating population of organisms. And suppose a genetic mutation occurs in that population (either dominant or recessive). How large a selective advantage would that mutation have to confer on its possessor for it to become fixed in that population in n generations? (Or alternatively, how many generations would it take for a trait of a given selection coefficent s to be eliminated from the population?)" This, and literally hundreds of other structurally similar questions, can be explored independently of information regarding the actual sources of selective advantage in particular cases (without confirmed *source laws*, to use Sober's terminology—Sober 1984b, 55–59). Similarly, such population genetical models have been crucial in the face of phenomena which seemed to be impossible to account for on Darwinian premises. Models of kin selection which account for so-called *"altruistic" behavior* (behavior which seems to confer a slight disadvantage on its performer and advantages on other members of the species) are good examples. It would seem that such behavior conflicts with neo-Darwinism because animals which behave this way should be eliminated by selection. Kin selection models, however, show that genes for altruistic behavior will be preserved by selection provided the danger of so behaving is slight and the organisms who benefit the most are those most closely related to the altruist. For there is a high probability that these "kin" will also have the gene in question. The use of population genetic models therefore shows that the phenomenon has a possible explanation within Darwinian theory. It still remains, of course, to *test* the solution on a diverse range of populations. But merely showing that an apparent anomaly is capable of solution is a valuable elaboration of the theory, and makes the inquiry into the proposed solution possible. Among our case studies, Cases 2 and 4 served as crucial support for ideas that were first worked out as theoretical possibilities using the mathematical techniques of the population geneticist. Such models, then, have important roles to play *both* within causal explanations in evolutionary biology, *and* as a means of exploring the explanatory possibilities open to Darwinism. They are valuable tools indeed.

7.4 NATURAL SELECTION AND TELEOLOGY

In the previous discussion it was suggested that, in order for evolutionary theory to be explanatorily powerful, it must distinguish, as Darwin did, questions of adaptation from questions of evolutionary change as a product of adaptation. The theory of natural selection cannot remain a theory about the causes of evolutionary change if it simply says that certain traits increase in frequency in a population because parents with those traits leave more offspring. For, while this claim is true, it fails to distinguish evolution by selection from evolution by drift, and it fails to provide an explanation for *why* those parents leave more offspring. It is claims about *better adapted* organisms tending to leave more offspring which both puts the theory at some risk and provides it with its explanatory guts.

Having insisted on this, some might say, is not only to put the theory at *scientific* risk. Because of the form such explanations take, we have put evolutionary theory at *philosophic* risk as well. This is because such explanations appear to be *teleological,* that is, to explain things by referring to the goals they achieve. Adaptive explanations take the form of statements that an organism has a certain trait *in order to* escape predation or *for the purpose of* camouflage. It is time to see whether in fact it is correct to say that such explanations are teleological, and if so, what that implies about evolutionary theory.

Questions about why certain classes of organisms have the traits they do are often answered by specifying the functional advantages these traits confer on the individuals of the classes which have them. That is, this particular ''why'' question is answered by identifying *what the trait is for, what its value is to its possessor.* This is an answer that we do not find in sciences interested only in inanimate phenomena. For example, astrophysics does not explain why Saturn has gaseous rings surrounding it by appealing to the value of such rings to Saturn.

Answers of this form are traditionally termed ''teleological explanations.'' Plato was the first philosopher to explicitly defend their use. He assumed that they were appropriate if and only if the fact being explained arose as a result of intelligent design. For Plato, it would be appropriate to explain the rings of Saturn this way if in fact our solar system were the product of an intelligent designer who sought to achieve some good by designing Saturn with rings. And it would be appropriate to explain the fact that horses have hooves teleologically if they have them because God determined that it would be good for them (or for any other reason good) to have hooves (Plato, *Phaedo* 97b8–99d1; *Timaeus* 46c–48b, 68d–69c; see Lennox 1985).

Aristotle, one of Plato's students, and the founder of the science of biology, also defended teleology, but did not think intelligent deliberation was at the core of such explanations. For Aristotle, certain features of organisms make functional contributions to the organism's life, and the organism comes to have those features *because* they make those contributions. Thus in order to fully explain such features, we need to identify what they are for, what activity they perform which contributes to the organism's life (Aristotle, *Parts of Animals* I 641b11–642a14, 645b14–646a2; *Physics* II 8; *Generation of Animals* II 742a17–742b18, V 778a29–778b20, 789b3–23).[3]

On the face of it, when an evolutionary biologist identifies a biological trait as an adaptation he says (1) that it makes a functional contribution to the life of an organism in a specific environment; and (2) that selection was *for* that trait, that is, the trait in question became common and remains common in a population *because* it makes that contribution. To use one of our examples: To say that **AS** hemoglobin is an adaptation is to say (1) that having **AS** hemoglobin reduces the risk of malaria where it exists, and (2) that it is common in certain human populations *because* it reduces the risk of malaria. Thus it does look as if the goal of the trait somehow explains it—the explanation *looks* teleological.

However, is this not simply shorthand for a very long historical explanation, one which would tediously mention all the physical interactions which gave rise to a population of animals with this trait? And by giving such an historical explanation would we not remove the appearance of teleology from our evolutionary explanations? To answer this, let us sketch a model of such an explanation and see if we can banish teleology from the natural sciences altogether.

The form of question we begin with is "Why do the members of species *S* have trait **t** in environment *E*?" To begin with, let us make the simplifying assumption that **t** is a phenotypic trait the production of which can be explained at the molecular and biochemical level. Once we give a genetic and biochemical and developmental explanation of **t**'s production, have we answered our initial question? No. We know that the way evolutionary biology provides understanding is by helping us to identify how natural selection "adapts" organisms to their environments. Of course, we say, genes will be initiating the production of **t** in each of them. But we want to know why individuals with **t** (and the appropriate biochemical machinery to produce it) were (and are) selected over those with an alternative trait **t***. The above explanation does not answer this question.

But we can go further. For simplicity's sake, let us suppose that we have access to the account of how the mutation which first produced **t** in some individual occurred. We thus begin by explaining how a certain mutation occurred in a particular organism at some moment in the past. The challenge for those who want to eliminate teleological explanations from biology comes precisely at this point. Now that an individual member of *S* has **t**, and the genetic material which codes for it, **t** becomes a potential target for selection. The challenge is to construct a selectionist explanation which does *not* say that the gene for **t** spread and became fixed in *S because of* its functional contribution to members of *S*. If a functional advantage conferred on individuals by **t** is required in order to fully explain why members of *S* have **t**, the explanation becomes teleological. (For important recent discussion of this question,

---

[3] For three different interpretations of Aristotle's teleology, see the papers of John Cooper (1987), David Balme (1987) and Allan Gotthelf (1987, 199–286).

see Brandon 1981; Rosenberg 1985; Sober 1984a, Chapters 18–21; and Wright 1976. The position defended here is indebted in important ways to Brandon and Wright, views which appear to be in different ways akin to Aristotle's.)

The same challenge can be made by once again remembering that not all traits are present in populations because of natural selection. In order to distinguish adaptations from other features of organisms, we need to express a selectionist explanation in a way which distinguishes it fundamentally from other explanations of evolutionary change. The way this is in fact done in evolutionary biology is by insisting that the term "adaptation" be confined in its application to traits which we have good grounds for thinking were selected because of the benefits they confer on their possessors.

Why would we want to banish to the hinterlands of the social sciences an explanatory form which seems so useful? (Some of course want to ban it from the hinterlands as well; see Chapter 11.) A common reason given is this. A sign of progress in the development of natural science is the gradual removal of anthropomorphic forms of thinking from it (see Chapter 1). Teleological explanations are sometimes argued to be just such "primitive" forms of thinking and thus progress in biology demands their elimination. That is, they seem to imply one of two things: Either genes, biosynthetic pathways, populations, or species possess something akin to purposeful deliberation, or future events (ends, goals) are somehow capable of causing prior events to take place. Modern biology has no room for either implication. Thus we must banish teleology in the name of scientific progress.

While all of the premises of this argument are correct, they do not imply the conclusion. Selection explanations, in particular, imply neither of the beliefs that worry teleology's detractors. They identify differences such as longer teeth, hooves, **AS**-hemoglobin or larger beaks as having *advantageous consequences* in specific environments. They go on to claim that having those advantageous consequences in a given environment is *causally relevant* to the organisms having those traits. To use Larry Wright's language, these explanations are *consequence etiologies*—causal explanations of a special kind (Wright 1976). Such explanations claim that what a trait, or a variation in a trait, allows an organism to do, or do better, in a given environment is causally relevant to an organism's having that trait. Organisms may have certain *functional advantages* in virtue of having certain traits. These functional advantages produce the biases in survival and reproductive ability that are referred to as natural selection.

Figuring out, by controlled field observation or experiment, *which* of the consequences of a trait's presence in an organism helps explain its presence is a crucial part of evolutionary inquiry. For example, in our Case 4, it seems odd that larger birds survived the drought better than smaller ones. On simple bioenergetic grounds it would seem that the smaller creatures, making more modest energy demands of the environment, would survive better. What ability did larger size confer on its possessors which turned out to be causally relevant to survival? Or, what consequence of being larger was relevant to the survival of finches in a drought environment?

On the presupposition that the very term "teleology" carries with it the above suspect implications, yet aware that adaptation explanations in evolutionary biology look teleological, some biologists have suggested solving the problem by changing their name, referring to them as "teleonomic" rather than teleological. This has nothing to recommend it, though, either from the standpoint of etymology or history. Explanations which answer questions of the form "Why does A have B?" in the form "A has B for the sake of C" have, since the eighteenth century, been termed "teleological." What such explanations commit us to philosophically has, from the time of Plato and Aristotle, been a question for philosophical debate. The challenge, then, is either to show that such explanations are in fact reducible to explanations by causal antecedents, or to demonstrate that reference to the functional consequences of a trait is an ineliminable aspect of explanation by natural selection.

An historical note: It is sometimes claimed that *On the Origin of Species* was the very document that rid biology of teleology once and for all (Ghiselin 1974, Hull 1974). Its author did not think so. In a review of Darwin's work in *Nature,* the distinguished American naturalist Asa Gray claimed that Darwin had brought back teleology to natural science. Darwin immediately sent a note to Gray, saying, "What you say about Teleology pleases me especially, and I do not think any one else has ever noticed the point. I have always said you were the man to hit the nail on the head" (Darwin [1892] 1958, 308). This ringing endorsement of Gray's point is underscored by Darwin's rich use of teleological explanation throughout his career. Darwin was an avowed enemy of *anthropomorphic* teleology in biology, but he recognized that explanations by means of natural selection were teleological.

7.5 THE LAW OF HIGGLEDY-PIGGLEDY

7.5.1 Nature Doesn't Always Select

John Herschel, a philosopher and mathematician much admired by Charles Darwin, described the theory put forward in *On the Origin of Species* as "the law of higgledy-piggledy" (Darwin 1887, 241), presumably because of the large role given to "chance" in the theory. It is common to point out that Darwin treated the origins of variation as a matter of chance, and we will look at the modern analogue of this claim momentarily. But apart from that, you will note that in Darwin's summary of the theory, with which we began, he talks about organisms with useful variations having the *best chance* of surviving and *tending* to produce the most offspring. Think back to our twin zebras on the African savannah and you can see why he spoke this way. Both zebras were equally well adapted to their normal environment, but one survived and reproduced while the other was swept away. Likewise, in any individual case, the better adapted of two *particular* organisms may not be the one to survive and reproduce. The theory depends on "the law of averages," on assuming that over the long term differently adapted organisms will be treated equally in the (nonselective) lottery of life. Events which affect survival and reproduction rates, but which have nothing to do with natural selection, are thus treated as essentially random. This aspect of Darwin's theory is seldom given the importance it deserves. It forced Darwin to think of the theory of evolution by natural selection as a theory of large populations, that is, as a *statistical* theory; and this was to change the face of biology forever.

7.5.2 From Chance Variation to Random Mutation

An evolutionary alternative to Darwin's theory which he worked hard to distance himself from was that of Jean Baptiste Lamarck (1774–1829). According to Lamarck's evolutionary theory, the physiologically registered needs of an organism would give rise to changes in habit, behavior and, ultimately, structure. At least some of these changes would then be inherited. Thus Lamarck thought that adaptive requirements played a direct causal role in the production of heritable variation. In such a theory it would be wrong to say that variation is random with respect to adaptation.

Darwin started out his evolutionary career in a somewhat Lamarckian spirit, but by the time the *Origin* was published he was vocal in his opposition. Yet when *he* talks of chance variation in the *Origin,* it is not primarily to stress his opposition to Lamarck. Rather, he usually intends "to acknowledge plainly our ignorance of the cause of each particular variation" ([1859] 1964, 131). Darwin, in discussing what we would call mutations, uses "chance" primarily to express *our inability to assign a specific cause* for each new variation.

Contemporary evolutionists do, however, use the concept of "chance" or "random" mutation to distance Darwinism from Lamarckian theories of the sources of inherited variations. Theodosius Dobzhansky, for example, describes mutation as "a random process with respect to the adaptive needs of the species" (1970, 65) The production of mutations is said to be "random" in the sense that the potential "usefulness" of a mutation does not affect the probability of its occurrence. Darwinian evolutionary theory presumes that no causal connection exists between the occurrence of a mutation and its potential usefulness. Within biochemistry, while the explanation of the occurrence of a mutation may be explained by a "statistical" theory and thus be a product of "chance" in the sense discussed in the opening subsection of 7.5, it is "by chance" or "random" in a very different sense within evolutionary theory. Here "random" means "unaffected by adaptive requirements," a product of "chance," not "design."

7.5.3 Chance and Population Genetics

If, at a given locus, there are two alleles (**A** and **a**), then there are three possible genetic combinations at that locus in the population. The Hardy–Weinberg equilibrium says that in the absence of various disruptive forces, the frequencies of each of those combinations will be the same from one generation to the next. That is because in the absence of disruptive forces the recombinations of alleles which occur during mating are a matter of chance. Ignoring population size (see the next subsection, "Genetic Drift"), the probability of getting **AA, Aa** or **aa** is a simple multiple of the number of A's and **a**'s in the initial gene pool.

As with all such "stochastic" laws, HWE cannot be used to predict with certainty any particular individual's genetic makeup (see the first subsection of 7.5 and Chapter 2); but it *can* be used to predict the genetic makeup of the next generation of a randomly mating population. And because evolutionary theory is concerned to explain the changes in the heritable characteristics of populations, this is useful indeed.

7.5.4 Genetic Drift

As already seen, one significant reason for distinguishing *reproductive success* from *adaptive success* is that changes in genotypic frequencies can be due to a number of factors other than natural selection. One of these is called "genetic drift." Ayala (1982) defines "random genetic drift" as "variation in gene frequency from one generation to another due to chance fluctuations" (p. 243). Because the combinations of genes in a given generation is always a sample of those present in the preceding population, the question naturally arises, is this current gene pool a *representative* sample of the previous one? The possibility of "sampling error" arises, and the smaller the reproductively active segment of the population, the more likely it is that a random sample will not be representative. In population genetics what counts as representative is HWE, the maintenance of the same genotypic frequencies across generations.

The extent to which major evolutionary trends have been affected by drift is currently much debated among evolutionists. A very few, genetically atypical members of a population may migrate away or otherwise become isolated and start a new colony (a phenomenon termed by Ernst Mayr the "founder principle" [1942, 237]), initiating an evolutionary change which is due to simple sampling error.

One might well ask what is random about such processes. To answer this question, consider the evolutionary theorist's initial questions. First of all, when studying a population through many generations, one of the first questions to answer is whether any long-term changes occur in the genetic makeup of that population. Is the population in something approaching HWE or not? If not, the next question is whether an explanation can be found for the changes taking place. If the answer is that the gene frequency changes are within the limits expected to result from sampling error for a population of that size, then this raises questions about whether the change is due to selection, migration or other possible causes (see Beatty 1984, 183–211).

The concepts of "chance" and "randomness" play a variety of roles within evolutionary theory. Some of these derive from the fact that, in the absence of various disruptive forces, the recombinations of genes which take place when populations of organisms reproduce occur according to the rules of probability, one of Mendel's initial insights. That further implies that recombinations in small interbreeding populations are subject to the usual sampling error effects.

We have also seen, however, that describing a process as chance or random is often intended to stress the fact that explanations by reference to adaptation are inappropriate. Here it is not combinations of independent elements that are referred to as products of chance, but rather particular events. In such cases the appropriate contrast is not "random" versus "determined," but "random" versus "designed." It is important to distinguish these two uses of the terms "chance" and "randomness" in evolutionary biology.

7.6 CONFIRMATION

In an insightful review which Darwin greatly admired, a mathematically trained engineer named Fleeming Jenkin levelled the following methodological broadside at the evolutionist.

> He can invent trains of ancestors of whose existence there is no evidence; he can marshal hosts of equally imaginary foes; he can call up continents, floods, and peculiar atmospheres, he can dry up oceans, split islands, and parcel out eternity at will; surely with these advantages he must be a dull fellow if he cannot scheme some series of animals and circumstances explaining our assumed difficulty quite naturally. (Jenkin [1867] 1983, 319)

This criticism predates Popper's similar claims about evolutionary biology by more than a century. Furthermore, as a criticism of *On the Origin of Species* it has some force—virtually all of Darwin's examples of explanation by appeal to natural selection were "thought experiments," imaginary illustrations, as he called them. We can see whether this criticism is still valid by referring back to our four cases and asking how evidence is brought to bear on their truth or falsity. Before doing so, however, there are two important preliminaries.

First, it is important to distinguish between *testing specific hypotheses* (or evaluating specific explanations) which use the Darwinian pattern of explanation, on the one hand, and *evaluating the overall success of the theory* of evolution itself, on the other. (That is, we need to distinguish questions of explanation and confirmation discussed in the first two chapters of this book and questions of evaluating the overall strength of evolutionary biology as a research program discussed in Chapter 4.) Jenkin allows the possibility that a hypothesis invoking a particular combination of processes might be "falsified"—the problem he raises is that the evolutionist has such an arsenal of alternatives. He notes that, in accounting for geographical distribution, "Darwin calls in alternately winds, tides, birds, beasts, all animated nature, as the diffusers of species, and then a good many of the same agencies as impenetrable barriers. . . . With these facilities of hypothesis there seems to be no particular reason why many theories should not be true" (Jenkin 1867 in Hull 1983, 342; see Kitcher 1985a, 154–168). Thus, while it is important to show that specific evolutionary explanations can be supported or rejected on the basis of evidence, it is also important to see whether biologists habitually adopt strategies which insulate evolutionary theory from more general evaluation in the ways suggested by Jenkin and, more recently, by Popper.

Secondly, we must again take seriously the idea that every explanation is an answer to a specific question (see Chapter 1). To evaluate the success of an explanation we need to know not just the fact the explanation is concerned with, but the *specific question* about that fact which the explanation seeks to answer. There is an old joke about a child getting a long and detailed account of human sexual reproduction when she asked where she had come from, after which she replied, "Oh, I thought I came from Canada."

Cases 1–4 cover a wide spectrum of types of evolutionary questions and explanations. Let us consider how these are in fact supported by evidence, and how they might be evaluated with respect to their truth.

Case 1 asks a question about an evolutionary trend in an historical lineage of animals. The question to be answered is "Why do we see a rapidly accelerating change in the height-length ratio of the molars in one lineage of the *Equidae* and not in others?" The specific fact about which we are asking the question is itself not established beyond reasonable doubt—we could imagine rich fossil finds which could, for example, dramatically change our estimations of the acceleration rates in the two lineages being compared. As trends supported by fossil evidence go, however, this one is well supported. But how strong is the evidential support for processes mentioned in the explanation?

The explanation supposes that this trend reflects a series of mutations, at one or a related set of genetic loci, producing changes in teeth structure which provided slight advantages to their possessors given a newly available nutritional resource. That is, it depends on premises about the genetics of these long extinct organisms, and about the adaptiveness of the changing dentition. Unlike Case 2, we have no *direct* way of testing these genetic assumptions. Indirect methods, however, are available. The mammals in question are ancestors of a currently existing genus. If genetically based variability in dentition in current populations of *Equus* is of the degree and rate required, there is good reason to trust the assumptions. In fact George Simpson, the originator of a careful biometric study of this data, used arguments of this kind to support the conclusion that the rates observed in the fossil record were well within the ranges observed today. If that had not been the case, the explanation would have been considerably weakened.

The adaptational assumption is supported by paleoecological data, data suggesting that the grasses were evolving at this time, providing a new niche to be exploited, and that a subpopulation of the *Equidae* is associated with grazing environments while another is associated with browsing environments; and that the acceleration of evolutionary change in relative tooth height is associated with one and not with the other. Each of these claims can be *independently* tested, that is, tested outside the context of this particular explanation, and again could be considerably weakened by future fossil finds. The history of this case up to now is that new fossil evidence has tended to strengthen previously weakly supported claims.

Other kinds of support not previously alluded to are important to mention because of their prominence in historical explanations of this sort. First, there may be correlated evolutionary changes in the same lineage which can be explained by reference to the same or related processes. In this case, there are in fact many, of which only two will be mentioned (for more, see Maynard Smith 1972, 261–270). First, the well-known change in this lineage to a "hoofed" creature from a four-toed creature, and a vast array of associated morphological changes, is consistent with these creatures shifting to an open, grassland environment (see Case 3). These changes occur only in the lineage which took up grazing, not in those which retained a browsing lifestyle. Furthermore, this lineage has gone through both relatively long periods of little change and a number of periods of rapid change. The periods of rapid change in dentition are closely correlated with those in limb structure. Second, along with the rapid change in tooth dimensions is a correlated change in the surface ridges of the teeth, again restricted to the grazers as our hypothesis would predict.

Finally, if optimality considerations lend any support to the claim that "well-designed" features are products of natural selection, then an engineering analysis which shows that certain types and relative sizes of teeth are superior in dealing with sources of nutrition high in silicates will lend further support to the above explanation. Such evidence is considered in discussing Case 3.

Enough has been said to indicate both the vulnerability of such explanations to tests, and the vast array of resources available to provide inductive support, or confirmation, for them.

Case 2 is, as mentioned earlier, a textbook case of an explanatory success for population genetics. In cases like this one, we begin with a question about the characteristics of a current population, and we want to know what causal mechanisms are responsible for the population having those characteristics. This case, however, involves a special problem as well—on the assumption that natural selection works ruthlessly to weed out disadvantageous genes, it is puzzling that a population should possess a lethal recessive gene at high frequencies. Recall that population geneticists had already deduced, from their mathematical models, the *possibility* that heterozygotes could have an advantage over either of the homozygotes in a population, under certain special conditions. We can generate any percentage frequency of such organisms in a population with such models simply by varying the relative fitness values of the alternative genotypes. This example, and a number of other classic studies, established that heterosis actually occurs in natural populations, and provided a selection explanation which identifies the adaptive advantage responsible for the maintenance of the potentially lethal gene.

The strength of this case flows from the following features: (1) The functional value of the "mixed hemoglobin" phenotype (**AS**) in a malarial environment is clearly established, as is the pathophysiology associated with the "sickling" of the red blood cells due to **S**-hemoglobin; (2) the genetics is simple, even at the molecular level, because the sickling of the red blood cells is due to a single amino acid substitution; (3) the role of the environment in establishing the adaptive advantage of the heterozygote is clear; (4) the adaptive advantage of the trait in question is not merely an assumption in this case, as it often is—it has been carefully established by epidemiological studies. The primary weakness of this explanation is that the precise mechanisms by which the blood type which is slightly *less* successful at oxygen transport manages to be *more* successful at counteracting malarial infection are not fully understood. That is, while it has been established *that* the heterozygote is at an adaptive advantage, it has not been established *why* this is so.

Once again, the genetics and physiology of these populations are independently supported by evidence unrelated to their role in this particular explanation. Such explanations draw selectively on an enormous amount of background knowledge in molecular genetics and biochemistry. If biologists consistently failed in such cases to establish heterozygote superiority, this would call into question certain assumptions which are currently taken for granted in explanations of this kind. Enough such failures might perhaps lead to doubts about certain principles of population genetics which are currently taken as established.

Case 3 is part of an attempt to provide a general explanation for an extremely extensive set of facts regarding animal locomotion. Different land animals have a variety of structural and behavioral differences related to locomotion. Are they all attempts on the part of natural selection to solve the same basic problem of locomotive energy efficiency? Specifically, is the horse's peculiar limb structure bioenergetically optimal? The use of optimality arguments of this kind is common in evolutionary biology, and has been the target of some serious criticism. In order to evaluate these criticisms, we must distinguish between three uses of the concept of "optimal design" in evolutionary explanations:

(a) an assumption is made that a certain trait is optimal relative to available alternatives as *part of* an evolutionary explanation;
(b) an optimality *hypothesis* is put forward, and is then tested in some way, but no claim is made about natural selection;
(c) a structure is taken to be optimally designed for a certain function, and this fact is taken to establish natural selection as the cause of that trait's presence in a population.

In this and similar cases, it is arguments of type (c) that have been criticized. The questionable assumption is that, at least for certain types of traits, establishing them to be, by engineering or game theoretic standards, the best of available alternatives is prima facie evidence that the trait was a product of a selection history. In many cases, this is the *only* evidence we have for natural selection. Thus two questions must be asked. First of all, can the optimality claim itself be tested? Second, supposing it can, and supposing it passes rigorous tests, does it support the *further* claim that the trait in question is an adaptation, that is, a product of natural selection?

Regarding the first question, here is the kind of objection that can be raised. It is Jenkin's objection, as you will see. Suppose Maynard Smith predicts that for mammals of the same size there should be a direct correlation between increased speed and time spent off the ground. And suppose in 50 percent of the animals studied, he finds no such correlation. Will he give up his optimality claim? Not at all, the critic says. He will simply add a subsidiary hypothesis, or a series of them, for the exceptions, claiming that other problems besides efficiency of locomotion entered the picture in these cases. For example, Maynard Smith might now hypothesize that in the 50 percent of mammals where the correlation fails, physiological constraints make more rapid limb movement the optimal solution to the problem of locomotive efficiency.

Notice that nothing *in principle* is wrong with this suggestion—as Jenkin pointed out long ago. The problem is, how can we test it? Is the introduction of such subsidiary hypotheses a legitimate search for the truth, or an ingenious con job? The answer depends centrally on the status of the subsidiary hypotheses. Two questions should be asked of them:

1. Do they refer to processes or entities of kinds which are known to have causal relevance to evolutionary change in other, similiar cases?
2. Is the presence of these processes and entities capable of *independent confirmation?* Independence, remember, simply means that there must be evidence for the truth of the subsidiary hypothesis apart from its ability to help explain the phenomenon currently under investigation.

In the case in question, for example, there are built-in historical constraints on adaptational "solutions," so nothing is inherently implausible about the proposed additional hypothesis appealed to. And if the imagined hypothesis has a chance of being acceptable, the 50 percent which were not as the initial hypothesis predicted should all share the same sorts of physiological constraints—they might, for example, be part of the same historical lineage, which could independently be established from the fossil record or by measures of biochemical similarity.

Suppose now that such an optimality hypothesis has successfully explained the similarities and differences among a class of organisms with respect to some feature. Does this by itself count as support for the hypothesis that these differences are due to natural selection? As mentioned in presenting this example, a number of assumptions have to be made to establish this connection: about the genetics related to the traits in question, about the existence of genetical alternatives, and about the absence of compelling reasons to suppose the trait would have been there anyway, selection or not. In principle, each of these further assumptions can be confirmed. In practice, however, optimal design arguments are often used precisely where confirmation for such assumptions is lacking (see Sober 1984a, Chapters 15–17 and Kitcher 1985b, Chapter 7 for further discussion).

Case 4 shows how it is possible to establish precise claims about relative adaptedness independently of population genetic measurements of actual reproductive contributions. Sophisticated statistical methods were used to establish hypotheses about causal relationships between changing features of the environment and changes in specific traits of the population in question (for example, correlations between availability of seeds of various sizes and nutritional values and differential survival rates of finches). As a number of classic studies have shown, evidence of selection more direct than, and independent of, measures of actual reproductive success are possible (see Endler 1986). This is crucial for the viability of the picture of evolutionary theory as a causal theory developed in this chapter.

This case also illustrates one other important feature of particular evolutionary explanations—their relevance to the actual existence in nature of various *postulated* models of speciation. Because of the ability to specify quantitative values for fitness differences, and then to solve problems regarding the effects of such differences on populations breeding along Mendelian lines, it is easy to imagine all sorts of ways that new species *could* be produced by various combinations of evolutionary mechanisms. Evolutionary theory's main task, however, is to help us understand life on earth as it actually was and is. Thus it is important to question whether these speculative models are realistic—are they likely to be instantiated in the natural world? In this case the idea of an "evolutionary bottleneck" seems to be supported, and in addition it suggests some potentially surprising consequences from such bottlenecks.

Enough has been said to indicate how specific applications of evolutionary theory to specific sorts of questions about specific facts are confirmed and rejected on the basis of evidence. Perhaps the best way to think about whether, nonetheless, Jenkin's worry still holds for the theory in general is to carefully study its historical development.

We speak of the theory which emerged with the integration of population genetics, paleontology, biogeography and ecology in the 1940s and 1950s as the "Neo-Darwinian Synthesis." The primary reason for this is that the central plank of Darwin's theory was taken to be dead on, after a long period of skepticism: *Evolution is predominantly due to the action of natural selection acting on small, random, heritable variations*. Prior to the synthesis, no serious biologist doubted that evolution had occurred, but there were nearly as many accounts of its causes as there were individuals worrying about the question.

As mentioned in the first section of this chapter, specific theories of inheritance and variation, and of population dynamics and geology, had to be true in order for a *Darwinian* theory of evolution to be true. Had things kept turning out otherwise, while "Darwinians" kept on insisting that their view must be correct somehow, it would now be viewed, to use a Lakatosian turn of phrase, as a degenerating research program (see Chapter 4).

On the other hand, neo-Darwinism is not Darwinism. It has rejected Darwin's view of the peripheral importance of geographic isolation in speciation, of inheritance (insofar as he had one), of the heritability of variations due to "use and disuse." Even the meanings of the theory's central concepts such as "fitness" and "random variation" have shifted to facilitate their use in population genetics. What is more, the theory includes, at least in its "ideal" form, principles such as the Hardy-Weinberg principle, and concepts such as "heterosis" which cannot even be formulated in Darwin's terminology. A careful study of the development of some of the key concepts within this still developing theory would raise interesting questions for the various models of scientific change discussed in Chapter 4. Certainly by any of the measures of scientific progress there discussed, Darwinian evolutionary theory is progressive. Unfortunately, given the models that have so far been developed to make such measures, that is a rather modest claim.

DISCUSSION QUESTIONS

In answering the questions below, it will be helpful to consult the bibliography listed in Section A of the Suggested Readings.

1. In Section 7.3 David Sloan Wilson is quoted as formulating "the basic theory of natural selection" in the following manner: "Organisms vary in a heritable fashion; some variants leave more offspring than others; their characteristics, therefore, are represented at a greater frequency in the next generation." What is wrong with this statement of the theory?

2. It is common sense to say that an insect appears like the leaves of the plants it rests on *in order to* or *for the sake of* avoiding its predators. But is it good science? Why are such statements (commonly termed "teleological") explanations? Explain what role they play in evolutionary biology. How can such explanations meet the requirements for scientific explanation discussed in Chapter 1?

3. What is an "optimality explanation?" Discuss what role such explanations play within evolutionary theory and provide a convincing argument that such explanations are empirically testable.

4. Evaluate one of the models of scientific change discussed in Chapter 4 in terms of its ability to account for the development of evolutionary biology from 1859–1959.

5. What do critics mean when they say evolutionary theory rests on a tautology? Provide an argument that either supports or undermines this objection to the theory.

6. Answer question 1 at the end of Chapter 2 with respect to one of the biological examples discussed in this chapter.

7. Which of the four types of explanations referred to in Table 1.1 of Chapter 1 does Case 2 in this chapter exemplify? Depending on which you decide on, evaluate this explanation in the light of the problems for that type of explanation discussed in Chapter 1.

SUGGESTED READINGS

A. Evolutionary Biology.   All these works are accessible to the nonspecialist. Reading them in the order listed makes some pedagogical sense.

DARWIN, CHARLES ([1859] 1964), *On the Origin of Species: A Facsimile of the First Edition.* Cambridge, MA: Harvard University Press. The founding document of evolutionary biology, it remains one of the subject's masterpieces.

HANSON, EARL D. (1981), *Understanding Evolution.* New York: Oxford University Press. A clear, comprehensive university-level textbook.

MAYNARD SMITH, JOHN (1972), *The Theory of Evolution.* 2nd ed. Baltimore: Penguin. The best nontechnical introduction to evolutionary biology, by a leading theorist.

MAYR, ERNST (1970), *Populations, Species, and Evolution: An Abridgment of* Animals Species and Evolution. Cambridge, MA: Belknap Press of Harvard University Press. More advanced than Maynard Smith and Hanson, and focused more narrowly on the mechanisms producing new species.

SHEPPARD, P.M. (1975), *Natural Selection and Heredity.* 4th ed. London: Hutchinson. A classic introduction to "ecological genetics," by one of its leading practitioners.

WILLIAMS, GEORGE C. (1966), *Adaptation and Natural Selection: A Critique of Some Current Evolutionary Thought.* Princeton: Princeton University Press. A tough-minded critique of some of evolutionary biology's excesses. Many philosophers of biology cite this as the book that first interested them in the subject.

B. Philosophy of Biology.   These sources collect a wide range of philosophical discussion of evolutionary biology, including issues not discussed in this chapter.

BRANDON, ROBERT (1990), *Adaptation and Environment*. Princeton: Princeton University Press. A biologically well-informed philosophical analysis of evolutionary theory and its explanations.

BRANDON, ROBERT N. AND RICHARD M. BURIAN (eds.) (1984), *Genes, Organisms, Populations: Controversies over the Units of Selection*. Cambridge, MA: MIT Press. A collection of the most important papers on the "levels" and "targets" of natural selection.

Philosophy of Science. (1984), Volume 51, number 2. *Special Issue: Philosophy of Biology*. This issue of the "official journal" of the Philosophy of Science Association was devoted entirely to philosophy of biology, and contains a number of fine essays. (In addition to *Philosophy of Science*, which regularly publishes papers on topics in philosophy of biology, the journal *Biology and Philosophy* is devoted exclusively to it.)

ROSENBERG, ALEXANDER (1985), *The Structure of Biological Science*. Cambridge, England: Cambridge University Press. The most up-to-date introduction to philosophy of biology, uncompromisingly reductionist in approach.

SOBER, ELLIOTT (ed.) (1984a), *Conceptual Issues in Evolutionary Biology: An Anthology*. Cambridge, MA: Bradford/MIT press. The best collection of essays available covering a wide range of topics in philosophy of biology.

————. (1984b), *The Nature of Selection: Evolutionary Theory in Philosophical Focus*. Cambridge, MA: Bradford/MIT Press. A detailed, advanced study of the central explanatory concept in evolutionary biology.

C. History of Evolutionary Biology.

BOWLER, PETER J. (1989), *Evolution: The History of an Idea*. Revised edition. Berkeley and Los Angeles: University of California Press. A comprehensive study of evolutionary thinking from the eighteenth century to the present.

RUSE, MICHAEL (1979), *The Darwinian Revolution*. Chicago: University of Chicago Press. A lively history of Darwin's impact on the nineteenth century.

Eight

PHILOSOPHY OF MEDICINE

Kenneth F. Schaffner

Philosophical analyses of medicine have a long pedigree (see Engelhardt 1986; Pellegrino 1976, 1986; Pellegrino and Thomasma 1981, Chapter 1). Nonetheless, only in the last dozen years or so have professional philosophers and philosophically sensitive physicians paid any sustained attention to a philosophy of medicine. In part this renewed interest can be traced to the burgeoning impact of medical ethics—a branch of philosophy of medicine broadly conceived. Part of this interest is also dependent on the dynamic growth of studies in the philosophy of biology which furnish insights into the sciences that are traditionally viewed as lying at the core of medicine.

In this chapter we will examine two related topics in the philosophy of medicine: the scientific status of medicine and the nature of reduction in the biomedical sciences. A satisfactory answer to the question whether medicine is a science will ultimately depend heavily on the account given of the nature of reduction, whence the rationale for the inclusion of this topic.

8.1 IS MEDICINE A SCIENCE?

We begin by considering a general question: "What is the nature of 'medicine'?" and in particular focus on the more specific question: "Is medicine a *science?*" By attempting to answer these questions we have an opportunity to examine both the reach of science as well as its limitations.

It should be clear to all without reciting any extensive number of examples that medicine is *at least in part grounded on* the biomedical sciences. The development of the germ theory of disease, the extent to which micro- and molecular biological techniques are used in the development of new antibiotics, and the manner in which biochemical tests are employed in diagnosis are testimony to the scientific nature of modern medicine. Moreover, significant progress has been made in rationalizing medical decision making through the application of Bayesian probabilistic reasoning techniques to clinical problems (Zarin and Pauker 1984), as well as other important advances in computerizing medical diagnosis (Miller, Pople, and Myers 1982; Schaffner 1985). However, medicine is not simply biology, as every patient will recognize. The doctor-patient interaction is a human interchange, and as Engelhardt has noted:

> One cannot effectively treat patients without attending to their ideas and values. One cannot humanely treat patients without recognizing that their lives are realized within a geography of cultural expectations, of ideas and images of how to live, be ill, suffer, and die. (1986, 3)

The great nineteenth-century pathologist Virchow (1849) observed that medicine may best be conceived of as a *social science*, in part because of these types of complexities. But other philosophers have argued in recent years that medicine is not—in its essence—a science at all. A review of some of these arguments and an elaboration of what we have come to learn about the intricate nature of medicine by discussing a brief example of a particular patient problem will help us understand the character of the subject better.

8.2 PROBLEMS WITH CHARACTERIZING MEDICINE AS A SCIENCE

Continuous debates have been in the pages of *The Journal of Medicine and Philosophy* and in its sister publication *Theoretical Medicine* about the nature of medicine as a science. In addition a vigorous debate has taken place within one medical discipline, psychiatry, about the scientific status of that subject (see Grünbaum 1985). In his article, Munson (1981) challenged Forstrom's (1977) earlier claim that medicine (in general) was a science by proposing we examine universally recognized sciences such as physics, chemistry and biology along three fundamental dimensions and compare medicine with these paradigm sciences. Munson suggested that we consider the fundamental *aims, criteria of success,* and *principles regulating the disciplines* of recognized sciences and medicine. In each of these three categories Munson claimed that significant differences exist between the recognized sciences and medicine, and then, on the basis of an additional further argument based on construing medicine "as an *enterprise"* (1981, 183), concluded that medicine is not, and cannot be, a science" (1981, 189). Let us examine Munson's three rubrics for comparing medicine and other sciences.

The first important dimension along which science and medicine are contrasted is the *aim* of the discipline. For Munson, the basic (internal) aim of science is *"the acquisition of knowledge and understanding of the world and the things that are in it."* (1981, 190) For medicine, however, the basic aim is *"to promote the health of people through the prevention or treatment of disease"* (1981, 191). Acquiring knowledge and understanding of diseases is subservient to disease prevention and treatment. The second rubric in Munson's comparison involves the *criteria of success* of the disciplines. Science is successful when it achieves true (or approximately true) knowledge, whereas medicine is successful when it achieves its aim of preventing or ameliorating disease regardless of the truth of its cognitive content. Finally, Munson compares the basic moral commitments of science and medicine. Science must give *honest* reports of observations or experimental results; without such honesty science as a cooperative endeavor cannot succeed. Medicine in contrast is primarily committed to "promoting the health of any individual accepted as patient" (1981, 196). This is not to assert that medicine as an enterprise can encourage dishonesty, nor that physicians ought not be honest with their patients, but only that health promotion for the physician's individual patient is primary and essential.

These are useful and largely correct points made by Munson and they suggest at the minimum that important *differences* exist between the traditional sciences and medicine. By appealing to one further argument, however, Munson believes that he can demonstrate that medicine cannot *ever* be *reduced to*—in the sense of replaced by—the biomedical sciences, which are so heavily employed by medicine. This additional argument is important since, as we see in the later parts of this chapter, *apparent differences* between sciences cannot be taken as compelling evidence that a reduction cannot be accomplished between those sciences. Thus the differences Munson specifies are but the first of two steps he must take to demonstrate that medicine cannot be a science.

Munson appears to accept something like the general reduction model to be described in Section 8.6 as prescribing the requirements for achieving a reduction. These requirements, which are considered in more detail later, essentially demand that we define all the terms in medicine in vocabulary that belongs to the biomedical *sciences,* and that we then be able to derive medicine as a special case from the biomedical sciences. On the basis of that type of account, however, Munson contends that medicine cannot be reduced to its conceptual content—the biomedical sciences—because it is an *enterprise*—an inherently social activity—as well as a discipline. Though the *content* of medicine might someday be reducible to biology, it would be "inappropriate to talk about reducing *medicine* to biology" (1981, 203). Munson provides an analogy:

> It is inappropriate in the same way it would be to talk about reducing hopscotch to physics. As a game, hopscotch must be understood in terms of rules and aims. The pattern of activity displayed in playing the game gains its significance from a network of social conventions. (1981, 203)

Munson's new argument rests on two new claims: (1) medicine has an essentially social character to it; and (2) as such, it is irreducible to biology because networks of social conventions (and interactions) are not reducible to biology. Munson's argument here bears some similarities to a view urged by Pellegrino and Thomasma, "medicine is not reducible to biology, physics, chemistry, or psychology; . . . we argue that medicine is a form of unique relationship" (1981, xiv). This relationship—if we understand it correctly—is that between the healer and the patient in which the patient participates as subject and object simultaneously. In the next section we examine the nature of the social aspect of medicine, and in the conclusion re-examine Munson's second claim, relating it to the material in the earlier sections of this chapter.

8.3 MEDICINE AND A BIOPSYCHOSOCIAL MODEL

As noted in Section 8.1, it was Virchow who first suggested that medicine should be conceived of as a "social science." Other scholars have proposed similar ideas, but the writer who has probably been the most articulate proponent of significance of the social—and psychological—components of medicine has been George Engel. Engel has propounded the notion of a "biopsychosocial" model for medicine to be contrasted with a more reductionistic "biomedical" model (1977, 1981).

In his writings, Engel suggests that the striking successes of twentieth-century medicine such as the discovery of insulin and the antibiotics, as well as new breakthroughs in molecular biology, can result in tunnel vision on the part of the physician, nurse, or other health-care professional. Such successes incline healers toward a biomedical vision of medicine, in which the patient is "nothing but" a bag of chemicals. For such a doctor or nurse, sensitivity to factors which can have marked influences on illness, such as psychosocial stresses, becomes weakened. Such healers may overlook important predisposing causes of disease, or may fail to incorporate significant psychosocial dimensions into their healing art.

Engel (1981) recounts the story of a patient suffering from a heart attack who is being examined in a hospital's emergency room. The patient has an inexperienced intern who attempts to take a sample of his arterial blood as part of a standard diagnostic workup for presumptive heart attack patients. This arterial puncture is a simple procedure which, however, can become painful if it is not performed deftly. Repeated unsuccessful attempts generate pain and anxiety which result in a second heart attack for the patient. Reassuring discussion with the patient, and a request for assistance with the arterial puncture, could have achieved a significantly better outcome for this patient. Figure 8.1 is taken from Engel's essay and graphically depicts the hierarchy of perspectives that are found in medicine and the complex ways that these "levels" interact in one time-slice of a patient's encounter with the health-care system.

Engel's suggestion is that better health care would be delivered by healers mindful of the psychosocial as well as the biological dimensions of illness. The appreciation of the complexity of a human being's reality and a joint analysis of the interactions among various levels of causation, from the molecular through the organ-level to the intellectual, emotional, familial, and economic, will both permit a better understanding of how illness arises, as well as provide a richer armamentarium for the physician. This richer, more complex perspective is what Engel terms the *biopsychosocial model of medicine*. Two fundamental questions to be addressed are whether such suggestions (1) are sound; and (2) count against the reducibility of medicine to biology.

Engel's proposals that medicine needs to be conceptualized as a complex multi-level science receive support from recent studies examining the interactions of the neurological and immunological systems. Thus Engel's suggestions can be viewed as

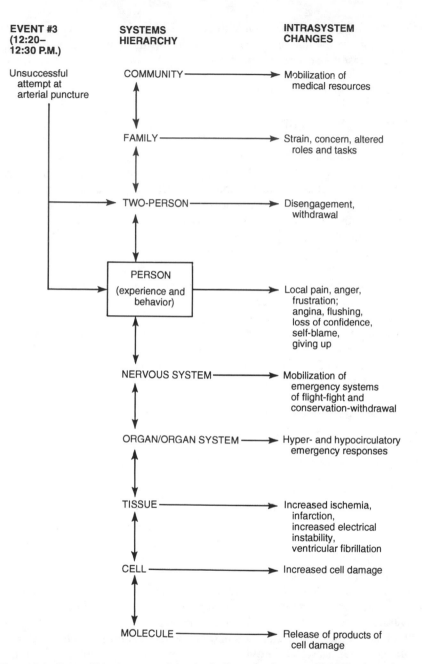

Figure 8.1 Event #3 in the course of a patient's diagnosis of a myocardial infarction or heart attack, depicting the interlevel causal interactions. See text for details. From Engel 1981, with permission.

sound, and also can envisage an important role for the social dimension of medicine at the core of its fundamental aim as a healing practice. We are still left, however, with the question whether the acceptance of these "levels" as working heuristics in a sense—and this seems to be all that we are committed to thus far—is any argument against in-principle reducibility. In an important sense, we may stop temporarily at many levels in the course of bringing about a reduction; we see how this works in the account of Kandel and his associates' research discussed in later sections. Another example to be examined explores the interaction between neurological and immunological systems—and the endocrine system as well—which may assist us in understanding how a "biopsychosocial" model may be implemented in the framework of a particular disease, such as breast cancer, and whether including such concepts as "emotion" into our medical ontology counts for or against the prospect of reduction.

8.4 REDUCTION IN THE BIOMEDICAL SCIENCES

Reduction for biomedical scientists often suggests that biological entities are "nothing but" aggregates of physicochemical entities. This can be termed "ontological reductionism." A closely related but more *methodological* position is that sound scientific generalizations are available only at the level of physics and chemistry. The contrary of the first *ontological* position is antireductionism, which often is further distinguished into a strong *vitalistic* position and a second weaker *emergentist* view. From the perspective of ontological antireductionism, the biomedical sciences are autonomous. The contrary of the second *methodological* reductionist position is a kind of methodological emergentism. Here too we can distinguish a strong position taken by such distinguished biologists as Simpson (1964) and Weiss (1969), and a weaker form that attributes the necessity to work with higher level entities to current gaps in our knowledge (Grobstein 1965).

The vitalistic thesis, championed earlier in this century by Hans Driesch, proposed the existence of special forces and irreducible causes which were peculiar to living entities. The vitalist thesis led to no testable hypothesis and is essentially a dead issue in contemporary biology.

Analysis of the emergentist position requires further distinctions. One type of emergentism, *in principle emergentism,* holds that the biomedical sciences cannot, regardless of progress in biology and the physicochemical sciences, be reduced to physics and chemistry. Generally this view is based on what its proponents perceive to be certain general timeless *logical* features of biomedical entities and modes of explanation. A similar thesis has been applied to other areas of experience, and we have seen such a view proposed for "medicine" as an enterprise by Munson (1981) and by Pellegrino and Thomasma (1981).

The weaker sense of "emergentism" refers to the *current* status of the relation between the biomedical and the physicochemical sciences (or between a science of social interactions and the biomedical sciences). It should be clear from the views developed thus far that not all biomedical entities, such as cells, are *at present* reducible to molecular terms and chemical theories. However, some *portions* of the biomedical sciences *are* so reduced. Such reductions, partial though they may be, are important accomplishments in the biomedical sciences and constitute the subject matter of biophysics, biochemistry, and parts of "molecular" biology. *Why* the accomplishments are important, and specifically, what conditions are and should be met in reductions will occupy us in the next few sections.

A major component of reductionism to be elucidated is often referred to as *intertheoretic* reductionism. Much of the discussion by scientists and philosophers has focused on this notion, though the term "intertheoretic" is somewhat misleading. Essentially intertheoretic reduction is the *explanation* of a theory initially formulated in one branch of science by a theory from another discipline. Sometimes this type of reduction is referred to as *branch* reduction when it is felt that emphasis on *theories* is not appropriate (Kemeny and Oppenheim 1950). This is a useful corrective, since what can be captured by general high-level abstract theory in the biomedical sciences is often considerably different from what can be so represented in physics. In the next section, a notion of branch reduction is developed and then discussed in the context of the various senses of reduction and emergence outlined above.

8.5 REDUCTION AND THE HUMAN MIND

An analysis of a complex concept such as "reduction in the biomedical sciences" will not be fruitful unless it can be shown how—at least in a schematic way—it works in practice in those sciences. This suggests that we will need to look at some substantive area in which reduction has occurred or may be occurring. There are a number of plausible candidates, given molecular biological developments in the past decade, including genetics and the neurosciences. Hull (1974, 1976) and Schaffner (1967, 1969, 1976, 1977) have engaged in a debate over the reducibility of classical genetics to molecular genetics, with important contributions to the discussion by Wimsatt (1976) and Kitcher (1984). Rosenberg (1985) has discussed both sickle-cell anemia and recombinant DNA in the context of reductionism, and has also developed a position on the relation of classical and molecular genetics. Also see Balzar and Dawe (1986) who have recently examined the reduction of classical to molecular genetics.

An examination, however, of the way in which reductionistic explanations of learning behavior are being developed by the neuroscientists is more relevant to the themes in this chapter. The explanation of mental processes through appeals to physical or biological events has traditionally been controversial, and remains so today. What has altered in the past decade are significant advances in the application of molecular biology to the neurosciences. In previous sections, "medicine" has been characterized as possessing important psychological (and social) dimensions, and it has been indicated that to adequately address these issues we needed an account of reduction which would be sufficiently strong to give promise of reducing such psychosocial dimensions to the biomedical sciences. Furthermore, significant advances are occurring in neurobiology which have immediate clinical applications, for example, neural tissue transplants for Parkinson's disease. Accordingly, it may be most fruitful to explore reduction at the leading edges of molecular neurobiology. Thus an account of reduction in the biomedical sciences is developed in the context of neuroscientists' molecular explanations of learning behavior and, ultimately, of the mind.

Questions concerning the substantial nature of the mind are as old as philosophy (and religion). Today the ontological status of "the mental" is coming under renewed scrutiny in the neurobiological sciences. Some of the major contributors to the science have argued that *molecular biological approaches* are superior to alternative means of studying neurobiology. In the Preface to a recent volume on *Molecular Neurobiology in Neurology and Psychiatry* (1987b), Eric R. Kandel wrote:

> This volume reflects the impact of molecular biology on neural science and particularly on neurology and psychiatry. These new approaches have accelerated the growth of neurobiology. The resulting increase in knowledge has brought with it two unanticipated consequences that have changed the ways in which clinical researchers and practitioners can now view the findings that came from basic science.
>
> The first consequence is a new unity, a greater coherence, in biology as a whole, as studies move from the level of the cell to that of the molecule. . . . [T]he second consequence of our increased knowledge . . . [is that a]s science becomes more powerful it becomes more ambitious—it becomes bolder. . . . Many molecular biologists now frankly admit that the ultimate object of their interest is not simply the system with which they work. It is not simply *lambda*, the T4 phages, or *E. coli*. It is not even *C. elegans, Drosophila*, or *Aplysia*. It is human biology. And some biological researchers are so bold as to see their ultimate interest as the function of the human mind. (1987, vii–viii)

In Section 8.7 we look closely at the nature of some of these molecular biological explanations of neurobiological phenomena, and most specifically at learning and at memory. In what follows we will present a brief historical sketch of some of the main features of philosophical analyses of reduction, and then go on both to criticize those accounts and to sketch the outlines of what we take to be a more adequate perspective on reduction. We believe that the account to be a proposed has direct application in molecular neurobiology. In point of fact we will elaborate a detailed example from molecular neurobiology, but we also think that its field of application is broader. We think the analysis to be presented in the present chapter covers not only neurobiology but also the biological, psychological, and social sciences, and that it can be applied and tested in a wide domain.

8.6 A BRIEF HISTORICAL ACCOUNT OF REDUCTION IN PHILOSOPHY OF SCIENCE

As mentioned in Section 8.4, much of the debate about reductionism has focused on the relationship between *theories*. The traditional analysis of a scientific theory in philosophy of science understood a theory to be a collection of a few sentences, typically each of them scientific laws. These sentences could be taken to be the basic *axioms* or leading premises from which theorems could be proven, much as in any high school geometry text. These axioms had both logical terms, such as the equal sign, and non-logical terms, such as "photon" and "gene." Philosophers of science also believed that casting those sentences into a logical formalism would aid in the clarification of the nature of a scientific theory, especially since the explicit relations between the axioms and the theorems—other "derived" scientific laws or descriptions of experimental results—could be examined with formal precision. More recently philosophers of science have thought of scientific theories in different ways, such as viewing them as "predicates" (see Chapter 3). This is often referred to as a "semantic" conception of scientific theories, as distinguished from the older "syntactic" conception. Such newer views are compatible with the older notion of a theory as a small collection of sentences. In fact, simply conjoining the collection of sentences (the axioms) together is sufficient to define the predicate of interest. (See Chapter 3.)

The *locus classicus* of philosophical analyses of theory reduction can be found in the work of Ernest Nagel (1947, 1961). Nagel envisaged reduction as a relation between *theories* in science, and also assimilated it to a generalization of the classical Hempel deductive-nomological model of explanation (see Chapter 1). A theory in biology, say, was reducible to a theory in chemistry, if and only if (1) all the nonlogical terms appearing in the biological theory were *connectable* with those in the chemical theory, for example, gene had to be connected with DNA; and (2) with the aid of these *connectability assumptions,* the biological theory could be *derived* from the chemical theory (with the additional aid of general logical principles). Later connectability came to be best seen as representing a kind of "synthetic identity," for example, gene = DNA sequence, (Schaffner 1967, Sklar 1967, Causey 1977).

An interesting alternative and more semantic approach to reduction was developed by Suppes and Adams:

> Many of the problems formulated in connection with the question of reducing one science to another may be formulated as a series of problems using the motion of a representation theorem for the models of a theory. For instance, the thesis that psychology may be reduced to physiology would be for many people appropriately established if one could show that for any model of a psychological theory it was possible to construct an isomorphic model within physiological theory (1967, 59)

Another example of this type of reduction is given by Suppes when he states:

> To show in a sharp sense that thermodynamics may be reduced to statistical mechanics, we would need to axiomatize both disciplines by defining appropriate set-theoretical predicates, and then show that given any model T of thermodynamics we may find a model of statistical mechanics on the basis of which we may construct a model isomorphic to T. (1957, 271)

This model-theoretic approach allows a somewhat complementary but essentially equivalent way to approach the issues, to be discussed.

In contrast to Suppes and Adams (and thus also later work by Sneed, Stegmuller, and their followers), most writers working on reduction dealt with the more syntactic requirements of Nagel: connectability and derivability. The *derivability* condition (as well as the connectability requirement) was strongly attacked in influential criticisms by Popper (1957a), Feyerabend (1961, 1962), and Kuhn (1962). Feyerabend, citing Watkins, suggested but did not agree with the proposal that perhaps some form of reduction could be preserved by allowing approximation (1962, 93). Schaffner (1967) elaborated a modified reduction model designed to preserve the strengths of the Nagel account but flexible enough to accommodate the criticisms of Popper, Feyerabend, and Kuhn. The model, termed the *general reduction model,* has been criticized (Hull 1974, Wimsatt 1976, Hooker 1981), defended (Schaffner 1976, Ruse 1976), further developed (Wimsatt 1976, Schaffner 1977, Hooker 1981), and recriticized recently by Kitcher (1984) and Rosenberg (1985). A somewhat similar approach to reduction was developed by Paul Churchland in several books and papers (1979, 1981, 1984).

An approach close to that of the original general reduction model has quite recently been applied in the area of neurobiology by Patricia Churchland (1986). She offers the following concise statements of that general model:

> Within the new, reducing theory T_B, construct *an analogue* T_R^* of the laws, etc., of the theory that is to be reduced, T_R. The analogue T_R^* can then be logically deduced from the reducing theory T_B plus sentences specifying special conditions (e.g., frictionless surfaces, perfect elasticity). Generally, the analogue will be constructed with a view to mapping expressions of the old theory onto expressions of the new theory, laws of the old theory onto sentences (but not necessarily *laws*) of the new. Under these conditions the old theory reduces to the new. When reduction is successfully achieved, the new theory will explain the old theory, it will explain why the old theory worked as well as it did, and it will explain much where the old theory was buffaloed. (Ibid., 282–283).

Churchland goes on to apply this notion of reduction to the sciences of psychology (and also what is termed "folk psychology") which are the sciences *to be reduced,* and to the rapidly evolving neurosciences, which are the *reducing sciences.* In so doing she finds she needs to relax the model even further, to accommodate, for example, cases not of reduction but of replacement, and of partial reduction. She never explicitly reformulates the model to take such modifications into account, however, and it would seem useful, given the importance of such modifications, to say more as to how this might be accomplished. The *most* general model must also allow for those cases in which the T_R is NOT modifiable into a T_R^* but rather is REPLACED by T_B or a T_B^*. Though not historically accurate, a reduction of phlogiston theory by a combination of Lavoisier's oxidation theory and Dalton's atomic theory would be such a replacement. Replacement of a demonic theory of disease with a germ theory of disease, but with retention, say of the detailed observations of the natural history of the diseases and perhaps preexisting syndrome clusters as well, is another example of reduction with replacement (see Schaffner 1977).

In the simplest replacement situation we have the essential *experimental arena* of the previous T_R (but not the theoretical premises) directly connected via new correspondence rules associated with T_B (or T_B^*) to the reducing theory. A correspondence rule is here understood as a telescoped causal sequence linking (relatively) theoretical processes to (relatively) observable ones. (A detailed account of this interpretation of correspondence rules can be found in Schaffner 1969.) Several of these rules would probably suffice, then, to allow for further explanation of the experimental results of T_R's subject area by a T_B (or T_B^*). In the more complex but realistic case, we also want to allow for partial reduction, that is, the possibility of a partially adequate component of T_R being maintained together with the entire *domain* (or even only part of the domain) of $T_R{}'$. (The sense given to "domain" here is that of Shapere (1974): a domain is a complex of experimental results which either are accounted for by T_R and/or *should be* accounted for by T_R when (and if) T_R is or becomes completely and adequately developed.) Thus, the possibility arises of a continuum of reduction relations in which T_B (or T_B^*) can participate. (In those cases where only one of T_B or T_B^* is the reducing theory, we use the expression T_B^*.) To allow for such a continuum, T_R must be construed not only as a completely integral theory but also as a theory dissociable into weaker versions of the theory, and also associated with an experimental subject area(s) or domain(s). Interestingly the Suppes-Adams model might lend some additional structure through the use of model-theoretic terminology to this notion of partial reduction, though it will need some (strengthened) modifications to that original schema, to be discussed.

We may conceive of "weaker versions of the theory" either (1) as those classes of models of the theory in which not all the assumptions of the theory are satisfied or (2) as a restricted subclass of all the models of the reduced and/or reducing theory. The first weakening represents a restriction of assumption, the second a restriction in the applied scope of the theory. As an example of the first type of restriction consider a set of models in which the first and second laws of Newton are satisfied but which is silent or which deny the third law. The application to reduction is straightforward, since in point of fact there are models of optical and electromagnetic theories in the nineteenth century which satisfy Newton's first two laws but which violate the third (see Schaffner 1972, 65). As an example of the second type of restriction, consider a restriction of scope of statistical mechanical models that eliminates (for the purpose of achieving the reduction) those peculiar systems (of measure zero) in which entropy does not increase, for example, a collection of particles advancing in a straight line. This second type of restriction is unfortunately ad hoc.

Under some reasonable assumptions, the general reduction model introduced in the quotation from Patricia Churchland above could be modified into the general reduction-replacement model, characterized by the conditions given in the following text box. (Text in boxes as well as all the figures with the exception of Figure 8.1 are optional material.) These conditions are of necessity formulated in somewhat technical language, but the concepts involved should be reasonably clear from the discussion above. (The italicized *ors* should be taken in the weak, inclusive sense of "or," i.e., and/or.)

Such a model has as a limiting case what we have previously characterized as the general reduction model, which in turn yields Nagel's model as a limiting case. The use of the weak sense of *or* in conditions (1), (2) and (4) allows the "continuum" ranging from reduction as subsumption to reduction as explanation of the experimental domain of the replaced theory. Though in this latter case we do not have intertheoretic reduction, we do maintain the "branch" reduction previously mentioned. This flexibility of the general reduction-replacement model is particularly useful in connection with discussions concerning current theories that may explain "mental" phenomena.

T_B—the reducing theory/model
T_B^*—the "corrected" reducing theory/model
T_R—the original reduced theory/model
T_R^*—the "corrected" reduced theory/model

Reduction in the most general sense occurs if and only if:

(1)(a) All primitive terms of T_R^* are associated with one or more of the terms of $T_B^{(*)}$, such that:
    (1) T_R^* (entities) = function ($T_B^{(*)}$ (entities))
    (ii) T_R^* (predicates) = function ($T_B^{(*)}$ (predicates))

or

(1)(b) The domain of T_R^* be connectable with $T_B^{(*)}$ via new correspondence rules. (Condition of generalized connectability.)

(2)(a) Given fulfillment of condition (1)(a), that T_R^* be derivable from $T_B^{(*)}$ supplemented with (1)(a)(i) and (1)(a)(ii) functions.

or

(2)(b) Given fulfillment of condition (1)(b) the domain of T_R be derivable from $T_B^{(*)}$ supplemented with the new correspondence rules. (Condition of generalized derivability.)

(3)   In case (1)(a) and (2)(a) are met, T_R^* corrects T_R, that is, T_R^* makes more accurate predictions. In case (1)(b) and (2)(b) are met, it may be the case that $T_B^{(*)}$ makes more accurate predictions in T_R's domain than did T_R.

(4)(a) T_R is explained by $T_B^{(*)}$ in that T_R and T_R^* are strongly analogous, and $T_B^{(*)}$ indicates why T_R worked as well as it did historically.

or

(4)(b) T_R's domain is explained by $T_B^{(*)}$ even when T_R is replaced.

Though these are useful expansions of the traditional model, we have yet to see a reexamination of one of the basic premises that reduction is best conceived of as a relation between *theories* (though we have allowed a reduction to occur between a theory and the *domain* of a previous theory). Furthermore, nowhere has it yet been seriously questioned whether the notion of *laws* or a collection of laws is the appropriate reductandum and reductans.

Wimsatt (1976) criticized Schaffner's earlier approach to theory reduction (Schaffner 1967, 1969, 1976) for among other things as not taking "mechanisms" as the focus of the reducing science. He seems to have been motivated in part by some of Hull's (1974) observations, and especially by Salmon et al.'s (1971) model of explanation. Wimsatt construes one of Salmon et al.'s important advances as shifting our attention away from statistical *laws* to statistically relevant *factors* and to underlying *mechanisms* (Wimsatt 1976, 488). Wimsatt's suggestions are very much to the point. The implications of them are considered further below in terms of the causal gloss placed on reductions as explanations.

The question of the relevance of "laws" in reductions in the biomedical sciences in general, and in connection with genetics in particular, was also raised more recently by Philip Kitcher. In his paper on reduction in molecular biology, Kitcher (1984) argued that the Nagel model of reduction and all analogous accounts suffered from several problems. One of those problems was that when one applies such accounts to biology in general, and to genetics in particular, it is hard to find collections of sentences which are the "laws" of the theory to be reduced.

Kitcher notes that finding many laws about genes is difficult, as opposed to the way that one finds various gas laws peppering the gas literature, or the extent to which there are explicit laws in optics, such as Snell's law, Brewster's law, and the like. Why this is the case is complex (see Schaffner 1980 and 1986), but it has interesting implications for reduction and thus requires some discussion. As Kitcher adds, however, two principles or laws have received that term and are extractable from Mendel's work and its rediscoverers. These are the famous laws of segregation and independent assortment. They continue to be cited even in very recent genetics texts such as Watson et al. (1987), and a statement of them can be found in Chapter 7. These laws were the subject of a searching historical inquiry by Olby (1966).

Olby believes that Mendel's laws are important and, in their appropriate context, still accurate. In answer to the question whether Mendel's law of independent assortment still holds, Olby writes:

> yes; but, like any other scientific law, it holds only under prescribed conditions. Mendel stated most of these conditions, but the need for no linkage, crossing-over, and polyploidy were stated after 1900. (1966, 140)

In his essay, Kitcher gleans a different lesson from the post-1900 discoveries of the limitations of Mendel's law. It is not that one could not save Mendel's law by specifying appropriate restrictions or articulating a suitable approximation and reducing it, but rather that "Mendel's second law, amended or unamended, simply becomes irrelevant to subsequent research in classical genetics" (Kitcher 1984, 342). What seems to concern Kitcher is that Mendel's second law is only interesting and relevant if it is embedded in a cytological perspective which assigns Mendelian genes to chromosomal segments within cells:

> What figures largely in genetics after Morgan is the technique [of using cytology], and this is hardly surprising when we realize that one of the major research problems of classical genetics has been the problem of discovering the distribution of genes *on the same chromosome,* a problem which is beyond the scope of the amended law. (1984, 343)

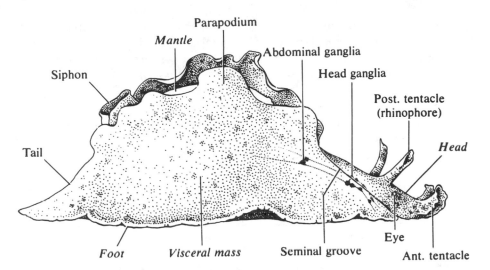

Figure 8.2 Gross structure of *Aplysia californica*, also known as the "sea hare." The gill structure is not visible in this view, but is located adjacent to the mantle behind the Parapodium and in front of the siphon. From *Cellular Basis of Behavior*. By E.R. Kandel copyright © 1976 by W.H. Freeman and Company, Reprinted with permission.

We think Kitcher is on the right track here but that he has not fully identified the units that are at work in actual scientific reductions. His later elaboration of what he terms "practices"—which is a kind of analogue of a Kuhnian paradigm, a Lakatosian research program, or a Laudanian research tradition (see Chapter 4)—though interesting in its own right does not really address the issue of *reduction,* but is more suited to clarify problems in the domain of scientific progress. What we take from Wimsatt's arguments and Kitcher's discussion is that a focus on the notion of a theory as a collection of general laws may be problematic in the biological sciences. Any biomedical generalization needs to be embedded in a broader context of similar and overlapping theories in order to adequately capture the nature of reduction in these sciences. We will see how this works in neurobiology in Section 8.8.

What is the import of this view for reduction? To assess that, it will be appropriate to turn to an extended example of reduction in neurobiology, after which we can in the light of the philosophical discussion just presented, *generalize* the example and examine what reduction in neurobiology might look like from a philosophical point of view.

8.7 SHORT-TERM AND LONG-TERM LEARNING IN *APLYSIA*

In a series of critically important papers in neurobiology, Kandel, Schwartz, and their colleagues have been deciphering the complex neurobiological events underlying the primitive forms of learning which the marine mollusc *Aplysia* exhibits. This invertebrate organism, sometimes called the "sea hare" (see Figure 8.2 for a diagram of its gross structure and note its similarity to a rabbit) exhibits simple reflexes which can be altered by environmental stimuli. The nervous system of *Aplysia* is known to consist of discrete aggregates of neurons called ganglia containing several thousand nerve cells. These nerve cells can be visualized under a microscope, and can have microelectrodes inserted into them for monitoring purposes. Individual sensory as well as motor nerve cells have been identified, and are essentially identical within the species. This has permitted the tracing of the synaptic connections among the nerve cells as well as the identification of the organs which they innervate. "Wiring diagrams" of the nervous system have been constructed from this information.

It has been a surprise to many neuroscientists that such simple organisms as shellfish, crayfish, and fruit flies can be shown to exhibit habituation, sensitization, classical conditioning, and operant conditioning. Both short-term and long-term memory by these organisms can be demonstrated and serve as the basis for model approaches to understanding the molecular basis of such learning and memory. Though it is not expected that a *single universal* mechanism for learning will be found that holds for all organisms including humans, it is felt that basic mechanisms will have certain fundamental relationships to one another, such as reasonably strong analogies.

1. Sensitization and Short-Term Memory. Sensitization is a simple form of nonassociative learning in which an organism such as *Aplysia* learns to respond to a usually noxious stimulus (e.g., a squirt of water from a water pik) with a strength-

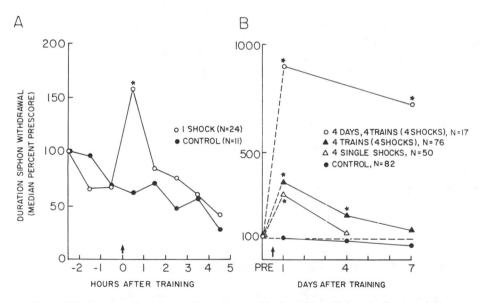

Figure 8.3 A and B. Short-term and long-term sensitization of the withdrawal reflex in *Aplysia*, after Kandel and others (1987), p. 117, with permission. The arrow indicates a single strong electric shock to the tail or the neck of *Aplysia*, resulting in the heightened response as shown in the open circles in the graphs. Filled-in circles are controls.

Philosophy of Medicine

ening of its defensive reflexes against a previously neutral stimulus. A well-known defensive reflex in *Aplysia* known as a *gill-siphon withdrawal reflex* has been studied intensively by Kandel and Schwartz and their colleagues. When the siphon or mantle shelf of *Aplysia* is stimulated by light touch, the siphon, mantle shelf, and gill all contract vigorously and withdraw into the mantle cavity (see Figure 8.2 for these structures). Short-term and long-term sensitization of the withdrawal reflex is shown in Figures 8.3A and 8.3B (from Kandel et al. 1987).

The nerve pathways that mediate this reflex are known, and a simplified wiring diagram of the gill component of the withdrawal reflex is shown in Figure 8.4. The sensitizing stimulus applied to the sensory receptor organ (in this case the head or the tail) activates *facilitatory* interneurons that in turn act on the follower cells of the sensory neurons to increase neurotransmitter release. Three types of cells apparently use different neurotransmitters; serotonin (= 5-HT) and two small interrelated peptides known as SCP_A and SCP_B are identified transmitters in two of the cell types. This explains sensitization at the cellular level to some extent, but Kandel and his

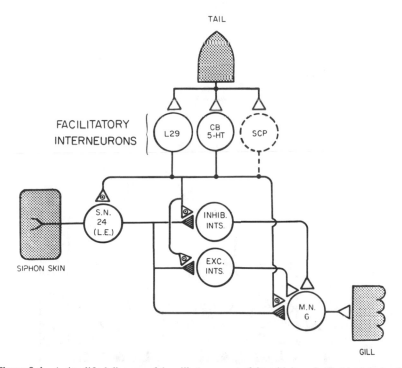

Figure 8.4 A simplified diagram of the gill component of the withdrawal reflex in *Aplysia* after Kandel and others (1987), p. 118, with permission. The 24 mechanoreceptor neurons transmit the information from the siphon skin to the inhibitory and excitatory interneurons, as well as to the gill motorneurons (M.N.). Stimulation of the tail or the neck excites some of the facilitory interneurons increasing the strength of the signal between the sensory neurons and the cells they stimulate. The facilitory interneurons use three different types of neurotransmitters, two of which have been identified as shown.

colleagues have pushed this investigation to a deeper level, that of the molecular mechanisms involved.

For short-term sensitization, all the neurotransmitters have a common mode of action: each activates an enzyme known as adenylate cyclase which increases the amount of cyclic AMP—a substance known as the second messenger—in the sensory neuronal cells. This cyclic AMP then turns on (or turns up) the activity of another enzyme, a protein kinase, which acts to modify a *"family* of substrate proteins to initiate a broad cellular program for short-term synaptic plasticity" (Kandel et al. 1987, 118). The program involves the kinase that phosphorylates (or adds a phosphate group onto) a K^+-channel protein, closing one class of K^+ channels that normally would restore or repolarize the neuron's action potential to the original level. This channel closing increases the excitability of the neuron and also prolongs its action potential, resulting in more Ca^{++} flowing into the terminals, and permitting more neurotransmitter to be released. (There may also be another component to this mechanism involving a movement of a C-kinase to the neuron's membrane where it may enhance mobilization and sustain release of neurotransmitter.) The diagram in Figure 8.5 should make this sequence of causally related events clearer.

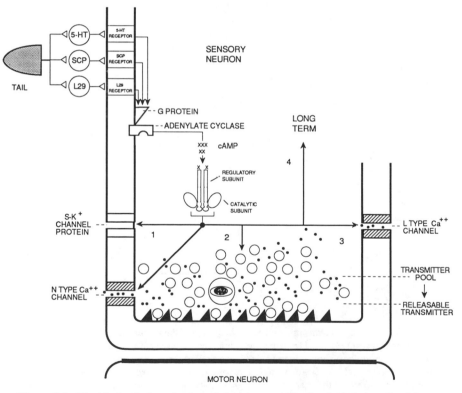

Figure 8.5 The biochemical mechanism of short-term sensitization in *Aplysia* updated from Kandel and others (1987), p. 119, with permission.

2. Long-Term Memory in **Aplysia.** A similar type of explanation can be provided for long-term memory for sensitization, but we will not go into it in any detail here, save to point out that the mechanism can best be described as exhibiting important *analogies* with short-term memory. Suffice it to say that the long-term mechanism is somewhat more complex, though it shares similarities with the short-term mechanisms such as the adenylate cyclase—cyclic AMP—protein kinase cascade. Since protein synthesis occurs in connection with long-term memories, additional different mechanisms need to be invoked, including, almost certainly, new genes being activated. Figure 8.6 provides us with one of at least two possible overviews of this more complex—and at present more speculative—mechanism.

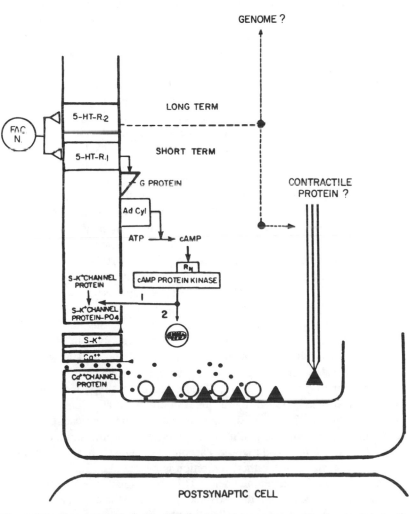

Figure 8.6 Speculative biochemical mechanism of long-term sensitization in *Aplysia* after Kandel and others (1987), p. 126, with permission.

Additional Complexity and Parallel Processing in Aplysia. The above description of Kandel and his associates' work focuses on an account which has "had the advantage of allowing a relatively detailed analysis of the cellular and molecular mechanisms underlying one component important for both short- and long-term memory for sensitization" (Frost et al. 1988, 298). As these authors note in their more recent analysis, however, "the siphon withdrawal response exhibits a complex choreography with many different components . . . (ibid.)." Further work on both the gill-component and the siphon component of this reflex has suggested that memory for sensitization in *Aplysia*—as well as other organisms—appears to involve "parallel processing," an expression which is "similar" to a "view, called 'parallel distributed processing' [which] has emerged from theoretical studies in artificial intelligence and cognitive psychology and is based on the idea that common computational processes recur at many sites within a network . . ." (Frost et al. 1988, 297–298).

In this chapter we will not have an opportunity to explore the complex interactions which Kandel and his colleagues have found even in this comparatively simple form of short-term memory. Suffice it to say that they have found it involves at least four circuit sites, each involving a different type of neuronal plasticity, which are shown in Figure 8.7: (1) presynaptic facilitation of the central sensory neuron connections as described earlier in this section; (2) presynaptic inhibition made by L30 onto circuit interneurons including L29 shown; (3) posttetanic potentiation (PTP) of the synapses made by L29, onto the siphon motor neurons, and (4) increases in the tonic firing rate of the LFS motor neurons, leading to neuromuscular facilitation (Frost et al. 1988, 299).

Frost et al. (1988) anticipated such complexity but were also surprised to find that though the PTP discovered at L29 was a homosynaptic process, the other three components appeared to be coordinately regulated by a common modulatory transmitter, serotonin, and that the common second-messenger system, cyclic AMP, was involved in each (pp. 323–324). Additional modulatory transmitters and other second-messenger systems are explicitly not ruled out, however (see 1988, 324).

In some other recent work, Kandel has in fact reported that tail stimuli in *Aplysia* can lead to transient inhibition (in addition to prominent facilitation). Studies of this inhibitory component indicate that the mechanism is presynaptic inhibition in which the neurotransmitter is the peptide FMRFamide, and that the "second messenger" in this component is not cyclic AMP; rather, inhibition is mediated by the lipoxygenase pathway of arachidonic acid. This "unexpected richness" as Kandel has characterized the existence of two balancing pathways, constitutes still further evidence for the philosophical views to be discussed below.

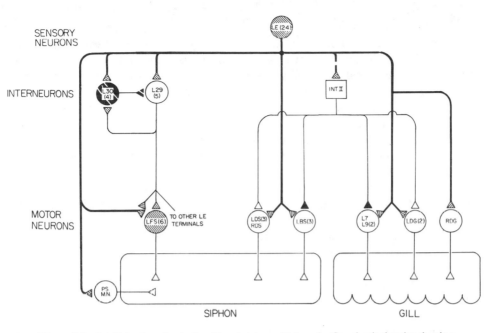

SENSORY
NEURONS

INTERNEURONS

MOTOR
NEURONS

SIPHON   GILL

Figure 8.7 Parallel processing in the gill and siphon withdrawal reflex circuit showing the sites found to be modified by sensitizing stimuli (shaded structures). From Frost, W. N., Clark, G. A. and Kandel, E. R. (1988), ''Parallel Processing of Short-term Memory for Sensitization in Aplysia'' in *Journal of Neurobiology* 19:291-334. Copyright © 1988 John Wiley & Sons, Inc. Reprinted by permission of John Wiley & Sons, Inc. See textbox 8.2 for details.

8.8 IMPLICATIONS OF THIS EXAMPLE FOR REDUCTION IN NEUROBIOLOGY

Several points are important to note in the account given in Section 8.7 which gives a molecular explanation of a behavioral, in a sense ''psychological,'' explanandum. First, the GRR model will be discussed in the light of the specific *Aplysia* example, then in the following section more general issues regarding the GRR model will be considered, ultimately leading to *two* forms of this account, one (comparatively) *simple,* and one both *partial* and more *complex.*

When we examine the *Aplysia* sensitization exemplar, we do not see anything fully resembling the ''laws'' we find in physics explanations. The generalizations that exist are of varying generality—for example, protein kinase enzymes act by phosphorylating molecules, and cyclic AMP is a (second) messenger—which we could extract from this account, though they are typically left implicit. Such generalizations are, however, introduced explicitly in introductory chapters in neuroscience texts. In addition, however, there are usually also generalizations of narrow scope, such as the ''two balancing pathways'' found in Kandel's very recent work. Further, these generalizations are typically borrowed from a very wide-ranging set of models (such as protein synthesis models, biochemical models, and neurotransmitter models). The set in biology is so broad it led Morowitz's (1985) committee to invoke the notion of a many-many modeling in the form of a complex biomedical ''matrix.'' But even when such generalizations, narrow or broad, are made explicit, they need to be supplemented with the details of specific connections in the system in order to constitute an explanation, such as the linkage of three-fold L-29—SCP—5-HT receptor to the G protein shown in Figure 8.5. Are these connections playing the role of "initial conditions" in traditional philosophy of science? This might not be the case since we do not have a set of powerful generalizations such as Newton's laws of motion to which we can add minor details concerning the force function and the initial position and momentum, and generate explanations across a wide range of domains. The generalizations we can glean from such molecular mechanisms have a variable scope, and typically need to be changed to *analogous* mechanisms as one changes organism or behavior. The brief account given of the relation between short-term and long-term mechanisms for sensitization in *Aplysia* is a case in point, as is the further complexity found in short-term sensitization in the more recent "parallel processing" account previously given.

What we appear to have are intricate *systems* using both broad and narrow **causal generalizations** which are typically *not* framed in purely biochemical terminology, but which are characteristically *interlevel*. Such systems often *explain* by providing a temporal (and often causal) sequence as part of their models (see Schaffner 1980, 1986, and forthcoming).

This view of the explaining system as a combination of broad and narrow generalizations that may be framed at a variety of different levels of aggregation suggests that *theory reduction* needs to consider not only general summary statements of entities' behavior which might be said to function at what we call the "γ (for very general) level of theory axiomatization" (e.g., the general definition of learning by sensitization), but also needs to note the significance of the finer-structured elements of a theory—for example, the cytological or what we term the "σ level of specification." (This σ, level of specification represents a means of *realizing* a higher level γ principle.) One can also in addition introduce a more fine-structured level than σ, that we call the "δ-level," which can represent molecular level details realizing in turn σ-level mechanisms. We reexamine the implications of these levels of specification for reduction further below in Section 8.9. The introduction of these levels of specification or realizations (γ, σ, and δ) very roughly track the general process, cell implementation, and molecular implementation levels, but are more intended to capture the level of *generality* than the level of aggregation (though empirically they *may* track both simultaneously.) For the utility of such levels as parts of "extended theories" in accounting for scientific change see Schaffner (forthcoming, Chapter 5).

What we appear to have in the neuroscience example is an instance of a multilevel system employing causal generalizations of both broad and narrow scope. Moreover, as an explanation of a behavioral phenomenon like sensitization is given in molecular terms, one maps the phenomenon into a neural vocabulary. Sensitization becomes not just the phenomenon shown earlier in Figure 8.3, but it is *reinterpreted as* an instance of neuronal excitability and increased transmitter release, that is as *enhanced synaptic transmission*. Thus something like Nagel's condition of connectability is found. Is this explanation also in accord with explanation by derivation from a theory? Again, the answer may be yes, but this time with some important differences. As argued above, we do not have a very general set of sentences (the laws) which can serve as the premises from which we can deduce the conclusion. Rather, what we have is a set of causal sentences of varying degrees of generality, many of them specific to the system in question. In some distant future all of these causal sentences of narrow scope, such as "This phosphorylation closes one class of K^+ channels that normally repolarize the action potential" (Kandel et al. 1987, 120–121), may be explainable by general laws of protein chemistry, but it is not the case at present. In part this is because we cannot even fully infer the three-dimensional structure of a protein like the kinase enzyme or the K^+ channels mentioned from a complete knowledge of the amino acids which make up the proteins. Fundamental and very general principles will have to await a more developed science than we will have for some time. Thus the explanans, the explaining generalizations in such an account, will be a complex web of interlevel causal generalizations of varying scope, and will typically be expressed in terms of an idealized system of the type shown in Figure 8.5, with some textual elaboration on the nature of the casual sequence leading through the system.

This then is a kind of **partial model reduction with largely implicit generalizations,** often of narrow scope, licensing the temporal sequence of causal propagation of events through the model. It is not unilevel reduction, that is, to biochemistry, but it *is* characteristically what is termed a *molecular biological explanation*. The model effecting the "reduction" is typically interlevel, mixing different levels of aggregation from cell to organ back to molecule, and the reducing model may be further integrated into another model, as the biochemical model is integrated into or seen as a more detailed expansion of the neural circuit model for the gill-siphon reflex. The model or models also may not be robust across this organism or other organisms; it may well have a narrow domain of application in contrast to what we typically encounter in physical theories.

In some recent discussion on the theses developed in the present essay, Wimsatt (personal communication) has raised the question of why not refer to the reduction as being accomplished by "mechanisms" in the sense of Wimsatt (1976), and simply forget about the more syntactic attempt to examine the issues in terms of "generalizations"? Though Wimsatt's suggestion (and his position) are fruitful, "mechanism" should not be taken as an unanalyzed term. Furthermore, it seems that we do (and, as will be argued, *must*) have "generalizations" of varying scope at work in these "molecular biological explanations" which are interlevel and preliminary surrogates for a unilevel reduction, and that it is important to understand the varying scope of the generalizations and how they can be applied "analogically" within and across various biological organisms. This point of view relates closely to the question of "theory structure" in the biomedical sciences, an issue mentioned earlier (also see Morowitz's National Academy of Science Report, Committee on Models for Biomedical Research 1985 and Schaffner 1987). This said, however, there is no reason that the *logic* of the relation between the explanans and the explanandum cannot be *cast* in deductive form. Typically this is not done because it requires more formalization than it is worth, but some fairly complex engineering circuits, such as a full adder shown in Figure 8.8 and described in the following text box, can effectively be represented in the first-order predicate calculus and useful deductions made from the premises, which in the adder example number some 30.

The picture of reduction that emerges from any detailed study of molecular biology as it is practiced is not an elegant one. A good overall impression of reduction

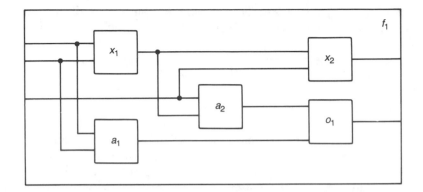

Figure 8.8 A circuit diagram of the full adder described in text box, after © 1987 Morgan Kaufmann publishers. Reprinted with permission from Geneserth and Nilsson, "Logical Foundations of Artificial Intelligence."

in practice can be obtained if one reads through such a text as Watson et al.'s version of his classic *Molecular Biology of the Gene* (1987) or Lewin's recent *Genes-IV* (1990). An alternative similar overall impression can be found in those intermediary metabolism charts that hang on laboratory walls, displaying wheels and cycles of complex reactions feeding into more cycles and pathways of complex reactions (Bechtel 1986). This material found in chapters of Watson et al. (1987), in Lewin (1990), and in the intermediary metabolism charts plays the role of the explaining theories and initial conditions of molecular biology and biochemistry. Unfortunately, as already remarked, it is not possible to usefully separate out a small core of general laws and add a list of initial conditions to the core to generate an explanation.

The extensive complexity found in molecular biology has tended to generate pessimism regarding the reduction of such theories as Mendelian genetics by molecular genetics (Hull 1974, Kitcher 1984, Rosenberg, 1985), though both Kitcher (1989) and Rosenberg (1985) acknowledge that different forms of knowledge representation may alleviate this pessimism, Rosenberg writing that "[T]he 'cannot' [involved in obtaining a derivation about a regularity in transmission genetics] is not a logical one; it has to do with the limitations on our powers to express and manipulate symbols . . ." (1985, 106). Kitcher's solution is to develop his notion of "practice" in some detail so that we might see how scientific research programs are accomplishing molecular explanations ("reductions") (Kitcher 1989). Rosenberg's approach is perhaps more sensitive to the complexity of molecular explanation than is Kitcher's, but his own solution, appealing to "supervenience," (1985, 111–117) appears to sidestep the real issue of knowledge representation and derivation (see Hull 1981, 135).

8.9 THE RELATION BETWEEN THE CAUSAL-MECHANICAL ACCOUNT OF REDUCTION AND THE GRR MODEL; TWO VERSIONS OF THE GRR MODEL—SIMPLE AND COMPLEX

In the previous section, the discussion is strongly focused on various causal features involved in (what is at least a partial) reduction. At several points, however, we addressed the relations of that account to Nagel's conditions of connectability and

The full adder shown in Figure 8.8 is an integer processing circuit consisting of five subcomponents called "gates." There are two exclusive "or" (*xor*) gates x_1 and x_2, two "and" gates a_1 and a_2, and an inclusive "or" gate o_1. There are input ports on the left side of the box and output ports on the right as well as ports into and out of the subcomponents. The universe of discourse thus consists of 26 objects: 6 components or subcomponents and 20 ports.

The structure and operation of this circuit can be captured in first order predicate logic (FOL), and a theorem prover used with the logical formulas to generate answers to questions about the device's operations. A full treatment is beyond the scope of this chapter (but can be found in Genesereth and Nilsson's (1987, 29–32; 78–84). Here we will give only a few of the axioms used to characterize this circuit in FOL. After the vocabulary is introduced, some 30 logic sentences are needed for the full description.

Vocabulary examples: Xorg(x) means that **x** is an *xor* gate; **I(i,x)** designates the ith (**1, 2,** or **3**) input port of device **x**; **Conn (x,y)** means that port **x** is connected to port **y**; **V (x,z)** means that the value of port **x** is **z**; **1** and **0** designate the high and low signals respectively.

Connectivity and behavior of components examples:

. . .

(7) Conn (I(1,F1), I(1,X1))
(8) Conn (I(2,F1), I(2,X1))

. . .

(26) $\forall x \, \forall n \, (\text{Org}(x) \wedge V(I(n,x),1) \Rightarrow V(O(1,x),1))$

. . .

(30) $\forall x \forall y \forall z \, (\text{Conn } (x,y) \wedge V(x,z) \Rightarrow V(y,z))$
(As noted, these assumptions represent only 4 of the 30 needed.)

---

derivability which continue to figure prominently in the GRR model introduced earlier in this chapter. In the present section, we consider those relations more systematically.

It appears that the GRR account, which might be termed following Salmon's (1989) account as a more "epistemic" approach to reduction, may *best* capture the relations that exist in a "clarified" science where a *uni*level reductans has been fully developed. We might term this a *simple interpretation* of the GRR model; it allows us to construe it as constituting a *set of conditions* which must be met if a *complete* reduction is to have been judged as effected. Under this interpretation the GRR model falls into what might be termed the justificatory or evaluative sphere. The requirements of connectability and derivability are seen in this light to be both reasonable and benign: connectability *must* be established if two languages are to be related, and derivability is just the implementation of a well understood truth-preserving mode of inference—one that carries impeccable credentials.

This *simple interpretation* of the GRR model does not, however, without elaboration and interpretation of its tenets along the lines proposed in Section 8.6, adequately capture the heavy reliance of scientific explanation on causal mechanisms of the type that Wimsatt (1976) and Hull (through personal communication) have stressed. In terms of active scientific investigation and of the language in scientific research reports (and textbooks), a causal-mechanical analysis (to use Salmon's [1989] term) *seems* much more suitable.

Along similar lines to those made by Salmon (1989) on explanation about the possibility of a "rapprochement" between causal and unificatory analyses, it is thus *tempting* to think of the causal reduction analysis presented above (in terms of interlevel causal generalizations that become progressively better characterized at the detailed δ levels of aggregation and specificity) as perhaps *another sense* of reduction, by analogy with Carnap's probability$_1$ and probability$_2$ (and especially Salmon's explanation$_1$ and explanation$_2$). On such a view, there would be a reduction$_{grr}$ and a reduction$_{cm}$ standing for the epistemic "*general reduction-replacement*" model at one extreme and for the ontic "*causal mechanism*" at the other, perhaps with some overlap (mixed) analyses between these two extreme types.

Though tempting, this view might be mistaken without keeping in mind four additional clarificatory comments. First, any appeal to "mechanisms" also requires appeals to the "*laws* of working" (the term is Mackie's 1974) that are implemented in the mechanisms and which provide part of the analysis and grounding of the "causal" aspect of the causal-mechanical approach (see Schaffner forthcoming, Chapter 6 for details). Second, if we appeal to *any* general notion of a theory (i.e., to a semantic or a syntactic analysis) it would seem that we would have to deal with "generalizations" and with connecting the generalizations' entities and predicates, as in the GRR model. For example, if we consider the semantic conception of a scientific theory as it might be involved in reductions, such "theories" are set-theoretic (or other language) predicates which are essentially constituted by component generalizations. Thus if we were to think about attempting to characterize a "causal-mechanical" approach to reduction with the aid of semantic construal of theories, we would still require recourse to traditional generalizations. Third, appeals to either a semantic or a syntactic notion of theories still requires that entity and predicate terms be used (in addition to the generalizations) to express the theories which we consider. Thus some form of connectability assumptions relating terms such as "sensitization" and "enhanced synaptic transmission" will be required to effect a reduction. Fourth and finally, in those cases in which we wish to verify the *reasoning* supporting some *causal consequence* of a reducing theory (whether it be a general result or a particular one), we can do no better than utilize the principles of *deductive* reasoning, other forms of logic and reasoning being significantly more suspect. These four points support the characterization of reduction presented in the GRR account previously provided.

If we keep these caveats in mind, however, it does seem that we *can* effect a "rapprochement" of sorts between the two CM and GRR senses of reduction by construing reductions as having two aspects: (1) ongoing advances that occur in *piecemeal* ways, for example, as some features of a model are progressively elaborated at the molecular level or perhaps a new "mechanism" is added to the model, and (2) *assessments* of the explanatory fit between two theories viewed as a *collection* of models or even as two reasonably complete *branches* of science. Failure to keep these two aspects distinct has led to confusion about the adequacy of the GRR model (by, for example, Wimsatt 1976). Aspect (1) is akin to what Mayr (1982) has termed explanatory reduction, a notion that has been further elaborated by Sarkar (1989, forthcoming). Aspect (2) is the more traditional notion of intertheoretic reduction.

As scientists further develop reducing theories in a domain, they frequently elaborate various aspects of the theories both by (a) modifying a δ assumption or proposing a δ-level assumption that accounts for (part of) a σ-level process, and (b) describing these new assumptions in *causal* terms. Thus the result of such advances are typically complex interlevel connections with causal processes appealed to as part of the connections. (Part-whole or constituent relationships may also be appealed to in such situations.) This task is typically done by scientists, and reflects what Salmon (1989) calls the *ontic* dimension. This ongoing process will establish connections at several different levels, and, prior to arriving at something like a unilevel theory in a clarified science, many weblike and bushy connections can be expected. These reticulate relations however do not count *against* a reduction; they can be thought of as part of the process leading *towards* a "simplified" unilevel reduction. Something very much like this progressive process is described in Culp and Kitcher's (1989) recent essay on the discovery of enzymatic RNA, though they prefer to characterize the process more as an "embedding of a problem-solving schemata of one field of science in those of another" (1989, 479) than as a complex partial reduction.

These various theoretical developments and attendant connections—bushy and weblike though they may be—frequently have the ability to "explain" at least partially some features of a higher level domain, for example, a phenomenon such as sensitization in *Aplysia*. The more *traditional* question of reduction, however, is a question of the adequacy of the explanatory relationship in a more *global* or gross sense which appeals (usually) to preexisting disciplines or fields (or theories). Working out this type of relation is typically done by the philosopher or by a scientist interested in reviewing very broad questions of domain relations. Such a relation(s) is much more *systematic* than the ontic dimension(s) referred to in the previous paragraph.

In connection with this global question, the various features of the simple GRR model can be appealed to as providing the *logical* conditions that must be satisfied, but usually satisfied in a post hoc evaluative manner. Thus one looks for (1) synthetic identities, which introduce basic ontological economies, and (2) derivability of the corrected reduced science (with the help of causal generalizations in the reducing science), which insures that security of relation that only truth-preserving nonampliative inference can provide. A representation of intertheoretic relations using the GRR model thus is a kind of "executive summary" of much scientific work that has gone on to establish interdisciplinary connections; such analysis also typically falls into the justificatory arena of philosophy of science. The GRR model in what may be considered its *simple* interpretation is thus useful in providing a kind of systematic summary and regulatory ideal, but it should not in general be confused with the *process* of establishing reductions and in the ongoing elaboration of the complex web of connections that typically unite reduced and reducing theories.

Though the GRR model, particularly in its limiting case as the general reduction model, can be conceived of as a summary and a regulatory ideal for unilevel reductions, it can also accommodate two *inter*field or *inter*level theories. This would *not* be the *simple interpretation* of the GRR model discussed above, but there is no reason in principle why a theory which was *interlevel but primarily cellular* could not be explained by a theory which was *interlevel but primarily molecular*. To an extent, this feature of the GRR model will also rely on its ability to accommodate *partial* reductions. Insofar as the explanation was adequate, this type of theory would be akin to a type of "homogeneous" but now interlevel reduction discussed by Nagel (1961). Perhaps it might better be termed a "complex homogeneous" reduction. The GRR model could accommodate this form of reduction, but its reduction functions will reflect the *mixed inter*level character of the theories. Such a *complex interpretation* of the GRR model may be more realistic than the *simple interpretation* of the GRR model discussed earlier where unilevel theories functioned as reductans and reductandum.

Certain general methodological principles are likely to govern the implementation of this *complex* GRR model. For example, it seems likely that

T_R^*'s γ-central hypotheses will be preserved unless the relation falls on the replacement side of the reduction-replacement continuum.

T_R^*'s σ hypotheses will either be replaced or amplified.

T_R^*'s δ hypotheses will either be replaced or amplified.

Kandel's group's recent amplifications of *Aplysia*'s cellular and molecular mechanisms accounting for sensitization appear to fit these suggestions. (In addition, in Schaffner forthcoming, Chapter 5, an extensive example that shows how these suggestions work in practice is provided in terms of the development of the clonal selection theory from 1957–1967. Current theories of antibody diversity generation and of T-lymphocyte stimulation (see Watson et al. 1987, Chapter 23) suggest that these types of changes are taking place in those fields studying the molecular biology of the immune response.)

The GRR model is also an appropriate framework in terms of which deeper logical questions, such as predicate analysis, the nature and means of establishing synthetic identities, and the metaphysical implications of such identities can be pursued (see Schaffner 1976, and forthcoming). As such, the GRR framework forces an actual ongoing reduction into a "rational reconstructive" mode. It should be added as a final point, however, that occasionally a focus on the satisfaction of even the simple unilevel interpretation of the GRR conditions can illuminate a scientific debate.

Accordingly the GRR model, in spite of its syntactic (epistemic?) flavor, and though it may prima facie appear to be intuitively less in accord with the pervasive causal flavor of middle range examples of theories (whether used in explanatory or reduction contexts) presented in this chapter is a strong and defensible account of reduction. Though the causal-mechanical alternative is a valuable approach representing an important aspect of reductions, it is not the whole story by any means.

It seems that deeper analysis of what a causal-mechanical alternative involves *leads us back* to just those assumptions which constitute the GRR model.

As noted, reduction in the biomedical sciences is complex, and causal and (re)interpretative connections are found at a number of loci. Before closing this section, it would be well to introduce one additional point of clarification to guard against the misinterpretation of interpretative connections as causal connections.

An examination of the system studied by Kandel and others (1987b) indicates that the **behavioral** level of analysis (represented by Figures 8.3A and 8.3B) deals with such entities as shocks administered to organisms, gross reflex descriptions, and a time course of hours or days. Comparison with the **neural network** level indicates that these entities are now viewed at a higher level of detail; this level of detail is further increased at the **biochemical** level. As one descends levels from the behavioral to the biochemical, one does *not* traverse a causal sequence in reverse, with behavior viewed as effects and biochemical mechanisms as causes, rather one traces a **perspectival** sequence. **Causal** sequences can be found at any level of aggregation *and between levels of aggregation,* but a description of a set of parts of a whole does not automatically entail a causal account relating the parts as causes to the whole as an effect.

What frequently is the case is that the causal sequence between cause and effect is best understood at the biochemical level. Thus stimuli can be **interpreted** as propagating action potentials, and neurotransmitters can be invoked as chemical messengers. Biochemical cascades can be examined as fine-structure connecting networks, leading to an event to be explained which can be **reinterpreted** so as to be identified with an original explanandum. What should not be done, however, is to conflate perspectival reinterpretation with causation for it can only lead to confusion in an already extraordinarily complex set of relations.

We thus finally have a response to Kandel's optimistic assessment of the impact of reductionism on neurobiology: The pursuit of reductionism in molecular biology, and that includes the molecular neurosciences, will result in systematicity and unity *only in comparison with* the even more bizarre complexity that biologists would have to deal with without some occasional broad generalizations obtainable from the molecular approach. Reduction is also bound to be patchy. Portions of learning theory as understood to be applied to specific types of model organisms will be reduced, but it is unlikely that any *general theory* will emerge which is universal and simple that does anything useful for neuroscientists—or for behavioral scientists. Thus *partial reductions* as well as some reductions which may well have the effect of *replacing* rather than explaining previous material is likely, given the complexity of the domains of interest and the lack of sound foundations. Thus we will most likely see progress in the neurosciences where features of the reduction-replacement model described are implemented implicitly, but with a rather baroque set of models playing the role of the reducing theories.

8.10 RELATIONS BETWEEN REDUCTION AND EXPLANATION

In the (comparatively) simple example from Kandel and his associates' analysis of learning in *Aplysia,* we appear to encounter a paradigm case of *both* a reduction *and* a scientific explanation. This suggests the question of how the account of explanation given in association with the Kandel example relates to extant models of scientific explanation discussed in Chapter 1 of this book.

The most useful approach may be to refer back to the contrast between two very general analyses of scientific explanation noted by Wesley Salmon in Chapter 1 as well as in his *Scientific Explanation and the Causal Structure of World* (1984). Salmon contrasts the "epistemic" approach to explanation with the "ontic" approach. The former, in its "inferential" interpretation, is best represented by the Hempel-Oppenheim model of explanation. The "ontic" approach, on the other hand, is one which Salmon has articulated and defended under the rubric of the "causal-mechanical" tradition. Salmon believes that the epistemic approach also characterizes the "theoretical unification" analyses of explanation developed by Friedman (1974) and Kitcher (1981, 1989).

Salmon was unwilling, however, to completely yield "unification" to the epistemic approach and suggested that it could be accommodated in a way within the causal-mechanical tradition:

> The ontic conception looks upon the world, to a large extent at least, as a black box whose workings we want to understand. Explanation involves laying bare the underlying mechanisms that connect the observable inputs to the observable outputs. We explain events by showing how they fit into the causal nexus. Since there seem to be a small number of fundamental causal mechanisms, and some extremely comprehensive laws that govern them, the ontic conception has as much right as the epistemic conception to take the unification of natural phenomena as a basic aspect of our comprehension of the world. *The unity lies in the pervasiveness of the underlying mechanisms* upon which we depend for explanation. (Salmon 1984, 276)

In Chapter 1 of the present book, Salmon introduces two briefly characterized biomedical examples to illustrate his approach. He mentions that "to understand AIDS, we must deal with viruses and cells. To understand the transmission of traits from parents to offspring, we become involved with the structure of the DNA molecule. . . . When we try to construct causal explanations we are attempting to discover the mechanisms—often hidden mechanisms—that bring about the facts we seek to understand" (p. 34).

In Chapter 1, Salmon adds to the contrast between the two major traditions that in "an important respect" these two traditions "overlap":

> When the search for hidden mechanisms is successful, the result is often to reveal a small number of basic mechanisms that underlie wide ranges of phenomena. The explanation of diverse phenomena in terms of the same mechanism constitutes theoretical unification. For instance . . . [t]he discovery of the double-helical structure of DNA . . . produced a major unification of biology and chemistry. (p. 34)

The biomedical sciences in general, and the neurosciences, AIDS virology, and molecular genetics in particular, tend, with a few important exceptions, to propose what Salmon terms causal-mechanical explanations. The Kandel example detailed above offers clear evidence to support his point. As such, the appeal to ''a small number of basic mechanisms that underlie wide ranges of phenomena'' is the way both explanation and reduction is achieved, though an important caveat is in order.

The caveat has to do with the expression ''small number of mechanisms.'' In point of fact, close analysis of the biomedical sciences discloses an extensive variety of mechanisms, some with comparatively narrow scope and some with almost universal scope. In a number of areas, a range of mechanisms which bear close analogies to each other are found. This, given the evolutionary backdrop, is not unexpected, but the subtle variation biologists continue to encounter requires them to be attentive to changes in biological processes as they analyze different types of organisms or even as they analyze different forms of entities in the same organism. Thus though the genetic code is (almost) universal, messenger RNA is processed quite differently in prokaryotes (e.g., bacteria) in contrast with eukaryotes (e.g., multicellular organisms such as *Aplysia*). Furthermore, variation can be found in the same organisms, for example, in the human where the actions of muscle fibers in skeletal muscle are regulated importantly differently than in cardiac muscle, and serious errors can result if the differences are not kept in mind.

Thus the biomedical sciences display a rather complex and partially attenuated theoretical unification: some mechanisms are nearly universal, many have evolutionarily fixed idiosyncrasies built into them, and some are highly individualized. Almost a spectrum of scope is encountered in the diversity of the life sciences. Nevertheless, generalizations and mechanisms are available to be used in explanations and to be tested in different laboratories: the variable scope breadth does not negate the nomic force of the generalizations and the mechanisms.

This picture of explaining-reducing generalizations of variable scope instantiated in a series of overlapping mechanisms is congruent with the account of theory structure proposed by Schaffner (1980, 1986), and further developed in Schaffner (1987). The analysis also comports with the GRR model of reduction developed in Section 8.6. It would take us beyond the scope of this introductory book to pursue these issues in any further depth. Suffice it to say, then, that reduction (and explanation) can be located in that ''overlap'' range discussed by Salmon where (comparatively) few ''hidden'' mechanisms account for a (comparatively) wide range of phenomena by exhibiting (complex) causal pathways.

8.11 THE NEURO-IMMUNE-ENDOCRINE CONNECTION AS EXEMPLIFYING THE BIOPSYCHOSOCIAL MODEL

We have now traced the details of how a reduction occurs in a part of neurobiology, and it remains to apply those results back to the issues raised in our treatments of Munson and Engel in Sections 8.2 and 8.3. Over the past half-dozen years, a number of biomedical scientists have discovered that important connections exist between the nervous system and the immune system in both animals and in humans. That the nervous system interacted with the endocrine system—the group of glands that produce hormones—has been appreciated for some time, but the mutual influence of the immune system and the neuroendocrine apparatus is a relatively new idea.

Scientists have found bidirectional communication between the systems. The cells of the immune system have been found to synthesize biologically active neural hormones and the immune system's cells have receptors on their surface for such hormones. In addition, the central nervous system can affect immune responses by inducing the release of adrenal and other hormones, which can substantially modulate immune reactivity since immune cells also have receptors for adrenal hormones on their surface. Thus each system can influence the other via a variety of chemical substances. A recently published book on *The Neuro-Immune-Endocrine Connection* (Cotman et al. 1987) demonstrates that well characterized *causal pathways* exist whereby certain forms of behavioral stress can be shown to result in a decreased immune response. Animals experiencing such stress—often referred to as an analogue of a human's helpless-hopeless situation—become susceptible to bacterial and viral illnesses, and also appear to have a higher susceptibility of developing cancers.

This research on the neuro-immune-endocrine connection is based in part on an idea that was at one time controversial, but which today seems well-founded. This is the notion that one of the functions of the body's immune system is to identify at an early stage those cells which have become cancerous and to eliminate them before they can divide and overwhelm the body's defenses. This hypothesis was championed by the late Dr. MacFarlane Burnet who termed it "immunological surveillance." The increase of cancers in patients who have been artificially immunosuppressed because of receiving organ transplants, as well as the appearance of a rare form of cancer known as Kaposi's sarcoma in patients with the Acquired Immunodeficiency Syndrome or AIDS, supports this idea.

Recent studies done at the Pittsburgh Cancer Institute at the University of Pittsburgh by Dr. Sandra Levy and Dr. Ronald Herberman and their colleagues also suggests that psychological status including emotions may play an important role in accounting for different prognoses of cancer patients. These scientists followed the immunological and psychological status of 75 breast cancer patients who had undergone surgery and chemotherapy-radiation therapy. An important predictor of these patients' prognoses was the activity level of their natural killer (NK) or cancer-fighting immunological cells. Drs. Levy and Herberman found that they could explain about 51% of the decreased NK activity on the basis of three distress indicators: the patient's poor adjustment, the lack of social support, and their fatigue-depression symptoms. A similar but less pronounced effect was also noted after a period of three months. More recent work by these investigators, carried out on a second sample of 125 early stage breast cancer patients, showed that the perception of high quality emotional support from the patient's spouse was a more potent predictor of NK activity than the endocrinological character of the tumor itself, as determined by its endocrine receptor status. These investigators concluded that these central nervous system mediated factors thus accounted for a significant effect on the cancer's prognosis.

Other attempts to correlate emotions and cancer prognosis have been less successful, but these linkages have been given a rational basis by the neuro-immunological-endocrine connection discussed. Human beings are very complex and it is very difficult to isolate the extensive interactions just indicated so as to produce unequivocal experimental conclusions. Research in cancer immunology is proceeding at an increasing pace as investigators realize that the more traditional forms of chemotherapy have likely gone as far as possible, and the new direction is likely to involve means of boosting or stimulating the immune system via interferons, interleukins, and other active immunological modifiers to help the body fight cancer. Augmentation and fine tuning of the immunological response via the neuroendocrine connection is in the process of exploration.

These considerations force us to take seriously that multiple perspectives, in the sense of multiple interacting systems, play significant roles in disease and in the healing process. To this point, however, the advances described both in the early sections and in the previous section suggest that biomolecular advances are discovering just *how* such interlevel causal connections as indicated in Figure 8.1 are implemented *at the biomolecular level*. Thus such progress as is occurring seems to support the reducibility of medicine to biology, and would confirm the characterization of medicine as a science. There is, however, one additional at least prima facie difficulty for such a reductionistic approach. As suggested in Engelhardt's remarks at the beginning of Section 8.1, and in Munson's comments about the essential moral features of medicine, medicine may contain intrinsic normative or ethical features that make a reductionistic approach difficult to defend. It is to this issue that we now turn.

8.12 THE ETHICAL DIMENSION OF MEDICINE

Ethical problems in medicine have become an important part of the public consciousness in recent years. As medicine has developed its technical prowess, it has generated hard choices concerning the use of its capacity to help individuals with their health. The costs associated with leading-edge innovations such as organ transplants pose difficult allocation decisions for individuals and for society. The potential to keep terminally ill and/or comatose patients alive for indefinite periods has led to a vigorous ethical and legal debate about active and passive euthanasia. These choices and debates are not external to medicine but in a very real sense are part of health-care delivery.

The situation does not change whether we focus our attention on so-called ''macro-'' issues involving societal determinations of what portion of the gross national product to commit to health care or on ''micro-'' problems such as whether to withdraw an artificial respirator from a particular patient. In both types of deliberation, normative principles need to be utilized in order to reach an ethical conclusion of what it is ''best'' to do. The literature of medical ethics has grown enormously over the past fifteen or so years, with no lack of approaches which offer normative frameworks to assist with these matters.

Many ethicists have found that what can be described as a ''principle'' approach to medical ethics is helpful in clarifying and resolving clinical ethical problems. Such a perspective can be both flexible and sufficiently philosophically eclectic that it does not generate additional problems of communication by dogmatic adherence to any one philosophical or religious system. The approach has its roots in several Federal Commission reports (including the *Belmont Report* and the many volumes published by the President's Commission for the Study of Ethical Problems in Medicine) and has been articulated systematically by Beauchamp and Childress (1989).

It will not be feasible in this chapter to develop any significant medical ethical analysis utilizing this approach. Suffice it to say that the "principle" approach does introduce several prima facie *values* such as individual "self-determination" and "well-being" which represent what is taken to be ethically desirable features to be maximized in any given situation. Unfortunately such values frequently conflict and may require a more general and comprehensive ethical theory such as "utilitarianism" or Rawlsian social contract theory to resolve them. Critical for our inquiry in this chapter is the question whether such normative principles could be in any interesting sense "reduced" to the biological sciences. A major barrier to such a reduction is an observation made by the philosopher David Hume several hundred years ago and sometimes referred to as the "is-ought" distinction.

Hume pointed out that no logical rule of inference exists by which a sentence expressing (moral) obligation, e.g., "You ought not steal," can be *derived* from a set of premises that are descriptive of the world or of human nature ([1739–1740] 1978). Such premises, including all standard scientific statements among them, state what *is* the case, not what *ought* to be the case. Recalling the conditions of the general reduction model of Section 8.6, what would be necessary to achieve a reduction of ethics would be a "connectability assumption" defining an ethical property in terms of a descriptive property. This, however, seems to be exactly against what Hume has cautioned us.

The situation is somewhat more complex though not more encouraging in terms of pointing toward a clear solution. A number of philosophers, John Stuart Mill (the utilitarian) among them, contend that the "good"—what is ethically desirable—*can* be specified in what are prima facie nonethical terms. For Mill, the "good" or *"summum bonum"* was "the greatest happiness" ([1861] 1979, 1,3). Mill offered a complex theory of happiness as well as a number of subtle arguments for his identifying the "good" with "the greatest happiness for the greatest number," none of which we can discuss here. Other philosophers, however, including Kant, have contended that such a reduction or *naturalizing* of ethical notions is faulty and incorrect. Thus we encounter in our discussion of the reducibility of medicine—itself a difficult question—one of the perennial philosophical problems about the nature of ethics.

While no major new insights will be offered here concerning this issue, the concluding section suggests a tentative position which has the virtue of permitting us to continue with further inquiry, and which may point toward a more satisfactory resolution of the problem.

8.13 SUMMARY AND CONCLUSION

In this chapter we have examined two interrelated problems in the philosophy of medicine. The first issue addressed was the status of medicine as a science and of the reducibility of medicine to its constituent biological sciences. There the significant and essential social (and psychological) dimensions of medicine were emphasized.

In the second part we examined the nature of reduction in the biomedical sciences in general, and reduction involving what has often been called "mental" to the "physical" in particular. Those sections showed both how complex such a reduction is, as well as how fragmentary the results have been in spite of major advances in molecular neurobiology. On the way to these conclusions, a complex model of reduction was also developed—the GRR account—which was suggested to have special utility in those areas involving the relation of the mental and the biological.

Finally, the inquiry closed by suggesting that though on the other hand molecular biology may be furthering ways to reduce the biopsychosocial intricacies of medicine to biochemistry, on the other hand the normative aspects of the biopsychosocial dimensions of medicine seemed to strongly resist any such reduction.

One way to come to a reasonable conclusion regarding these different tendencies is to acknowledge that *for the present and the foreseeable future* medicine will not be a reducible science. Whether medicine can be conceived of as as science at all will also wait on a resolution of the reducibility of the inherently normative components which constitute part of it. This is a position described in Section 8.4. as "weak emergentism." Close analysis of the example of *Aplysia* sensitization will affirm that even this paradigm of the "reduction" of a behavioral event to a set of molecular mechanisms involves conjoint appeals to multi-level concepts including the cellular as well as the molecular. On the other hand, the GRR model as well as the *Aplysia* example suggests that translation and reinterpretation—and *possible* replacement—of some fundamental entities, properties, or processes may occur as science advances. This is a thesis that Paul and Patricia Churchland have argued may even involve basic concepts in "folk psychology" such as "beliefs" and "desires" (see Churchland 1986, 1988). A very powerful molecular learning theory may well indicate why we possess and utilize such normative concepts as self-determination, and the role that such ethical analyses may play in individual and collective homeostasis. Would such a theory commit the "naturalistic fallacy" of attempting to derive an "ought" from an "is" that Hume ([1739–1740] 1978) and Moore ([1903] 1962) said was not possible, or might it be conceived more in the spirit of a Millean attempt at a rational reconstruction of a naturalistic foundation for ethics? Such possibilities are at present mere (and weak) speculations but as the molecular neurosciences advance rapidly, their salience—and importance—are highly likely to confront us with increased urgency.

DISCUSSION QUESTIONS

1. What is the difference between ontological and methodological reductionism? Discuss the subtypes within these two classes. Where does the doctrine of "vitalism" fit? Into what type or subtype(s) would (Cartesian) "dualism" fit?

2. What are the basic conditions of the Nagel Model (NM) of reduction? How does the Generalized Reduction model go beyond the NM?

3. What does it mean to claim that there are different "levels" of aggregation at which biological organisms can be studied? Define three such levels using Kandel and Schwartz's studies on *Aplysia*.

4. What conditions of the Nagel Model and/or the Generalized Reduction (or—a harder question—the GRR) model are satisfied in the reduction of short-term sensitization in *Aplysia* to biochemistry?

5. What are some problems which Munson detects in attempts to characterize medicine as a science?

6. What key argument must Munson make to show that in addition to medicine being *different* from typical sciences, it can *never* become a science? Discuss why Munson's argument is or is not sound.

7. Describe Engel's "biopychosocial" model of medicine. How might its adoption lead to changes in health care?

8. How might thoughts and emotions interact with biological entities? Is there any evidence that such studies can be scientific? What might be their implications for the reduction of mind to matter?

9. What did Hume mean by the "is-ought" distinction? What possible implications might it have for the reduction of medicine to the biological sciences? In your answer discuss whether you believe medicine, *per se,* contains "normative ethical" assumptions.

SUGGESTED READINGS

BEAUCHAMP, TOM L. and JAMES F. CHILDRESS (1989), *Principles of Biomedical Ethics.* 3rd ed. New York: Oxford University Press. This is a thorough introduction to contemporary problems in medical ethics.

CHURCHLAND, PATRICIA SMITH (1986), *Neurophilosophy: Toward a Unified Science of the Mind/ Brain.* Cambridge, MA: Bradford/MIT Press. A systematic treatment of philosophical issues in neuroscience, including reduction.

ENGEL, GEORGE L. (1981), "The Clinical Application of the Biopsychosocial Model," *The Journal of Medicine and Philosophy* 6: 101–123. One of Engel's most detailed developments of his influential biopsychosocial approach to medicine.

HULL, DAVID L. (1974), *Philosophy of Biological Science.* Englewood Cliffs, NJ: Prentice-Hall. A general introduction to philosophy of biology; Chapter 1 criticizes the Nagel model and its extensions.

KANDEL ERIC R. and JAMES H. SCHWARTZ (eds.) (1985), *Principles of Neural Science.* 2nd ed. New York: Elsevier. An excellent and extensive introduction to the contemporary neurosciences.

KITCHER, PHILIP (1984), "1953 and All That: A Tale of Two Sciences," *The Philosophical Review* 93: 335–373. A detailed criticism of the Nagel model of reduction and proposals for Kitcher's alternative approach.

MUNSON, RONALD (1981), "Why Medicine Cannot Be a Science," *The Journal of Medicine and Philosophy* 6: 183–208. Munson's detailed arguments against the possibility of medicine being a science.

NAGEL, ERNEST (1961), *The Structure of Science: Problems in the Logic of Scientific Explanation.* New York: Harcourt, Brace & World. A comprehensive treatment of the Nagel model of reduction appears in Chapter 11.

SALMON, WESLEY C. (1989), "Four Decades of Scientific Explanation," In Philip Kitcher & Wesley C. Salmon (eds.), *Minnesota Studies in the Philosophy of Science.* Volume 13, *Scientific*

Explanation. Minneapolis: University of Minnesota Press, pp. 3–219. An enormously detailed history of the development of philosophical analyses of scientific explanation up to the present.

SCHAFFNER, KENNETH F. (1977), ''Reduction, Reductionism, Values, and Progress in the Biomedical Sciences,'' in Robert G. Colodny (ed.), *Logic, Laws, and Life: Some Philosophical Complications*. Pittsburgh: University of Pittsburgh Press, pp. 143–171. A detailed account of the GRR model presented in the current chapter.

Nine

PHILOSOPHY OF PSYCHOLOGY

Peter Machamer

In many ways the philosophy of psychology is an old and venerable subject dating back to the Greeks. In another way philosophy of psychology only became possible in the late nineteenth century when psychology "split" from philosophy and became its own empirical discipline. Yet in still other ways philsophy of psychology was not treated as a major component in the modern professional philosophical community until the last twenty years. To understand why all these statements are true is to understand much about the different ways in which philosophy and psychology have been linked.

Psychology comes from a Greek root *psyche* meaning *soul,* which later became synonymous with *mind.* The goal of psychology was to study the mind. Looking at the work of Plato in *Timaeus* and *Sophist* and later Aristotle in *De Anima,* we see a curious mixture of empirical, descriptive material about how sensation, perception and thought occur intermingled with assumptions and arguments about the metaphysical principles and substances that underlie those descriptions—all leading toward an answer to the question of how humans come to know.

These texts—and they are similar in structure to those that were to follow through the eighteenth century—did not distinguish between issues that were metaphysical, epistemological or psychological. Although there is once again debate over whether these distinctions can be clearly or usefully drawn, at the beginning of the twentieth century most professional philosophers thought that the issue was settled and that these three types of inquiry were indeed separate and distinct. Science, including psychology, was one enterprise, and philosophy, done scientifically, was another type of intellectual activity. Neither had any place for metaphysics which was held to be meaningless. Scientific claims were to be empirically verifiable. All claims about philosophy of science were metalinguistic claims about the structure of science and its verifiable nature. Epistemology at this time was taken to be the analysis of the *logical* bases for claims to knowledge.

From the time of the Logical Positivists at the beginning of the twentieth century through the later Wittgenstein (mid-1950s), philosophy was thought to have little or nothing to do with psychology. (Ludwig Wittgenstein and the Logical Positivists associated with the Vienna Circle dominated much of early twentieth-century philosophical thinking. Their goal, briefly stated, was to make philosophy its own rigorous analytic discipline like science. In order to do this they held that philosophy and science were wholly distinct.) You can see this position and its tensions clearly in Norwood Russell Hanson's work *Patterns of Discovery* (1958) where in a very Wittgensteinian footnote he writes how philosophers would have much to learn from reading the Gestalt psychologists and then says, of course, what psychologists have to say, strictly speaking, is irrelevant to philosophy since psychologists deal with facts.

Epistemological questions for these thinkers dealt with how to analyze sentences like *"S knows that p"* into components, often held to be truth conditions, such as *"S believes p"*, *"S is justified in believing p"* and *"p."* Despite their scientific bent, it was as though science had nothing to tell us about knowing. Philosophy, and thus philosophy of psychology, were *sui generis* activities.

The bringing together again of psychological and epistemological inquiry dates back to the period around the publication of a widely read paper by W.V.O. Quine, entitled "Epistemology Naturalized" (1969a). Suddenly it became acceptable once more to use empirical truths garnered from the science of psychology to help establish claims about how humans know. It was all right to fill in the analysis of *"S knows that p"* with empirical sentences about how the psychological and physiological systems of human beings were thought to work. Quine gives a negative argument for this conclusion stressing that classical empiricism must fail in its attempt to base all knowledge upon sensations and to construct everything we know from that. He then concludes that the only alternative is to consider epistemology as just part of the history of psychology. Classical empiricism was just not a particularly insightful part of that history.

The earlier attacks on classical empiricism by Hanson, Kuhn and Feyerabend (see Chapter 4) have already shown why the positivist's program of reconstructing the logical structure of science and their concepts of observations and verifiability had to fail. But I believe the real popularity of Quine's paper had more to do with other contemporaneous movements in philosophy and psychology—notably the rise of cognitive science as the dominant approach to psychology and the increasing influence of computer models on the practice of philosophy and on thinking about the mind.

Prior to this reconnection with psychology, philosophy of psychology was largely confined to what is called philosophical psychology. Briefly, this discipline took affective or psychological concepts and attempted to analyze them in terms of their logical and, mostly, behavioral components. Numerous books and papers were published analyzing concepts of will, thinking or mind. Strangely, these thinkers, while often aware of, and even inspired by, Sigmund Freud, failed to see that his analysis of concepts such as jealousy or guilt did a far better job than their neo-Wittgensteinian offerings. Perhaps this is because the latter were essentially behavioristic while Freud dealt with the internal structure of affective (or emotional) states. Freud and behaviorism will be discussed later.

Yet another aspect of philosophy of psychology overlaps with the philosophy of mind. Philosophy of mind, at least since the seventeenth century and Descartes, when the subjective experiences of the knower (namely, the *cogito*) emerged as perhaps *the* major problem in philosophy, asks questions about the nature of mind. The chief questions are what kind of a thing (or substance) is the mind and what are its properties. Many attempts to answer questions about the nature of mind turn into questions about the adequacy of reducing mind (or talk about mind) to some other substance (or talk about some other substance). In the most popular philosophy of psychology the language to which mind-talk would be reduced is not the brain language of physiology or neurobiology (as discussed in Chapter 8) but to a language described as functional.

This claim is best understood by use of a computer analogy—which is where the theory of functionalism arose. A computer in one sense is nothing but bits of hardware (most often transistors)—switches that are on or off, open or closed. A network of switches can be completely modeled for information purposes as a system that instantiates Boolean algebra (or the propositional calculus.) This is the binary logic of the digital computer. (If we add to the Boolean algebra the probability calculus that was discussed in Chapter 2 and a few shorthand tricks about logarithms, we have the basis for Shannon-Weaver mathematical information theory.)

Now in addition to these hardware-based logical operations, computers also have routines which we call programs. These are complex operations that the machine can perform, such as adding numbers or solving differential equations. In solving these complex problems a computer cannot be doing anything in reality but concatenating (putting together) sets of off-on sequences. Mechanically that is all it can do. But many different sequences of off-on hardware responses can do the same job at the program level. When this is the case such sequences are functionally equivalent. Because of these equivalencies, the program level can be discussed independently from the hardware level. Daniel C. Dennett (1981, 1987) has called this level of activity the program level, and the hardware description, the machine level. (Dennett also argues that an even higher level of talk is necessary, the intentional level. This level mirrors talk about human intentions. Dennett claims this is necessary because complex problem solving machines cannot be described adequately without reference to intentional locutions. The question, if he is right, is: Does the need to use intentional talk originate from some characteristics concerning the structure of such machines or are they imported by humans in ways that reflect programmer designs of merely convenient and comfortable anthropomorphism?)

What is interesting is that sometimes *in practice the program cannot be totally specified (reduced) in terms of the machine level, though we know in principle that* it can be done. So, the functionalist hypothesis goes, the human mind is like a program which can be realized in many different mechanical and material forms including neural hardware. Furthermore, it may be that because of the mathematical complexity of the relations between the program and material levels that reduction of one to the other will never occur. This way of looking at the mind in functional terms allows a certain independence of psychology from physiology. It also helps explain why we tend in our commonsense folk psychology to treat the mind as independent from and behaving according to different laws than the body or brain. At present, the adequacy of functionalist theory is much debated (functionalism as a theory of psychology is treated in a section of Ned Block 1980). Additional interesting questions about the mechanical nature of mind arise in the debate about whether computing machines can think or be creative. (Machines as being possibly creative is a thesis of Herb Simon. Counter arguments are opposed by Dreyfus 1972 and Weizenbaum 1976).

9.1 FREUD

No account of the philosophy of psychology would be adequate if it did not discuss, at least in passing, the work of Sigmund Freud. Freud's work has had a special and tantalizing appeal for philosophers ever since the 1920s when he and psychoanalysis rose to prominence together. Despite some waning interest during the heightened interest in behaviorism, Freud has continued to be a fit subject for philosophical debate. Because of certain recent trends he may be more popular today than he ever was.

The contemporary influence of Continental thought views Freud as an adherent or progenitor of a hermeneutic approach to understanding people.[1] This view of Freud holds that psychoanalytic theory is not a scientific theory because it is not subject to the usual constraints concerning evidence and testability. Rather Freud is said to provide an interpretative schema for understanding human behavior. Interpretative schemas are judged according to how well they organize seemingly disparate bits of data. The criterion for "how well" is not usually articulated very clearly, but seems to concern whether the organization brought about by the schema leads to understanding or to some special, though usually unspecified, psychological state in the theorist or philosopher doing the interpreting.

Another line of contemporary interest in Freud was started by Karl Popper's (1963) critique. He argued that Freud's theory which postulated unconscious mental motivational mechanisms was unscientific because it was not falsifiable. No evidence, Popper argued, could be found that would show that a psychoanalytic hypothesis was false. Whatever a patient said or did could be explained by the theory. Therefore, psychoanalysis was unfalsifiable and therefore not a scientific theory.

More recently, Adolph Grünbaum (1984) has taken the Continental hermeneuticists, Popper, and Freud to task. Against the hermeneuticists and Popper, Grünbaum argues that Freud's theory is indeed scientific, as it was meant to be, and is testable. However, he continues, it is not very good science. Grünbaum argues that psychoanalysis is empirically unsupported on many of its central theoretical points.

---

[1] The Continental views of Freud tend to be in the hermeneutic—interpretative—tradition where they argue that psychoanalysis is not really a science like the natural sciences. Psychoanalysis is a humanistic way of "seeing" the world. Influential writers include Paul Ricoeur, Jacques Lacan, and Jurgen Habermas. This tradition is criticized soundly by Grünbaum (1984).

Still other current interests have brought Freud to prominence. Some feminist thinkers have chided Freud's theories, especially his views on femininity and women. Others have overtly accused him of blatant sexism, (for Freud and feminist critiques that in strange and interesting ways still depend upon Freud and utilize Freudian concepts see Lerman 1986, Gilligan 1982, van Herik 1982, and Williams 1977). Other people have criticized Freud because they believe his view of the fantasies of childhood trauma have helped to cover up the real and scary problem of child abuse. (Mason 1984, Malcolm 1984).

These current studies are all interesting and provide good reasons why Freud ought to be studied. However one aspect of Freud's work has not received much recent attention and is important for understanding what constitutes an adequate theory in psychology—his theory of the affective or emotional dimension of human personality.

Psychoanalysis arose as a means of treatment, primarily for patients suffering from various neuroses. From its inception psychoanalysis was regarded as primarily a clinical discipline, a therapeutic branch of medicine.

As Freud's study of neuroses grew, he began more and more to expand and elaborate theoretical concepts. These became the bases for Freud's theory of personality. More specifically, the theoretical elaboration was an attempt to present a causal theory of human affect or emotions. Concurrently with this theoretical work Freud continued to develop his ideas about effective therapy and the treatment of patients. Most of the evidence for Freud's claims were based on insights gleaned from his clinical, therapeutic experiences.

Despite their complex interrelations, it is both possible and important to draw a distinction between Freud's general theory of personality (his theory of human affective motivation) and his views on how therapy may be used as evidence. Such a distinction, however, does involve reconstructing Freud's work with a view toward isolating the influential and insightful theory of the human affective personality from the questions of the effectiveness of psychoanalytic therapy and the peculiar problems attendant upon the nature of clinical evidence.

Freudian theory about the affective human personality is not only fascinating but virtually unrivaled. The concepts of unconscious, repression, ego defenses, psychosexual development and, most importantly, the importance of the family complex in childhood development are but a few of the Freudian ideas that have pervaded almost all contemporary thought. Details regarding the structure of these concepts and how they are to be applied to understanding human behavior and actions are arguable, but that many people think in these terms is undeniable. Moreover, no detailed alternative theory of human affective states and their motivational functions to replace Freud's exists.

Freud's theory of personality is a mentalistic or cognitive theory that attempts to postulate functional parts of the mind in order to account for human emotions and (normal and abnormal) actions that arise from them.

More specifically, Freud's theory postulates basic instincts which are the motivational forces for human affective states and behaviors. These instincts, in the form of unconscious memories charged with affective power, are channeled and restructured by other mental mechanisms (internalized external influences, such as views of parents or siblings, and individual dispositions) into particular forms of behaviors and actions. In this way, Freud's view of the mind is of a set of internal processing mechanisms (functional subsystems) that act upon basic biological drives and instincts. By reforming those basic instincts with additional inputs form the ego and superego (self-preservation and external influences) he attempts to explain how certain human emotions function.

A classic example is Freud's analysis of jealousy. Jealousy, according to Freud ([1922] 1963a), is a normal feeling that comes from reaction to competition for a love object. The feeling is analyzed into its constitutive elements including a feeling of grief at the thought of losing the loved one, a concomitant narcissistic wound (self-pity), enmity toward the successful rival, and a degree of self-criticism as the ego is blamed for the loss. The basic instinct here is love for a particular person (though this itself ought to be analyzed further). More excessive forms of jealousy add a dimension of projection when faithlessness—real or imagined—in an individual's own life projects to the infidelity of the partner. The final most extreme form of jealousy is a form of paranoia in which the jealous person becomes delusional. In such cases paranoic distrust causes signs of infidelity to be seen everywhere.

This is just one example of a Freudian analysis of an affective state. His analysis of human emotions in terms that can be related to other psychological theorizing is very important. Indeed as psychologists began to realize and study the importance of the social dimensions of cognition and behavior, the role that emotions play becomes increasingly significant. Freud's theory is the best place to start in attempting an adequate theory of the emotional dimensions of human life.

9.2 PHILOSOPHY OF PSYCHOLOGY AS A SCIENCE

Finally let us introduce another conception of the philosophy of psychology which treats it as similar to the philosophy of physics, biology or medicine. On this view is a philosophical inquiry into the science of psychology (called experimental psychology by most academic departments) which attempts to characterize the scientific nature of psychological theories, methods of inquiry, and results.

Among the many ways to approach this topic, using problems in one area of psychology—perception—to illustrate the various theoretical positions may be most useful. In the history of twentieth-century psychology three basic types of theories have dealt with perception, thinking, and behavior: response-based (behaviorism), organism-based (cognitivism or constructivism) and stimulus-based (realism).

These three approaches have not always been kept distinct, nor have proponents always adhered exclusively to just one approach. Judicious simplifications, however, will better allow us to formulate the different theories and their corresponding positions on experiment and method.

9.3 BEHAVIORISM

The behaviorist approach originated in the United States with John W. Watson (1919) in the early twentieth century. In the 1930s and 1940s it was the most popular approach in American universities, with Clark Hull et al. (1940) and B.F. Skinner (1953) as its best known proponents. In what follows let us concentrate on Skinner's view because it is the methodologically purest and most easily accessible.

The basic concept in Skinnerian behaviorism is operant conditioning or how an organism learns specific behaviors. The idea is simple. A stimulus affects an organism which then responds with some behavior. The behavior elicited is then positively or negatively reinforced, either increasing or decreasing the probability that the same behavior will occur again under similar stimulus conditions. Reinforcement is brought about by environmental contingencies (effects) that impinge on the organism as, or shortly after, it exhibits the original behavior.

Schedules of reinforcement are designed in order to ascertain what temporal intervals for reinforcement are optimal and how often the reinforcement should be applied. We can draw a graph of what schedule of reinforcement elicits the greatest change in probability that a behavior is elicited. Learning consists in a change in the probability that a given behavior is elicited under similar stimulus conditions. One additional bit of information required to record the changes in probabilities is the state of the organism relevant to the reinforcer, that is, the level of deprivation of the organism.

The experimental paradigms (favored designs for experiments) associated with behaviorism are discrimination tasks and learning trials. The first tries to determine whether an organism can discriminate when a stimulus is present. Discrimination is measured by a behavioral response, such as a pigeon pecking at a new color. Learning trials set up situations in which reinforcers are coordinated with behaviors as when a rat is rewarded with food pellets for pressing a bar.

Part of the methodology of Skinner's most extreme form of behaviorism stresses that a science of behavior ought not speculate about what goes on internally within an organism. Perhaps someday physiology will describe what happens *in* organisms, but a science of behavior is independent of such internal concerns. A science of behavior merely formulates laws of discrimination and learning associated with different organisms and different stimulus and reinforcing conditions.

Even stronger, a behavioral scientist is enjoined upon pain of non-sense from using mental terms to explain behavior. The use of intentional or mental terms such as "will," "desire," "wish," "intend," "believe," and so forth are said to be wholly nonexplanatory. They are no better than obfuscating redundancies of the type ridiculed by Moliere when he described a doctor explaining why opium puts people to sleep by saying that opium has a *virtus dormitiva* (that is, a power to bring sleep). The language of mental states refers to what is unobservable in principle and so has no place in empirical science. All we can observe and measure are overt behaviors, which are the domain for good empirical science. This accords with a strict form of an operationalist thesis (see Chapter 1) which is a very strong form of the empiricist's doctrine of meaning.

Criticisms of behaviorism came fast and hard in the late 1950s. Most notable among the critics was the linguist Noam Chomsky (1959) who, in a review of Skinner's *Verbal Behavior,* pointed out that the operant conditioning model was insufficient to explain a human's ability to produce novel sentences in a natural language. Chomsky pointed out that each of us is capable of producing an infinite number of sentences, many of which had never been uttered before. Since uttering a novel sentence is behavior that has never before occurred, how could prior reinforcement explain how humans learned to produce such novel sentences?

Other criticisms of behaviorism were more diffuse but more devastating. First and foremost were queries about how to identify and "count" behaviors. It is easy to see what behavior is being reinforced in a controlled experiment where the behavior to be learned is specified in advance, but in more ecologically salient (real world) contexts behaviors are generally not isolable. Further, reinforcing contingencies are not easy to identify. In fact, in the behaviorist program reinforcers are defined post hoc as conditions which have brought about a change in the probability of a behavior's being elicited. This means one records the change in frequency of behavior and then looks for the reinforcer that is responsible for it. This may involve a vicious circularity. More significantly many different events and objects are always present and subsequent to every behavior, so how can we pick out which was the reinforcing one? Remember the only information that can be used to pick out behaviors according to the strict methodological principles are facts about spatial and temporal proximity.

The general problem is that instead of ending up with a neat one-to-one correspondence between a specifiable stimulus, an identifiable behavior and a clear reinforcing contingency, we have many-many correspondences. This means that laws describing behavior in a one-one mapping of physical stimulus conditions onto a specific behavioral response (the paradigm for a psychophysical law) do not exist.

Finally a different type of criticism was raised by the philosopher-psychologist Jerry Fodor. Fodor, and his colleague Bever, studied human perception of pause, (this perception of pause experiment is reported in Fodor, 1968, 79–89. Fodor uses it as an argument against behaviorism). This is the ability of humans to hear breaks or pauses between words as they are spoken in utterances. They point out that when we hear the name Bob Lees pronounced, for example, we hear a break or pause between the "Bob" and the "Lees." However, if we look at a sound wave representation of the vocalization of a normal utterance of this name, we see that the break in the sound curve actually comes between "BobL" and "ees." This means that the hearer is perceiving a pause which does not naturally correlate with the break in the sound wave pattern. The behavior of pause perception is not shaped by the physical characteristics of the sound waves. Fodor concludes that internal processing variables are necessary to bring back an explanatory correlation. This is an argument for internal processing, or a form of cognitivism.

9.4 COGNITIVISM (CONSTRUCTIVISM)

The importance of the organism's contribution to perception and cognition drew support from the computer revolution. After World War Two, with the introduction of digital computers, then called "thinking machines," scientists in psychology and physiology began to think seriously about internal processing models. The computer showed them that they could represent internal processes in logical form.

In perceptual theory, Jerome Bruner (1947, 1957) was one person who came to symbolize the break with behaviorism. The "new look" in perception had at least two important facets and many ramifications. Perceptions, it was held, went beyond the information given to the organism from the environment. The nature of the processes by which this "going beyond" occurred became the subject of inquiry. The question became: What is added by the organism to the stimulus information? One answer came as Bruner and his coworkers revived interest in the study of affective variables. They looked at how values, needs and expectancies influenced what was perceived, (see Haber 1969b).

The study of affective factors as causes for enhancing or inhibiting recognition and for influencing perceptual judgement was part of an interest in a topic called perceptual set. Theorists and experimenters asked, how does an organism's prior experience or training affect or make it "set for" what it perceives? Why does "13" in a set of numerals get seen as thirteen, while the same stimulus when surrounded by letters of the alphabet is called "B"? What role does expectation have in recognizing objects, and how can we model the role that perceptual readiness played in recognition tasks?

Though many of these studies were later to be criticized because of confusions between perception, attention, and responses, they served well to awaken psychologists to the importance of the organism's internal processing mechanisms, and opened new experimental ways of studying memory and cognition.

Another aspect that deserves mention, though it will receive scant attention here, is an increasing emphasis on developmental aspects. Bruner again had a role in the developmental studies of children's abilities, but the real force behind developmental studies in cognition and perception was the Swiss psychologist Jean Piaget. (The developmental literature by Jean Piaget and his colleagues is voluminous. An early classic that shows the method of inquiry and scope of interest is Piaget [1926] 1959. See also Bruner 1973.) Developmental studies describe how a person's perceptions and ways of thinking about the world develop and change as the child grows older. They attempt to describe the changing states that children go through on their way to an "adequate" view of the world. The internal mechanisms that characterize the different stages of development again called attention to the contribution of the organism to the form and character of its experience.

By concentrating on the function of prior experience and other internal variables, the problems in perception to be investigated by cognitive theorists were shifted from those of discrimination to those of identification and recognition. What past set of experiences allow a subject to identify or recognize a given stimulus object more quickly? The scientific (quantifiable) question is: Given different background conditions (training), how long does it take the subjects to identify or recognize a stimulus object? The length of time required then is correlated with the subject's prior training.

The experimental paradigm for identification and recognition tasks is to determine the minimal information and time required for a subject to identify or recognize something presented. Minimal information is guaranteed by short times, for example, tachistoscopic flashes, by unclear or ill-defined images, or by ambiguous stimuli (illusions). The rationale for this paradigm is that by providing a subject minimal stimulus information we isolate what must be added by the subject. Thus, the experimenter is allowed to study these additive or constructive processes and to postulate mechanisms or models by which these are accomplished. These new concerns with internal mechanisms illustrate how a shift in theoretical models leads to new sorts of problems to be investigated and this leads to new experimental paradigms.

The experimental paradigms of the constructivist, information-processing theories, are those of minimal information, and the associated measures are temporal. Reaction times are the basis for claims to being a quantitative science of psychology. In short, the computer model allowed psychologists to allay their theoretical scruples about the mental, and the measurement of reaction times allowed them a sense of scientific objectivity.

The constructivist approach to the study of perception is well exemplified by the work of Richard Gregory (1990). He believes that perceptual processes are best described as (abductive or inductive) unconscious inferences that augment incoming incomplete data. The processes result in the formulation of hypotheses about what is in the world.

The Gregory program starts with minimal information. His experimental stimuli typically are illusions or other informationally impoverished or ambiguous objects. (Bruner and Potter 1964 used one-of-focus slides. Richard Gregory has a machine that increases density of dots on a screen—the greater the density of these dots, the easier is recognition.) He notes that an organism tends to respond to these stimuli in ways that suggest that it is not in doubt as to what is in its environment. The organisms acts as if the world determines its behavior. Therefore, the organism must utilize the perceptual information to induce a hypothesis about what is really in the world. The rules of induction are the augmentive rules that constitute our unconscious inferences. The hypotheses formed also usually reflect prior experiences, for example, in the Müller-Lyer illusion:

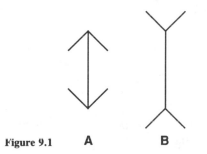

Figure 9.1   A   B

Gregory claims that an organism sees arrow *A* as shorter because it forms the hypothesis that the arrow is the projecting edge of a three-dimensional corner, while it hypothesizes the longer-looking arrow *B* is a farther away receding edge of a corner. If you look at the corner of a room, at the line where the walls join together in a room, they look somewhat like arrow *B* where the extensions are the lines formed by the ceiling and floor.

What Gregory is doing in this explanation is twofold. He is claiming that in perceiving the Müller-Lyer lines we are forming a hypothesis based on the visual input from the lines plus information gleaned what we have experienced about three-dimensional objects such as corners of rooms. This information about corners causes the organism, when stimulated by the lines, to hypothesize that the lines are really edges of three-dimensional (or representations of three-dimensional) objects. After this hypothesis is formed, the organism then further infers that since one set of lines *B* is more distant from it than the other, *A*, that *B* must be larger. So the organism unconsciously makes a size constancy correction to make line *B* look longer, and then experiences *B* as longer.

The missing premise here is that the organism in some sense unconsciously saw *A* and *B* as being the same size, and then under the three-dimensional hypothesis, applied a size-constancy distance correction function in order to experience *B* as longer. This supplementation of the details of the theory brings out the major fault of Gregory's constructive hypothesis formation theory. There seems to be no consistent or principled description about how such hypotheses are formed. We are never clear what hypothesis is being formulated or whether one hypothesis or a set of them is required.

Gregory's experimental designs most often start with a well-specified illusion or ambiguous object and a well-specified assumption about the organism's prior experiences or relevant behaviors. The hypothesis attributed to the organisms is defined, qualitatively, by what is necessary to map the stimulus so specified into the response so described. This smacks suspiciously of the mentalistic ad hoc method that so irritated Skinner. Indeed, many of Gregory's inferences and hypotheses are similar to Moliere's *virtus dormitiva*.

Though Gregory's work raises certain methodological problems, it still points to a new direction for seeking answers about questions of perception. The attempts to move from qualitative to quantitative studies about internal processing variables have brought some changes. The work of Fred Attneave (1971) and Julian Hochberg (1968, 1978) is similar to Gregory's in intention, though much more careful in execution. Both of these investigators have used experimental paradigms like those used by Gregory to isolate some of the detail concerning the organism's contribution. Hochberg has contributed many studies to the character of the organism's schemas, maps, and plans. Perception, in these studies, starts with the organism obtaining a set of successive glimpses. These glimpses (by encoding) are then fitted into a pre-existent schematic structure. The structure "fills in" the gaps between the glimpses and, even more importantly, provides an organizing principle by which the glimpses get encoded and stored. Some such principle is needed, so it is argued, to avoid chaotic or randomly distributed storage.

Common to all theorists who concentrate on the organism's contribution is their almost exclusive explanatory emphasis on the nature and function of the schemas and associated internal processing mechanisms. "Schemas" or some functional equivalent are by far the most important explanatory term in the constructivist's theory. Schemas get introduced to do every imaginable job. Remember that words like "schema," "set," "schematic map," "paradigm," "hypothesis," "script," "theory,' and "network" all serve to name internal processing variables.

A recent review of the literature shows that schemas and their kin were assigned abilities to perform the following functions:

1. Identification
2. Recognition
3. Comprehension
4. Categorical encoding
5. Memory storage and retrieval
6. Initial selection of inputs: attention, readiness, set
7. Inferences, entailments and implications
8. Expectations and predictions
9. Effector direction: action plans, behavior guidance systems
10. Coordinate intermodal information
11. Coordinate perception with the "language" system

(Representative and survey work on the concepts of schemas, scripts, and so forth can be found in Neisser 1976, Abelson 1981, Murphy and Medin 1985.)

In broader terms, schemas are the variables that are used to explain why incoming information is meaningful or significant for an organism. Somehow, meaning is given to the information by the role it plays in the schematic structures.

Other features that schema theories must have include:

12. showing how schemas are developmentally changeable;
13. showing how they are suitable for modification by learning in law-like ways by new inputs, reafference and internal introspection;
14. the ability to explain the conscious-unconscious distinction;
15. a way to handle the knowledge-belief (opinion) distinction.

Finally some cognitive theorists require that

16. only computational functions (computable) may link the elements (the nodes) of the schemas and those connecting various schemas.

Usually this last requirement means that schemas and the rules that describe their functioning can be modeled in some language that can be run on a digital computer.

The above list presents just some aspects of schemas that have been considered by psychologists. This varied list suggests that the nature and function of schemas are not clear. Often the detailed mechanism by which a schema can do any *one* of these tasks is also not clear. Moreover, it is unclear how any one mechanism could perform many or all of these roles. In some cases, properties that seem necessary for performing one task are antithetical to properties that seem necessary for performing another. Since what we attend to is not always what we do or can act upon, a schema's job to be both an attention selector and an action planner seems most difficult.

From where did these multiroled requirements for schemas come? They seem to stem from the rejection of behaviorism. Having noted the need for internal variables, constructivists, cognitivists, and information processors stressed the role of internal variables to the exclusion of all others. In their work, the stimulus affecting the organism, the environment in which it is located, and the behaviors it performs are theoretically unimportant. We have seen, for instance, the unrealistic degeneration of the stimulus to the tachistoscopic moment, the illusion, and the glimpse.

The behaviorists' single-minded focus on descriptions of overt behaviors has fallen from explanatory favor. For cognitivists, behaviors are explained in terms of schemas. Behavior, insofar as it has any special methodological status, remains the major evidential basis for schema attribution and thus for theory construction. More importantly, however, the class of behaviors that are of interest to psychologists has changed too. Now problem solving, thinking and reasoning are most important.

Much more can be said about the organism's contribution to perception. It seems reasonable to take internal processing and schematic variables as important, and in many cases, such as thinking and problem solving, to make them central. The detailed specification of these variables and the sorting out of the problems raised about schemas, such as whether schemas are imagistic, propositional or procedural, are the focus of the most work and the major debates in contemporary psychology.

9.5 REALISM

Behaviorists and cognitivists both neglect the stimulus and the environment. For behaviorists the stimulus was unproblematic since it was simply that which was discriminated by the behaving organism. Generally the experimental conditions under which the behavior was elicited were prescribed so carefully that there was no real question about what the discriminatory stimulus was. The target to be pecked or the bar to be pressed could be unproblematically isolated and its role described. The epitome of stimulus neglect is that in some learning paradigms, such as maze running, no stimulus was part of the design.

Cognitivists (constructivists) also found no problems with the stimulus since almost everything of psychological interest occurred internally. The stimulus was whatever in the environment that provided the trigger for internal processing. Experimental designs raised no problems about isolating or identifying stimuli. A subject in a darkened room, whose head is clamped into position, responds to a briefly flashed form. The experimenter has almost complete control over the artificially produced stimulus. The complexities of the real-world environments do not enter into these experimental designs.

Against this background of stimulus neglect we can best appreciate the work of J. J. Gibson and those he has influenced (Gibson, 1950, 1966, 1979). Gibson, beginning in the 1950s, emphasized the importance of the stimulus. He argued against any theory that held that percepts or perceptual experiences were constructed out of the data of sensations. Sensations, he held, played no role in perception.

Gibson concentrated on identifying significant or salient aspects and events in the environment of a perceiver. He looked at the ways in which information was available in the environment and he attempted to isolate those aspects of complex structures which elicit a response. Since no adequate vocabulary existed to describe these aspects Gibson introduced a new theoretical, descriptive vocabulary of "ecological science."

For example, Gibson argued that depth perception did not arise from triangulation based on nerve impressions (convergence), nor from calculation on binocular disparity. Gibson turned to the real world and noticed that surfaces therein were textured. A normal surface is broken into more or less regular elements of patterned color or brightness. Considering a textured surface stretched before a perceiver, regular changes in the sizes or densities of the elements constituting the texture of that surface (texture gradients) will be proportional to differing distances. So if perceivers in fact were attuned to these reasonably regular changes in the elements, they would directly perceive relative distance.

To illustrate the idea of a texture gradient (Gibson 1966, Rosinski 1977), let us deal only with static, monocular distance perception, which rules out the use of cues like binocular disparity or convergence as a basis for depth perception.

Consider an object sitting on a tile floor with a regular checkered pattern, typical of Renaissance buildings. In perspective from a perceiver's point of view it would look as follows in Figure 9.2.

The visual angle that each square, or element, makes with the perceiver's eye changes and gets smaller as the eye attends to more distant elements. A gradient is defined as a rate of change in the dimensions of these elements. Three different properties of the elements in our picture exhibit this rate of change of size. The height of each element decreases, the width of each element decreases, and the number of elements (density) included in the same visual angle increases.

The rate of change of these elements is proportional to the relative distance of the perceiver of the object. Thus, as the heights, widths and compressions of the elements increasingly diminish, the distance of the object from the perceiver is increasingly greater.

The perceiver is picking up information about the rate of change of these elements at the same time as perceiving the distant object. The information about distance is specified by the gradients. In experimental tests, perceivers performed with great accuracy in both relative and absolute distance judgements. The conclusion, therefore, was that perceivers obtained information about distance from directly attending to the variables of texture gradients. The texture gradients are instances of Gibson's complex variables that specify information in the environment. There is no computing or calculating from sensations; a perceiver merely picks up information about distances from the environment.

Reflection upon the Renaissance floor example brings some interesting theoretical aspects. First, the tiles are uniform in size. Under such conditions, the accuracy of distance perception is high. On other surfaces in the environment, we cannot expect such a neat, uniform distribution of elements. But many surfaces come close. Think of a grass lawn. For all practical purposes (i.e., within the limits of accuracy needed by most perceiving organisms in most conditions), the distribution of blades of grass is uniform. Texture gradients can be defined over that distribution.

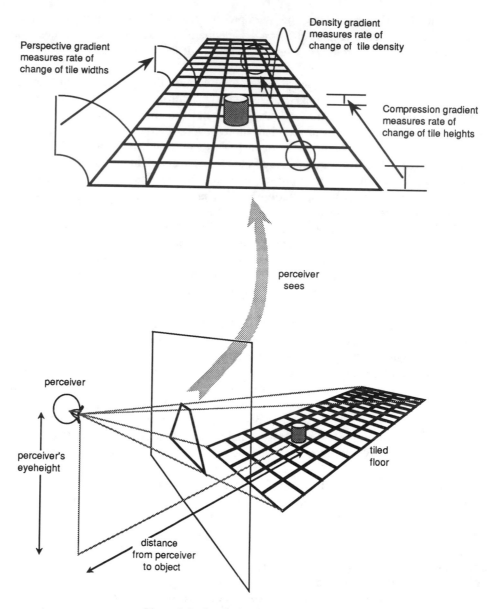

Figure 9.2 Perspective view of a tiled floor.

When elements are less uniform in size or their distribution on the surface is less regular, we expect less reliable distance perception. Such is in fact the case. Accuracy relates directly to the uniformity and regularity of the texture elements, (see Wohwill 1963, and Flock and Moscatelli 1964). Errors correspond directly to greater variability in the elements.

Perceptions of impending collisions or looming phenomena, another example of a higher-order complex variable, exhibit another important dimension of Gibson's theory of stimuli (Schiff 1965 and Rosinski 1978). Many of the variables that a perceiver directly picks up are defined over time.

As an owl swiftly swoops down to grab a field mouse, the mouse may become aware of a dark object of constantly increasing magnitude in its visual field. As an observer moves toward an object, or as an object moves toward an observer, the object looms large as the distance between the two decreases. The rate of change of a symmetric expansion in the perceiver's visual field determines a complex variable that the organism attends to in perceiving when that object will collide with it.

If $å_1$ is the angular size of the object at t_1 and $å_2$ is the angular size of the object at a later time, t_2, then the time until collision (TC) is shown by the formula:

$$TC = å_1/ (å_2 - å_1/t_2 - t_1),$$

for motion with a constant velocity. Roughly, holding the velocity constant, the differences in the apparent sizes determine a rate of change with respect to time. The organism responds to these rates in terms that specify when the distance between the object and itself are nil.

It has also been experimentally established that there are temporally differential avoidance responses to different rates of change. That is, organisms start to get out of the way more quickly if the object that will collide with them is moving more quickly. With chickens, an object pattern that expanded at a rate specifying a collision in 4 seconds led to avoidance behavior in 3–3.5 seconds. And expansion patterns specifying a collision in 1.75 seconds led to avoidance behavior in 1.4–1.6 seconds.

The expansion pattern that is perceived to specify looming or time to collision is also perceived as being symmetrical within the organism's visual field. If we center sight on the midpoint of the impending object, then its size will increase equally on both sides of the point. If it does not expand symmetrically, the object is veering off to one side or another. Experiments with human infants showed that they could perceive differences in looming between patterns that specified a collision and those that specified a near miss (Rosinski 1977).

The important aspect of the looming variable for an organism (its affordance) is that it is taken to mean "time to collision." Gibson and his followers hold that the perceiver responds directly to his TC information instead of registering putatively constituent bits and then computing time to collision. Gibson says the perceptual system of the organism is tuned to this complex variable and resonates, under proper conditions, to its presence in the environment. The variable is defined completely in physical terms and is taken to be an objective property of the environment.

The evolutionary significance of the information carried by the looming variable is obvious because it concerns predators and collisions. Similar information is used by pilots, baseball players and beach ball fanatics. However, in these latter cases responses must be learned. It does not seem that all perceptual cases, particularly those involving cultural categories, can fit easily into Gibson's direct perception framework.

The variables of texture gradients and looming are examples of Gibson's higher order environmental variables. He holds that perceiving organisms pick up information from their environment by extracting such variables from their environment. Thus organisms are directly affected by the environment, and the best description of what in the environment affects them is given by such analytically complex variables. However, the physical situations referred to by such variables are not really complex. They are just the structures in the environment that are significant to the perceiving organisms.

9.6 CONCLUSION

It should be obvious from this review of the major types of theories in perceptual psychology that an overemphasis on any one type of explanatory variable is too limiting for an adequate theory. Behaviorism's concentration on overt behaviors neglects the environment and the organism. Cognitivism neglects the environment and fails theoretically to account for the relevance of behaviors. Realism, with its direct pick-up mechanisms, neglects what the organism adds to its experiencing.

A straightforward attempt at a more adequate theory would include these variables in one comprehensive account and try to establish the relations that exist among them. Some psychologists have attempted to elaborate more comprehensive theories. However, at the present, no theoretical model incorporates all the relevant variables for perceptual psychology, let alone for the range of issues dealt with by other experimental psychologists.

Experimental psychologists often disdain their colleagues who go in for explicit theorizing. But as we have seen, experimental paradigms reflect most often the psychologist's implicit theory. Thus philosophers of psychology have a real opportunity to have an effective and desirable impact on the field of experimental psychology. To be effective, however, philosophers must both know the work in psychology and have philosophical sensitivity to issues of theory and experiment. With this knowledge and sensitivity they can contribute substantively to the development of an adequate theoretical model that incorporates the necessary variables, and they can contribute to the examination and development of experimental paradigms used to gather data and test theories.

For completeness it should be noted that in addition to the variables discussed in this chapter, an adequate theory must include a developmental model to show how different aspects of an organism's psychological functioning change over time. Moreover, an adequate theory will have to deal with the role of affect and emotions (after the manner discussed in Section 9.1) and explore how this dimension of human experience has its place in discussing issues of perception and cognition.

The student interested in philosophy of psychology should become aware of the many other areas that exist to be explored. Current fascinating and good work relates psychology to the information sciences and artificial intelligence, as well as to neurobiology and physiology. New approaches to theoretical and experimental issues are occurring in psycholinguistics (the psychology of language and reading), the study of reasoning and problem solving (both how children learn to solve problems in specific domains and the difference between naive and expert problem solvers), and the nature of the categories of thinking (or how people structure their world cognitively and linguistically). These are some of the most exciting research areas in a field that abounds with intriguing possibilities.

DISCUSSION QUESTIONS

1. Briefly describe and then critically assess the strengths and weaknesses of the three major types of psychological theories: behaviorism, cognitivism and realism.

2. Need it be the case that schemas, scripts, or the structures that carry on cognitive processing are linguistic in character? Another way of asking this is, is thought a language?

3. Gibson's realism seems to work well for certain cases such as looming or depth perception. These perceptual abilities also seem to be evolutionarily salient for most organisms. What perceptual abilities beyond those that might have been selected by evolution do you think a realistic theory can explain?

4. Discuss how a general model might be constructed that would incorporate the ''good parts'' of the three major types of psychological theories.

5. It is sometimes said that anger or jealousy makes a person see red. How would we begin to think of integrating the role of affect or emotion in perception or cognition?

6. How are explanations in psychology the same and/or different from those in biology or physics? from those in the social sciences?

7. Cognitivism, with its emphasis on what is constructed or added to perception and cognition by the human being, has been taken as evidence for an epistemological relativism. Different people with differing backgrounds will see things and think things differently. Therefore, any person's way of thinking or seeing is as good as any other person's. What is wrong with this argument, and what does this suggest about the limits of a purely constructivist theory?

SUGGESTED READING

Among the many fascinating recent books that have appeared treating these subjects I will recommend for further reading only a few:

BODEN, MARGARET A. (1988), *Computer Models of Mind: Computational Approaches in Theoretical Psychology*. Cambridge, England: Cambridge University Press.

CHURCHLAND, PATRICIA SMITH (1986), *Neurophilosophy: Toward a Unified Science of the Mind/Brain*. Cambridge, MA: Bradford/MIT Press.

FODOR, JERRY A. (1968), *Psychological Explanation: An Introduction to the Philosophy of Psychology*. New York: Random House.

JOHNSON-LAIRD, P.N. (1983), *Mental Models: Towards a Cognitive Science of Language, Inference, and Consciousness*. Cambridge, MA: Harvard University Press.

KOSSLYN, STEPHEN MICHAEL (1980), *Image and Mind*. Cambridge, MA: Harvard University Press.

LAKOFF, GEORGE (1987), *Women, Fire, and Dangerous Things: What Categories Reveal about the Mind*. Chicago: University of Chicago Press.

Ten

ANDROID EPISTEMOLOGY: COMPUTATION, ARTIFICIAL INTELLIGENCE AND THE PHILOSOPHY OF SCIENCE

Clark Glymour

Can a computer make scientific discoveries, design scientific experiments, and provide scientific explanations? When real scientists do these things do they use procedures that could be implemented in a computer program? Can computers be programmed to be *better* than human scientists at making some kinds of scientific discoveries? Until very recently, almost every philosopher of science gave a resounding "no" in answer to each of these questions. Philosophers of science frequently dismissed these possibilities by claiming that there is no "logic" of scientific discovery, or that there are no "rules" for discovery, or that there is no means by which a computer could introduce novel theoretical concepts not already specified in its program. However, over the last twenty-five years, work in the subjects of artificial intelligence and cognitive psychology has done a good deal to undermine these negative conclusions. Computer science is teaching us that computers can be programmed to do remarkably better scientific work than philosophers had thought possible, and cognitive psychology is teaching us that practicing scientists often do less well than philosophers had imagined.

Artificial intelligence is the attempt to design, implement and test computer programs that will solve problems intelligently. The problems may consist of playing winning chess, diagnosing appendicitis, answering questions about a story, discovering physical laws, extracting causal relations from large data bases, or any other cognitive task of interest. Contemporary computational cognitive psychology does something different but closely related. In cognitive psychology the aim is to form correct descriptions of the computational procedures by which people reason, think, remember, recognize, solve problems, learn or perform any other cognitive task.

The difference between artificial intelligence and cognitive psychology is that the second subject makes an empirical assumption and has empirical goals that artificial intelligence does not. Artificial intelligence can be pursued without any assumption that people think by computation, and without any claim that any program describes how people actually work. Nonetheless, artificial intelligence and computational cognitive psychology share a natural affinity because both involve thinking about how intelligent action *could be* brought about through computation.

The surprising and perhaps ironic fact is that the development of the digital computer owes a great deal to logic and the philosophy of mathematics, and furthermore much of the work in artificial intelligence which suggests that computers can do good scientific work was made possible by using ideas from the very philosophers of science who rejected the possibility. One result is that certain parts of contemporary computer science—especially the specialty known as *machine learning*—may properly be viewed as philosophy of science carried on by other means. This chapter reviews some of the philosophical history that led to the subject of artificial intelligence, and then describes a few problems and programs that should be of particular interest to philosophy of science.

10.1 SOME HISTORICAL BACKGROUND

Although the word derives from Arabic mathematicians, the notion of an *algorithm* seems as ancient as any idea in mathematics. Greek geometers thought of algorithms as procedures for constructing figures with rule and compass. Thus, in translation, the first proposition of Euclid's elements states *a procedure* using rule and compass for constructing an equilateral triangle on any given line segment in any given plane containing that line segment. But it was in the theory of arithmetic and in algebra that the idea of an algorithm became explicit. Current teaching of arithmetic in elementary schools consists in large part of the teaching of algorithms for computing simple arithmetic functions: addition, subtraction, multiplication, and division (although the availability of cheap electric calculators may make such skills obsolete). In algebra, much of the focus of research by Arab and then by Christian mathematicians was on finding algorithms for the solution of various classes of equations. Thus in algebra we learn an algorithm (or perhaps several algorithms) for solving linear equations and for solving systems of linear equations, algorithms for solving quadratic equations, and an algorithm for solving third-degree equations.

As we will see, the idea of an algorithm is not even yet fully understood, but roughly it means a mechanical or automatic procedure that will, at least ideally, compute something for any of an infinity of different possible cases. If the something is a function, an algorithm will compute the value of the function for any argument upon which the function is defined. Such an algorithm is always an algorithm for computing some specific function or other; there might be many ways to compute one and the same function, and so many distinct algorithms can exist that compute one and the same function.

The idea that machines can carry out procedures to help people reason about numbers is as old as civilization, but the idea that machines can help people reason about subjects other than numbers seems more recent. One early and influential formulation of the idea is due to Ramon Lull, a thirteenth-century Spanish monk. Lull developed a strategy that he thought would convert Moslems to Christianity. His idea, in effect, was that unbelievers need help with their reasoning processes. Lull thought that if followers of Mohammed were able to see the many, many combinations of Christ's virtues, they would be converted. Suppose, for example, that only 20 of Christ's virtues are considered; then an easy calculation shows that these 20 virtues can be formed into 190 distinct *pairs* of Christ's virtues and 1,080 distinct *triples* of Christ's virtues. Far too many combinations, Lull thought, for humans to recognize without the aid of a machine.

Lull's machines consisted of two or three disks of different radii, all rotating on the same axis or spindle. Each of the disks could be rotated independently of the others. The disks were divided into the same number of sections, each section labeled with a letter or name indicating one of the virtues. By rotating the disks different pairs or triples of virtues would line up with one another and could therefore easily be recognized. Lull made several trips to North Africa with his machines to convert Moslems to Christianity. On his third trip, past the age of 80, he was stoned to death.

In the seventeenth century, partly under the influence of the Lullian tradition, Gottfried Leibniz, the great philosopher, mathematician and diplomat, proposed the development of a kind of *calculus* in which every proposition on any subject could be stated formally and unequivocally, and for which an algorithm would determine the truth or falsity of every proposition that could be so stated. Of course Leibniz did not have such a calculus, or any such algorithm, but he envisioned their development as a principal goal of science. In the same period a remarkable idea began to emerge: Thought, any thought, is some form of computation. Thomas Hobbes articulated this idea clearly in the middle of the century. It remained for an eighteenth-century French physician and philosopher, Julien de La Mettrie, to give the clearest statement of the thesis that thought is mechanical. La Mettrie ([1815]1988) argued that consciousness is subject to physical laws, emphasized the similarities of men and animals, especially apes, and speculated that apes might be taught language.

The mechanical view of mind was influential in the nineteenth century, especially among physiologists. Nowhere did that view receive a clearer or more prescient formulation than in the writings of the young Sigmund Freud. After studying with an antimaterialist philosopher, Franz Brentano, and a materialist physiologist, Ernst Brucke, in 1895 Freud wrote a long essay entitled "Project for a Scientific Psychology" ([1895] 1966a). Freud speculated that the cells of the nervous system pass excitation from one to another through the synaptic connections, that the connections can be altered to make the passage easier or more difficult, that facilitation of the passage of excitation depends in part on how often the synaptic connection has previously passed activation, and that the entire system of nerve cells acts as a mechanical network that acts to minimize its long-run average activation. All of human reasoning, planning and desiring arises through the mechanics—the computations—of this system. Today very similar methods of computation are called *neural net* models, and are widely studied.

Freud speculated about the computational structure underlying the mind, but he had no definite idea about algorithms or programs. The first clear attempt at an algorithmic description of fundamental processes of mind came from another, equally surprising source: the young Rudolf Carnap. Unlike Freud's, the system Carnap ([1928] 1967) developed was not intended as an empirical description of the computations by which humans actually do form various categories and concepts. Rather, Carnap's aim was to show how a plethora of concepts—including those of color, sound, shape, object, space and time—*could be* constructed algorithmically from a list of facts of a very simple kind. Carnap's theory was a design for the knowledge categories of a possible sort of person. It was, we might as well say, epistemology for androids. The input or base to Carnap's system was a list of facts asserting that one entire moment of experience bore some similarity or other to another entire moment of experience. No other features of the ''elementary experiences'' were assumed by the system other than facts about their similarity relations to one another. Using Frege's logical methods, Carnap defined classes of elementary experiences that were similar to one other, classes whose members all contained some one common feature (''quality classes'' Carnap called them), series of qualities corresponding to various sensory modalities, and so on. Carnap was able to do so in some measure because he sought entirely structural features characteristic of various concepts, such as the concept of color. For the first few stages of this construction Carnap attached to every definition a procedure or ''fictive operation'' as he called it. The operations are perfectly computable, and the entire system can be viewed as an algorithm that constructs one after the other from an input list of facts about the similarities of a finite collection of ''elementary experiences.'' Carnap wrote the first artificial intelligence program.

Carnap had another influence on artificial intelligence through his students, Walter Pitts and Herbert Simon. In the 1950s Pitts was one of the people responsible for the revival of neural net models of computation. In roughly the same period, Simon began the project of trying to make digital computers behave intelligently, and in fact he is one of the people responsible for the term ''artificial intelligence.'' Before any of that could happen, though, electronic digital computers had to be invented, and before *that* could happen, people had to get a better understanding of the nature of computation itself. The understanding of computation arose partly from logic and partly from technical responses to problems in the philosophy of mathematics.

10.2 THE DEVELOPMENT OF THE THEORY OF COMPUTATION[1]

The theory of computation belongs as much to philosophy as to other disciplines. The modern theory of computation arose through the application of mathematical logic to philosophical issues concerning the foundations of mathematics. Moreover, at least a rudimentary understanding of the theory of computation is essential to understanding important areas of contemporary philosophical concern besides artificial intelligence, including theories of knowledge and philosophy of mind. Finally, the theory of computation forms one important limiting part of the modern theory of rationality.

---

[1] This section contains some challenging material. Later sections do not presuppose it.

In the nineteenth century Cantor developed the theory of sets and produced a number of perplexing results. Cantor's theory provided the *existence* of certain transfinite sets, and of infinite collections that are, in a well-defined sense, *larger* than the infinity constituted by the natural numbers. Cantor's mathematical work, and the theories that developed from it, set off a line of research that led to the modern theory of computation. Some mathematicians, notably Kronecker, did not believe Cantor's mathematical results, and did not even regard them as real mathematics. Although the criticisms of Cantor's work were sometimes clothed in nineteenth-century metaphysical jargon, one theme in the criticisms was as follows: *The principles of arithmetic are the most certain and sure foundation for mathematics, and any new mathematical theory must be proved consistent by the methods of arithmetic, and any new mathematical objects or functions should be computable from elementary arithmetic functions, that is, there should be arithmetic algorithms for any such novelties.* Kant's philosophy, which gave a special status to arithmetic, provided part of the support for this view.

At about the same time that the conflict between Cantor and Kronecker developed, Gottlob Frege had published the first presentation of modern logic. Frege's logic permitted the explicit formulation of nearly every mathematical theory, from number theory to geometry to set theory. The power of Frege's conception had already been exploited by many mathematicians and philosophers as early as the 1920s. David Hilbert, who ranked among the greatest of mathematicians of the late nineteenth and twentieth centuries, saw in formal logic a means of meeting Kronecker's objection to set theory, and indeed of reducing the question of the consistency of any mathematical theory of questions of elementary arithmetic.

10.2.1 Hilbert's Program

One of Hilbert's ideas can be derived from a simple reflection on how mathematics is applied in the physical sciences (perhaps it is relevant that Hilbert was also a great mathematical physicist, and, with Einstein, the codiscoverer of the general theory of relativity). When we apply mathematics, we count things, or we assign numbers to things or to states of things. We assign numerical measures to objects when, for example, we weigh them. By counting and assigning measures, we are able to make inferences about things *by doing arithmetic*.

Consider weighing things on a scale. We do so by adopting some *convention* that correlates states of the scale with numbers. The same number is also correlated with the body we have weighed on the scale. The convention we use may be built into the scale (as with modern chemical scales that have a digital readout) or we may have to mentally assign a number to the state of the scale according to some rule (as with an old fashioned pan balance). There are many different systems by which numbers can be associated with states of the scale—we can measure in grams or ounces or some other unit—but the important thing is to use one such system; it doesn't matter which.

When we use a scale and a standard of measurement to assign weights to objects, and we weigh first one object and then another on the scale, we use our measurement convention to assign a number to the state of the scale in each case. We call that number the *weight* of the object on the scale. Now the interesting and useful thing about measurement scales is this: We can use arithmetic to determine what the state of the scale will be when, for example, both objects are placed on it. To get the answer we need do only the following: (1) weigh each object separately; (2) add the numbers representing the weights of the two objects; (3) use our measurement convention to infer the state of the scale that is associated with the number that is the sum. Our measurement practice enables us to represent properties of weights as simple arithmetic relationships.

Hilbert's idea was that we can associate numbers with the *language of a mathematical theory* in such a way that properties of the mathematical theory, such as its consistency, are represented by arithmetical relationships among numbers associated with parts of the language. *Formal* properties of a theory, such as its consistency, then become equivalent to arithmetical properties of sets of numbers. We could then prove (or disprove) the consistency of a mathematical theory by using nothing but arithmetic. Since no one doubted arithmetic proofs, doubts about the consistency of various mathematical theories, such as set theory, could be resolved. (One important disanalogy should be noted. An *empirical* fact, established by experiment, is that weights are additive, that is, the weight of two objects together is the sum of their individual weights. By assigning numbers to weights, that empirical fact is represented by an arithmetic relation. In contrast, the *consistency* of a theory is a *logical* property, not an empirical property, and one hopes to be able to establish consistency by an a priori proof rather than by experiment.)

When a mathematical theory—set theory, for example—is completely formalized the vocabulary of the language of the theory can be assigned *natural numbers* in any arbitrary but mechanically computable way. The sentences of the language of the theory can then also be assigned numbers, because each sentence is just a sequence of vocabulary elements. Further, because we insisted that our formalized languages be constructed so that an algorithm determines whether any arbitrary string of vocabulary elements is a well-formed formula, some arithmetic algorithm determines the numbers that correspond to well-formed formulas of the language of the formalized mathematical theory under study.

Let us suppose that when formalized the mathematical theory we are considering is axiomatizable. The axioms of the theory are then in correspondence with some determinate set of numbers. Since there is an algorithm for determining the axioms of the theory, there is a numerical algorithm for determining the numbers that correspond to the axioms.

A proof, or derivation from the axioms of the formalized theory, then corresponds to a finite sequence of numbers, according to a proof theory such as Frege's. Since one of the requirements of proof theory is that there be an algorithm to determine whether any given sequence of formulas is a proof, there will also be a numerical algorithm that decides whether a sequence of numbers is the sequence of a proof. The number of the last sentence in a proof is a number of a theorem of whatever theory is being considered.

A theory is said to be **consistent** if and only if there is no sentence A such that both A and $\neg$ A ("denial of A") are theorems of the theory. Whether a specific theory, say set theory, is consistent thus becomes equivalent to a question that is purely about arithmetic: whether there are, or are not, pairs of sequences of numbers of the appropriate kinds corresponding to a proof of some sentence and its denial, both within the same theory.

If for a formalized theory there exists an algorithm that determines whether or not any given sentence is a theorem of the theory, then it would seem that the spirit of Kronecker's objections would be fully met. One could prove a theory consistent by purely arithmetic—or as Hilbert put it, by "finitary"—means. One could do so by

1. formalizing the theory;
2. arithmetizing its language;
3. proving the existence of a computable number theoretical function for deciding whether an arbitrary sentence is a theorem;
4. showing that the function has the property that no sentence and its denial are both theorems of the theory whose consistency is in question.

Hilbert called this program **metamathematics.** It is mathematics done on the *language* of mathematics.

10.2.2 Gödel's Theorems

In 1931 a young Viennese logician, Kurt Gödel, proved two theorems that were understood to mean that Hilbert's program as originally conceived could not succeed. Gödel's theorems used a generalization of Cantor's diagonalization strategy. Gödel applied Hilbert's program to arithmetic itself. If the goal of the Hilbert program were to succeed, then we could decide the question of consistency of any axiomatizable theory by using nothing more than arithmetic. So we could in principle formalize the theory of arithmetic itself, and in formalized arithmetic we could represent the sentences of the language of *any* formalized axiomatic theory—represent them as collections of numbers, or special functions on the natural numbers. Then in formal arithmetic we could give a formal proof of the consistency or inconsistency of the formalized theory, whatever it might be.

Gödel proved two extraordinary theorems. The first theorem implies that arithmetic itself cannot be represented as an axiomatizable formalized theory. That requires some explanation. In the nineteenth century Peano had developed an axiomatic system for arithmetic. In their first order formulation Peano's axioms are infinite in number, but they can be formalized and they form a set of sentences for which there is an algorithm to determine membership, so Peano's axioms could be the object of arithmetic study as in Hilbert's program. In fact, we can formalize Peano's axioms for arithmetic, and then in the language of that formal theory assign a numeral (a term in the language of the theory) to each sentence of that very language. To understand Gödel's first theorem it is necessary to recall the definition of a *complete* theory: A first order theory is complete if and only if, for every sentence s in the language of the theory, either s is the theory or $\neg s$ is in the theory.

Gödel's first theorem says the following:

Gödel's Theorem: No axiomatizable formal theory that is true in the natural numbers is complete.

What does this mean? Like any other structure, the structure determined by the natural numbers, $N = [N, +, x, s, <, =]$, determines a complete theory, call it T, namely, the set of all sentences true in the structure N. Theory T includes all of the sentences that are logical consequences of the formalized counterparts of Peano's axioms, but T contains much else besides. Gödel's theorem says something about theory T: It is not axiomatizable. To say that the theory is not axiomatizable is to say something about the *nonexistence of algorithms*. It is to say that there does not exist any possible algorithm that assigns 1 or 0 to sentences such that the set of sentences assigned the value 1 consists of a set of axioms that entail all of the sentences true in the natural numbers (that is, in the structure $[N, +, x, s, <, =]$), and of no other sentences.

Gödel's theorems actually say more than that the complete theory of arithmetic is not axiomatizable. They say that *there is an algorithm* that, given as input a finite description of any axiomatizable theory, T that is true in N and that entails Peano arithmetic, will produce as output a sentence that is true in N but is not a theorem of the theory.

Why should we care whether the theory of the natural numbers or other complete theories that extend Peano's arithmetic are axiomatizable? For at least two reasons. One is that Hilbert's program supposed we could axiomatize the theories that interest us; Gödel's result says that we can't always do so since the complete theory of the natural numbers is, for example, a theory that very much interests us. However, there is a second philosophical reason why we should care that the theory of the natural numbers cannot be axiomatized. *For if a formalized theory cannot be axiomatized then there exists no possible algorithm that will decide for each sentence of the language of the theory whether that sentence is in the theory.* For if there were such an algorithm we could let the set of all sentences which the algorithm says are in the theory be itself the axiomatization of the theory. Actually, something stronger is true. If a theory is not axiomatizable then there is not even an algorithmic means to *list* the theorems of the theory; there is no mechanical procedure that will, every now and then, output some sentence that is a theorem of the theory, never outputting a sentence that is not in the theory, and for every sentence of the theory, eventually output it. We have no effective means of specifying a theory that cannot be axiomatized.

We all know an algorithm that, for any natural numbers n, l and m, will determine whether or not $n + l = m$. The procedure is just the addition algorithm we learn in elementary school. We know algorithms that will determine the answers to other classes of arithmetic questions—for example, whether an arbitrary number is prime. So we might hope that there is an algorithm that will answer *every* question of arithmetic. Such an algorithm would decide the truth or falsity in the natural numbers of any proposition we might choose to put to it. Such an algorithm would be part of the fulfillment of Leibniz's dream. Gödel's theorem says that no such algorithm exists; it is not just that we have not found such an algorithm yet. Rather, *no such algorithm is logically possible.*

Now the notion of an "axiomatizable" theory was defined in terms of the notion of an "algorithm," or alternatively in terms of a *computable* function (i.e., a function for which there is an algorithm) assigning the number 1 to formulas that are in an axiom set and the number 0 to formulas that are not in the axiom set. As Gödel's theorem has been stated, it is therefore about the nonexistence of certain kinds of algorithms, or about the fact that certain functions are *not computable*. To prove his claims, Gödel therefore had to characterize the computable functions, and that effort, quite as much as his astonishing theorems, led to the development of computation theory. Rather than describe Gödel's characterizations of the computable functions as what are called the *recursive functions,* we instead consider the characterization provided shortly after by Alan Turing.

10.2.3 Turing Machines

Turing conceived of a kind of machine whose operations are so simple that to think them in any way mysterious would be absurd. Further, he wanted his machines to compute functions. The idea is that the machine somehow represents a function on the natural numbers, so that if in a fixed code we enter the representation of any number or finite sequence of numbers, the machine gives back the value of the function for that number or sequence of numbers. The coding of numbers can be done in many ways. Any number base provides a system of digits whose finite sequences encode all numbers. We are used to representing numbers in base 10, with 10 digits; but we could just as well represent numbers in base 2 or any other base. Given a finite sequence of digits in some specified base, Turing's machines would compute another sequence of digits in that base, representing the value of a function for the number input.

The machines Turing conceived can read a square of a tape on which some number has been written; the machine has a finite list of instructions that tell the machine, when it is in a particular state reading a particular square of the tape, to erase the digit on the square and write some other digit in its place, or to move on to the next square to the left, or to the right, on the tape, and to change its state. The tape is unbounded; that is, should the machine reach the end of the tape, more tape is always added. The machine starts with its reading and writing device over a blank square of tape, and all of the squares to one side of the read-write head, say to the left, are blank. A finite number of squares on the other side, to the right, contain symbols, representing the input to the machine, and the rest are blank. A computation is carried out with such a machine by starting the machine in its start state. The machine reads a square and does its work reading, writing, moving, sometimes to the right, sometimes to the left, changing its internal state as it does its work. Eventually the machine may stop, and when it does some sequence of symbols will be written on the tape. That sequence is the output of the machine for the number input.

Behaviorally, a Turing machine can do just three things:

1. It can erase a symbol on a square of tape and write another symbol in its place;
2. It can move the read-write head one square to the right;
3. It can move the read-write head one square to the left.

Internally, the machine can only do one kind of thing: *change its state.*

Such machines could be physically realized in many ways. The tape could be a magnetic tape, or even a section of paper tape; the read-write head of the machine could be a simple optical scanner connected to a printer in one movable unit. The instructions in each state could be implemented by cogs, pulleys, and ropes; by electronic tubes as in early digital computers; or by silicon chips.

There is nothing special about numbers save that certain functions on the numbers form our principal and clearest examples of computable functions. but we know that given any finite alphabet, the finite sequences of letters from that alphabet, or *words* as they are called, can be systematically coded as numbers, so functions from words to words can be represented as numerical functions. Thus we could just as well have Turing machines that have some finite vocabulary other than a system of digits, and compute functions from words on that vocabulary to words on that vocabulary.

We can think of the instructions, or program, of a Turing machine as a system of annotated points and lines: Represent each distinct state of the machine as a point or node, and draw an arrow from one node to another if some inscription on a tape square will cause the Turing machine when in the first state to go into the second state. Annotate the arrow with the symbol the machine must read when in the first state to go into the second, and with the symbol it writes or the direction it moves. Thus a machine that operates on the vocabulary {B, 0, 1}, where B represents a blank square of tape, and which changes every digit to 1, can be pictured as in Figure 10.1.

The machine starts at square 1, which is blank by convention, in state 1. Then it moves one square to the right, and changes to state 2. If the square to the right of square one is blank (no input) the machine *halts* and does nothing further. If that square has a 0, the machine writes a 1 in its place, and stays in state 2 and moves one square to the right. If that square has a 1, the machine moves to the right one square and stays in state 2.

To describe the sequence of states of the machine and tape, we let the first number represent the number of the state of the machine, the second number represent the number of the square the read-write head is over, counting the square underneath the read-write head initially as 1, and counting positively to the right, and then we let the remaining sequence of numbers be the sequence of digits on the tape. For example:

INPUT TAPE: B1001

SEQUENCE:

1,1,B1001B

2,2,B1001B

2,3,B1001B

2,3,B1101B

2,4,B1101B

2,4,B1111B

2,5,B1111B

2,6,B1111B

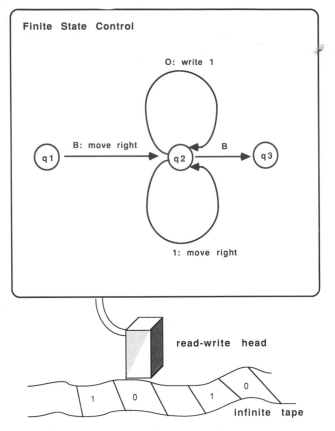

Figure 10.1 A Turing machine.

A Turing machine can also be described as a precise mathematical object, specifically as a finite set of quadruples of numbers. The first number of such a quadruple simply names the state of the machine corresponding to that four-tuple. (Since the quadruples each specify what the machine does when it is reading a square in a certain state, for most machine states and each possible digit that can occur on a square, there will be a distinct quadruple.) The second number of the quadruple is a possible digit that can occur on a square, the third number is a digit that can be written on a square or a number representing "move right" or a number representing "move left"; and the fourth number is again the number naming a state of the machine. Informally, $<n_1 \ n_2 \ n_3 \ n_4>$ is the instruction: "If in state n_1 reading digit n_2, erase n_2 and write n_3, or move one square to the right if n_3 equals the special value, or move one square to the left if n_3 equals the other special value, and go into state n_4.

By convention we specify that the state $n_1 \ = \ 1$ is the *start state* of the machine.

We can define the *computation* executed by a Turing machine given by a finite set of quadruples as a sequence of *instantaneous states of the machine* in the way illustrated in the previous example. Using the definition of a Turing machine as a finite set of quadruples, we could provide a precise inductive definition of a Turing machine computation, but that is left as a (difficult) exercise.

Turing proved that a single machine will compute every function that can be computed by any Turing machine. Currently such machines are referred to as *Universal Turing machines*. The idea is this. Since all Turing machines can be enumerated, we can form some encoding that assigns each Turing machine a number. Then we can design a Turing machine that interprets the first number on its tape as the number of a Turing machine, some conventional sequence of digits following that number as a space marker, and the following number on its tape as the input to the machine represented by the first number on its tape. Then our Universal Turing machine simulates the computations that the Turing machine named on its tape would do for the input number on its tape. So there is, in a sense, one universal algorithm that will do anything any algorithm can do.

10.2.4 Church's Thesis

It seems indisputable that any function computed by some Turing machine or other should count as a computable function. Besides Turing machines, we have many other ways in which we could try to model the notion of computation. For example, we could imagine a machine with an unlimited number of registers, in each of which a number can be written. Suppose we are allowed to write any finite sequence of instructions of any of the following kinds:

$r_i := 0$ (Set the number in the ith register equal to zero)

$r_i := r_j$ (Set the number in the ith register equal to whatever number is in the jth register)

$r_i := r_i + 1$ (Set the number in the ith register equal to its successor)

Goto (i,j,k) (If $r_i = r_j$ then jump to the kth instruction in the list; otherwise go to the next instruction.)

A program for such an *Unlimited Register Machine* (URM) consists of a finite set of instructions. Input is given by numbers in a finite set of the registers, with the convention that any register that does not have a number in it as input is assumed to have the value zero in it unless given another value in the course of the computation. Output is given by the number in the first register when the machine stops.

A computation with an URM proceeds from a first program line and the initial state of the registers. The registers are changed in accordance with the program line and a new state of the registers and program line results (which will be the next program line, or possibly some other program line if the present line has a "Goto" instruction). Thus, just as with Turing machines, an URM computation can be described as a sequence of finite lists where each list gives the program line and specifies for each i the value of r_i, the number in the ith (nonempty) register. As with Turing machines, some URM machines may never halt for certain inputs.

Theorem: A function is computable by some URM machine if and only if it is computable by some Turing machine.

The characterization of the computable functions as the recursive functions, as the functions computable by Turing machine and as the URM computable functions are all equivalent in the sense that exactly the same class of functions satisfies all three descriptions. The computable number theoretic functions have many other characterizations. The philosopher and logician Alonzo Church, who gave the very first characterization (one different from any of those we have described) of this class of functions, formulated the following thesis:

Church's Thesis: The computable number theoretic functions are the Turing computable functions.

Church's thesis is not a mathematical theorem; it is rather in the nature of a proposal. The proposal is that, in view of the coherence of several conceptually very different approaches to computability in characterizing one and the same class functions, in view of the evidence Turing's work provides that the functions in this class are indeed computable by very simple machines, and in view of the fact that every function on the natural numbers that anyone is sure is computable turns out to be Turing computable, we should simply regard the computable functions as those computable by some Turing machine.

Since we know how to reduce functions defined on finite sequences of objects of any specified collection of types to functions on numbers, Church's thesis has broad implications. It is not just about the computability of functions on the numbers; it is also about the computability of functions on whatever can be counted or enumerated by a computable one-to-one mapping or coding of objects to natural numbers. Through the device of characteristic functions which assign the value 1 to all members of a set and the value 0 to all members of its complement, the computability of (countable) sets can be reduced to the computability of numerical functions. Since the extensions of properties and relations are sets, the computability of such extensions, provided they are countable, is also reduced to the computability of numerical functions.

Church's thesis could be taken more broadly as the claim that whatever can be computed can be reduced to the computation of a recursive function. If the world is in some respects continuous, as our physical theories assume, then we can use physical systems to compute functions defined not just on natural numbers but on the system of real numbers, or at least the system of rationals. Again, we might find physical processes that compute characteristic functions for uncountable sets. There have been attempts to characterize the notion of computability for real valued functions without reducing such computations to computations on the natural numbers.

10.2.5 Decision Problems

Every Turing machine is a finite sequence of quadruples. The collection of all *finite* sequences of numbers can be enumerated, and so can be put in one-to-one correspondence with the natural numbers themselves. So if we fix a vocabulary for input and output, the collection of all Turing machines for that alphabet can be enumerated, and in fact the enumeration can be done in a way that is intuitively computable. This is one way to see something interesting about the set of all computable functions: Since, by Church's thesis, the set of computable functions is just the set of functions that can be computed by Turing machine, and since the set of Turing machines is countably infinite, the set of all computable functions on the natural numbers is countably infinite. Since by Cantor's results the set of *all* functions on the natural numbers is uncountably infinite, this means that the computable functions form only a tiny fragment of the set of all functions on the natural numbers. And that means something very important:

> *For many functions and for many sets and properties and relations for which we wish to have algorithms, it may be the case that no algorithms are possible.*

Let us consider an example of an uncomputable function. Consider the function $f(m,n)$ that assigns 1 to the pair (m,n) if m is the number of a Turing machine that halts on input n, and assigns 0 otherwise. Is this function computable? It can be proved that it is not.

The halting problem gives us one example of a function that is not computable. There are many others. Consider the problem of deciding where an arbitrary first order formula is valid. A decision procedure for that problem is some algorithm for computing a function that assigns 1 to a formula if and only if it is a valid first-order formula, and 0 otherwise. Clearly there is such a function, but is it a function for which there exists an algorithm? Is it a computable function? The answer is no. *There is no algorithm that will determine for us whether an arbitrary first order formula is valid.* That means also that there is no algorithm that will determine for us whether an arbitrary first order *argument* is valid: We cannot have a mechanical procedure that takes as input a finite list of formulas and determines for us whether the set consisting of all but the last member of the list entails the last member of the list.

The philosophical consequences of this result, in combination with Church's thesis, are enormous. It means, for example, that the idea of solving all mathematical problems by an algorithm is hopeless; it means that Leibniz's vision of a calculus in which all of knowledge can be expressed and its consequences obtained algorithmically must fail. It also means that a rich and difficult question emerges which only makes sense given that first order validity is not decidable: Which first order theories *are* decidable? That is, which sets of sentences are such that there is an algorithm that will determine, for any sentence, whether it is a consequence of the set. Put another way, theories are deductively closed collections of sentences. Which such collections have a computable characteristic function? These questions form what is known in logic as *the decision problem*.

Consider the theory of numbers. Part of what Gödel proved is that the set of theorems of Peano arithmetic is not decidable. Consider the set of valid Boolean formulas. That set is computable.

Consider any first-order theory that is both axiomatizable and complete. The set of sentences in that theory is computable, and the theory is said to be **decidable.** It is easy to see that a complete axiomatizable theory is decidable, for given any sentence in the language of the theory, either that sentence or its denial is in the theory. Since the theory is axiomatizable, there is an algorithm that will decide whether a sentence is in an axiom set for the theory. We can also effectively enumerate all finite sequences of sentences in the language and computably determine, for any such sequence, whether it is a proof of axioms of the theory. For any sentence in the theory such a proof will exist. Thus we can computably enumerate all of the proofs from axioms of the theory until we find a proof either of a given sentence or of its denial. If we find a proof of its denial our procedure reports that the sentence is not in the theory; if we find a proof of the sentence our procedure reports that the sentence is in the theory. So while much that we might wish to compute is not computable, many important functions, properties and relationships are computable. The theory of elementary Euclidean geometry, for example, is axiomatizable and complete, and hence there is an algorithm that will decide for us for any sentence in the language of geometry whether the sentence is a theorem of Euclidean geometry.

These examples scarcely touch the intricate structure of the decision problem for first-order theories, a problem that is still an active area of research.

10.2.6 What Is a Computation?

The theory of computability began with two questions that have not been answered: What is a computation? What is an algorithm? Rather than answering either one of those questions, the development of the theory of computation proceeded by providing *specific* computational systems, and then characterized the computable as whatever can be computed in any one of these several systems. But this strategy does not give us any *general* characterization of a *computational system,* and so it does not give any general characterization of the notion of a computation. Nor does it give us a general account of the notion of an *algorithm*. Each specific computational system gives us the notion of a *program*—for example, the list of instructions of an URM program, or the first number on the input to a Universal Turing Machine—but we don't have any characterization of when it is that two programs in *different* computational systems—say an URM program and a Universal Turing Machine program—are or are not implementations of one and the same algorithm.

We can get a taste of the variety of alternative computational systems by considering a few examples. The programming languages, PASCAL and LISP, when implemented on a computing machine, form a computational system. Moreover they are computational systems that are capable of computing any recursive function provided the machine memory can be increased whenever more is needed.

Consider what can be done with Turing machines. Rather than have a single tape on which the input, the intermediate work, and the output must all be done, we could consider a machine designed like a Turing machine but having several tapes: one for input, say, one for output, any number for intermediate computations. For any number of tapes k the class of k-tape Turing machines can compute exactly the Turing computable functions, that is, the class of functions computable by a one-tape machine.

A different sort of computational system is obtained if we introduce probabilities into Turing machines. The fourth entry in each quadruple in the mathematical description of a Turing machine is the name of a specific machine state. The Turing machine goes from one state to another specific state at each stage. We can make a *stochastic Turing machine* by replacing the fourth entry in each quadruple by a *probability distribution* over machine states. (We could also make what a machine *writes* in a given state undetermined). The idea is that a stochastic Turing machine, when in a certain state reading a certain digit, writes or moves and then throws a die or does something else random to determine what state to go into next. (We can say that a machine of this kind computes a given function f provided that for every number n in the domain of the function, the machine gives $f(n)$ with probability greater than $1/2$.) A surprising fact is that the class of functions computable by stochastic Turing machines is exactly the class of Turing computable functions.

Not every imaginable computational system will be capable of computing the Turing computable functions. To the contrary, many systems can only compute a more restricted class of functions. For example, consider computational systems called *finite state automata*. A finite state automaton looks like the graph representing the machine state transitions of a Turing machine. There are a set of nodes, including a unique distinguished initial node, and a nonempty set of "final nodes." Each node has a certain number, say k, of arcs from it to other nodes (including possibly to itself), where k is the same number for all nodes, and where the k-arcs out of a node are given k distinct labels.

A finite state automaton executes a computation by taking a finite string of labels as input; the automaton begins in the initial state, and then follows the arc with label corresponding to the first element of the input string, then the arc with label corresponding to the second element of the input string, and so on. The automaton is said to *accept* the input string s if it ends up in one of the final states. Accepting string s is the same as computing a function f that assigns 1 a string if it is accepted and 0 to it otherwise. In representations of finite state automata a final state is represented by a double ring. (See Figure 10.2.) Some Turing computable functions cannot be computed by any finite state automaton.

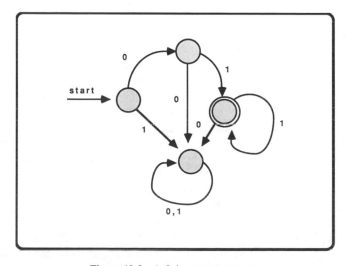

Figure 10.2 A finite state automaton.

Computational structures such as finite state automata have an important role in representations of problems that humans might have to solve. Consider, for example, the task of making a telescope lens from a glass blank. The blank must be ground, polished and aluminized in order to make it into a lens. These actions—grind, polish, and aluminize—each affect the state of the glass. Moreover, in order to obtain a suitable lens, the operations must be applied in the correct order. If you aluminize the glass, then polish it, then grind it, the result will not be a telescope lens but simply a glass ground to the correct shape. In a problem such as this, the various states of the glass are represented by nodes of a finite state automaton, and the alternative actions are each represented by a label (for example, g, p and a). There is a start state, representing the untreated glass blank, and a final state, representing the desired state achieved when the appropriate sequence of actions is taken. Each node or state has three arcs coming out of it, one arc for each action that could possibly be applied to the glass blank in the state represented by the node.

We can also consider *stochastic finite automata,* in which each label corresponds to a probability function connecting one node with a set of other nodes.

Other computational systems are modeled, at least crudely, on the neural linkages of the brain. For example, *parallel distributed processing* or *connectionist* machines consist of a system of nodes connected by arcs. Each node can be on or off. Whether a node is on is, save for input nodes, a stochastic function of the on-off state of the nodes immediately connected to that node. Some of the nodes can record input; that is, the external environment determines whether they are on or off. Other nodes are designated as output nodes, and their final or equilibrium state after the system has been given some input is the output of the system.

Clearly the range of systems that can intelligibly be called "computational" is very large and diverse; we have no characterization that we can argue provides necessary and sufficient conditions for something to be a computational system.

10.2.7 Complexity

If you are given a problem to solve, it may be easy or hard, depending on your abilities and your knowledge. If you happen already to know the answer to the problem (say you saw the answer sheet), then the problem is very easy. If you don't, solving the problem may require a lot of work. So the intuitive, informal notion we have of the difficulty of a problem makes difficulty a *relation* between persons and tasks; a task isn't easy or hard in itself, but only *for* some person.

Now there is an obvious notion of the difficulty of a computational task, a notion that is likewise relational. Suppose, for example, that we have some total function and a Turing machine that can compute that function. If we give a number to the Turing machine as input, we will get an output. In computing the output, the Turing machine will go through a certain number of steps; more exactly, if we write down the sequence of instantaneous descriptions of the Turing machine as it goes through the computation, that sequence will have some definite number of members. We can use that number, whatever it is, as a measure of the effort the computation required of the machine. If each step requires the same amount of time, the measure can be thought of as a measure of the time required for the computation.

Evidently, for one and the same input, different Turing machines may require different numbers of steps. Even machines that compute the very same function may differ in the number of computational steps they require. Given any specific Turing machine, we can always form another Turing machine that computes the very same function as the first one but requires more steps for some inputs. In fact, we can always find a machine that requires more steps for *every input*. We simply have to tack on extra initial states and state transitions that do nothing.

To say roughly how difficult it is for a particular Turing machine to compute its output for a specific input is easy. But we can say something more general? Suppose we have two Turing machines that compute one and the same total function f. Is there some way to *compare* the difficulty they have in computing the function? There is.

For each machine T we look at how the number of computational steps the machine takes *varies* with the size of the input given to the machine. Suppose, for example, we are representing numbers in some base, so each number is represented by a sequence of digits. We can measure the *size* of an input by the number of digits it contains. For any given size value, only a finite number of inputs of that size are possible.

For inputs of size 1, any particular machine T will require a number of computational steps. Machine T might require 10 steps if the input is 0, 15 if the input is 1. Again, for inputs of size 2, T will require a number of computational steps, say 12 for 00, 13 for 10, 15 for 11 and 30 for 01. In principle we could make a table, Table 10.1, listing the possible inputs of each size, and the number of computational steps that T requires for each input:

TABLE 10.1

Input	# of computational steps required by T
0	10
1	15
00	12
10	13
01	30
11	15

Let us get the pieces straight. We are considering a Turing machine T which computes a function $f(x)$; we have a measure $s(x)$ of the size of inputs; we have a measure $c(T,x)$ of the computational cost for T to compute a value from input x.

Now if we have two different Turing machines, say M and N, each of which compute the same function, we can compare the two functions $c(M,x)$ and $c(N,x)$. It might be, for example, that for every input size, $c(M,x) > c(N,x)$. Or it might be that for all but a finite number of inputs, $c(M,n) > c(N,x)$. Or it might be that $c(N,x)$ is never greater than $c(M,x)$, but $c(M,x) > c(N,x)$ infinitely often. Each of these conditions is distinct, but in each case we would be inclined to say that the computation is more difficult for machine N than for machine M. Of course, it might turn out that none of these conditions obtain, and that we can say no more than that for some inputs M has it easier and for other inputs N has it easier.

Our measure of the computational cost is not very precise. It might be that in a real machine some steps require more time than others; it might be that in a real machine some steps get faster (or slower) if they are repeated. But we have at least a crude way of comparing the time requirements of different machines, and thus of comparing the difficulty they have in computation.

Here is a more difficult issue. We have made some sense of the notion of the difficulty of a problem as a relation between a task and a problem solver. We even made some sense of the difficulty or complexity for the infinite set of tasks involved in the capacity to compute a function: Each input presents what we might call a problem instance, and the problem as a whole is to be able to compute the value of a given function on every input. That task may be more difficult for some Turing machines that compute the function than for others. We have not yet characterized the *intrinsic difficulty* of a problem. We know that some functions are simply not computable, but of computable functions we have as yet found no way to say that some are intrinsically more difficult to compute than others. We have not yet found a way to clarify difficulty as a *property* of computational tasks themselves rather than as a relation between a problem and a problem solver. We can do so.

For each input size there is an input for which the number of computational steps that T requires is the largest, or at least as large as any other. For inputs of size 1, the biggest value of the number of computational steps T requires for inputs of that size is 15; for inputs of size 2 the largest value is 30. So we could construct another table:

TABLE 10.2

Size of input	Largest number of steps T requires for input of that size
1	15
2	30

and we have a function, call it $W(T,s)$ which measures the computational cost for T of the most difficult input of size s. The function $W(T,s)$ enables us to begin to talk more systematically about the difficulties of computation. In general, $W(T,s)$ will increase as s increases, but not always.

Now for a given Turing machine, $W(T,s)$ may be a ragged function that is not easy to describe. But we can ask about functions that *bound* $W(T,s)$. That is, we can ask whether a given well-behaved function of s is always greater than $W(T,s)$. For example, we can ask whether a function $g(s) = as + c$, where a and c are constants, is such that $g(s) > W(T,s)$ for all s. If that were true for some a and c, then we would say that $W(T,s)$ is *linearly bounded*. Or we might ask: Is there *any* polynomial function of s, call it $P(s)$ such that for all s, $P(s) > W(T,s)$? If there is, we say that $W(T,s)$ is *polynomially bounded*. Or, we might ask: Is there any exponential function, call it $E(s)$, such that $E(s) > W(T,s)$ for all s? If so, $W(T,s)$ is *exponentially bounded*.

If $W(T,s)$ is roughly a polynomial function, then it will be **polynomially bounded**, and if it is roughly exponential than it will be **exponentially bounded** but not polynomially bounded.

Android Epistemology

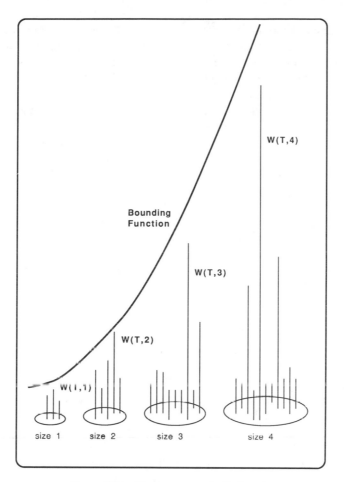

Figure 10.3 Worst-case complexity bounds.

Recall that the issue is whether there is a way to describe how difficult it is to compute a computable function; we are interested in a notion of difficulty that is not a relation between the function to be computed and the machine that computes it, but is instead a property of the function itself. We know that if we have a Turing machine that computes a function with a worst-case computational difficulty given by the function $W(T,s)$, then we can always build a Turing machine that has a harder time of it. (See Figure 10.3). By adding extra steps to T that do nothing useful, we can always find a machine T' such that $W(T,s) < W(T',s)$ for all s. *But there is no guarantee that for an arbitrary f we can find a Turing machine such that $W(T,s)$ is polynomially bounded.* Whether such a Turing machine exists depends on the function f. Similarly, whether there is a Turing machine that computes f such that $W(T,s)$ is polynomially bounded depends on f.

We say that a computable function is *computable in polynomial time* if there is a Turing machine T that computes f and such that $W(T,s)$ is polynomially bounded.

We say that a computable function is computable in exponential time if there is a Turing machine T that computes f and such that $W(T,s)$ is exponentially bounded. Clearly every function computable in polynomial time is also computable in exponential time, but the converse is not true. This classification of computable functions orders them by their intrinsic difficulty.

If a function is computable in exponential but not in polynomial time, then we may expect that no Turing machine will offer a feasible means of computing the function. Every Turing machine that computes such a function will for some inputs require exponentially increasing time as the problem instances become larger. We can see what happens with a simple example. Suppose $W(T,s)$ is of the order of 10^S. Then for the most difficult instances of each size the time required increases as in Table 10.3 and Figure 10.4.

Are there any interesting functions that are computable but that are of exponential class? A great many functions whose computation is of enormous practical importance *so far as we know* are in this class. Since our interest is principally in the theory of rationality, consider an example germane to that theory. We know that there is an algorithm that, given any well-formed Boolean formula S compounded of n variables, will determine whether S is satisfiable; that is, no assignment of truth and falsity to the variables of S will have the effect of assigning truth to S. In principle we could implement any such algorithm on a Turing machine. But every known algorithm for this problem requires a number of steps that increases exponentially as n increases.

Something more remarkable is true. Suppose we fix the number of simple sentences or sentential variables at any number $k > 2$, and suppose we measure the size of a problem by the length (that is, the number of symbols occurring) of a sentential formula. Then every known algorithm that decides the consistency of Boolean formulas (that is, whether there exists an assignment of truth values to the variables of a formula that makes the formula true) requires computational time that, in the worst case, increases exponentially with the size.

There are thousands of other problems for which every known algorithm is worst-case exponential (if implemented on a Turing machine). Some of them are simple. A **graph** is any nonempty collection of nodes or vertices, some pairs of which may be connected by lines. A pair of vertices that is connected is said to be *adjacent*. Consider the following problem: Determine for any graph whether each of its vertices

TABLE 10.3

s	$W(T,s) = 10^S$
1	10
2	100
3	1,000
4	10,000
5	100,000
6	1,000,000
7	10,000,000
8	100,000,000
9	1,000,000,000

Android Epistemology

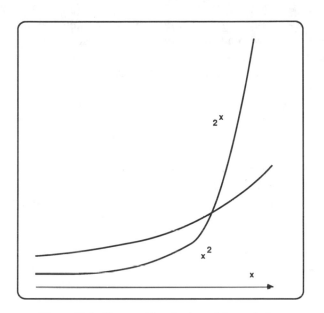

Figure 10.4 Exponential and polynomial complexity.

can be given one of three colors in such a way that no two adjacent vertices have the same color. So far as we know in the worst case, the number of steps any Turing machine will require to solve this problem increases exponentially.

We have ignored an important question: What is special about Turing machines? That a function is of exponential class may be a property of that function, but it is an uninteresting function if some computational system other than Turing machines can compute the function in polynomial time. What about multitape Turing machines, or Turing machines with several read-write heads, or URM machines, or random access memory machines like ordinary computers? The particular computational bound (for the worst case) for a function depends on the class of machines considered. For example, if a function is computed by a two-tape Turing machine with a worst-case time bound $W(T,s)$, then some one-tape Turing machine will compute the same function with a worst-case time bound of $W(T,s)^2$. Moreover, there are functions such that any Turing machine that computes them requires (up to a multiplicative constant) the square of the time a two-tape machine requires.

The fact that different kinds of computational systems will have different worst-case bounds for one and the same function suggests that there is no intrinsic measure of the complexity of a function that is independent of the computational system considered. While that is true, it is not as serious as it seems, since some important distinctions *do* appear to be invariant. For example, whether a function is computable in polynomial time seems to be invariant over all familiar computational systems. We cannot claim this invariance as a mathematical fact since we do not know exactly what the class of computational systems includes, but every computational system we know of seems to have this property.

Another important question we have ignored has to do with our focus on the most difficult case in assessing the complexity of a computational task. Some algorithms used every day without difficulty are worst-case exponential. The practical success of these algorithms is not due to the fact that the inputs given to them are small, but rather to the fact that the computationally difficult cases of any large size are *very rare*.

Probability and decision theory suggest an alternative way to measure the intrinsic difficulty of computing a function. For a given function f and a given size measure s, and a Turing machine T that computes f, let us consider the *average number of computational steps T requires for inputs of size s.*

Denote the average number of computational steps T requires for inputs of size s by $E(T,n)$, and call it the *expected complexity* of T for inputs of size n.

Let $E(T,n)$ be regarded as a function of n. Now we can ask about any computable function, f, and any function $g(n)$ (e.g., a linear function or an exponential function, and such), whether there exists a Turing machine such that $E(T,n) < g(n)$. This mathematical apparatus enables us to compare expected complexity and worst-case computational complexity for one and the same function. If a function has a very low expected complexity it may in practice be feasible to compute the function even though its worst-case complexity is very high. In fact, that sometimes turns out to be the case.

Consider again the graph 3-coloring problem discussed earlier. Every known algorithm for deciding the problem is worst-case exponential. But there is an algorithm for solving the problem that has *constant* expected complexity. The constant is not even very large: 192. What happens is that as the size of graphs increases, the proportion of graphs whose three colorability is difficult to assess decreases exponentially. While there are always hard cases, they become increasingly rare.

10.3 ARTIFICIAL INTELLIGENCE

Immediately after World War Two the theory of computation was given a practical realization with the development of stored-program electronic digital computers. It was obvious to the designers of the first digital computers that the machines could be used to simulate and in some respects surpass human intellectual abilities. The first attempts at computer programs that would learn and solve problems appeared in the 1950s. Currently artificial intelligence is a diverse subject with more or less established subareas of study, almost all of which have some relation to the philosophy of science. One large area of study is computer vision, which aims to transform images into information about the environment. Another important area of research is called "knowledge representation." Philosophers of science have long been concerned with how to make scientific theories as clear and explicit as possible. The study of knowledge representation is concerned with the structures in which beliefs should be represented and organized in order to make feasible fast, computable, relevant inferences from those beliefs. Still another area of research is computer planning: If the computer must design an experiment or plan a procedure to make a telescope, certain actions must be taken in a certain order, and which actions should be taken may depend on circumstances not entirely under the computer's control. How can such planning be carried out in a computer? The area of artificial intelligence most closely connected with philosophy of science is *machine learning,* the subject which tries to find algorithmic procedures for discovery. Machine learning has a purely formal counterpart: computational learning theory. Rather than trying to give an overview of all of artificial intelligence, or even of machine learning, let us concentrate on just a few programs for scientific discovery.

10.3.1 The BACON Programs

One common sort of scientific inference consists of fitting curves to data. The curve graphically depicts a functional dependency among the measured variables, and that functional dependency often incorporates an empirical law, or approximate law. Boyle's law was obtained in this way; so were Kepler's laws, the Rutherford-Soddy law of radioactive decay, and many other empirical regularities of physics, chemistry, and biology. Sometimes a scientist working on a particular subject already knows the form of the functional dependency. Thus electrochemists expect, on theoretical grounds, to get "S" shaped curves when they do their measurements. We speak to the form of a functional dependency when we mean that the functional dependency is known to belong to a parametrized class of dependencies. For example, $y = ax + b$ is a parametric representation of the class of linear dependencies of y on x, with a and b as parameters; for any real numerical values we give to a and b, $y = ax + b$ becomes a linear equation. If the scientist knows that a functional dependency belongs to a certain parametric family, then experiments or observations need only determine the values of the parameters. That kind of problem is the subject of statistics. The scientist's problem is more difficult if the functional form is unknown. That was Boyle's problem, and Kepler's as well. Their task was not so much to estimate parameters as to figure out, first, that parameters needed to be estimated.

The BACON programs, developed at Carnegie-Mellon University by Pat Langley and Herbert Simon, address the problem of discovering a functional dependency when the form of the functional dependency is not known; Kepler's third law provides an example. The law asserts that for any two planets, the ratio of the cube of their (mean) distance from the sun to the square of their periods is the same. Kepler did not know initially that the dependency was of this form; he inferred the conclusion from the data. Let us leave aside the historical issue of how Kepler did it, and consider how, by examining distances and periods of planets, this regularity could be discovered.

One way we might discover Kepler's third law from data about the planets is as follows. For each planet, we have measurements of its distance D and its period P. So for each planet we calculate the ratio D/P. Now, since we have values for D/P and for D, for each planet we can multiply the value of D/P by the value D and obtain a value of D^2/P. Now, since we have values of D/P and of D^2/P for each planet, we can multiply them together and obtain values of D^3P^2. If we do these steps, with data from the planets, we find that for each planet the value of D^3/P^2 obtained is nearly the same as for every other planet. That is Kepler's third law. Table 10.4 gives an illustration.

TABLE 10.4

Planet	D	P	D/P	D^2/P	D^3/P^2
A	1.0	1.0	1.0	1.0	1.0
B	4.0	8.0	0.5	2.0	1.0
C	9.0	27.0	0.333	3.0	1.0

This is all well and good, but it is not a procedure: How would we know which combinations of D and P to multiply together so that we are led to Kepler's law?

The simplest BACON program, BACON.1, uses the following search procedure in which the program is told initially which variables are independent and which are dependent. In the Kepler example, D and P are each functions of the planet's identity, so they, and any terms formed from them, are counted as dependent variables:

1. If the absolute values of a term X increase as the absolute values of a term Y increase, and the X and Y values are not linearly related, compute the values of the term X/Y;

2. If the absolute values of a term X increase as the absolute values of a term Y decrease, and the X and Y values are not linearly related, compute the values of the term XY';

3. If the values for a term X and the values for a term Y are linearly related with the slope m and intercept b, output $Y = mX + b$;

4. If a dependent term X has a constant value c output $X = c$.

One of the lessons to be learned from the discussion of scientific realism is that we ought to consider the reliability of any proposed discovery procedure. Where will BACON.1 succeed and where will it fail? We can see immediately that it is confined to discovering algebraic dependencies that can be expressed as linear functions of no more than two terms. Each of these terms must be a product of powers of measured variables. So BACON.1 can find only those functional dependencies among quantities $Q_1, \ldots, Q_k$ that are mathematically equivalent to an equation of the form:

$$aQ_1^n Q_2^m \ldots Q_k^p + bQ_1^r Q_2^s \ldots Q_k^t = c$$

where a, b, and c and any real numbers and n, m, p, r, s, t, and so on are integers, either positive or negative. (In fact, the implementation of BACON.1 limits k to 2). Of course many functional dependencies occur in the natural sciences that are not of these forms. The Rutherford-Soddy law, for example, involves exponential functions; many laws involve logarithmic or trigonometric or other functions.

Consider the ideal gas law, $PV = nRT$, where P is pressure, V is volume, T is absolute temperature, n is the number of moles of gas and R is the gas constant. If one tried to discover this law from measured values of P, V, T, and n, using the BACON.1 heuristics, the following difficulty would arise. The values of P and V, because they depend on n and T, neither increase nor decrease systematically with one another. So BACON.1 would not form the product PV, and so would not find the gas law. A natural remedy for this limitation suggests itself. Suppose the program is modified so that it groups together values of some variables when values of other variables are fixed. For example, the program could look at values of P and V when n and T are given a pair of fixed values, and it could look at another set of values of P and V when n and T are given a different set of fixed values. Suppose these various subsets of measured values are treated as separate discovery problems. So for fixed values of n and T, but varying values of P and V, the program uses BACON.1 heuristics to look for a relationship between P and V. What the program would find is an instance of Boyle's law, $PV = K_1$, where K_1 is a constant. On another set of values of P and V, with different fixed values of n and T, the program would find another instance of Boyle's law, $PV = K_2$, where in general K_1 would not equal K_2. So the program would discover certain laws, and could then use these laws in discovering others. How could that be done? The program treats the value K_1, K_2, and so on, as values of a new quantity K. Then, when the program looks at data in which P, V, and K vary, but T is fixed, the term PV/K will be formed, and then found to be equal to T (using BACON.1 heuristics). Further, comparing the values of K and n, the program can in the same way discover that $K = nR$, where R is a constant.

What is happening in the process we have just described is a procedural version of the bootstrapping relation discussed in the chapter on confirmation. The procedure locates certain laws in specialized cases, and then uses them to permit the calculation of new quantities that in turn lead to the discovery of further, more general, laws. This procedure is implemented in the BACON.3 program. It permits the discovery of laws that involve several distinct variables, and its scope, the range of problems in which it will discover the truth, is more difficult to analyze exactly.

A later version of the program, BACON.4, introduces two novel kinds of discovery problems. One of those problems is simple to illustrate. Ohm's law says that the voltage (V) in an electrical circuit is proportional to the product of the current (I) and the resistance (R) in the circuit. Suppose you have a device (an ammeter) for measuring the current in any circuit, and you have a supply of wires of different compositions and lengths, and a supply of different kinds of batteries. You have no way to measure resistance and you have no way to measure voltage. Worse, you do not even know that there are such properties. How might you nonetheless discover Ohm's law? Suppose for concreteness that you have three batteries A, B and C, and three wires, X, Y and Z.

You could do the following. Choose a battery, say A, and hook it up to each of the three wires in turn. You find that the three circuits contain different currents:

TABLE 10.5

Battery	Wire	Current	(I)
	A	X	3.4763
	A	Y	4.8763
	A	Z	3.0590

What varies in this situation is the wires and the current, not the battery. So it is some difference in a property of the wires that produces the differences in current.

Call this property c. You could let the current itself (when the wires are hooked up to battery A) measure this property of the wires. Then the wires will (you hope) have property c even when they are not hooked up to battery A, indeed even when they are hooked up to battery B or battery C. The assumption is that the property measured by hooking them up to battery A is a dispositional property of the wire itself. In fact, adopt the following heuristic: Whenever a dependent quality X varies systematically with a "nominal" variable V (that is, a nonnumerical variable such as battery, or wire or planet), when a set of other variables is held constant, let the values of X measure a property of V measured under those conditions, provided no such property of V has already been defined.

TABLE 10.6

Battery	Wire	Current (I)	c
A	X	3.4763	3.4763
A	Y	4.8763	4.8763
A	Z	3.0590	3.0590
B	X	3.9781	
B	Y	5.5803	
B	Z	3.5007	
C	X	5.629	
C	Y	7.8034	
C	Z	4.8592	

Suppose now you apply the heuristics of BACON.3 to the subproblem given by the first three rows of this table. Then you will find that I and c are linearly related, with slope $v = 1.0$ and intercept $b = 0$. Of course this is no news because you defined the measure of c so that it would be true. But if you apply the same heuristics to the subproblem given by the second three rows of Table 10.6 (which is an exercise to fill in) you will find that I and c are again linearly related with $v = 1.1444$, and similarly the subproblem consisting of the final three rows gives a linear relation with slope 1.6003.

So, proceeding as BACON.3 does, you record the values of v as a new quantity, and obtain Table 10.7:

TABLE 10.7

Battery	Wire	Current (I)	c	v
A	X	3.4763	3.4763	1.0
A	Y	4.8763	4.8763	1.0
A	Z	3.0590	3.0590	1.0
B	X	3.9781	3.4763	1.1444
B	Y	5.5803	4.8763	1.1444
B	Z	3.5007	3.0590	1.1444
C	X	5.629	3.4763	1.6003
C	Y	7.0834	4.8763	1.6003
C	Z	4.8592	3.0590	1.6003

Now if you fix the wire, say to X, you find that v (but not c) varies systematically with the nominal variable "Battery" and is constant if the value of "Battery" does not change. So assume that v is a property of batteries. You have established the law, $I = vc$. If you identify c as the "conductance" or reciprocal of resistance of a wire, the law is equivalent to $V = IR$, or Ohm's law.

The procedure we have sketched is carried out automatically by the BACON.4 program. It is a procedural combination of ideas from the theory of measurement together with ideas about testing hypotheses. Moreover, the last problem we considered, discovering Ohm's law without measures of voltages or current, is an example of an interesting sort of discovery problem that is common in the sciences.

Consider any real (or rational, or integer as the case may be) valued function of n-tuples of nominal variables. In the circuits considered previously, for example, current I is a function of each pair of variables for the nominal pair (battery, wire). In general we have $F(X_1, \ldots, X_n)$. Let F be equal to some compositions of functions and subsets of the nominal variables. For example, I(battery, wire) = V(battery)*R(wire), where * is multiplication. A discovery problem consists of a set of functions on subsets of tuples of nominal variables, and for each tuple and set of functions, a function that is a composition of (i.e., some function of) that set. The learner's task is to infer the decomposition of values of the composite function.

Much of clever science consists in solving instances of problems of functional decomposition, and thus discovering important but initially unmeasured properties. The properties of discovery problems of this kind, and of algorithms for solving them, are almost completely unstudied.

The BACON.4 program has another interesting procedure that is associated with a novel kind of discovery problem. If BACON.4 finds that a series of values (or a series of reciprocal values) for measures of a property of a nominal variable all have a common divisor under appropriate conditions, then the program will associate that division, whatever it is, with the nominal property. Consider what happened in the history of chemistry with Cannizzaro and vapor densities. About 1860, Cannizzaro found that the fractional weight of hydrogen in every vaporized unit volume of any hydrogen compound is divisible by one-half the density of hydrogen vapor. He thus concluded that one-half the density of hydrogen vapor measures a property of hydrogen, the atomic weight. From parallel data, he drew similar conclusions about other elements. Computer program BACON.4 could reproduce Cannizzaro's reasoning.

10.3.2 The Significance of the BACON Programs

What general morals should we draw from the Bacon programs? They certainly are not descriptions of all of the methods that physical scientists use in finding regularities in their data. The programs have a limited scope. They will not find many of the forms of equations that commonly occur in the physical sciences, for example, a simple logarithmic dependency. Many of the functional forms that occur in physics are most naturally viewed as solutions to differential equations. The natural logarithm, *ln x*, for example, arises in the equation $y = ln\ x$ that is one solution of the differential equation:

$$dy/dx\ -\ 1/x\ =\ 0$$

Furthermore, in physics at least, the mathematical component of a theory is often a differential equation, or system of differential equations; Newton's theory of gravitation, for example, is a parametrized system of differential equations. The DELPHI system, developed by John Norton of the Department of History and Philosophy of Science at University of Pittsburgh, infers differential equations directly from data. Rather than taking particular values of *x* and *y* and inferring, say, a logarithmic dependency, the DELPHI procedure is to take the particular values and infer a differential equation directly. The procedure uses standard statistical techniques and mathematical properties of differential equations and their solutions.

The BACON programs are not of great practical value in autonomously discovering new relationships because they are very sensitive to errors in the data. In effect, the programs cannot make very informed decisions about how "noisy" the data are, and they do not calculate how error is propagated through calculations of the value of one quantity from others. Improvements in these respects have been undertaken by Jan Zytkow, a philosopher and computer scientist.

We might ask the following question: If the BACON procedures are not a description of how scientists actually discover laws, and if they are not of practical use in discovering new laws, what good are they? What is their point? Their importance lies in demystifying scientific discovery, in refuting some common claims about that process, and in suggesting a change in perspective in the philosophy of science. First, the programs suggest that an important part of the process of scientific discovery can be mechanized; even if human scientists do not use the BACON heuristics, or do not use them in the way the program does, the success of the programs lends credence to the notion that in theoretical reasoning scientists are using some sort of computable heuristics. Second, Carl Hempel, the distinguished philosopher of science, has claimed that no computer program could discover laws or theories which required novel properties or entities not already described in the data or in the computer's program. But the BACON.4 program does that very thing in using regularities it finds to postulate new, unmeasured properties of objects and in proposing measures of these new properties. Third, in carrying out a sort of automated bootstrapping, the BACON procedures suggest that the fundamental thing in understanding science as a cognitive activity is not confirmation or explanation, but *discovery*. Rather than focusing on confirmation relations or explanation relations, those who want to understand the power and limits of science might do well to focus on procedures for forming conjectures, and on the reliabilities of those procedures. Of course determining confirmation relations or explanatory relations might be a *stage* of a discovery process—a cog in the machine. The BACON procedures, for example, make use of bootstrapping ideas, and we will see that philosophical ideas about scientific explanation have played a role in other artificial intelligence systems.

10.4 EXPERT SYSTEMS AND BAYESIAN NETWORKS

In many practical cases we want human experts to make some sort of inference based on their knowledge. We want physicians to make diagnoses and recommend treatments to patients, chemists to determine the structure of unknown compounds, clinical psychologists to predict whether candidates for parole are likely to commit new crimes, and so on. One important idea about the use of artificial intelligence is that machines can be made to serve as substitutes for human experts. By putting the knowledge of human experts into the computer and giving the computer rules for making inferences, machines might be made to diagnose, identify compounds, predict recidivism or do other inference tasks as reliably and more cheaply than human experts. Programs of this kind are called *expert systems*.

10.4.1 The DENDRAL Program

One of the first and most impressive expert systems is the DENDRAL program. The aim of the program is to identify the structure of unknown chemical compounds. Many pairs of distinct chemical compounds are *isomers;* a molecule of one isomeric compound has the same number of each kind of atom as does a molecule of the other compound, but the atoms are differently arranged. For example, both butane and isobutane have the same molecular formula, C_4H_{10}, but the atoms are differently arranged in the two molecules. In butane, the four carbon atoms are arranged in a line, but in isobutane one carbon atom is bonded to all three of the others, like a star.

One technique chemists use to determine which isomeric structure describes an unknown compound is *mass spectrometry*. In a mass spectrometer a compound is bombarded by an electron beam. By breaking chemical bonds within the molecules, the beam breaks many of the molecules of the compound into pieces. After bombardment, the material is passed through a magnetic field. While the original compound has no net electrical charge, the pieces formed by breaking bonds may be charged, although only rarely with more than a unit positive or negative charge. Elementary physics teaches that a charged particle moving through a magnetic field will be subject to a force which will accelerate the particle. From Newton's second law, the greater the mass of the particle, the smaller its acceleration by any given force. In the mass spectrometer, the acceleration of a particle causes it to move in a curved trajectory. The result is that charged fragments having a small mass tend to move one way and charged fragments having a large mass tend to move another way. By recording the intensity of fragments emerging in different places, the mass spectrometer tells the chemist the relative proportions of fragments of various masses formed when the electron beam hits the molecules of the original compound.

How can the chemist use this information? Consider the example of butane and isobutane again. It seems comparatively unlikely that any one molecule has more than one chemical bond broken by the electron beam. In butane there are three bonds between the four carbon atoms. Breaking either of the bonds on the end will leave a fragment with one carbon atom and a fragment with three carbon atoms; but breaking the middle bond will leave two fragments each with two carbon atoms. In isobutane, however, as long as only one bond between carbon atoms is broken, no matter which, the result will be a fragment with one carbon atom and a fragment with three carbon atoms.

$$
\begin{array}{rcl}
\text{C} \mid \text{C - C - C} & \rightarrow & \text{C} + \text{C - C - C} \\
\text{C - C - C-} \mid \text{C} & \rightarrow & \text{C - C - C} + \text{C} \\
\text{C - C} \mid \text{C - C} & \rightarrow & \text{C - C} + \text{C - C} \\
\end{array}
$$

$$
\begin{array}{ccc}
\quad\ \overset{\displaystyle \text{C}}{\underset{\displaystyle |}{}} & & \overset{\displaystyle \text{C}}{\underset{\displaystyle |}{}} \\
\text{C} - \text{C} - \mid - \text{C} & \rightarrow & \text{C - C} + \text{C}
\end{array}
$$

By considering the relative proportions of two carbon fragments to one and three carbon fragments, the chemist can make a good guess about whether the unknown compound is butane or isobutane.

Real problems are often much more complicated. In principle, the chemist has to figure out every possible isomer that is consistent with the molecular formula and with whatever else the chemist knows about the compound. Then the ratios of fragments of various masses to be expected from each isomer must be calculated (or guessed) and compared with the data from mass spectroscopy. In practice chemists probably do not literally carry out this procedure in full detail except in very simple cases. The DENDRAL program does. The procedure it uses involves a general algorithm for finding the isomers of any molecular formula from organic chemistry. The program includes a theory about which bonds will break in an electron beam, depending on the kind of atoms involved in the bond and the nearby atoms. The theory is applied to a candidate isomer to derive predictions about the mass fragments that should be seen, and the isomer is given a confirmation score according to how well the predictions fit the mass spectroscopic data. The procedure makes use of Hempel's ideas about explanation and of hypothetico-deductive ideas about confirmation. This is no wonder since one of the principal workers on the project, Bruce Buchanan, was trained in the philosophy of science.

10.4.2 Bayesian Networks

The DENDRAL program is an example of what is often called a *rule-based* expert system. The program applies an algorithm for generating an exact class of structures, and after that it makes deductive inferences using a theory and assigns scores. No probabilities are calculated, and no probabilistic inferences are made. Another way of constructing expert systems uses probabilistic calculations and depends on applications of Bayes's Theorem.

The Bayesian conception of inquiry supposes that the investigator has probabilities for all propositions of interest. These probabilities include conditional probabilities for any hypothesis, given any evidence. Recall that the conditional probability of hypothesis H on evidence e is defined in terms of unconditional probabilities by the following formula:

$$\text{Prob}(H/e) = \text{Prob}(H\&e)/\text{Prob}(e)$$

In the Bayesian concept of inquiry, the investigator starts with an initial probability assignment (or probability distribution) to all propositions of interest. Upon making observations, investigators learn something new, say that e is the case. Upon learning e, investigators revise their probability assignment by giving each proposition the conditional probability of that proposition on e. As more evidence is obtained, the probability assignment is revised, and so on. Eventually, we hope investigators will give more and more probability to the truth, and less and less to whatever propositions are in fact false.

In this simple form, the Bayesian conception of inquiry does not produce conclusions. The Bayesian theorist does not typically conclude that one hypothesis is true and all others false (although in especially simple cases such conclusions are possible). In some sense, however, we often produce conclusions through action. A medical doctor must decide whether to run expensive, uncomfortable, and perhaps risky tests on a patient; the doctor must sometimes make a decision as to the correct diagnosis and proceed to treat the patient based on that diagnosis. How does the Bayesian conception take account of such things? In two ways.

First, some (but not all) Bayesians allow that conclusions may be detached. If the probability of a hypothesis is high enough, and if appropriate further conditions are met, these methodologists allow that one may treat the hypothesis as though it has probability 1.

Second, many Bayesians place the study of procedures of inquiry within a larger conception called decision theory. The idea is that we must sometimes make decisions about action—for example, a physician must sometimes decide which of a set of alternative treatments to give a patient, or which of a set of alternative tests to run on the patient. The consequences of any actions will depend on which hypothesis is true. Giving someone a particular antibiotic may cure infections and cause no harm; but if the patient is allergic to the antibiotic, that treatment may cause a great deal of harm. Bayesian decision theory assumes that we know what the consequences of each possible action will be if any particular hypothesis is true, but we do not know which particular hypothesis is true. Each possible consequence of the action has a value which is measured by a utility function. The utility of not having an allergic reaction is, for example, higher than the utility of having one. For each possible hypothesis about the state of things there is a probability, and for each outcome there is a utility. So for each alternative action, say action A_i, and each alternative hypothesis, say H_j, there is a utility $U(O_{ij})$ for the outcome that will result from the action if the hypothesis is in fact correct.

The expected utility of an action is the weighted average of the utilities of that action under the various hypotheses. Suppose the alternative hypotheses are H_1, H_2, and H_3. The expected utility of action A_i is defined as:

$$\text{Exp}(A_i) = \text{Prob}(H_1)U(O_{i1}) + \text{Prob}(H_2)U(O_{i2}) + \text{Prob}(H_3)U(O_{i3}).$$

The fundamental principle of Bayesian decision theory is that a decision maker should do whatever action has the highest quality expected utility.

In the 1960s, Bayesian methods were used in expert systems for medical diagnosis. Gorry and Barnett (1968) developed a computer program that performed diagnosis and made decisions about whether to conduct further laboratory tests on a patient. The program was applied to problems of diagnosing congenital heart defects. A variety of related programs were developed about the same time. Gorry and

Barnett's program works as follows. Thirty-five different congenital heart disease conditions were considered, and 53 different possible findings, including 25 distinct kinds of heart murmurs, 11 X-ray findings, 7 EKG findings, 6 photocardiographic tracing findings, and three age groups. Clearly some of these findings would appear as data only if specific tests, such as an X-ray, were ordered. The program started with probabilities for each of the heart disease conditions, and with conditional probabilities for the findings given each of these hypotheses. The program was provided with values of the utility (or disutility) of a misdiagnosis and the costs (or disutilities) of various tests.

In running the program, the system was presented with initial findings about the patient. These initial findings were used to compute a new conditional probability for the various hypotheses. As soon as the program calculates the new probabilities from the initial evidence, it is faced with a decision; it must either report one of the 35 possible diagnoses or it must run an additional test. The program is also provided with some arbitrary utilities: The cost of a misdiagnosis is set at 1000 for all misdiagnoses. The program can compute for each of 35 conditions the expected utility of making the diagnosis that the condition obtains. By a straightforward calculation, the program can also compute, for each possible test, the expected utility of making a diagnosis after it has obtained the results of such a test. The program takes whatever action has the highest utility. Thus depending on the initial probabilities and findings, the program will typically ask for some specific further tests, but not for others, because at each stage it calculates whether the improvement in confidence of the diagnosis that would be obtained with the further finding is worth more than the cost of the test. Eventually, however, the program reaches a stage in which the cost of further tests is not worth the value of the information they would provide. In that case, the program gives out a diagnosis based on the expected utility principle—but since all misdiagnoses have the same utility, when making the decision the program always makes the diagnosis that has the highest probability after conditioning on the evidence and the test results.

Early Bayesian programs have largely been replaced by an expert system design known as a *Bayesian network*. In a Bayesian network various quantities or properties of interest are represented as nodes. The quantities may have any of several values, and even the properties have at least two values, representing the presence or absence of the property in the system. In a Bayesian network two nodes may be connected by a directed edge, or arrow, from one to the other, or they may be unconnected. So for each node N there is a set of nodes with edges directed into N, called the *parents* of N. Often the directed edges are understood to mean that one variable causes another, but sometimes they are not. (See Figure 10.5.)

The computational rule for the probabilities in Bayesian networks is very simple: Starting with the terminal nodes which have no directed edges out of them, the computer is given a probability for each value of the terminal node *conditional* on each set of values of all of its parents. For each parent of a terminal node, the computer is given the probability of each of its values conditional on each set of values for all of *its* parents, and so on until nodes are reached that have no parents. Nodes with no parents are just assigned probability numbers for each of their possible values. The probability of any set of values for any of the nodes can be calculated from these numbers.

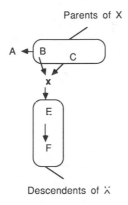

Parents of X

Descendents of X

Figure 10.5

A Bayesian network for variables relevant to emergency medicine is shown in Figure 10.6. (Beinlich et al. 1989). Called the ALARM network, it was put together by a team of physician computer scientists, who also assigned probabilities and conditional probabilities to the various nodes.

Such networks have many uses. If for a particular person or unit the values of some of the variables are known, the network can be "updated" for that individual by using Bayes's rule to compute the conditional probabilities of all of the other variables given the values of the known variables. The task of updating is known to be very computationally demanding in the worst case (under a widely held assumption, its worst-case complexity is at least exponential), but it is feasible

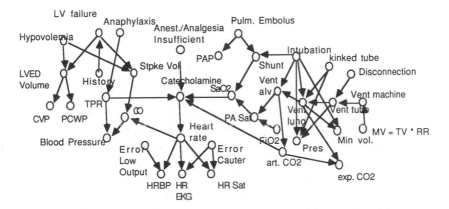

Figure 10.6 CO: cardiac output. CVP: central venous pressure. LVED volume: left ventricular endiastolic volume. Lv failure: left ventricular failure. MV: minute ventilation. PA Sat: pulmonary artery oxygen saturation. PAP: pulmonary artery pressure. PCWP: pulmonary capillary wedge pressure. Pres: breathing pressure. RR: respiratory rate. TPR: total peripheral resistance. TV: tidal volume.

when the network is sufficiently sparse so that most pairs of variables are not directly connected with one another. When the edges in the network accurately represent causal dependencies, the network can also be used to predict the probability distributions of some variables when other variables are manipulated or forced to have specific values.

10.4.3 The Limitations of Experts

The original idea in expert system design was to extract a theory, or rules, or probabilities from human experts and put them into a computer as a rule-based system or Bayesian network. That strategy is still widely followed, but it faces a fundamental problem: Human experts, it turns out, are often not as expert as they think, and for many areas in which we wish to make predictions there are really no experts at all, or what the experts really know (as distinct from what they may think they know) is very limited.

Study after study has found that at many tasks medical and psychological diagnosticians do not perform well when their predictions can be checked objectively. In a test widely used for organic brain damage, Ph.D. psychologists with several years of experience with the procedure were found to discriminate between patients with and without organic brain damage only slightly better than chance: we would get slightly worse predictions by flipping a coin. A study of radiologists diagnosing malignant ulcers from X-rays found that the correlations of nine radiologists' judgements with one another were negative. Various tests of trained polygraph operators have found that they discriminate liars from those who tell the truth either no better, or only slightly better, than we would do by flipping a coin.

Many studies have shown that when judgements or predictions must be made under uncertainty, whether in medicine or psychology or elsewhere, in many cases better predictions are obtained simply by collecting relevant statistics correlating the phenomenon to be predicted with easily determined properties of individuals or systems and then using the most elementary statistical procedures for prediction. Statistical prediction is more reliable than human prediction at predicting parole violations, recidivism, educational success, and many diagnostic tasks.

Another disappointing finding of cognitive psychology is that when we ask experts to make judgements under conditions of uncertainty, their responses do not conform to the theory of probability. This is not any defect of the theory of probability, but instead a reflection of the fact that because intensive computations are required to make a large body of judgements in accord with the theory, people use heuristics that are inconsistent with probability theory. If we try to elicit probability judgements from experts, we typically find that they are incoherent.

Further, there are many subjects in which one might reasonably doubt there are any experts of the right kind. Social scientists have collected vast amounts of data on poverty, criminology, economics, drug use, sexual behavior, education, war, and many other broad areas of concern. The same is true in many areas of medicine and epidemiology. Such data are typically not reports of controlled experiments; they are at best reports of random observations. Experts in sociology, economics, political science, education, and other subjects sometimes use such data banks to construct "causal models" of the processes that generate the data. The causal models are very much like Bayesian networks, and they are often used to influence public policy, or decisions made by private companies. Many people think that little credence should be given to causal models produced in this way. They object that these "causal models" are arbitrary, and that many alternative models could fit the data equally well unless there are experimental controls, and no one knows what all the alternatives may be; when the models are based on social scientific theories, the objection continues, those theories themselves are often without any adequate empirical support.

In some measure, these problems with Bayesian networks and social scientific causal models have been overcome by artificial intelligence techniques that draw from ideas in philosophy of science. Some computational procedures can build Bayesian networks from appropriate data alone, or from statistical data combined with human knowledge. The same procedures build causal models in the same way, exactly because the Bayesian networks they construct *are* causal models. The surprising thing is that under precise and very general conditions, the causal conclusions obtained in this way must be correct, and when a unique causal structure cannot be determined, the procedure will describe for us the set of alternatives.

10.5 DISCOVERING CAUSAL STRUCTURE: THE TETRAD PROGRAMS

Hans Reichenbach proposed some simple relations between causal relationships and probability relationships. Let us represent any supposed causal structure by a network (or technically, by a *directed graph).* Reichenbach's conditions connecting causal structure and probability can be stated generally and uniformly by a condition that is now called the *Markov* condition.

> **Markov Condition:** If a set *A* of variables contains no descendent of a variable *B,* then *A* is independent of variable *B* conditional on the set of all parents of *B.*

The Markov condition means that if the mechanism by which *A* affects *B* is entirely mediated by another variable *C,* then if *C* does not change, variation in *A* will not produce any variation in *B.* Reichenbach would have said that *C screens-off A* from *B.* The Markov conditions also says that if, in another sort of case, *A* and *B* are both caused by variable *C* but are otherwise causally unconnected, then conditional on *C, A* and *B* are independent. Reichenbach called this assumption the *principle of the common cause.* Related ideas have been proposed by a number of other philosophers of science, notably Patrick Suppes.

The Markov condition appears to be violated in certain quantum mechanical experiments where the recordings of spatially remote sensors are correlated, but there is no prior state of the system on which the recordings are conditionally independent. Quantum mysteries aside, the problem with the Markov condition is that there are conditional *dependence* relations that ought to be required but which the principle does not ensure. Suppose for example that your car will not start and you know that it must be either because the battery is dead or because the gasoline tank is empty, but you do not know which. Suppose you assign some probability $P(D)$ to the proposition that the battery is dead, and some probability $P(E)$ to the proposition that the gas tank is empty. There is, so far as you know, no connection between having an empty gas tank and having a dead battery, so the two are independent: $P(D\&E) = P(D)P(E)$. Now suppose, finally, that on further investigation you find that the gas tank is not empty. Presumably this new information would change the probability you give to the proposition that the battery is dead. If so, this shows that while B and D are independent, they are *not* independent conditional on the fact that the car will not start. *Causal variables that are independent of one another are dependent conditionally on any common effect.* The Markov condition does not guarantee this relationship, but a stronger condition does:

> **Markov Completeness Condition:** Given a set of variables with causal relations represented by a directed graph, all of the conditional independence relations among the variables are consequences of the Markov condition applied to the graph of causal relations.

The Markov condition and the Markov completeness conditions imply that a great deal of causal structure is determined by probability relations. If we knew the facts about independence and conditional independence for the variables in an unknown causal structure, we could in principle determine a great deal about the causal structure. In some cases we could determine the causal structure uniquely from the probability relations, but in other cases alternative causal structures are consistent with the same probability relations. Mathematical investigations have characterized these indistinguishability facts exactly. That is, given any arbitrary directed graph (without any sequence of directed edges that cycles back on itself) we are able to characterize the set of alternative causal structures that cannot be distinguished by probability relations among variables. For example, according to the Markov Completeness condition the structures

$$X \rightarrow Y \rightarrow Z \text{ and } X \leftarrow Y \leftarrow Z$$

both entail that X, Z are independent conditional on Y and do not entail any other independence relations.

In real data we are not told the independence facts. We have to decide them from the properties of finite samples. Using statistical tests decisions can be made about whether variables are independent or conditionally independent. Provided the results of such decisions agree with the Markov principle, there are algorithms that will recover the causal graph from the independence facts; when two or more graphs are indistinguishable and fit the independence facts found, the algorithm will output all of the alternative causal structures. These algorithms have been implemented in a computer program, the TETRAD II program, where they can be applied to systems with a hundred or more variables.

How good are automated procedures for generating Bayesian networks? Consider the ALARM network again. In the network there are 37 variables and 46 directed edges. The network designers used the network and the probabilities they had assigned to it to generate (by computer) data for a sample of 10,000 units. The data represent a computer simulation of the data we might collect from 10,000 patients if the ALARM network correctly described the structural relations among the features of each patient. From these data alone, the TETRAD II program reconstructed the entire ALARM network up to statistical indistinguishability with only two errors. Extensive simulation tests of the procedures on many other networks show that on sample data they are very reliable.

One of the ancient conundrums in the philosophy of science is the "problem of theoretical terms." One version of that problem is this: How can we determine when unmeasured causes are producing associations between variables we do measure? Mathematical investigations of the Markov Completeness condition reveal that a variety of probability relations among measured variables suffice to indicate the presence of unmeasured common causes. Under special conditions the causal relations among the unmeasured variables can even be determined from probability relations among the measured variables.

10.6 CONCLUSION

A great deal of philosophical debate about artificial intelligence has focused on the thesis that people are really biological computers. Some critics of artificial intelligence deny that the brain is a computer at all. Other critics emphasize issues of meaning, interpretation and understanding, and claim that even if we are biological computers that does not explain anything about human understanding, including scientific understanding. These debates will undoubtedly continue, but it is important to realize that they form a marginal part of the connection between philosophy, especially philosophy of science, and artificial intelligence. The more elaborate and substantial connections are that philosophy of science provides both techniques and issues for artificial intelligence research, and, reciprocally, artificial intelligence research reveals heuristics and algorithmic connections between theory and evidence that ought to be of great interest to those concerned with how science works or can work. And most importantly, work in machine learning reemphasizes the central traditional question in philosophy of science: How is it possible to make discoveries?

DISCUSSION QUESTIONS

1. Gödel's second theorem, also proved in 1931, is this:

Gödel's second theorem: There is an algorithm that, given a finite description of any consistent, axiomatizable theory that entails formalized Peano arithmetic, outputs a sentence asserting the consistency of the theory, and that sentence cannot be proved in the theory.

Ronald Fisher, the late and distinguished statistician, remarked in connection with Hardy's observation that a contradiction implies anything, but Gödel's result ought not to have been a surprise to anyone: "After all, suppose a Ph.D. student came, breathless with excitement, and said, 'Sir, I have *proved* that this system of axioms is free from all contradictions.' You'd say, 'Did you prove it using only axioms?' He might say, 'Yes, I have here written out a chain of propositions which demonstrate that these axioms are free from all contradiction.' Well, perhaps you'd look at him with mild surprise, and you might say, 'I suppose you know that if this system of axioms *did* contain a contradiction, you could prove exactly those same propositions,' and so you have the situation that certain propositions which purport to prove the truth, the truth of the theorem, could be equally well demonstrated by the ordinary rigorous processes of deductive reasoning if they were false. And I don't know how much we would give, then, for the chain of theorem which purported to prove that the system of axioms was free from contradictions. It would seem a little absurd to imagine that such a thing is possible."

Consider carefully whether this passage is a sound attack on Hilbert's program and on the significance of Gödel's results.

2. An algorithm that lists all of the theorems of a theory does something intuitively less difficult than does an algorithm that, for every sentence, correctly decides whether the sentence is a theorem of a theory. Suppose, however, that you had a procedure that would list all of the theorems of a theory T, and that you also had another procedure that would list all of the sentences that are not theorems of that same theory T. Explain how these two listing procedures could be used together to form a procedure that, for every sentence, decides whether the sentence is a theorem of T.

3. Write annotated directed graphs representing Turing machines that
   a. Change all 0s to 1s and erase the last 1.
   b. Change all 1s to 0s and add one further 0.
   c. For some input, never halt.

4. For exercises 1 and 2, give a trace of the sequence of instantaneous states of the Turing machine you described for input 1001.

5. Describe the set of strings the finite automation illustrated will accept.

6. Describe a finite state automaton that will accept any input that is a sequence of an *even* number of 1s, but no other input.

7. Describe a finite state automaton for the telescopic lens problem.

8. Verify that applying rules 1, 2, 3 and 4 in appropriate sequence will give Kepler's law with the data in Table 10.4.

9. Use BACON heuristics 1, 2, 3, and 4 to find the law relating distance and time for bodies falling near the surface of the Earth from the following data:

Time (T)	Distance (D)
0.1	0.098
0.2	0.392
0.3	0.882
0.4	1.568
0.5	2.450
0.6	3.528

10. Fill in the rest of the "c" column for the first battery example, Table 10.6.

11. Explain why the Markov completeness condition guarantees that in the automobile example, "out of gas" and "dead battery" are not conditionally independent given that the car will not start.

SUGGESTED READINGS

GARDNER, MARTIN (1958), *Logic Machines and Diagrams*. New York: McGraw-Hill. A delightful introduction to Lull's background and ideas as well as to other attempts to design machines for reasoning.

LANGLEY, PAT; HERBERT A. SIMON; GARY L. BRADSHAW and JAN M. ZYTKOW (1987), *Scientific Discovery: Computational Exploration of the Creative Process*. Cambridge, MA: MIT Press. This book describes the BACON programs.

SPIRTES, PETER; CLARK GLYMOUR; and RICHARD SCHEINES (1992), *Causality, Prediction and Search*. New York: Springer-Verlag. A systematic advanced discussion of computerized causal inference.

WILF, HERBERT S. (1986), *Algorithms and Complexity*. Englewood Cliffs, NJ: Prentice-Hall. An accessible introduction to aspects of complexity and computation.

Eleven

PHILOSOPHY OF THE SOCIAL SCIENCES

Merrilee H. Salmon

Is it possible to study the activities of people, their social relationships, and the structures of the societies they live in with the same techniques and methods that have brought us so much knowledge in the physical and biological sciences? Can there be a *science* of human behavior modeled on the natural sciences? This is the major philosophical question concerning the social sciences. Accordingly, this chapter looks at the so-called social sciences with particular attention to whether we can investigate human behavior in the way scientists study the rest of the natural world.

Because scientific studies are so centrally concerned with causal relationships, a question closely related to our main theme is how to understand causation in the social world. For example, are we to regard individual humans as primary causal forces with the ability to shape social customs and practices to suit their individual and social needs? Or are we rather to understand humans as themselves the products of historical and social forces which constrain their beliefs and actions? Social theorists who lean toward the first position are called "individualists" while those who support the latter are called "collectivists" or "holists." Neither side in the dispute (actually, there are many different versions of both sides) denies the obvious causal interplay between individuals and societies. Nevertheless, individualists and collectivists disagree about the ultimate causes of human behavior. Correlatively, individualists and collectivists differ significantly in their accounts of human nature and of how to bring about social change. Human free will as a force for change in society looms large for many individualists, but collectivists tend to disregard or deny the efficacy of individual choice in shaping human affairs. In addition, according to many collectivists, we must be aware of the historical development of social, economic, and cultural institutions before we can change them. In contrast, many individualists pay little attention to how a situation developed historically when their primary interest is in changing it.

11.1 THE NATURALISTIC TRADITION

Conflict about whether human social behavior can or should be studied scientifically is reflected in uneasiness about how to categorize the disciplines of sociology, economics, political science, anthropology, psychology, and history. In many universities, these are all classified as "social sciences." During the past forty years, many departments that were formerly called "government" changed their name to "political science" to signal their allegiance to the aims and methods of science. Experimental psychologists, many of whom have extensive training in physiology and neurology, see their field as one of the "life sciences." However, some historians and anthropolgists protest this labeling for they see their fields of study—or at least their own work in these fields—as more closely allied with typical "humanistic" studies, such as literature or philosophy.

Clearly, the label "scientific" carries a certain cachet in our society. Scientists are respected, and their work is heavily funded by public and private agencies. More importantly, the use of scientific methods is believed by many people to be the best way to obtain genuine (though not infallible) knowledge about the world. Despite this widespread admiration for science and its methods, it is not easy to say exactly what science or the scientific method is. The nature of science has been considered and discussed in the first four chapters of this text. This chapter does not attempt to offer a definition of science since it is likely that by now the reader has formed a satisfactory, if somewhat vague, notion of what science is. Instead, let us assume that science involves such features as laws, testability, prediction, and explanations as described in earlier chapters.

Given the honorific force of "science" and "scientific," why would anyone hesitate to try to attain scientific understanding of human social life? One of the oldest arguments raised against the possibility of a social *science* is that there can be no *science* of individuals with free will. Since humans are not constrained in their voluntary actions by the sorts of laws that operate in the physical world, critics say, we cannot *predict* human behavior with anything like the accuracy that is required in a respectable science. Human beings, unlike atoms and amoebas, are rational beings. They make and carry out plans, reflect on progress toward their goals, revise original plans in the light of such reflection, replace old goals with different ones, and even change their minds "for no apparent reason." Humans, if they are informed of predictions of their behavior, sometimes can and do thwart those predictions. No one likes being thought of as utterly predictable. For example, if you predict in my hearing that I will order a vanilla ice cream as usual, I may order chocolate just to refute you although I would have preferred vanilla. Being predictable seems to conflict with our sense of autonomy and spontaneity. For pragmatic reasons, businesses and governments do not want their competitors to regard them as too predictable, and so those in power employ many strategies to keep the opposition guessing. Obviously, the value humans place on spontaneity can interfere with successful prediction.

Without denying the force of the ability to thwart predictions or the possibility of free will, we must notice that our ordinary everyday social interaction with others involves a great deal of successful prediction. To call someone "reliable" does not carry the same negative connotation as to call the person "predictable," but the behavior described by the two words is the same. When we ask a close friend for help, we can count on a generous response. We expect classes and movies to begin more or less on time. Successful restaurants can predict roughly how much food they will need to feed their clients on a given night. A batter who hits a tie-breaking run in the final inning can count on the cheers of teammates and fans. Even the natural desire of humans to overturn predictions about themselves can be exploited in the service of successful prediction—as when opinion polls take account of "bandwagon effects" or "underdog effects." If people were completely unpredictable, social life would break down altogether.

John Stuart Mill, a nineteenth-century philosopher who strongly believed that a science of human behavior could be modeled on the physical sciences, tried to deal with the difficulty of making reliable predictions about human behavior without directly addressing the issues of whether humans have free will. Mill (1874) points out that even in the physical sciences precise predictions are not always possible. He distinguishes between "exact" sciences—those capable of accurate prediction—and "inexact" sciences, those in which predictions are much less precise. Astronomy is Mill's example of an exact science. Astronomers make reliable predictions regarding the positions of planets at various times of the year, eclipses of the sun and moon, and even the paths of comets.

Tidology—the science of tidal movements, in contrast, is far less accurate in its predictions. Scientists since Newton have understood the lawful relationship between tides and the phases of the moon, and have known that tides are affected in regular ways by the movement of winds and the shapes of shorelines and ocean floors. Although the laws that govern the movements of the tides are well-known, because of shifting winds and irregularities in shorelines and ocean floor, precise predictions of tidal movements are not always possible. Predicting tides is not so risky as predicting weather, but because so many minor causal forces counteract and reinforce one another, we cannot be sure exactly what is going to happen. Still, Mill says, tidology can make imprecise predictions, and no one denies that tidology is a science even though it is less exact than astronomy. Mill believes that if we would use the same methods as the physical sciences to discover the causes of human behavior, a *science* of human behavior could be developed that would be at least as exact as tidology. Thus he admits that *precise* prediction of human behavior is not possible, but he denies that this prevents the development of social science.

Mill's claim that absolutely precise predictions are not required for science is further supported by contemporary scientists' recognition that astronomy and physics are not so exact as Mill believed. Ironically, as more precise instruments have been developed for measuring such features as the exact location of physical bodies, the accuracy of predicting their exact locations has decreased. With better instruments of measurement, scientists see that even in the so-called exact sciences, sensitive dependence on initial conditions undermines the possibility of precise prediction (see Chapter 6).

Mill believes that human actions are caused by the thoughts (beliefs and desires) of the actors. He thinks that just as in the rest of the natural world, observation and—to some extent—experimentation are the ways to learn about the casual regularities that connect thoughts and behavior. Of course, experiments must not treat humans as if they were insentient beings; limits are imposed by humane considerations. However, limits on experimentation occur in physical sciences too; astronomy and geology are primarily based on observation rather than experimentation, but they are no less "scientific" for that reason.

Mill's approach to the study of human behavior can be called naturalistic, for he sees human thoughts and feelings as well as the actions they give rise to as part of the natural world. The special problems that we find in the projected human science come about, he believes, because of complexity of the subject matter, not because the study of humans must be intrinsically different from the study of biology, chemistry, and physics. Mill believes that scientific generalizations ranging over both individuals and social units are possible, and that with these generalizations satisfactory explanation and reasonably accurate predictions of individual and social behavior are possible.

Following the naturalistic tradition of Mill, the contemporary philosopher C. G. Hempel argues that explanation and prediction in the social sciences have the same *logical* structure as in the physical and biological sciences (see Chapter 1). In other words, the actions of humans, like other natural phenomena, are to be explained by subsumption under laws, and one of the aims of social science is to discover suitable laws.

Many students of human behavior agree with Mill's and Hempel's assessment of the possibility of human science. The eminent sociologist George Homans, for example, while admitting that the results of social science are not so impressive as those of modern physics, says:

> What makes a science are its aims, not its results. If it aims at establishing more or less general relationships between properties of nature, when the test of the truth of a relationship lies finally in the data themselves, and the data are not wholly manufactured—when nature, however stretched out on the rack, still has a chance to say "No!"—then the subject is a science. (1967, 4)

As mentioned earlier, not everyone agrees with this naturalistic position. Mill's answers to those worried about the lack of predictability in the social sciences did not silence all of his critics. Doubts about the compatibility of free will and human science persist and other problems are raised as well. Contemporary critics of the view that studies of human behavior are or can be scientific fall into three categories. The first group, called *interpretivists,* claim that explanations of human behavior are structured entirely differently from explanations of the behavior of physical objects since human behavior, they say, consists of actions done for *reasons* rather than events resulting from *causes*. The second group, called here *nomological skeptics,* do not deny that human behavior is subject to causal laws, but doubt that it will ever be possible to find laws of human behavior that are similar in power and scope to those in physical science. The third group, called *critical theorists,* claim that it is inappropriate even to try to explain human behavior in terms of laws of cause and effect because to do so denies the value of human autonomy (free will). In addition, they say, any attempt to construct a social science on the model of physical sciences promotes unethical manipulation of humans and discourages any attempts to improve the conditions of social life. The alternatives to naturalism will be discussed in the following sections.

11.2 INTERPRETIVISM

R. G. Collingwood (1946), a philosopher of history, presents a forceful exposition of the interpretive point of view. Collingwood uses the term "history" to refer to all (and only those) studies that are primarily concerned with voluntary actions of humans—history, parts of anthropology, sociology, political science, psychology, and economics. He contrasts "history" or "historical thought" with "natural science" which studies only physical aspects of the world.

Events studied by history, in contrast to those studied in natural sciences, Collingwood says, have both an "inside," which must be described in terms of *thoughts* of the agents responsible for the events, and an "outside," which can be described in terms of bodies or their movements. Events studied by natural sciences have only an outside.

Collingwood insists that an adequate *description* of human voluntary actions requires an account of the beliefs and desires of human agents. Without knowing the thought that informs an action, it is impossible to say what action occurred. From a description of the "outside" of one person cutting another with a knife (i.e., a description of bodies and their movements), for example, we cannot tell whether the action is an operation to correct a medical problem or an attempt at murder. Additional details about the "outside," such as the setting in which the cutting occurred can provide clues to the beliefs and desires of the person who did the cutting, but unless we know the intention of the cutter, we cannot say what kind of action was performed.

According to Collingwood, the relationship between intentions (reasons) and actions is different from the relationship between causes and their effects in the physical world. Causes and their effects are distinct events or distinct parts of a process. As separate realities, there is no *logically* necessary connection between their occurrences. Furthermore, a description of the effect does not have the cause as part of its *meaning*. We can understand, for example, that a bridge has collapsed without being aware of the cause of the collapse. In contrast, we cannot know that a murder was committed without being aware that the perpetrator intended to commit a crime, for such an intention must be present if we are to classify the action as murder rather than manslaughter or an accident. (We may, of course, be unaware who the murderer was or why the murderer formed the intention to kill the victim, but those are separate issues.) Causes in the natural world are thought of—at least since Hume (see Chapter 2) as events distinct from their effects; there is no *logical* connection between events that are identified as causes and the events identified as effects of those causes. However, Collingwood argues, the intention of the agent is logically inseparable from an action's being the sort of action it is—an action cannot be murder without the murderous intent of the agent; change the intent, and the action may be manslaughter or accidental homicide rather than murder. Collingwood says that because the intention gives meaning to the action, and makes it the kind of action it is, the relationship between intentions (reasons) and the actions that express those reasons cannot be similar to the causal relationship that holds between events with only an outside.

Peter Winch (1958), a contemporary British philosopher, agrees with Collingwood that because reasons and the actions that reasons are invoked to explain are logically intertwined with one another, causal laws of nature are not appropriate for explaining human behavior. Following Hume's dictum that there is no logical connection between causes and their effects, Winch concludes that reasons cannot be the causes of actions.

Winch, a disciple of the influential philosopher Wittgenstein, adopts the latter's emphasis on the importance of *rules* in social behavior. Voluntary behavior is best understood, Winch says, as behavior that follows some rule, not behavior that is *caused* by reasons. "Rule," in this context, refers not only to formal regulations, such as "Drive on the right," but also to unstated cultural norms like conventions that govern the appropriate distance between speakers in a face-to-face conversation. The notion of rule is also extended to cover practices and institutions, such as language, banking, religion, and various forms of government. Consider the action of writing checks to pay bills. Writing checks would not be possible without a system of banking governed by sets of rules specifying how monies are to be deposited, held, and transferred. Some of these rules are regulations, such as the rule requiring checks to be dated; others, however, specify what it means to be a bank. Without these latter constitutive rules, there would be no institution of banking and the meaning of writing amounts of money on a little printed piece of paper, dating, and signing the paper would not be what it now is.

Consider also a social institution that recognizes some objects as sacred and also prescribes special behavior toward such objects. Only within such a context can someone commit a *sacrilege*. In this example we can understand that it is the rule (that is to say, practice or institution which gives meaning to the notion of a sacred object)—not merely the intention of the agent—that makes the act of sacrilege what it is. Interpretivists admit that when we have knowledge of the rules of a society and an understanding of how the rules are applied, we can often predict what people will do. However, interpretivists insist that reliable prediction is not the goal of studies of social behavior and that insofar as behavior is rule-governed, it is not subject to causal laws. Rules, according to interpretivists, are standards or norms of behavior. They give meaning to behavior, they do not cause behavior. After all, it is obvious that humans can violate or disregard the rules of their societies; they can and do shape new institutions, conventions, and regulations. In contrast, the causal laws that govern the movement of bodies, such as the law of gravity, cannot be changed, suspended or amended by human intervention. Humans discover and formulate statements that describe relationships that hold in the physical world, but the relationships—expressed in natural laws—obtain independently of human agency. Rules, in contrast, are made by and for humans and can be abided by, ignored, or changed by them.

The interpretivist's concern with rules is shared by many social scientists. Ward Goodenough (1957), a cognitive anthropolgist, for example, defines the crucial anthropological concept of *culture* in terms of rules. He says that culture consists of the forms of organization of things, people, behavior, and emotions in the minds of people. The task of the cognitive anthropologist, he says, is to uncover those forms (i.e., rules) by observing the behavior of subjects and listening to what they say about how they organize the various aspects of their world. An anthropolgist comes to understand another culture by learning its rules *and* how they are applied in various situations.

Clifford Geertz (1975), a symbolic anthropologist, who disagrees with many points in Goodenough's cognitive approach to anthropology, nevertheless also recognizes the value of the interpretivist position. In fact, he insists that the task of anthropology is not to find covering laws that would allow us to construct the "'scientific'' explanations favored by Mill and Hempel, and to predict people's behavior, but rather to *interpret* the flow of social discourse.

To interpret, Geertz says, is to inscribe or record the flow of social discourse, to "fix" it so that it can be shared and reexamined at a later time. He calls this activity "ethnography," or "thick description." A major problem for the ethnographer is how to describe or classify observed behavior. To classify behavior correctly, Geertz says (in agreement with cognitive anthropologists) we need to be aware not only of the agent's intentions, but also of the rules (social conventions, norms, and institutions) of the society and the way in which actors apply those rules. Geertz differs from the cognitive anthropologists in his belief (shared by Winch) that the rules which constitute culture are *public* rather than hidden in the minds of the actors. For example, suppose that the anthropologist who is studying some group of people is aware that meetings are taking place at odd hours with great concern for secrecy, and that political issues are being discussed in low voices. The anthropologist tries to figure out whether these activities should be described as "fomenting a rebellion" or in some less inflammatory way. By observing behavior, including of course linguistic behavior, the anthropologist tries to discern whether the meetings are secret merely because it is considered impolite to discuss political arrangements publicly or for some other reason. When the ethnographer can *describe* what is going on in terms of the purposes of the actors and the standards of the society, Geertz says, no further explanation by subsumption under covering laws is necessary, for then in the very act of giving a "thick description" (Geertz borrows the term from the philosopher Gilbert Ryle) of the action, we have told *why* it was done, and thus *explained* it. Geertz's thick descriptions of human behavior are obviously very similar to Collingwood's descriptions that include the inside as well as the outside of human behavior.

Geertz's interpretivism differs from Winch's, however, since Geertz's chief objection to lawful explanation in anthropology seems not to be the complaint that a logical or "meaningful" relation between intention and action precludes a causal relationship between them, but rather that any true statements that are general enough to serve in covering-law explanations in anthropology are either trivial or hopelessly vague. This objection is also that of the nomological skeptic, to be discussed after examining a criticism of interpretivism (Section 11.3) and an attempt to reconcile interpretivism with the naturalistic tradition (Section 11.4).

11.3 REASONS, ACTIONS, AND CAUSES

The interpretivists' claim that there can be no causal connection between reasons and actions because the two are logically related has been severely undercut by the work of Donald Davidson (1980). Davidson points out that causal relationships that hold between events in the world (such as a bridge's understructure rusting away and the bridge's collapse) can be described in various sentences, some of which express logically necessary truths whereas others express contingent truth (i.e., truths that might have been otherwise). For example, (1) "A rusty understructure caused the bridge to collapse," does not express a logical truth, for the sentence could be false. For example, weak concrete rather than a rusty understructure might have caused the collapse. In contrast, (2) "The cause of the bridge's collapse caused the bridge to collapse," is logically true. Nevertheless, if the first (contingent) sentence happens to be true, "the cause of the bridge's collapse" and "the rusty understructure" refer to the same state of affairs so that sentence (1) and sentence (2) are different ways of describing the same situation. The real-world relationship between the cause of the bridge's collapse (the rusty understructure) and the event of the collapse holds independently of what sentences are used to represent it. Davidson's point is that two *descriptions* of the same situation may differ in their logical character, but that any causal relationship that obtains in the situation holds independently of the sentences used to characterize it. The mistake made by interpretivists is to confuse relationships between descriptions of events with relationships between events themselves. If we can accept Davidson's analysis of this mistake, then we can see that although in our usual descriptions of human behavior an agent's intentions are logically related to the action, this does not mean that there can be no genuine causal connection between the reason for acting and the action. We may not know how to describe the reason in an appropriate way, but the fact of its occurring can be causally related to the resulting action. Davidson says there must be such a relationship if it is true that the action was done for a reason (and not as a result of some other type of cause).

Davidson says not only that reasons are causes of actions, but also that there are *causal laws* connecting reasons and actions, for he agrees with Hempel that "if A causes B, there must be descriptions of A and B which show that A and B fall under a law" (1980, 262). However, the difficulty is that we hardly ever, if ever, know what that empirical law is. Davidson believes that our ignorance results from our inability to describe reasons in the nonpsychological terms appropriate for framing causal laws. He further argues that the unknown laws cannot be *psychophysical;* that is to say, they cannot have the form of stating a regular connection between a psychological cause (such as a belief or desire) and an action.

For example, suppose it is true that Jason ran in a marathon because he wanted to prove his worth. His reason (the desire to prove his worth) caused him to run the marathon, but there is obviously no general law of the form "Whenever someone desires to prove his or her worth, that person runs in a marathon." There is not even a law that "Whenever Jason desires to prove his worth, he runs a marathon." However deficient our understanding of the relationships between reasons and actions, it is rich enough for us to see that neither of these claims is even a good candidate for a law of human behavior. Moreover, even when generalizations such as these are subjected to any number of qualifications, the results are open to obvious objections. On the basis of these and other reasons, Davidson argues that genuine laws connecting reasons and actions must have some different, but as yet unknown, form.

Even though Davidson refutes a major point of interpretivism, he cannot be clearly identified with the naturalist position. He agrees with Winch that Mill was misguided in searching for explanatory laws that have the form of generalizations connecting descriptions of mental events with descriptions of behavior. Moreover, although Davidson says that his arguments reinforce Hempel's view "that reason explanations do not differ in their general logical character from explanation in physics or elsewhere" (1980, 274), his insistence on the special character of explanations in terms of reasons and his arguments against the possibility of covering laws that connect descriptions of psychological states and behavior in such explanations represent a significant departure form the tradition of Hempel and Mill.

11.4 THE DECISION-THEORETICAL APPROACH TO EXPLAINING HUMAN BEHAVIOR: AN ATTEMPT TO RECONCILE INTERPRETIVIST AND NATURALIST VIEWS

Appeals to an agent's reasons—beliefs and desires (values)—to explain the agent's actions form the common-sense basis for the theory of naturalistic explanations of human behavior. When commonsense explanations are presented, however, they usually are incomplete in the sense that no laws are explicitly stated. Davidson, as we have seen, has emphasized the difficulty of finding appropriate laws connecting reasons with behavior. We now consider an approach to naturalistic explanations of human behavior that claims that all that is required is a law which credits agents with *rationality*.

Many contemporary social scientists, especially economists, characterize actions that arise from a particular balance of the agent's beliefs and values as *rational*. Their definition of rational action requires the following conditions: agents act independently, on the basis of their own sets of beliefs and desires (values), are capable of ordering their preferences, and have sufficient partial knowledge of the possible outcomes of the contemplated actions to assign probabilities (see Chapter 2) to those outcomes. In such circumstances, agents are said to act *rationally* just in case they choose that action which maximizes *expected utility* (desirability or value).

Consider, for example, a high-school senior who excels both at academic subjects and at athletics and who must choose between no-strings-attached scholarships at two universities, State and Out-of-State. The student wants both to play volleyball and to receive a superior education. She figures that the probability of making the team at OSU is 0.75. The OSU academic program is weaker than at State, so a degree from OSU is worth less than one from State. At either school, the student is sure of earning a degree. The probability of making the team at State, however, is only 0.50. On a scale of 1 to 10, she assigns 10 units of value to playing ball at State and earning a degree there, 8 units of value to playing ball at OSU and earning a degree there, 6 units of value to going to State without playing ball, and 4 units of value to going to

OSU without playing ball. How should the student choose between going to State and going to OSU?

Two outcomes are possible for each action: earning a degree and playing ball or earning a degree but not making the team. Assigned to each of these four outcomes is a probability and a value, as shown in Table 11.1.

TABLE 11.1

| | Outcomes | |
Actions	Degree + Play Ball	Degree − Play Ball
State U.	Pr. 0.50 Val. 10	Pr. 0.50 Val. 6
Out-of-State U.	Pr. 0.75 Val. 8	Pr. 0.25 Val. 4

The expected utility of an action is calculated by multiplying the probability and value of each of its possible outcomes and summing the total. Thus the expected utility of going to State is $(0.50 \times 10) + (0.50 \times 6) = 8$; and the expected utility of going to OSU is $(0.75 \times 8) + (0.25 \times 4) = 7$. Following the rule of maximizing expected utility, the rational student chooses to go to State. Maximizing expected utility just means choosing the action with a sum that is not lower than the sum of any other action open to the agent.

The decision-theoretical account of rationality does not specify what beliefs or values are reasonable for agents or how to assign values or probabilities. Two high-school seniors, for example, might be ranked equally as ball players for the same position by the high-school athletic association. Nevertheless, if both are facing the choice of going to State or OSU, they might assign different probabilities to their chances of making the State team. A more self-confident player could assign 0.95, for example, and the less confident player, 0.50, or an even lower probability. Neither assignment is considered irrational by the decision-theoretical account of rationality. More obviously, values may also differ. One player could assign a value of 10 to playing at State, while the other may assign a lower value. Moreover, the decision-theoretical account of rationality does not require agents to actually perform a calculation as in the example above. All the decision-theoretical criterion of rationality requires is that an agent choose those acts that maximize expected utility on the basis of the agent's own beliefs and values without regard to how well-founded or appropriate these beliefs and values are.

David Papineau (1978) adopts the decision-theoretical approach to rationality in his attempt to reconcile the interpretivist and naturalist approaches to explaining human behavior. According to Papineau, the law-like generalization that underlies covering-law explanations of human behavior is that agents always perform those actions with greatest expected desirability [utility] (1978, 88). In other words, if two high-school scholar-athletes facing a choice between State and OSU choose differently, it is because their beliefs (assignments of probability to the various outcomes) or desires (assignments of values to the various outcomes) or both differ. Descriptions of those beliefs and desires are part of the explanation of an action. Following Hempel's model of explanation (see Chapter 1), Papineau says that statements of the agent's beliefs and desires belong to the set of initial conditions. From these, in conjunction with the law of maximizing expected desirability, a description of the action can be derived and thus explained.

Agreeing with the interpretivist point that rules do not *cause* behavior, Papineau says rules "influence" the desires and beliefs that agents have. By using the weaker term "influence" rather than "cause," Papineau acknowledges that an individual's beliefs and desires can clash with prevailing cultural norms. Regardless of whether the norm is followed or rejected in a particular action, however, the agent usually is aware of the norm and reacting to its presence. Nevertheless, according to Papineau, although statements of individuals' beliefs and desires constitute initial conditions in explanations, information about *how agents come to have their particular beliefs and desires* supplements, but is not part of, explanations of behavior. Once the beliefs and values can be attributed to an agent—regardless of how the agent came to have the beliefs or values—the action can be explained as arising from the particular beliefs and values in conjunction with the principle of maximizing expected desirability. Norms thus do not cause behavior, but they do "influence" the formation of beliefs and desires that are the causes of behavior.

A serious objection to Papineau's account of the explanation of human behavior is the apparent lack of empirical content of the law of maximizing utility. This "law" is problematic because observing an agent's behavior (including verbal behavior) is the only way to determine what an agent's beliefs and desires are. For example, suppose one of your friends accepts a party invitation with great enthusiasm, but fails to come to the party. We may explain this action by saying that the friend forgot, or had another commitment, or found something better to do, or even that the response to the invitation was insincere. In other words, we can attribute some different set of beliefs and values on the basis of your friend's failing to appear than we attributed when the invitation was accepted. We have no other basis than what people say and do for assigning beliefs and desires to them. Papineau tries to defend his "law" by saying that if we attribute extraordinary desires or beliefs to agents, we must be prepared to give some account of the circumstances that will lead to independently testable claims. For example, in support of "He forgot," we might give examples of his general tendency to forget: "He's so forgetful that he didn't show up for his own birthday party." Nevertheless, it seems that with enough ingenuity we can always attribute an "appropriate" set of beliefs and desires to make any action count as rational. However outlandish an action, it is always open to us to say, "It seemed (to the agent) a good idea at the time." Such a possibility weakens the claim that "Agents always perform those actions with greatest desirability" has any empirical content. In the words of Karl Popper, the alleged law is "unfalsifiable" (see Chapter 2). Since the naturalistic position, as presented by Mill and Hempel, and subscribed to by Papineau, requires empirical content in explanatory covering laws (Chapter 1), Papineau's attempt to reconcile interpretivist considerations with a naturalistic account of social sciences is defective.

If we try to retain empirical content in Papineau's proposed law by refusing to supply agents with a set of beliefs and desires that would make all their actions rational, it is not difficult to find examples in which persons perform actions that fail to maximize expected utility. For example, a person who has completed careful consumer's research concerning repair records and performance statistics before choosing to buy a particular brand of automobile may reject all that evidence in the face of a single report of a "lemon" of the same brand bought by a friend of a friend. Recent studies by Tversky (Kahneman et al. 1982) and other psychologists have shown that such actions are not random occurrences, but follow systematic patterns. In examples like this, vividness of specific information tends to diminish the force of the statistical knowledge. Thus, it seems that empirical content for the principle that agents always perform those actions with greatest expected desirability can be salvaged only at the price of rendering the principle false. In either case, the principle is not a suitable law to ground scientific explanations.

11.5 NOMOTHETIC SKEPTICISM

The problem that plagues the principle "Agents always act so as to maximize expected utility" arises with varying degrees of severity in many proposed laws in the social sciences. The alleged laws seem either to lack any real content or to be obviously false. Our apparent inability to come up with laws that are nontrivial and yet general enough to explain human behavior is discouraging to many social scientists.

The rejection of the possibility of scientific (lawful) explanation in social sciences because of the unavailability of nontrivial laws differs from the interpretivists' complaint that the relationship between reasons and actions is a logical rather than a causal relationship. The interpretivist raises a logical problem about the meaning of sentences that link reasons with actions, whereas the nomothetic skeptic focuses on the more pragmatic issue of whether we can ever capture the variety and complexity of behavior in laws that are at once nonvacuous and widely applicable. In the physical sciences, the discovery of laws extends our knowledge and allows us to see relationships that go far beyond our commonsense understanding of the phenomena. So often in the social sciences, in contrast, generalizations seem merely to summarize in a less rich form the insights already grasped by common sense. Many social scientists, following Mill, agree that because human behavior is complicated and subject to so many subtle causal influences, framing appropriate laws is far more difficult than in the physical sciences. Nevertheless, they hope that with enough time, we will eventually develop a stock of useful laws that can ground scientific explanations of behavior. Philip Converse, of the Institute for Social Research at the University of Michigan, for example, says that given the complexity of the data, "it would not surprise me if social science took five hundred years to match the accomplishment of the first fifty years of physics" (1986, 48).

Alasdair MacIntyre, a philosopher, does not share the optimism of Converse. He claims that the salient fact about the social sciences is "the absence of any law-like generalizations whatsoever" (1984, 88). MacIntyre acknowledges that social scientists have offered some interesting generalizations that are supported by confirming instances. One example he cites is Oscar Newman's generalization that "the crime rate rises in high-rise buildings with the height of a building up to a height of thirteen floors, but at more than thirteen floors levels off" (ibid., 90). Nevertheless, MacIntyre says that all of these generalizations share features that distinguish them from law-like generalizations:

1. They coexist in their disciplines with recognized counterexamples;
2. They lack both universal quantifiers and scope modifiers (i.e., they contain unspecified ceteris paribus clauses);
3. They do not entail any well-defined set of *counterfactual conditionals*. (See MacIntyre 1984, 90–91.)

To answer the first criticism, we can refer back to Mill's characterization of social sciences as inexact. Mill focused his attention on the imprecise character of predictions in social science, whereas we comment on the statistical or probabilistic form of laws in the social sciences. Laws of social science tell us what *usually, typically,* or *rarely* happens rather than what always, without exception, or never happens. MacIntyre acknowledges that generalizations in the social sciences are statistical rather than universal, but says that does not account for their defectiveness. He claims that the statistical laws found in quantum physics (see Chapter 6), for example, express universal claims about *sets* of atoms rather than individual atoms, and are subject to refutation in just the same way as nonprobabilistic laws.

However, MacIntyre misunderstands the nature of statistical laws. Statistical laws are not universal generalizations about sets; instead, they state some probability (greater than zero and less than 1) that an event of a certain type will occur together with or be followed by an event of another type. Nor are statistical generalizations confirmed or disconfirmed in just the same way as universal generalizations. A tentative universal law can be overthrown by a single genuine counterexample that cannot be accommodated by a suitable revision of auxiliary hypotheses (see Chapter 2). In contrast, any distribution in a given sample is compatible with a statistical generalization. To take a simple example, suppose I have a deck of cards, but I am not sure whether it is a standard deck or some odd combination of several decks, so I do not know the actual distribution of red and black cards. I hypothesize that one half of the cards are red. In the first sample of ten cards that I draw, nine are red and one is black. We know that a sample of this type will not occur very often if the deck really is half red, but it can happen, and the preponderance of red cards in the sample cannot by itself refute the hypothesis that half of the cards in the deck are red. Consider in contrast a test of the universal hypothesis "All of the cards in the deck are red." The occurrence of a single black card can refute this hypothesis.

MacIntyre is correct in saying that statistical laws in physics are more impressive than those in social science, but this is because the physical theories that such laws belong to have been tested and supported by far more elaborate and conclusive evidence than any theory of the social sciences. The question, however, is not whether theories in the social sciences are as advanced as those of physics—clearly they are not. The question is whether there can be laws of social science. If it turns out that the only laws possible for social science are statistical, this is no great defect. Laws are an indispensable part of science, but genuine scientific laws may be either statistical or universal.

The claim that generalizations in the social sciences differ from those in physical sciences because the intended domain of application of generalizations in the social sciences is only vaguely indicated has received a lot of attention from philosophers concerned to point out differences between the social and physical sciences (see Scriven 1959b). In the physical sciences, the exact conditions under which a law is supposed to apply are presumably spelled out in detail. The ideal gas laws, for example, apply only under conditions of moderate temperature and pressure, and these limits can be specified precisely, whereas proposed laws in the social sciences are usually qualified with some vague claim such as "under normal conditions" or "all things being equal."

MacIntyre says that such qualifications are necessary in the social sciences because there are ineliminable sources of unpredictability in human affairs. Hempel has recently pointed out the widespread need for similar scope modifiers in the physical sciences. In discussing the theory of magnetism, for example, Hempel (1988) notes that the laws of magnetism do not take account of the possibility (though it surely exists) of initial conditions such as strong air currents or strong magnetic fields of suitable direction that would prevent two bar magnets, suspended by fine threads close to one another at the same level, from arranging themselves in a straight line as the theory predicts. Yet, he says, such ceteris paribus clauses are implicit in the theory. Moreover, it would not be feasible to spell out all of the circumstances that might interfere with the operation of a physical law.

Ceteris paribus clauses are invoked when information about the conditions under which a law is applicable is lacking. In this way, laws can be stated tentatively, while research goes on in an effort to refine the domain of application of the law. Such tidying occurs in physical sciences as well as social sciences. Critics who despair of ever finding laws in the social sciences comparable to those in the physical sciences often have an exaggerated notion of the precision of physical sciences. Hempel's work indicates that the absence of scope modifiers does not constitute the major distinction between physical and social sciences, although in the physical sciences, ceteris paribus clauses are rarely acknowledged explicitly.

MacIntyre's complaint that generalizations in the social sciences fail to support counterfactuals raises the complicated and unresolved issue of how to distinguish genuine laws from coincidentally true generalizations, such as the true claim that "No gold spheres weigh more than 5000 pounds." One difference that has been noted is that genuine scientific laws support counterfactuals (see Chapter 1). This means that a genuine law (whether statistical or universal) is supposed to describe not only what actually happens but also what would happen if certain conditions obtained even though in fact they do not. For example, we can appeal to the law of gravity to say what *would* happen *if* a typewriter were dropped from the top of a ten-story building in New York City. Only when generalizations will support counterfactuals can we use them confidently in scientific explanation and prediction. Regrettably, however, we do not have any satisfactory account of the nature of counterfactuals that is independent of our understanding of laws. Therefore, MacIntyre's objection that generalizations in the social sciences will not support counterfactuals is just another way of saying that they are not genuine scientific laws.

How well justified is nomothetic skepticism? We have already discussed the difficulty of finding causal laws that connect reasons with actions. However, social scientists are also interested in laws that do not refer to any beliefs or desires of individuals. Consider for example the laws proposed by the anthropologist George P.

Murdock (1949). He tried to relate various systems of kinship in different societies to other features of those societies, such as their forms of marriage, patterns of post-marital residence, rules of descent, and forms of the family. Murdock's generalizations are not candidates for causal laws since they do not attempt to assign temporal priority to either kinship systems or the other related features, nor do they invoke any mechanisms for the regularities described. Moreover, the generalizations Murdock proposes are not *rules* or norms. Although the proposed laws state that some rules in societies are regularly associated with other rules (such as the rule of marriage with the rule of descent), they do not themselves have the form of normative rules. Noncausal generalizations such as Murdock's are sometimes called *structural* generalizations. In the physical sciences, "Copper conducts electricity" is an example of a structural law. Other examples of proposed structural laws in social science are the law of evolutionary potential: The more specialized the system, the less likely it is that evolution to the next state will occur (Sahlins and Service 1960) and the law of cultural diffusion: the greater the distance between two groups in time and space, the more unlikely is diffusion to take place between them (Sanders and Price 1968). Structural generalizations could lead to the discovery of causal laws by focusing our search for requisite mechanisms (e.g., what is it about copper that makes it a conductor?).

Descriptions of well-established social regularities are sometimes so puzzling or surprising that they require rather than support explanations. Newman's generalization about crime in high-rise buildings belong in this category, as do those well-supported generalizations that associate being an eldest child or eldest male sibling with intelligence and achievement. Correlations such as these, especially when they are documented cross-culturally, stimulate social scientists to search for deeper regularities to explain them (Converse 1986). The deeper regularities, if discovered, might be causal laws.

Consider Newman's generalization connecting crime and high-rise buildings. We would like to know how well this correlation would stand up in new situations. Is it just a coincidence or does it seem to support counterfactuals? Can we frame some interesting and testable hypothesis about *why* this correlation exists? What sorts of causal mechanisms can be operating? Interesting correlations stimulate the acquisition of new data and framing of new hypotheses. Ultimately, the process could lead to laws that are very different in form—not merely in the refinement of their scope modifiers—from the original generalization.

Social theorists often raise skeptical doubts concerning laws relating reasons and actions, but laws of human behavior need not take that form. Mill, who represents the individualist position regarding the causes of human behavior, believed that the ultimate explanatory principles would be laws of human thought. Emile Durkheim ([1897] 1951), a holist who is regarded by many as the founder of modern sociology, argued against individualists that *social facts* (such as the suicide rate in a particular society) require *social causes* for their explanation. Durkheim did not deny that individuals commit suicide for specific and highly personal reasons, but felt that the explanations afforded by those reasons were inadequate. According to the records that Durkheim studied, the individual reasons offered for suicides were similar in Catholic and Protestant countries. Yet proportionately greater numbers in Protestant countries committed suicide for those reasons. Durkheim believed that the difference in *rates of* suicide therefore could not be accounted for entirely in terms of individual reasons. He considered various other possible causal factors, such as climatic and seasonal influences, and concluded that none of these could explain the difference in rates.

On the basis of his investigations, Durkheim argued that some *social fact* was the cause of the observed difference in suicide rates. He claims that the lower suicide rate among Catholics could be explained by their greater degree of social integration. Degree of social integration, for Durkheim, was a primary example of a social fact, that is to say, a fact of human behavior that cannot be entirely explained in terms of individual beliefs and desires. Durkheim argued that social facts were causal forces, and that it was appropriate, and indeed necessary, to appeal to such forces when we frame causal explanations of social phenomena.

Durkheim's holism (or collectivism)—the belief that there are facts about societies that cannot be entirely accounted for in terms of facts about the individual members who make up the society or the relationships that hold among individuals— has many adherents among social scientists, as does the opposing position of individualism. Although space is not taken to discuss the arguments for and against various versions of collectivism and individualism in this chapter,[1] it is interesting to note that Durkheim's holism does not preclude his agreement with Mill about the possibility of a social science. Durkheim accepts—and indeed urges students of human behavior to adopt—the aims and methods of science. Although he believes that the ultimate causes of human behavior are social norms and other social facts, rather than Mill's individualistic "Laws of Thought," Durkheim and Mill agree that behavior is governed by causes. Durkheim believes that many apparently voluntary actions are dictated by social pressures of which we are unaware. He does not deny that humans can "break the rules" of society, but he insists that they do so only at some social cost, which can range from personal feelings of guilt to mild disapproval, or more severe penalties, from others. Thus, whether or not we comply with the rules of society, its influence on us cannot be avoided.

11.6 CRITICAL THEORY

We began this chapter with doubts about the possibility of the scientific study of creatures with free will. After raising the question, we put it aside, since Mill, the first representative of Naturalism that we considered, regarded the problem as one of predictability, and he believed that human behavior, while not predictable with complete accuracy, was predictable enough to allow the development of an inexact science. Mill did not deny that humans have free will. He believed that humans act freely when they act without coercion, that is to say, when their acts are caused by their own beliefs and desires instead of being constrained by other forces. This position, called "compatibilism" (see Chapter 6), attempts to reconcile the claim that

[1] Social facts are discussed in Chapter 4. The main question is whether social facts can be reduced to facts about individual members of a society in every case. See Chapter 8 for a discussion of reductionism. For further discussion of social facts and references, see Papineau (1978, Chapter 1).

human actions can be caused and yet free. It represents a compromise that appeals to many philosophers and social scientists.

However, compatibilism does not really come to grips with the problem of the causes of beliefs and desires themselves. On a compatibilist account, for example, even if an agent's beliefs and desires are completely molded (caused) by social forces, for example, the acts arising from those beliefs and desires are regarded as free. Yet, the situation just described hardly seems to afford a satisfactory context in which to talk about human freedom.

"Freedom" has sometimes been explained as a forensic notion. That is to say, an agent's actions are said to be free insofar as we are willing to hold the agent morally, legally, or socially responsible for those actions. On such an analysis of freedom, an agent who had been brainwashed to hold certain beliefs might not be held responsible for actions arising from those beliefs, though this would depend on a judgement about how effective the brainwashing was. Freedom and responsibility are certainly closely linked in our distribution of blame and praise for human actions. If we know that a person managed to accomplish something *purely by accident,* we usually would not praise or blame the person for the action. In making judgements about specific actions, sorting out accidental or mitigating conditioning factors can be very difficult. On the whole, however, despite some lack of clarity in the compatibilist position, it seems reasonable to link the notion of human freedom with that of "internal" (arising from an agent's own beliefs and desires) as opposed to "external" causation, instead of maintaining that free behavior is simply uncaused.

Unlike compatibilists, social scientists who are strict behaviorists in the tradition of the psychologist B. F. Skinner regard human freedom as an illusion, and a dangerous one at that. They believe that the illusion of freedom is dangerous because it can hinder us from learning the true causes of human behavior and thus prevent us from using such knowledge to improve society. Human behavior, according to Skinner (1972), occurs in response to various stimuli and schedules of reinforcement, that is, rewards and punishments (see Chapter 9). Humans undeniably have feelings of resolve, guilt, and the like, but, according to behaviorists, these feelings do not represent accurately what is happening. Once we understand what really causes behavior, behaviorists say, we can control it by providing the appropriate stimulation and reinforcement.

Behaviorism is not as popular as it once was, mainly because it has failed to live up to its promise of accounting for complex human behavior in behavioristic terms. The extreme behaviorists have not been able to translate convincingly all mental terms (such as belief, desire, hope, intelligence, and fear) into descriptions of physical or linguistic behavior. Even the less extreme behaviorists, who claim only that behavior can be described satisfactorily in terms of the mechanism of stimulus-response-reinforcement, without any appeal to mental concepts (beliefs and desires), have had major successes with simple patterns of animal behavior in controlled laboratory studies. Although some forms of behavioral therapy are recognized as effective, much complex human behavior seems to resist a behaviorist analysis. For example, on a behaviorist account, it is difficult to understand how very young children acquire facility with language, including the ability to frame sentences to which they have never been exposed (see Chapter 10).

Despite the relative lack of enthusiasm for strict behaviorism in the social sciences, an influential group of social theorists, who call themselves *critical theorists,* regard behaviorism's program for manipulating and controlling behavior as a serious threat to human autonomy. Critical theorists also believe that behaviorism is representative of what is wrong with the whole enterprise of social sciences. These theorists, who are associated with the Frankfurt School, include Horkheimer, Adorno, Marcuse, Habermas, Apel, and others. Although they disagree among themselves on many points, all of them regard any attempt to model social science on the pattern of the physical sciences as a denial of human freedom, and thus both mistaken and immoral.

Like the interpretivists, critical theorists believe that the attempt to provide covering-law explanations in social science involves a fundamental confusion between causal laws and normative rules. In addition, they point out that explanations in the physical sciences are deliberately ahistorical, a position that is inappropriate for human sciences.

Not everyone agrees that physical sciences are all ahistorical. Evolutionary biology, geology, and cosmology, they point out, are *historical* sciences. In addition, even in physics, certain irreversible processes give rise to physical laws that embody historical and contextual features, so it is simply incorrect to say that all explanations in physics are ahistorical (Grünbaum 1984). Although critical theorists acknowledge that this is so, they nevertheless insist that the requisite sense of "historical" is not applicable to such laws or to the so-called historical sciences. For example, Apel says:

> It is true, I think that physics has to deal with irreversibility in the sense of the second principle of thermodynamics . . . But, in this very sense of irreversibility, physics may support nature's being definitely determined concerning its future and thus having no history in a sense that would resist ontological objectification.
>
> Contrary to this, social science . . . must not only suppose irreversibility—in the sense of a statistically determined process—but irreversibility, in the sense of the advance of human knowledge influencing the process of history in an irreversible manner. (1979, 20)

Apel's remarks suggest that when critical theorists refer to the special historic character of social science, they are using the term "history" in a special way, namely, in Collingwood's sense. That is to say, they recognize that humans—unlike mere physical bodies—are often able to use their knowledge of what has happened to redirect the course of events, and to change the outcomes. Since humans are conscious agents with purposes of their own, they shape history in ways not possible for any nonthinking part of nature.

Like Collingwood, critical theorists want to reserve the term "history" to refer to accounts of autonomous human behavior. Recall that Collingwood does not apply the term "historical" to any processes of nature that do not involve human intentions—not even to geological and evolutionary processes which are usually called historical, for, as he says:

> The processes of nature can therefore be properly described as sequences of mere events, but those of history cannot. They are not processes of mere events but processes of

actions, which have an inner side, consisting of processes of thought; and what the historian is looking for is these processes of thought. All history is the history of thought. (Collingwood 1946, 215)

The broader critical program with its emphasis on history becomes clearer when we consider that critical theorists are convinced that most contemporary social systems are oppressive, unjust, and in need of reform. They believe that trying to model social sciences on the pattern of the physical sciences is immoral because it stands in the way of positive social change. Critical theorists say that although the regularities that are observed in our present social system are the result of unfortunate historical circumstances, when social scientists attempt to present these regularities as *laws*, they suggest an analogy to unchangeable laws of nature. Furthermore, explanations of the present corrupt situation that invoke these "laws" suggest that a better social arrangement is not possible (since it would violate the laws). Critical theorists fear that adopting the physical scientists' model of explanation for social phenomena will serve to maintain the status quo of unjust and oppressive social institutions.

Contrary to what critical theorists suppose, however, scientists do not regard every observed regularity as a law. No social scientist is prevented by a commitment to science from distinguishing between accidental connections or passing trends and genuine laws. Nor does the existence of laws imply the absence of social change. A change in conditions can result in changes for better or worse. Thus a social science that is committed to providing scientific explanations is not thereby committed to serve the ends of regimes that want to maintain their dominance by making any existing social arrangements seem inevitable outcomes of natural laws.

Because critical theorists believe that the primary goals of the physical sciences are the prediction and control of the physical environment, they think that a social science that is similar in its methods and aims must have the prediction and control of human behavior as its goal. Such a human science, they say, would be inherently manipulative and thus ethically repugnant.

Although, as we have seen, some of the extravagant claims of behaviorism might lend support to the view that prediction and control of behavior is the aim of human science, this view is inadequate. The picture presented of natural science as mere technique is false. The primary aim of science is the acquisition of knowledge about the world. This is not to say that science is morally neutral. The acquisition of knowledge can raise moral questions. For example, if subjecting humans to certain forms of torture is the only way to learn how they will respond to such treatment, moral considerations may prevent obtaining that knowledge. How knowledge is obtained, how it is used, and what resources should be committed to its acquisition are all questions that raise important moral issues that must be addressed. Legitimate moral concerns are not advanced, however, by misrepresenting or misunderstanding the aims of science. Moreover, the critical theorists' view that science is directed primarily at the manipulation and control of nature is at odds with their own opinion that science lends support to the status quo and is insensitive to possibilities of change and reform.

Even though the critical theorists' assessments of the character of physical science is inaccurate, their ethical worries, like those of MacIntyre, concerning a social science may be well founded. Clearly any scientific study of humans must be sensitive to the values and fundamental human rights of the subjects (see the discussion of medical ethics in Chapter 8). It may be considered immoral, for example, to subject humans to experimental drugs without obtaining their informed consent and taking every precaution to avoid bad consequences. It might also be deemed immoral to cause psychological stress and discomfort for the purpose of gaining knowledge of how humans handle such stress.

In both the physical and the social sciences, it may be reasonable to hold that moral considerations should guide the choice of research problems and the expenditure of funds to support such research. With limited resources, not every research program can be supported, and although it would be imprudent to reject studies just because there is no ''immediate payoff'' in human benefits, surely the promise of improvement in the human condition should be an important if not overriding factor in making research decisions.

In addition, if social scientists are asked to offer ''expert advice'' to lawmakers empowered to enact recommendations that affect the quality of human life, the social scientists have a strong moral obligation not to overstate the reliability of their results. For example, directing primary-school students into various academic and vocational tracks on the basis of early testing can unfairly deprive those students of their right to pursue a career of their own choosing unless there is a recognition of the fallibility of such tests, along with provisions for reevaluation and changes in status. Social scientists would be acting immorally if they were to use the aura of ''science'' to put forth the present findings of their disciplines in the same light as well-founded physical theories. When the implementation of one's theories (in social or in physical sciences) can gravely affect human life, we have the moral responsibility to admit the fallibility of the theories and to proceed with care in their implementation.

Most of the above considerations apply to the conduct of physical scientists as well as social sciences. The claim that facts and values can be neatly separated and that science is concerned only with facts and not with values is insupportable, as the above examples and many others show, even with strictures and qualifications with respect to the physical sciences. In addition, since the nature of values and value judgements are such a central concern of human social life, a concern with understanding those values would seem to be an integral component of the social sciences.

11.7 CONCLUSION

The problem with which this chapter began was whether a *science* of human behavior could be modeled on the natural sciences. In the course of the chapter we have considered some positive and negative responses to this question, but have shown most sympathy to the naturalistic position. The criticisms offered by interpretivists and critical theorists of the possibility of social sciences suffer from some misunderstanding of the aims and methods of the physical sciences.

At the same time, we are confronted with the reality of very slow progress in the social sciences and the naturalists' inability to explain this in a satisfactory way. The answer that social sciences are much newer and just have not had time to progress is weak, because humans have been concerned with these issues as long or longer than they have been concerned with the nature of the physical world. The answer that the subject matter is more complex is similarly questionable, because the physical world is certainly not lacking in complexity. What does seem to be true, however, is that in the physical sciences, everyone agrees that some of the complexities can be ignored in the pursuit of law-like generalizations. Consider, for example, the decision, widely adopted in the seventeenth century (though not without some dissent), to focus on the "primary" qualities of the material world, such as mass and velocity, and to ignore for the purposes of scientific knowledge such features as color, taste, smell, sound, and tactile qualities. This decision allowed a great surge of progress in science. The primary qualities lent themselves more easily to measurement with standard instruments, and this in turn brought to light relationships that could be expressed in mathematical terms.

With the advance of statistical techniques, social scientists have been able to measure and quantify features of social life, and to study important relationships that hold between various quantities, but we are still far from agreement about what features of social life can be ignored, even temporarily, in choosing features to quantify. We are never so sure in the social sciences as we are in the physical sciences that what we are counting or measuring is what is essential for understanding. The selection of variables is closely tied to the purposes that we have for studying human behavior, and these purposes are themselves varied and complex, so we often wonder whether we have captured the important variables. Judgements of value are notoriously difficult to quantify, and human preferences do not always follow the rigid criteria set by the economists' standard of rationality. Our uncertainty about these matters may be the major reason for our uneasiness about framing laws in social science. Furthermore, the social scientists' proposals for laws often appear to be vague or trivial because of the caution they rightly exercise in their selection of variables.

Since the seventeenth century, the ability to state relationships in quantitative terms has been an intrinsic part of what we mean by scientific knowledge. Why do we place so much emphasis on quantification in science? This too is an extremely difficult and inadequately investigated question, but part of the answer seems to lie in the ease of transmission of quantitative information. Scientific knowledge is a cooperative enterprise, and we expect the results and discoveries of individual workers to be communicated and understood by others with minimal ambiguity. But, as even Mill (1874, Book 6, Chapter 7) recognized, the progress of social sciences can be hindered by mindless attempts to apply the techniques that have worked so well in the physical sciences to human behavior before we have decided what it is we really want to measure and what we are willing to ignore. Until we make further progress in solving this difficult problem, the goal of naturalistic social science will be hard to attain.

Philosophy of the social sciences, like the area it studies, is much less well developed than the philosophy of the physical sciences and the philosophy of the biological sciences. The underdevelopment of the field is at least partly a result of less attention to it by philosophers. Ironically, however, philosophical work may be more important for the social sciences than for more advanced sciences, for philosophical reflection on human capacities, needs, and values will certainly be necessary to make informed decisions about reducing complexity in characterizations of the social world while preserving what we most want to know about ourselves as social beings. The possibility of a social science built on sound philosophical foundations exists and is worthy of the immense effort required to transform the possibility to a reality.

DISCUSSION QUESTIONS

1. Criticize or defend the Interpretivist position that the relationship between reasons and actions is not a causal relationship.

2. What is meant by the claim that social sciences are less well developed than the physical sciences?

3. What do you think is the best explanation for the less well developed state of the social sciences?

4. One of the chief tenets of modern philosophy of science (i.e., Logical Positivism, Logical Empiricism) is that science is value neutral. Can there be a value neutral philosophy of the social sciences? Defend your answer against possible criticisms.

5. In *Aspects* (1965b) and in "Explanation in Science and History" (1962b), Hempel defends the position that explanations of human behavior have the form of covering-law explanation. Critically discuss his account, with special attention to the question of whether the suggested laws in such explanations have empirical content.

6. Explain why critical theorists fear that a science of human behavior would threaten autonomy.

7. Discuss why there can (or cannot) be a science that has as its subject matter individuals who have free will. In your discussion, be sure to explain what having free will means.

8. Critically discuss the following argument against individualism: "People do not think and behave in the same way when in social groups as they would as isolated individuals. So the properties of social situations cannot derive solely from the thoughts and actions of single individuals" (Papineau 1978, 8).

SUGGESTED READINGS

In recent years, a number of introductory books addressed to a nonspecialist audience on the philosophy of social science have been published. A selection of books which contain useful bibliographies, is listed below:

BRAYBROOKE, DAVID (1987), *Philosophy of Social Science*. Englewood Cliffs, NJ: Prentice-Hall. Braybrooke discusses in considerable detail three different approaches to social science: naturalism, interpretivism, critical theory.

LITTLE, DANIEL (1990), *Varieties of Social Explanation: An Introduction to the Philosophy of Social Science*. Boulder: Westview Press. Little examines many actual examples of explanation in various fields of social science.

PAPINEAU, DAVID (1978), *For Science in the Social Sciences*. New York: St. Martin's Press. Papineau defends the naturalist view against the criticisms of interpretivists and critical theorists in his discussion of some central problems in the social sciences.

ROSENBERG, ALEXANDER (1988), *Philosophy of Social Science*. Boulder: Westview Press. An excellent introduction to the field that relates issues in the social sciences to central concerns of contemporary philosophy.

BIBLIOGRAPHY

ABELSON, ROBERT P. (1981), "Psychological Status of the Script Concept," *American Psychologist 36:* 715–729.

ADAMS, ERNEST W. (1959), "The Foundations of Rigid Body Mecahnics and the Derivation of Its Laws from Those of Particle Mechanics," in Henkin, Suppes and Tarski, pp. 250–265.

ALLÉN, STURE (ed.) (1989), *Possible Worlds in Humanities, Arts and Sciences: Proceedings of Nobel Symposium 65.* Berlin: de Gruyter.

APEL, KARL-OTTO (1979), "Types of Social Science in the Light of Human Cognitive Interests," in Brown, pp. 3–50.

ARISTOTLE (1970), *Aristotelis Opera.* Volume 1. Edited by I. Bekker. Berlin: de Gruyter.

ARNTZENIUS, FRANK (1990), "Casual Paradoxes in Special Relativity," *The British Journal for the Philosophy of Science 41:* 223–243.

ASQUITH, PETER D. and THOMAS NICKLES (eds.) (1983), *PSA 1982.* Volume 2. East Lansing, MI: Philosophy of Science Association.

ATTNEAVE, FRED (1971), "Multistability in Perception," *Scientific American 225* (6): 62–71.

AUSTIN, D. F. (ed.) (1988), *Philosophical Analysis.* Dordrecht: Kluwer Academic.

AYALA, FRANCISCO J. (1982), *Population and Evolutionary Genetics: A Primer.* Menlo Park, CA: Benjamin/Cummings.

AYER, A. J. (1952), *Language, Truth, and Logic.* New York: Dover.

––––––– (1956), *The Problem of Knowledge.* New York: St. Martin's Press.

––––––– (ed.) (1959), *Logical Positivism.* New York: The Free Press.

BACHELARD, GASTON (1984), *The New Scientific Spirit.* Translated by Arthur Goldhammer. Boston: Beacon Press.

BALME, D. M. (1987), "Teleology and Necessity," in Gotthelf and Lennox, pp. 275–285.

BALZER, W. and C. M. DAWE (1986), "Structure and Comparison of Genetic Theories: (1) Classical Genetics; (2) The Reduction of Character-Factor Genetics to Molecular Genetics," *The British Journal for the Philosophy of Science 37:* 55–69 and 177–191.

BARKER, PETER and CECIL G. SHUGART (eds.) (1981), *After Einstein: Proceedings of the Einstein Centennial Celebration at Memphis State University.* Memphis, TN: Memphis State University Press.

BARNES, BARRY (1974), *Scientific Knowledge and Sociological Theory.* London: Routledge & Kegan Paul.

——— (1977), *Interests and the Growth of Knowledge.* London: Routledge & Kegan Paul.

BAUMRIN, BERNARD (ed.) (1963), *Philosophy of Science: The Delaware Seminar.* Volume 2. New York: Interscience.

BEATTY, JOHN (1984), "Chance and Natural Selection," *Philosophy of Science 51:* 183–211.

BEAUCHAMP, TOM L. and JAMES F. CHILDRESS (1989), *Principles of Biomedical Ethics.* 3rd ed. New York: Oxford University Press.

BECHTEL, WILLIAM (ed.) (1986a), *Integrating Scientific Disciplines.* Dordrecht: Martinus Nijhoff.

——— (1986b), "The Nature of Scientific Integration," in Bechtel, pp. 3–52.

BEINLICH, I.; H. SUERMONDT; R. CHAVEZ; and G. COOPER (1989), "The Alarm Monitoring System," in *Proceedings of the Second European Conference on Artificial Intelligence in Medicine.* London.

BERGER, PETER L. and THOMAS LUCKMANN (1967), *The Social Construction of Reality: A Treatise in the Sociology of Knowledge.* Garden City, NY: Anchor Books.

BEROFSKY, BERNARD (ed.) (1966), *Free Will and Determinism.* New York: Harper & Row.

BLACK, MAX (ed.) (1950), *Philosophical Analysis: A Collection of Essays.* Ithaca: Cornell University Press.

——— (1954), *Problems of Analysis: Philosophical Essays.* Ithaca: Cornell University Press.

BLAKEMORE, COLIN and SUSAN GREENFIELD (eds.) (1987), *Mindwaves: Thoughts on Intelligence, Identity and Consciousness.* Oxford: Blackwell.

BLOCK, NED (ed.) (1980), *Readings in Philosophy of Psychology.* Volume 1. Cambridge, MA: Harvard University Press.

BLOOR, DAVID (1976), *Knowledge and Social Imagery.* London: Routledge & Kegan Paul.

——— (1981), "The Strengths of the Strong Programme," *Philosophy of the Social Sciences 11:* 199–213.

——— (1983), *Wittgenstein: A Social Theory of Knowledge.* New York: Columbia University Press.

BOAG, PETER T. and PETER R. GRANT ([1981] 1982), "Intense Natural Selection in a Population of Darwin's Finches (*Geospizinae*) in the Galápagos," in Maynard Smith, pp. 175–181. Originally appeared in *Science 214:* 82–85.

BODEN, MARGARET A. (1988), *Computer Models of Mind: Computational Approaches in Theoretical Psychology.* Cambridge, England: Cambridge University Press.

BORN, MAX (1971), *The Born-Einstein Letters: Correspondence between Albert Einstein and Max and Hedwig Born from 1916 to 1955.* Translated by Irene Born. London: Macmillan.

BOURDIEU, PIERRE (1975), "The Specificity of the Scientific Field and the Social Conditions of the Progress of Reason," *Social Science Information 14* (6): 19–47.

——— (1977), *Outline of a Theory of Practice.* Translated by Richard Nice. Cambridge, England: Cambridge University Press.

BOWLER, PETER J. (1989), *Evolution: The History of an Idea*. Revised edition. Berkeley and Los Angeles: University of California Press.

BRAITHWAITE, R. B. (1953), *Scientific Explanation: A Study of the Function of Theory, Probability and Law in Science*. Cambridge, England: Cambridge University Press.

BRANDON, ROBERT N. (1978), "Adaptation and Evolutionary Theory," *Studies in History and Philosophy of Science 9:* 181–206. Reprinted in Sober (1984a), pp. 58–82.

———— (1981), "Biological Teleology: Questions and Explanations," *Studies in History and Philosophy of Science 12:* 91–105.

———— (1985), "Adaptation Explanations: Are Adaptations for the Good of Replicators or Interactors?" in Depew and Weber, pp. 81–96.

———— (1990), *Adaptation and Environment*. Princeton: Princeton University Press.

BRANDON, ROBERT N. and JOHN BEATTY (1984), "The Propensity Interpretation of 'Fitness'—No Interpretation Is No Substitute," *Philosophy of Science 51:* 342–347.

BRANDON, ROBERT N. and RICHARD M. BURIAN (eds.) (1984), *Genes, Organisms, Populations: Controversies over the Units of Selection*. Cambridge, MA: MIT Press.

BRAYBROOKE, DAVID (1987), *Philosophy of Social Science*. Englewood Cliffs, NJ: Prentice-Hall.

BRIDGMAN, P. W. (1927), *The Logic of Modern Physics*. New York: Macmillan.

BROAD, CHARLIE DUNBAR (1926), *The Philosophy of Francis Bacon*. Cambridge, England: Cambridge University Press.

BROOKS, DANIEL R. and E. O. WILEY (1986), *Evolution as Entropy: Toward a Unified Theory of Biology*. Chicago: University of Chicago Press.

BROWN, S. C. (ed.) (1979), *Philosophical Disputes in the Social Sciences*. Atlantic Highlands, NJ: Humanities Press.

BRUNER, JEROME S. (1957), "On Perceptual Readiness," *Psychological Review 64:* 123–152.

———— (1973), *Beyond the Information Given: Studies in the Psychology of Knowing*. New York: Norton.

BRUNER, JEROME S. and CECILE C. GOODMAN (1947), "Value and Need as Organizing Factors in Perception," *The Journal of Abnormal and Social Psychology 42:* 33–44.

BRUNER, JEROME S. and MARY C. POTTER (1964), "Interference in Visual Recognition," *Science 144:* 424–425.

BURIAN, RICHARD M. (1983), " 'Adaptation'," in Grene, pp. 287–314.

BURKS, ARTHUR W. (1953), "The Presupposition Theory of Induction," *Philosophy of Science 20:* 177–197.

———— (1977), *Chance, Cause, Reason: An Inquiry into the Nature of Scientific Evidence*. Chicago: University of Chicago Press.

CAHN, STEVEN M. (ed.) (1977), *Classics of Western Philosophy*. Indianapolis: Hackett.

CARNAP, RUDOLF (1936), "Testability and Meaning," *Philosophy of Science 3:* 419–471.

———— (1937), *The Logical Syntax of Language*. Translated by Amethe Smeaton. New York: Harcourt, Brace.

———— (1950), *Logical Foundations of Probability*. Chicago: University of Chicago Press.

———— (1956a), "Empiricism, Semantics, and Ontology," in Carnap (1956 b), pp. 205–221.

———— (1956b), *Meaning and Necessity: A Study in Semantics and Modal Logic*. 2d ed. Chicago: University of Chicago Press.

———— ([1950] 1962), *Logical Foundations of Probability*. 2d ed. New Preface. Chicago: University of Chicago Press.

—— (1963), "Replies and Systematic Expositions," in Schilpp, pp. 859–1013.

—— ([1928] 1967), *The Logical Structure of the World: Pseudoproblems in Philosophy.* Translated by Rolf A. George. Berkeley and Los Angeles: University of California Press.

CARTWRIGHT, NANCY (1983), *How the Laws of Physics Lie.* Oxford: Clarendon Press.

CAUSEY, ROBERT L. (1977), *Unity of Science.* Dordrecht: Reidel.

CHISHOLM, RODERICK M. (1982), "Human Freedom and the Self," in Watson, pp. 24–35.

CHOMSKY, NOAM (1959), Review of *Verbal Behavior,* by B. F. Skinner. *Language 35:* 26–58.

CHRISTENSEN, DAVID (1983), "Glymour on Evidential Relevance," *Philosophy of Science 50:* 471–481.

CHURCHLAND, PATRICIA SMITH (1986), *Neurophilosophy: Toward a Unified Science of the Mind/ Brain.* Cambridge, MA: Bradford/MIT Press.

—— (1988), "The Significance of Neuroscience for Philosophy," *Trends in Neurosciences 11:* 304–307.

CHURCHLAND, PAUL M. (1979), *Scientific Realism and the Plasticity of Mind.* Cambridge, England: Cambridge University Press.

—— (1981), "Eliminative Materialism and the Propositional Attitudes," *The Journal of Philosophy 78:* 67–90.

—— (1984), *Matter and Consciousness: A Contemporary Introduction to the Philosophy of Mind.* Cambridge, MA: Bradford/MIT Press.

COHEN, R. S.; C. A. HOOKER; A. C. MICHALOS; and J. W. VAN EVRA (eds.) (1976), *PSA 1974.* Dordrecht: Reidel.

COLLINGWOOD, R. G. (1946), *The Idea of History.* Oxford: Oxford University Press.

COLLINS, H. M. (1985), *Changing Order: Replication and Induction in Scientific Practice.* Beverly Hills and London: Sage Publications.

COLODNY, ROBERT G. (ed.) (1962), *Frontiers of Science and Philosophy.* Pittsburgh: University of Pittsburgh Press.

—— (ed.) (1965), *Beyond the Edge of Certainty: Essays in Contemporary Science and Philosophy.* Englewood Cliffs, NJ: Prentice-Hall.

—— (ed.) (1970), *The Nature and Function of Scientific Theories.* Pittsburgh: University of Pittsburgh Press.

—— (ed.) (1977), *Logic, Laws, and Life: Some Philosophical Complications.* Pittsburgh: University of Pittsburgh Press.

COMMITTEE ON MODELS FOR BIOMEDICAL RESEARCH (1985), *Models for Biomedical Research: A New Perspective.* Washington, DC: National Academy Press.

CONVERSE, PHILIP E. (1986), "Generalization and the Social Psychology of 'Other Worlds'," in Fiske and Shweder, pp. 42–60.

COOPER, JOHN M. (1987), "Hypothetical Necessity and Natural Teleology," in Gotthelf and Lennox, pp. 243–274.

COPERNICUS, NICOLAUS (1976), *On the Revolutions of the Heavenly Spheres.* Translated by A. M. Duncan. New York: Barnes & Noble.

COTMAN, CARL W.; ROBERTA E. BRINTON; ALBERT GALABURDA; BRUCE McEWEN; and DIANA M. SCHNEIDER (eds.) (1987), *The Neuro-Immune-Endocrine Connection.* New York: Raven Press.

CULP, SYLVIA and PHILIP KITCHER (1989), "Theory Structure amd Theory Change in Contemporary Molecular Biology," *The British Journal for the Philisophy of Science 40:* 459–483.

CUSHING, JAMES T. and ERNAN MCMULLIN (eds.) (1989), *Philosophical Consequences of Quantum Theory: Reflections on Bell's Theorem*. Notre Dame: University of Notre Dame Press.

CUTLAND, NIGEL (1980), *Computability: An Introduction to Recursive Function Theory*. Cambridge, England: Cambridge University Press.

DARDEN, LINDLEY and NANCY MAULL (1977), "Interfield Theories," *Philosophy of Science 44:* 43–64.

DARWIN, CHARLES (1887), *The Life and Letters of Charles Darwin, Including an Autobiographical Chapter*. 3rd ed. Volume 2. Edited by Francis Darwin. London: Murray.

——— ([1892] 1958), *The Autobiography of Charles Darwin and Selected Letters*. Edited by Francis Darwin, New York: Dover.

——— ([1872] 1962), *The Origin of Species by Means of Natural Selection: or, The Preservation of Favoured Races in the Struggle for Life*. 6th ed. New York: Collier.

——— ([1859] 1964), *On the Origin of Species: A Facsimile of the First Edition*. Cambridge, MA: Harvard University Press.

DAVIDSON, DONALD (1980), *Essays on Actions and Events*. Oxford: Clarendon Press.

——— (1984), *Inquiries into Truth and Interpretation*. Oxford: Clarendon Press.

DAWES, ROBYN M. (1988), *Rational Choice in an Uncertain World*. San Diego: Harcourt Brace Jovanovich.

DAWKINS, RICHARD (1982), *The Extended Phenotype: The Gene as the Unit of Selection*. Oxford: Freeman.

DENNETT, DANIEL C. (1981), *Brainstorms: Philosophical Essays on Mind and Psychology*. Cambridge, MA: MIT Press.

——— (1987), *The Intentioinal Stance*. Cambridge, MA: Bradford/MIT Press.

DEPEW, DAVID J. and BRUCE H. WEBER (eds.) (1985), *Evolution at a Crossroads: The New Biology and the New Philosophy of Science*. Cambridge, MA: Bradford/MIT Press.

DEVITT, MICHAEL (1984), *Realism and Truth*. Princeton: Princeton University Press.

DEWITT, BRYCE S. and NEILL GRAHAM (eds.) (1973), *The Many-Worlds Interpretation of Quantum Mechanics*. Princeton: Princeton University Press.

DOBZHANSKY, THEODOSIUS (1970), *Genetics of the Evolutionary Process*. New York: Columbia University Press.

DREYFUS, HUBERT L. (1972), *What Computers Can't Do: A Critique of Artificial Reason*. New York: Harper & Row.

DUHEM, PIERRE ([1906] 1954), *The Aim and Structure of Physical Theory*. Translated by Philip P. Wiener. Princeton: Princeton University Press.

DUMMETT, MICHAEL (1978), *Truth and Other Enigmas*. Cambridge, MA: Harvard University Press.

DUNN, J. MICHAEL and GEOFFREY HELLMAN (1986) "Dualling: A Critique of an Argument of Popper and Miller," *The British Journal for the Philosophy of Science 37:* 220–223.

DURKHEIM, EMILE ([1897] 1951), *Suicide: A Study in Sociology*. Translated by John A. Spaulding and George Simpson. New York: The Free Press.

EARMAN, JOHN (ed.) (1983), *Minnesota Studies in the Philosophy of Science*. Volume 10, *Testing Scientific Theories*. Minneapolis: University of Minnesota Press.

——— (1986), *A Primer on Determinism*. Dordrecht: Reidel.

——— (1989), *World Enough and Space-Time: Absolute versus Relational Theories of Space and Time*. Cambridge, MA: Bradford/MIT Press.

EARMAN, JOHN and JOHN NORTON (1987), "What Price Spacetime Substantivalism: The Hole Story," *The British Journal for the Philosophy of Science 38:* 515–525.

EARMAN, JOHN; CLARK GLYMOUR; and JOHN STACHEL (eds.) (1977), *Minnesota Studies in the Philosophy of Science.* Volume 8, *Foundations of Space-Time Theories.* Minneapolis: University of Minnesota Press.

EDIDIN, ARON (1983), "Bootstrapping without Bootstraps," in Earman, pp. 43–54.

EELLS, ELLERY (1985), "Problems of Old Evidence," *Pacific Philosophical Quarterly 66:* 283–302.

EINSTEIN, ALBERT ([1916] 1952a), "The Foundation of the General Theory of Relativity," in Lorentz, Einstein, Minkowski and Weyl, pp. 109–164.

——— ([1905] 1952b), "On the Electrodynamics of Moving Bodies," in Lorentz, Einstein, Minkowski and Weyl, pp. 35–65.

———([1921] 1954a), "Geometry and Experience," in Einstein (1954b), pp. 232–246.

——— (1954b), *Ideas and Opinions.* New York: Bonanza.

——— ([1917] 1954c), *Relativity: The Special and the General Theory.* 15th ed. Translated by Robert W. Lawson. New York: Crown.

——— ([1922] 1956), *The Meaning of Relativity.* 5th ed. Translated by Edwin Plimpton Adams, Ernst G. Straus and Sonja Bargmann. Princeton: Princeton University Press.

——— ([1948] 1971), "Quantum Mechanics and Reality," in Born, pp. 168–173.

EINSTEIN, A.; B. PODOLSKY; and N. ROSEN (1935), "Can Quantum-Mechanical Description of Reality Be Complete?" *Physical Review 47:* 777–780. Reprinted in Wheeler and Zurek (1983), pp. 138–141.

ELSTER, JON (1983), *Explaining Technical Change: A Case Study in the Philosophy of Science.* Cambridge, England: Cambridge University Press.

ENDLER, JOHN A. (1986), *Natural Selection in the Wild.* Princeton: Princeton University Press.

ENGEL, GEORGE L. (1977), "The Need for a New Medical Model: A Challenge for Biomedicine," *Science 196:* 129–136.

——— (1981), "The Clinical Application of the Biopsychosocial Model," *The Journal of Medicine and Philosophy 6:* 101–123.

ENGLEHARDT, JR., H. TRISTRAM (1986), "From Philosophy *and* Medicine to Philosophy *of* Medicine," *The Journal of Medicine and Philosophy 11:* 3–8.

EUCLID (1956), *The Thirteen Books of Euclid's Elements.* 2d ed. 3 volumes. Translated by Thomas L. Heath. New York: Dover.

FAUST, DAVID (1984), *The Limits of Scientific Reasoning.* Minneapolis: University of Minnesota Press.

FEIGL, HERBERT (1950), "De Principiis Non Disputandum . . . ?" in Black, pp. 119–156.

——— (1953), "Notes on Causality," in Feigl and Brodbeck, pp. 408–418.

FEIGL, HERBERT and MAY BRODBECK (eds.) (1953), *Readings in the Philosophy of Science.* New York: Appleton-Century-Crofts.

FEIGL, HERBERT and GROVER MAXWELL (eds.) (1962), *Minnesota Studies in the Philosophy of Science.* Volume 3, *Scientific Explanation, Space, and Time.* Minneapolis: University of Minnesota Press.

FEYERABEND, PAUL K. (1962), "Explanation, Reduction, and Empiricism," in Feigl and Maxwell, pp. 28–97.

———— (1963), "How to be a Good Empiricist—A Plea for Tolerance in Matters Epistemological," in Baumrin, pp. 3–39.

———— (1965), "Problems of Empiricism," in Colodny, pp. 145–260.

———— (1970), "Problems of Empiricism, Part II," in Colodny, pp. 275–353.

———— (1978), *Against Method*. London: Verso Press.

FEYNMAN, RICHARD (1985), *Surely You're Joking, Mr. Feynman*. New York: Norton.

FIELD, HARTRY (1973), "Theory Change and the Indeterminacy of Reference," *The Journal of Philosophy 70:* 462–481.

———— (1980), *Science Without Numbers: A Defence of Nominalism*. Princeton: Princeton University Press.

FINE, ARTHUR (1981), "Einstein's Critique of Quantum Theory: The Roots and Significance of EPR," in Barker and Shugart, pp. 147–158.

———— (1986), *The Shaky Game: Einstein, Realism, and the Quantum Theory*. Chicago: University of Chicago Press.

———— (forthcoming), "Indeterminism and Freedom of the Will," to appear in a *Festschrift* for Adolf Grünbaum. University of Konstanz Press and University of Pittsburgh Press.

FINE, ARTHUR and JARRETT LEPLIN (1989), *PSA 1988*. Volume 2. East Lansing, MI: Philosophy of Science Association.

FINE, ARTHUR and PETER MACHAMER (eds.) (1987), *PSA 1986*. East Lansing, MI: Philosophy of Science Association.

FISKE, DONALD W. and RICHARD A. SHWEDER (eds.) (1986), *Metatheory in Social Science: Pluralisms and Subjectivities*. Chicago: University of Chicago Press.

FLECK, LUDWIK ([1935] 1979), *Genesis and Development of a Scientific Fact*. Edited by Thaddeus J. Trenn and Robert K. Merton. Translated by Fred Bradley and Thaddeus J. Trenn. Chicago: University of Chicago Press.

FLOCK, HOWARD R. and ANTHONY MOSCATELLI (1964), "Variables of Surface Texture and Accuracy of Space Perceptions,"*Perceptual and Motor Skills 19:* 327–334.

FODOR, JERRY A. (1968), *Psychological Explanation: An Introduction to the Philosophy of Psychology*. New York: Random House.

———— (1987), *Psychosemantics: The Problem of Meaning in the Philosophy of Mind*. Cambridge, MA: Bradford/MIT Press.

FORSTROM, LEE A. (1977), "The Scientific Autonomy of Clinical Medicine," *The Journal of Medicine and Philosophy 2:* 8–19.

FOUCAULT, MICHEL (1978), *The History of Sexuality*. Volume 1. Translated by Robert Hurley. New York: Vintage.

FREGE, GOTTLOB (1972), *Conceptual Notation*, Edited by Terry Bynum. Oxford: Oxford University Press.

FREUD, SIGMUND ([1922] 1963a), "Certain Neurotic Mechanisms in Jealousy, Paranoia and Homosexuality," translated by Joan Riviere, in Freud (1963 b), pp. 160–170.

———— (1963b), *Sexuality and the Psychology of Love*. Edited by Philip Rieff. New York: Collier Books.

———— ([1895] 1966a), "Project for a Scientific Psychology," in Freud (1966 b), pp. 295–359.

———— ([1886–1899] 1966b), *The Standard Edition of the Complete Psychological Works of Sigmund Freud*. Volume 1. Edited and translated by James Strachey. London: Hogarth Press.

FRIEDMAN, MICHAEL (1974), "Explanation and Scientific Understanding," *The Journal of Philosophy 71:* 5–19.

———— (1983), *Foundations of Space-Time Theories: Relativistic Physics and Philosophy of Science.* Princeton: Princeton University Press.

FROHLICH, CLIFF (1980), "The Physics of Somersaulting and Twisting," *Scientific American 242 (3):* 154–164.

FROST, WILLIAM N.; GREGORY A. CLARK; and ERIC R. KANDEL (1988), "Parallel Processing of Short-term Memory for Sensitization in *Aplysia," Journal of Neurobiology 19:* 297–334.

FULLER, STEVE (1988), *Social Epistemology.* Bloomington, IN: Indiana University Press.

———— (1989), *The Philosophy of Science and Its Discontents.* Boulder: Westview Press.

GAIFMAN, HAIM (1979), "Subjective Probability, Natural Predicates and Hempel's Ravens," *Erkenntnis 14:* 105–147.

GAIFMAN, HAIM and MARC SNIR (1982), "Probabilities Over Rich Languages, Testing and Randomness," *The Journal of Symbolic Logic 47:* 495–548.

GALISON, PETER (1987), *How Experiments End.* Chicago: University of Chicago Press.

GARBER, DANIEL (1983), "Old Evidence and Logical Omniscience in Bayesian Confirmation Theory," in Earman, pp. 99–131.

GARDINER, PATRICK (ed.) (1959), *Theories of History: Readings from Classical and Contemporary Sources.* New York: The Free Press.

GARDNER, MARTIN (1958), *Logic Machines and Diagrams.* New York: McGraw-Hill.

GEERTZ, CLIFFORD (1975), *The Interpretation of Cultures: Selected Essays.* London: Hutchinson.

GENESERETH, MICHAEL R. and NILS J. NILSSON (1987), *Logical Foundations of Artificial Intelligence.* Los Altos, CA: Morgan Kaufmann.

GEROCH, ROBERT (1978), *General Relativity from A to B.* Chicago: University of Chicago Press.

GERVER, JOSEPH L. (1984), "A Possible Model for a Singularity without Collisions in the Five Body Problem" *Journal of Differential Equations 52:* 76–90.

GHISELIN, MICHAEL T. (1969), *The Triumph of the Darwinian Method.* Berkeley and Los Angeles: University of California Press.

GIBBS, J. WILLARD ([1902] 1960), *Elementary Principles in Statistical Mechanics, Developed with Especial Reference to the Rational Foundation of Thermodynamics.* New York: Dover.

GIBSON, JAMES J. (1950), *The Perception of the Visual World.* Boston: Houghton Mifflin.

———— (1966), *The Senses Considered as Perceptual Systems.* Boston: Houghton Mifflin.

———— (1979), *The Ecological Approach to Visual Perception.* Boston: Houghton Mifflin.

GIERE, RONALD N. (1988), *Explaining Science: A Cognitive Approach.* Chicago: University of Chicago Press.

GILLIES, DONALD (1986), "In Defense of the Popper-Miller Argument," *Philosophy of Science 53:* 110–113.

GILLIGAN, CAROL (1982), *In a Different Voice: Psychological Theory and Women's Development.* Cambridge, MA: Harvard University Press.

GLEICK, JAMES (1987), *Chaos: Making a New Science.* New York: Viking.

GLYMOUR, CLARK (1977), "Indistinguishable Space-Times and the Fundamental Group," in Earman, Glymour and Stachel, pp. 50–60.

———— (1980), *Theory and Evidence.* Princeton: Princeton University Press.

———— (1983), "Revisions of Bootstrap Testing," *Philosophy of Science 50:* 626–629.

—— (1991), "The Mathematics of Discovery and the Hierarchies of Knowledge," *Minds and Machines 1:* 75–95.

GLYMOUR, CLARK and KEVIN KELLY (forthcoming), *Logic, Computation and Discovery*. Cambridge, England: Cambridge University Press.

GLYMOUR, CLARK; RICHARD SCHEINES: PETER SPIRTES; and KEVIN KELLY (1987), *Discovering Causal Structure: Artificial Intelligence, Philosophy of Science, and Statistical Modeling*. Orlando: Academic Press.

GOOODENOUGH, WARD H. (1957), "Cultural Anthropology and Linguistics," in *Georgetown University Monograph Series on Language and Linguistics 9:* 167–173.

GOODMAN, NELSON (1955), *Fact, Fiction, and Forecast*. 1st ed. Cambridge, MA: Harvard University Press.

—— (1978), *Ways of Worldmaking*. Indianapolis: Hackett.

GORRY, G. ANTHONY and G. OCTO BARNETT (1968), "Experience with a Model of Sequential Diagnosis," *Computers in Biomedical Research 1:* 490–507.

GOTTHELF, ALLAN (1987), "Aristotle's Conception of Final Causality," in Gotthelf and Lennox, pp. 204–242.

GOTTHELF, ALLAN and JAMES G. LENNOX (eds.) (1987), *Philosophical Issues in Aristotle's Biology*. Cambridge, England: Cambridge University Press.

GREGORY, RICHARD L. (1990), *Eye and Brain: The Psychology of Seeing*. 4th ed. Princeton: Princeton University Press.

GRENE, MAJORIE (ed.) (1983), *Dimensions of Darwinism: Themes and Counterthemes in Twentieth-Century Evolutionary Theory*. Cambridge, England: Cambridge University Press.

GROBSTEIN, CLIFFORD (1965), *The Strategy of Life*. San Francisco: Freeman.

GROSS, LLEWELLYN (ed.) (1959), *Symposium on Sociological Theory*. Evanston, IL: Row, Peterson.

GRÜNBAUM, ADOLF (1973), *Philosophical Problems of Space and Time*. 2d ed. Dordrecht: Reidel.

—— (1984), *The Foundations of Psychoanalysis: A Philosophical Critique*. Berkeley and Los Angeles: University of California Press.

GUYER, PAUL (1987), *Kant and the Claims of Knowledge*. Cambridge, England: Cambridge University Press.

HABER, RALPH NORMAN (ed.) (1968), *Contemporary Theory and Research in Visual Perception*. New York: Holt, Rinehart & Winston.

—— (ed.) (1969a), *Information-Processing Approaches to Visual Perception*. New York: Holt, Rinehart & Winston.

—— (1969b), "Nature of the Effect of Set on Perception," in Haber, pp. 326–339.

HACKING, IAN (1975), *The Emergence of Probability: A Philosophical Study of Early Ideas about Probability, Induction and Statistical Inference*. Cambridge, England: Cambridge University Press.

—— (1981a), "Lakatos's Philosophy of Science," in Hacking, pp. 128–143.

—— (ed.) (1981b), *Scientific Revolutions*. Oxford: Oxford University Press.

—— (1983), *Representing and Intervening: Introductory Topics in the Philosophy of Natural Science*. Cambridge, England: Cambridge University Press.

HAMILTON, EDITH and HUNTINGTON CAIRNS (eds.) (1961), *The Collected Dialogues of Plato Including the Letters*. Translated by Lane Cooper and others. New York: Pantheon Books.

HANSON, EARL D. (1981), *Understandinig Evolution*. New York: Oxford University Press.

HANSON, NORWOOD RUSSELL (1958), *Patterns of Discovery: An Inquiry into the Conceptual Foundations of Science*. Cambridge, England: Cambridge University Press.

HARDY, G. H. ([1908] 1959), "Mendelian Proportions in a Mixed Population," in Peters, pp. 60–62.

HEIDEGGER, MARTIN ([1927] 1962), *Being and Time*. Translated by John Macquarrie and Edward Robinson. New York: Harper.

HELLMAN, GEOFFREY (1982a), "Einstein and Bell: Strengthening the Case for Microphysical Randomness," *Synthese 53:* 445–460.

———— (1982b), "Stochastic Einstein-Locality and the Bell Theorems," *Synthese 53:* 461–504.

HEMPEL, CARL G. (1942), "The Function of General Laws in History," *The Journal of Philosophy 39:* 35–48. Reprinted in Hempel (1965b), pp. 231–243.

———— (1945), "Studies in the Logic of Confirmation," *Mind 54:* 1–26, 97–121. Reprinted in Hempel (1965b), pp. 3–46, with a 1964 Postscript added.

———— (1959), "The Logic of Functional Analysis," in Gross, pp. 271–307. Reprinted in Hempel (1965b), pp. 297–330.

———— (1962a), "Deductive-Nomological vs. Statistical Explanation," in Feigl and Maxwell, pp. 98–169.

———— (1962b), "Explanation in Science and in History," in Colodny, pp. 7–33.

———— (1965a), "Aspects of Scientific Explanation," in Hempel, pp. 331–496.

———— (1965b), *Aspects of Scientific Explanation and Other Essays in the Philosophy of Science*. New York: The Free Press.

———— (1966), *Philosophy of Natural Science*. Englewood Cliffs, NJ: Prentice-Hall.

———— (1988), "Provisoes: A Problem Concerning the Inferential Function of Scientific Theories," *Erkenntnis 28:* 147–164.

HEMPEL, CARL G. and PAUL OPPENHEIM (1948), "Studies in the Logic of Explanation," *Philosophy of Science 15:* 135–175. Reprinted in Hempel (1965b), pp. 245–290, with a 1964 Postscript added.

HENKIN, LEON; PATRICK SUPPES; and ALFRED TARSKI (eds.) (1959), *The Axiomatic Method, with Special Reference to Geometry and Physics*. Amsterdam: North-Holland

HERSCHEL, JOHN ([1830] 1987), *A Preliminary Discourse on the Study of Natural Philosophy*. Chicago: University of Chicago Press.

HILBERT, DAVID ([1908] 1971), *Foundations of Geometry*. Translated by Leo Unger. Lasalle, IL: Open Court.

HINTIKKA, JAAKKO and PATRICK SUPPES (eds.) (1966), *Aspects of Inductive Logic*. Amsterdam: North-Holland.

HOBART, R. E. (1966), "Free Will as Involving Determination and Inconceivable Without It," in Berofsky, pp. 63–95.

HOCHBERG, JULIAN E. (1968), "In the Mind's Eye," in Haber, pp. 309–331.

———— (1978), *Perception*. 2d ed. Englewood Cliffs, NJ: Prentice-Hall.

HOLMES, FREDERIC LAWRENCE (1985), *Lavoisier and the Chemistry of Life: An Exploration of Scientific Creativity*. Madison, WI: University of Wisconsin Press.

HOMANS, GEORGE C. (1967), *The Nature of Social Science*. New York. Harcourt, Brace & World.

HOOKER, C. A. (1981), "Towards a General Theory of Reduction. Part I: Historical and Scientific Setting. Part II: Identity in Reduction. Part III: Cross-Categorial Reduction," *Dialogue 20:* 38–59, 201–236, 496–529.

HORWICH, PAUL (1982), *Probability and Evidence*. Cambridge, England: Cambridge University Press.

———— (1987), *Asymmetries in Time: Problems in the Philosophy of Science*. Cambridge, MA: Bradford/MIT Press.

———— (1990), *Truth*. Oxford: Blackwell.

HOWSON, COLIN (1984), "Bayesianism and Support by Novel Facts," *The British Journal for the Philosophy of Science 35:* 245–251.

HUGHES, R. I. G. (1989), *The Structure and Interpretation of Quantum Mechanics*. Cambridge, MA: Harvard University Press.

HULL, CLARK L.; CARL I. HOVLAND; ROBERT T. ROSS; MARSHALL HALL; DONALD T. PERKINS; and FREDERIC B. FITCH (1940), *Mathematico-Deductive Theory of Rote Learning: A Study in Scientific Methodology*. New Haven: Yale University Press.

HULL, DAVID L. (1974), *Philosophy of Biological Science*. Englewood Cliffs, NJ: Prentice-Hall.

———— (1976), "Informal Aspects of Theory Reduction," in Cohen, Hooker, Michalos and van Evra, pp. 653–670.

———— (1981a), "Reduction and Genetics," *The Journal of Medicine and Philosophy 6:* 125–143.

———— (1981b), "Units of Evolution: A Metaphysical Essay," in Jensen and Harré, pp. 23–44.

———— (1983), *Darwin and His Critics: The Reception of Darwin's Theory of Evolution by the Scientific Community*. Chicago: University of Chicago Press.

———— (1988), *Science as a Process: An Evolutionary Account of the Social and Conceptual Development of Science*. Chicago: University of Chicago Press.

HUME, DAVID (1748), *An Enquiry Concerning Human Understanding*. Many editions available.

———— ([1739–1740] 1978), *A Treatise of Human Nature*. Edited by L. A. Selby-Bigge. 2d ed. Oxford: Clarendon Press.

HUMPHREYS, PAUL (1981), "Aleatory Explanations," *Synthese 48:* 225–232.

JACQUARD, ALBERT (1974), *The Genetic Structure of Populations*. Translated by D. and B. Charlesworth. New York: Springer-Verlag.

JARRETT, JON P. (1984), "On the Physical Significance of the Locality Conditions in the Bell Arguments," *Noûs 18:* 569–589.

JEFFREY, RICHARD C. (1969), "Statistical Explanation vs. Statistical Inference," in Rescher, pp. 104–113. Reprinted in Salmon, Jeffrey and Greeno (1971), pp. 19–28.

———— (1983), "Bayesianism with a Human Face," in Earman, pp. 133–156.

———— (1984), Response to "A Proof of the Impossibility of Inductive Probability," by Karl R. Popper and David Miller. *Nature 310:* 433.

JEFFREYS, SIR HAROLD (1957), *Scientific Inference*. 2d ed. Cambridge: Cambridge University Press.

JENKIN, FLEEMING ([1867] 1983), Review of *On the Origin of Species*, by Charles Darwin. In Hull, pp. 303–350.

JENSEN, U. J. and R. HARRÉ (eds.) (1981), *The Philosophy of Evolution*. Brighton, Sussex: Harvester Press.

JOHNSON-LAIRD, P. N. (1983), *Mental Models: Towards a Cognitive Science of Language, Inference, and Consciousness*. Cambridge, MA: Harvard University Press.

JOHNSTON, RICHARD F. (ed.) (1978), *Annual Review of Ecology and Systematics*. Volume 9. Palo Alto, CA: Annual Reviews.

JONSEN, ALBERT R. and STEPHEN TOULMIN (1988), *The Abuse of Casuistry: A History of Moral Reasoning*. Berkeley and Los Angeles: University of California Press.

The Journal of Medicine and Philosophy. Dordrecht: Reidel.

KAHNEMAN, DANIEL; PAUL SLOVIC; and AMOS TVERSKY (eds.) (1982), *Judgment Under Uncertainty: Heuristics and Biases*. Cambridge, England: Cambridge University Press.

KANDEL, ERIC R. (ed.) (1987a), *Molecular Neurobiology in Neurology and Psychiatry*. New York: Raven Press.

—— (1987b), "Preface: Molecular Neurobiology and the Proper Study of Humankind," in Kandel, pp. vii–ix.

KANDEL, ERIC R. and JAMES H. SCHWARTZ (eds.) (1985), *Principles of Neural Science*. 2d ed. New York: Elsevier.

KANDEL, ERIC R.; VINCENT F. CASTELLUCCI; PHILIP GOELET; and SAMUEL SCHACHER (1987), "Cell-Biological Interrelationships between Short-Term and Long-Term Memory," in Kandel, pp. 111–132.

KANT, IMMANUEL ([1787] 1865), *Critique of Pure Reason*. Translated by Norman Kemp Smith. New York: St. Martin's Press.

KELLY, KEVIN (forthcoming), *The Logic of Reliable Inquiry*.

KELLY, KEVIN and CLARK GLYMOUR, "Inductive Inference from Theory Laden Data," *Journal of Philosophical Logic*.

KEMENY, JOHN G. and PAUL OPPENHEIM (1956), "On Reduction," *Philosophical Studies 7:* 6–19.

KITCHER, PATRICIA (1990): *Kant's Transcendental Psychology*. New York: Oxford University Press.

KITCHER, PHILIP (1976), "Explanation, Conjunction, and Unification," *The Journal of Philosophy 73:* 207–212.

—— (1981), "Explanatory Unification,"*Philosophy of Science 48:* 507–531.

—— (1984), "1953 and All That. A Tale of Two Sciences," *The Philosophical Review 93:* 335–373.

—— (1985a), "Darwin's Achievement," in Rescher, pp. 127–189.

—— (1985b), *Vaulting Ambition: Sociobiology and the Quest for Human Nature*. Cambridge, MA: MIT Press.

—— (1989), "Explanatory Unification and the Causal Structure of the World," in Kitcher and Salmon, pp. 410–505.

KITCHER, PHILIP and WESLEY C. SALMON (eds.) (1989), *Minnesota Studies in the Philosophy of Science*. Volume 13, *Scientific Explanation*. Minneapolis: University of Minnesota Press.

KOESTLER, ARTHUR and J. R. SMYTHIES (eds.) (1969), *Beyond Reductionism: New Perspectives in the Life Sciences*. Boston: Beacon Press.

KÖRNER, S. (ed.) (1957), *Observation and Interpretation: A Symposium of Philosophers and Physicists*. London: Butterworths Scientific Publications.

KOSSLYN, STEPHEN MICHAEL (1980), *Image and Mind*. Cambridge, MA: Harvard University Press.

KUFFLER, STEPHEN W.; JOHN G. NICHOLLS; and A. ROBERT MARTIN (1984), *From Neuron to Brain: A Cellular Approach to the Function of the Nervous System*. 2d ed. Sunderland, MA: Sinauer Associates.

KUHN, THOMAS S. (1962), *The Structure of Scientific Revolutions*. Chicago: University of Chicago Press.

———— (1970), *The Structure of Scientific Revolutions*. 2d ed. Chicago: University of Chicago Press.

———— (1983), "Commensurability, Comparability, Communicability," in Asquith and Nickles, pp. 669–688.

———— (1989), "Possible Worlds in History of Science," in Allén, pp. 9–32.

LAKATOS, IMRE (1970), "Falsification and the Methodology of Scientific Research Programmes," in Lakatos and Musgrave, pp. 91–195.

———— (1981), "History of Science and Its Rational Reconstructions," in Hacking, pp. 107–127.

LAKATOS, IMRE and ALAN MUSGRAVE (eds.) (1970), *Criticism and the Growth of Knowledge*. Cambridge, England: Cambridge University Press.

LAKOFF, GEORGE (1987), *Women, Fire, and Dangerous Things: What Categories Reveal about the Mind*. Chicago: University of Chicago Press.

LA METTRIE, JULIEN OFFRAY DE ([1915] 1988), *Man a Machine*. Lasalle, IL: Open Court.

LANGLEY, PAT; HERBERT A. SIMON; GARY L. BRADSHAW; and JAN M. ZYTKOW (1987), *Scientific Discovery: Computational Explorations of the Creative Process*. Cambridge, MA: MIT Press.

LAPLACE, PIERRE-SIMON (1820), *Théorie Analytique des Probabilités*. 3rd ed. Paris: V. Courcier.

———— ([1814] 1951), *A Philosophical Essay on Probabilities*. Translated by Frederick Wilson Truscott and Frederick Lincoln Emory. New York: Dover.

LATOUR, BRUNO (1987), *Science in Action: How to Follow Scientists and Engineers Through Society*. Cambridge, MA: Harvard University Press.

———— (1990), "Postmodern? No, Simply Amodern! Steps Towards an Anthropology of Science," *Studies in History and Philosophy of Science 21*: 145–171.

LATOUR, BRUNO and STEVE WOOLGAR (1986), *Laboratory Life: The Construction of Scientific Facts*. Princeton: Princeton University Press.

LAUDAN, LARRY (1977), *Progress and Its Problems: Toward a Theory of Scientific Growth*. Berkeley and Los Angeles: University of California Press.

———— (1981), "A Problem-Solving Approach to Scientific Progress," in Hacking, pp. 144–155.

———— (1984), *Science and Values: An Essay on the Aims of Science and Their Role in Scientific Debate*. Berkeley and Los Angeles: University of California Press.

LEHNINGER, ALBERT L. (1971), *Bioenergetics: The Molecular Basis of Biological Energy Transformations*. 2d. ed. Menlo Park, CA: Benjamin/Cummings.

LEIBNIZ, GOTTFRIED WILHELM (1970), *Philosophical Papers and Letters*. 2d ed. Edited and translated by Leroy E. Loemker. Dordrecht: Reidel.

LENNOX, JAMES G. (1985), "Plato's Unnatural Teleology," in O'Meara, pp. 195–218.

LEPLIN, JARRETT (ed.) (1984), *Scientific Realism*. Berkeley and Los Angeles: University of California Press.

LERMAN, HANNAH (1986), *A Mote in Freud's Eye: From Psychoanalysis to the Psychology of Women*. New York: Springer Verlag.

LEVINS, RICHARD (1966), "The Strategy of Model Building in Population Biology," *American Scientist 54*: 421–431. Reprinted in Sober (1984a), pp. 18–27.

LEVY, S.; R. HERBERMAN; M. LIPPMAN; and T. D'ANGELO (1987), "Correlation of Stress Factors with Sustained Depression of Natural Killer Cell Activity and Predicted Prognosis in Patients with Breast Cancer," *Journal of Clinical Oncology 5*: 348–353.

LEVY, SANDRA M.; JERRY LEE; CAROLINE BAGLEY; and MARC LIPPMAN (1988), "Survival Hazards Analysis in First Recurrent Breast Cancer Patients: Seven-year Follow-up," *Psychosomatic Medicine 50*: 520–528.

LEWIN, BENJAMIN (1990), *Genes IV*. Oxford: Oxford University Press.

LEWIS, CLARENCE IRVING ([1929] 1956), *Mind and the World-Order: Outline of a Theory of Knowledge*. New York: Dover.

LINDSAY, R.K.; B. BUCHANAN; E. FEIGENBAUM; and J. LEDERBERG (1980), *Applications of Artificial Intelligence for Organic Chemistry: The DENDRAL Project*. New York: McGraw-Hill.

LITTLE, DANIEL (1990), *Varieties of Social Explanation: An Introduction to the Philosophy of Social Science*. Boulder: Westview Press.

LLOYD, CHRISTOPHER (1986), *Explanation in Social History*. New York: Blackwell.

LOCKE, JOHN ([1690] 1924), *An Essay Concerning Human Understanding*. Abridged and edited by A. S. Pringle-Pattison. Oxford: Clarendon Press.

LORENTZ, H. A.; A. EINSTEIN; H. MINKOWSKI; and H. WEYL ([1923] 1952), *The Principle of Relativity: A Collection of Original Memoirs on the Special and General Theory of Relativity*. Translated by W. Perrett and G. B. Jeffery. New York: Dover.

LYNCH, MICHAEL (1985), *Art and Artifact in Laboratory Science: A Study of Shop Work and Shop Talk in a Research Laboratory*. London: Routledge & Kegan Paul.

MACINTYRE, ALASDAIR C. (1984), *After Virtue: A Study in Moral Theory*. 2d ed. Notre Dame: University of Notre Dame Press.

MACKIE, J. (1974), *The Cement of the Universe: A Study of Causation*. Oxford: Clarendon Press.

MALAMENT, DAVID (1977a), "Casual Theories of Time and the Conventionality of Simultaneity," *Noûs 11:* 293–300.

——— (1977b), "Observationally Indistinguishable Space-times," in Earman, Glymour and Stachel, pp. 61–80.

MALCOLM, JANET (1984), *In the Freud Archives*. New York: Knopf.

MANNHEIM, KARL (1936), *Ideology and Utopia: An Introduction to the Sociology of Knowledge*. Translated by Louis Wirth and Edward Shils. London: Paul, Trench, Trubner & Co.

MASSON, JEFFREY MOUSSAIEFF (1984), *The Assault on Truth: Freud's Suppression of the Seduction Theory*. New York: Farrar, Straus & Giroux.

MAXWELL, GROVER and ROBERT M. ANDERSON, JR. (eds.) (1975), *Minnesota Studies in the Philosophy of Science*. Volume 6, *Induction, Probability, and Confirmation*. Minneapolis: University of Minnesota Press.

MAXWELL, JAMES CLERK ([1876] 1952), *Matter and Motion*. New York: Dover.

MAYNARD SMITH, JOHN (1972), *The Theory of Evolution*, 2d ed. Baltimore: Penguin.

——— (1978), "Optimization Theory in Evolution," in Johnston, pp. 31–56. Reprinted in Sober (1984a), pp. 289–315.

——— (ed.) (1982), *Evolution Now: A Century After Darwin*. San Francisco: Freeman.

——— (1983), "Current Controversies in Evolutionary Biology," in Grene, pp. 273–286.

MAYR, ERNST (1942), *Systematics and the Origin of Species from the Viewpoint of a Zoologist*. New York: Columbia University Press.

——— (1970), *Populations, Species, and Evolution: An Abridgment of* Animal Species and Evolution. Cambridge, MA: Belknap Press of Harvard University Press.

——— (1982), *The Growth of Biological Thought: Diversity, Evolution, and Inheritance*. Cambridge, MA: Belknap Press of Harvard University Press.

MCGEER, PATRICK L.; JOHN ECCLES; and EDITH G. MCGEER (1987), *Molecular Neurobiology of the Mammalian Brain*. 2d ed. New York: Plenum Press.

McLAUGHLIN, ROBERT (ed.) (1982), *What? Where? When? Why? Essays on Induction, Space and Time, Explanation*. Dordrecht: Reidel.

MENDEL, GREGOR ([1865] 1966), "Experiments on Plant Hybrids," translated by Eva R. Sherwood, in Stern and Sherwood, pp. 1–48.

MILL, JOHN STUART (1874), *A System of Logic, Ratiocinative and Inductive: Being a Connected View of the Principles of Evidence and the Methods of Scientific Investigation*. 8th ed. New York: Harper & Brothers.

——— ([1861] 1979), *Utilitarianism*. Edited by George Sher. Indianapolis: Hackett.

MILLER, RANDOLPH A.; HARRY E. POPLE, JR.; and JACK D. MYERS (1982), "*Internist-I*, an Experimental Computer-Based Diagnostic Consultant for General Internal Medicine," *The New England Journal of Medicine 307:* 468–476.

MILLER, RICHARD W. (1987), *Fact and Method: Explanation, Confirmation and Reality in the Natural and the Social Sciences*. Princeton: Princeton University Press.

MINKOWSKI, H. ([1908] 1952), "Space and Time," in Lorentz, Einstein, Minkowski and Weyl, pp. 73–91.

MITCHELL, SANDRA D. (1987), "Competing Units of Selection?: A Case of Symbiosis," *Philosophy of Science 54:* 351–367.

MOORE, GEORGE EDWARD ([1903] 1962), *Principia Ethica*. Cambridge, England: Cambridge University Press.

MORGENBESSER, SIDNEY (ed.) (1967), *Philosophy of Science Today*. New York Basic Books.

MUNSON, RONALD (1981), "Why Medicine Cannot Be a Science," *The Journal of Medicine and Philosophy 6:* 183–208.

MURDOCK, GEORGE PETER (1949), *Social Structure*. New York: Macmillan.

MURPHY, GREGORY L. and DOUGLAS L. MEDIN (1985), "The Role of Theories in Conceptual Coherence," *Psychological Review 92:* 289–316.

NAGEL, ERNEST (1949), "The Meaning of Reduction in the Natural Sciences," in Stauffer, pp. 97–135.

——— (1961), *The Structure of Science: Problems in the Logic of Scientific Explanation*. New York: Harcourt, Brace & World.

THE NATIONAL COMMISSION FOR THE PROTECTION OF HUMAN SUBJECTS OF BIOMEDICAL AND BEHAVIORAL RESEARCH (1978), *The Belmont Report: Ethical Principles and Guidelines for the Protection of Human Subjects of Research*. Washington, DC: U.S. Government Printing Office.

NEISSER, ULRIC (1976), *Cognition and Reality: Principles and Implications of Cognitive Psychology*. San Francisco: Freeman.

NEWTON, ISAAC ([1687] 1962), *Mathematical Principles of Natural Philosophy*. Volume 1. Translated by Andrew Motte; revised by Florian Cajori. Berkeley and Los Angeles: University of California Press.

NORTON, JOHN (1989), "The Hole Argument," in Fine and Leplin, pp. 56–64.

NOZICK, ROBERT (1981), *Philosophical Explanations*. Cambridge, MA: Harvard University Press.

OLBY, ROBERT C. (1966), *Origins of Mendelism*. London: Constable.

O'MEARA, DOMINIC J. (ed.) (1985), *Platonic Investigations*. Washington, DC: Catholic University of America Press.

OSHERSON, DANIEL N.; MICHAEL STOB; and SCOTT WEINSTEIN (1986), *Systems That Learn: An Introduction to Learning Theory for Cognitive and Computer Scientists*. Cambridge, MA: Bradford/MIT Press.

PAPINEAU, DAVID (1978), *For Science in the Social Sciences*. New York: St. Martin's Press.

PAULI, W. ([1921] 1958), *Theory of Relativity*. Translated by G. Field. New York: Pergamon Press.

PELLEGRINO, EDMUND D. (1976), "Philosophy of Medicine: Problematic and Potential," *The Journal of Medicine and Philosophy 1:* 5–31.

———— (1986), "Philosophy of Medicine: Towards a Definition," *The Journal of Medicine and Philosophy 11:* 9–16.

PELLEGRINO, EDMUND D. and DAVID C. THOMASMA (1981), *A Philosophical Basis of Medical Practice: Toward a Philosophy and Ethic of the Healing Professions*. New York: Oxford University Press.

PETERS, JAMES A. (ed.) (1959), *Classic Papers in Genetics*. Engelwood Cliffs, NJ: Prentice-Hall.

Philosophy of Science (1984). Volume 51, number 2. *Special Issue: Philosophy of Biology*.

PIAGET, JEAN ([1926] 1959), *The Language and Thought of the Child*. 3rd ed. Translated by Marjorie and Ruth Gabain. London: Routledge & Kegan Paul.

PIANKA, ERIC R. (1974), *Evolutionary Ecology*. New York: Harper & Row.

PICKERING, ANDREW (1984), *Constructing Quarks: A Sociological History of Particle Physics*. Edinburgh: Edinburgh University Press.

PITT, JOSEPH C. (ed.) (1988), *Theories of Explanation*. New York: Oxford University Press.

PLANCK, MAX (1932), *Der Kausalbegriff in der Physik*. Leipzig: Johann Ambrosius Barth.

PLATO (1900–1903), *Platonis Opera*. Volumes 1–5. Edited by J. Burnet. Oxford: Oxford University Press.

POINCARÉ, HENRI (1907), *Science and Hypothesis*. Translated by William John Greenstreet. London: Walter Scott.

POPPER, KARL R. (1957a), "The Aim of Science," *Ratio 1:* 24–35.

———— (1957b), "The Propensity Interpretation of the Calculus of Probability, and the Quantum Theory," in Körner, pp. 65–70.

———— ([1935] 1959), *The Logic of Scientific Discovery*. New York: Basic Books.

———— (1960), "The Propensity Interpretation of Probability," *The British Journal for the Philosophy of Science 10:* 25–42.

———— (1963), *Conjectures and Refutations: The Growth of Scientific Knowledge*. London: Routledge & Kegan Paul.

———— (1972a), "Conjectural Knowledge: My Solution of the Problem of Induction," in Popper, pp. 1–31.

———— (1972b), *Objective Knowledge: An Evolutionary Approach*. Oxford: Clarendon Press.

———— (1982), *The Open Universe: An Argument for Indeterminism*. Totowa, NJ: Rowman & Littlefield.

POPPER, KARL R. and DAVID MILLER (1983), "A Proof of the Impossibility of Inductive Probability," *Nature 302:* 687–688.

PROVINE, WILLIAM B. (ed.) (1971), *The Origins of Theoretical Population Genetics*. Chicago: University of Chicago Press.

PUTNAM, HILARY (1965), "Trial and Error Predicates and the Solution to a Problem of Mostowski," *The Journal of Symbolic Logic 30:* 49–57.

———— (1981), *Reason, Truth, and History*. Cambridge, England: Cambridge University Press.

QUINE, W.V. (1960), *Word and Object*. Cambridge, MA: MIT Press.

———— (1961), *From a Logical Point of View: 9 Logico-Philosophical Essays*. 2d ed. Cambridge, MA: Harvard University Press.

——— (1969a), "Epistemology Naturalized," In Quine, pp. 69–90.

——— (1969b), *Ontological Relativity and Other Essays*. New York: Columbia University Press.

RAILTON, PETER (1981), "Probability, Explanation, and Information," *Synthese 48:* 233–256.

REICHENBACH, HANS (1938), *Experience and Prediction: An Analysis of the Foundations and the Structure of Knowledge*. Chicago: University of Chicago Press.

——— (1949), *The Theory of Probability: An Inquiry into the Logical and Mathematical Foundations of the Calculus of Probability*. 2d ed. Translated by Ernest H. Hutten and Maria Reichenbach. Berkeley and Los Angeles: University of California Press.

——— (1956), *The Direction of Time*. Edited by Maria Reichenbach. Berkeley and Los Angeles: University of California Press.

——— ([1928] 1957), *The Philosophy of Space and Time*. Translated by Maria Reichenbach and John Freund. New York: Dover.

——— ([1924] 1969), *Axiomatization of the Theory of Relativity*. Translated and edited by Maria Reichenbach. Berkeley and Los Angeles: University of California Press.

RESCHER, NICHOLAS (ed.) (1969), *Essays in Honor of Carl G. Hempel*. Dordrecht: Reidel.

——— (ed.) (1985), *Reason and Rationality in Natural Science: A Group of Essays*. Lanham, MD: University Press of America.

ROBB, ALFRED A. (1914), *A Theory of Time and Space*. Cambridge, England: Cambridge University Press.

——— (1921), *The Absolute Relations of Time and Space*. Cambridge, England: Cambridge University Press.

RORTY, RICHARD (1979), *Philosophy and the Mirror of Nature*. Princeton University Press.

ROSENBERG, ALEXANDER (1985), *The Structure of Biological Science*. Cambridge, England: Cambridge University Press.

——— (1988), *Philosophy of Social Science*. Boulder: Westview Press.

ROSINSKI, RICHARD R. (1977), *The Development of Visual Perception*. Santa Monica, CA: Goodyear.

RUBEN, DAVID-HILLEL (1985), *The Metaphysics of the Social World*. London: Routledge & Kegan Paul.

RUSE, MICHAEL (1976), "Reduction in Genetics," in Cohen, Hooker, Michalos and van Evra, pp. 633–651.

RUSE, MICHAEL (1979), *The Darwinian Revolution*. Chicago: University of Chicago Press.

RUSSELL, BERTRAND (1948), *Human Knowledge: Its Scope and Limits*. New York: Simon & Schuster.

——— (1956a), *Logic and Knowledge: Essays, 1901–1950*. Edited by Robert Charles Marsh. New York: Macmillan.

——— ([1918] 1956b), "The Philosophy of Logical Atomism," in Russell, pp. 177–281.

——— ([1912] 1959), *The Problems of Philosophy*. New York: Oxford University Press.

——— ([1914] 1960), *Our Knowledge of the External World*. 2d. ed. New York: New American Library.

RUSSELL, BERTRAND and ALFRED NORTH WHITEHEAD (1925), *Principia Mathematica*. 1st ed. Cambridge, England: Cambridge University Press.

SAHLINS, MARSHALL D. and ELMAN R. SERVICE (eds.) (1960), *Evolution and Culture*. Ann Arbor: University of Michigan Press.

SALMON, WESLEY C. (1967), *The Foundations of Scientific Inference*. Pittsburgh: University of Pittsburgh Press.

—— (1973), "Confirmation," *Scientific American 228 (5)*: 75–83.

—— (1975), "Confirmation and Relevance," in Maxwell and Anderson, pp. 3–36.

—— (1977), "The Philosophical Significance of the One-Way Speed of Light," *Nôus 11*: 253–292.

—— (1978a), "Unfinished Business: The Problem of Induction," *Philosophical Studies 33*: 1–19.

—— (1978b), "Why Ask 'Why'? An Inquiry Concerning Scientific Explanation," *Proceedings and Addresses of the American Philosophical Association 51*: 683–705.

—— (1980), *Space, Time, and Motion: A Philosophical Introduction*. 2d ed. Minneapolis: University of Minnesota Press.

—— (1981), "Rational Prediction," *The British Journal for the Philosophy of Science 32*: 115–125.

—— (1982), "Comets, Pollen and Dreams: Some Reflections on Scientific Explanation," in McLaughlin, pp. 155–178.

—— (1984), *Scientific Explanation and the Causal Structure of the World*. Princeton: Princeton University Press.

—— (1989), "Four Decades of Scientific Explanation," in Kitcher and Salmon, pp. 3–219.

—— (1990), *Four Decades of Scientific Explanation*. Minneapolis: University of Minnesota Press.

SALMON, WESLEY C.; RICHARD C. JEFFREY; and JAMES G. GREENO (1971), *Statistical Explanation and Statistical Relevance*. Pittsburgh: University of Pittsburgh Press.

SANDERS, WILLIAM T. and BARBARA J. PRICE (1968), *Mesoamerica: The Evolution of a Civilization*. New York: Random House.

SARKAR, SAHOTRA (1989), "Reductionism and Molecular Biology: A Reappraisal." Ph.D. Dissertation, University of Chicago.

—— (forthcoming), "Models of Reduction and Categories of Reductionism," *Synthese*.

SCHAFFNER, KENNETH F. (1967), "Approaches to Reduction," *Philosophy of Science 34*: 137–147.

—— (1969), "Correspondence Rules," *Philosophy of Science 36*: 280–290.

—— (1972), *Nineteenth-Century Aether Theories*. Oxford: Pergamon Press.

—— (1976), "Reductionism in Biology: Problems and Prospects," in Cohen, Hooker, Michalos and van Evra, pp. 613–632.

—— (1977), "Reduction, Reductionism, Values, and Progress in the Biomedical Sciences," in Colodny, pp. 143–171.

—— (1980), "Theory Structure in the Biomedical Sciences," *The Journal of Medicine and Philosophy 5*: 57–97.

—— (ed.) (1985), *Logic of Discovery of Diagnosis in Medicine*. Berkeley and Los Angeles: University of California Press.

—— (1986), "Exemplar Reasoning about Biological Models and Diseases: A Relation Between the Philosophy of Medicine and Philosophy of Science," *The Journal of Medicine and Philosophy 11*: 63–80.

—— (1987), "Computerized Implementation of Biomedical Theory Structures: An Artificial Intelligence Approach," in Fine and Machamer, pp. 17–32.

——— (forthcoming), *Discovery and Explanation in Biology and Medicine*. Chicago: University of Chicago Press.

SCHIFF, WILLIAM (1965), "Perception of Impending Collision: A Study of Visually Directed Avoidant Behavior," *Psychological Monographs: General and Applied 79 (11)*: 1–26.

SCHILPP, PAUL ARTHUR (ed.)(1963), *The Philosophy of Rudolf Carnap*. Lasalle, IL: Open Court.

SCRIVEN, MICHAEL (1959a), "Explanation and Prediction in Evolutionary Theory," *Science 130:* 477–482.

——— (1959b), "Truisms as the Grounds for Historical Explanations," in Gardiner, pp. 443–475.

——— (1962), "Explanations, Predictions, and Laws," in Feigl and Maxwell, pp. 170–230.

SEARLE, JOHN (1984), *Minds, Brains, and Science*. Cambridge, MA: Harvard University Press.

——— (1987), "Minds and Brains Without Programs," in Blakemore and Greenfield, pp. 208–233.

SHAPERE, DUDLEY (1974), "Scientific Theories and Their Domains," in Suppe, pp. 518–565.

SHAPIN, STEVEN and SIMON SCHAFFER (1985), *Leviathan and the Air-Pump: Hobbes, Boyle, and the Experimental Life*. Princeton: Princeton University Press.

SHELDRAKE, RUPERT (1981), *A New Science of Life: The Hypothesis of Formative Causation*. Los Angeles: Tarcher.

SHEPPARD, P. M. (1975), *Natural Selection and Heredity*. 4th ed. London: Hutchinson.

SIMON, HERBERT A. (1977), *Models of Discovery and Other Topics in the Methods of Science*. Boston: Reidel.

SIMPSON, GEORGE GAYLORD (1964), *This View of Life: The World of an Evolutionist*. New York: Harcourt, Brace, & World.

SKINNER, B. F. (1953), *Science and Human Behavior*. New York: Macmillan.

——— (1972), *Beyond Freedom and Dignity*. New York Bantam/Vintage.

SKLAR, LAWRENCE (1967), "Types of Inter-Theoretic Reduction," *The British Journal of the Philosopohy of Science 18:* 109–124.

——— (1974), *Space, Time, and Spacetime*. Berkeley and Los Angeles: University of California Press.

SNEED, JOSEPH D. (1971), *The Logical Structure of Mathematical Physics*. Dordrecht: Reidel.

SOBER, ELLIOTT (ed.) (1984a), *Conceptual Issues in Evolutionary Biology: An Anthology*. Cambridge, MA: Bradford/MIT Press.

——— (1984b), *The Nature of Selection: Evolutionary Theory in Philosophical Focus*. Cambridge, MA: Bradford/MIT Press.

SPIRTES, PETER LAURENCE (1981), "Conventionalism and the Philosophy of Henri Poincaré." Ph.D. Dissertation, University of Pittsburgh.

SPIRTES, PETER; CLARK GLYMOUR; and RICHARD SCHEINES (forthcoming), *Causality, Prediction and Search*. New York: Springer-Verlag.

STAUFFER, ROBERT C. (ed.) (1949), *Science and Civilization*. Madison, WI; University of Wisconsin Press.

STERN, CURT and EVA R. SHERWOOD (eds.) (1966), *The Origins of Genetics: A Mendel Source Book*. San Francisco: Freeman.

STIGLER, STEPHEN M. (1986), *The History of Statistics: The Measurement of Uncertainty Before 1900*. Cambridge: MA: Belknap Press of Harvard University Press.

STRAWSON, P. F. (1952a), "Inductive Reasoning and Probability," in Strawson, pp. 233–263.

——— (1952b), *Introduction to Logical Theory*. London: Methuen.

Suppe, Frederick (ed.) (1974), *The Structure of Scientific Theories*. Urbana: University of Illinois Press.

———— (1989), *The Semantic Conception of Theories and Scientific Realism*. Chicago: University of Chicago Press.

Suppes, Patrick (1957), *Introduction to Logic*. Princeton: Van Nostrand.

———— (1966), "A Bayesian Approach to the Paradoxes of Confirmation," in Hintikka and Suppes, pp. 198–207.

———— (1967), "What Is a Scientific Theory?" in Morgenbesser, pp. 55–67.

———— (1970), *A Probabilistic Theory of Causality*. Amsterdam: North-Holland.

Taylor, Edwin F. and John Archibald Wheeler (1966), *Spacetime Physics*. San Francisco: Freeman.

Theoretical Medicine. Dordrecht: Reidel.

van Fraassen, Bas C. (1980), *The Scientific Image*. Oxford: Clarendon Press.

———— (1983), "Theory Comparison and Relevant Evidence," in Earman, pp. 27–42.

———— (1988), "The Problem of Old Evidence," in Austin, pp. 153–165.

van Heijenoort, Jean (ed.) (1967), *From Frege to Gödel: A Source Book in Mathematical Logic*. Cambridge, MA: Harvard University Press.

van Herik, Judith (1982), *Freud on Femininity and Faith*. Berkeley and Los Angeles: University of California Press.

van Inwagen, Peter (1982), "The Incompatibility of Free Will and Determinism," in Watson, pp. 46–58.

Venn, John (1866), *The Logic of Chance: An Essay on the Foundations and Province of the Theory of Probability, with Especial Reference to its Application to Moral and Social Science*. 1st ed. London: Macmillan.

———— ([1888] 1962), *The Logic of Chance*. 4th ed. New York: Chelsea.

Virchow, R. (1849), "Wissenschaftliche Methode und therapeutische Standpunkte," *Archiv für pathologische Anatomie und Physiologie und für klinische Medicin* 2: 37.

Vision, Gerald (1988), *Modern Anti-Realism and Manufactured Truth*. London: Routledge.

Wainwright, S. A.; W. D. Biggs; J. D. Currey; and J. M. Gosline (1982), *Mechanical Design in Organisms*. Princeton: Princeton University Press.

Wald, Robert M. (1984), *General Relativity*. Chicago: University of Chicago Press.

Watson, Gary (ed.) (1982), *Free Will*. Oxford: Oxford University Press.

Watson, James D., Nancy H. Hopkins; Jeffrey W. Roberts; Joan Argetsinger Steitz; and Alan M. Weiner (1987), *Molecular Biology of the Gene*. 4th ed. Two volumes. Menlo Park, CA: Benjamin/Cummings.

Watson, John B. (1919), *Psychology from the Standpoint of a Behaviorist*. Philadelphia: Lippincott.

Weber, Bruce H.; David J. Depew; and James D. Smith (eds.) (1988), *Entropy, Information, and Evolution: New Perspectives on Physical and Biological Evolution*. Cambridge, MA: Bradford/MIT Press.

Weiss, Paul A. (1969), "The Living System: Determinism Stratified," in Koestler and Smythies, pp. 3–42.

Weizenbaum, Joseph (1976), *Computer Power and Human Reason: From Judgment to Calculation*. San Francisco: Freeman.

WEYL, HERMANN ([1949] 1963), *Philosophy of Mathematics and Natural Science*. Translated by Olaf Helmer. New York: Atheneum.

WHEELER, JOHN ARCHIBALD and WOJCIECH HUBERT ZUREK (eds.) (1983), *Quantum Theory and Measurement*. Princeton: Princeton University Press.

WHEWELL, WILLIAM (1837), *History of the Inductive Sciences*. London: Parker.

WIGNER, EUGENE P. (1970), "On Hidden Variables and Quantum Mechanical Probabilities," *American Journal of Physics 38:* 1005–1009.

WILF, HERBERT S. (1986), *Algorithms and Complexity*. Englewood Cliffs, NJ: Prentice-Hall.

WILLIAMS, DONALD CARY (1947), *The Ground of Induction*. Cambridge, MA: Harvard University Press.

WILLIAMS, GEORGE C. (1966), *Adaptation and Natural Selection: A Critique of Some Current Evolutionary Thought*. Princeton: Princeton University Press.

WILLIAMS, JUANITA H. (1977), *Psychology of Women: Behavior in a Biosocial Context*. New York: Norton.

WILSON, DAVID SLOAN (1984), "Individual Selection and the Concept of Structured Demes," in Brandon and Burian, pp. 272–291.

WIMSATT, WILLIAM C. (1976), "Reductive Explanation: A Functional Account," in Cohen, Hooker, Michalos and van Evra, pp. 671–710.

WINCH, PETER (1958), *The Idea of a Social Science and Its Relation to Philosophy*. London: Routledge & Kegan Paul.

WINNIE, JOHN A. (1970), "Special Relativity Without One-Way Velocity Assumptions," *Philosophy of Science 37:* 81–99, 223–238.

——— (1977), "The Causal Theory of Space-time," in Earman, Glymour and Stachel, pp. 134–205.

WITTGENSTEIN, LUDWIG (1953), Philosophical Investigations. Translated by G. E. M. Anscombe. Oxford: Blackwell.

——— ([1922] 1955), *Tractatus Logico-Philosophicus*. London: Routledge & Kegan Paul.

WOHWILL, JOACHIM F. (1963), "Overconstancy in Distance Perception as a Function of the Texture of the Stimulus Field and Other Variables," *Perceptual and Motor Skills 17:* 831–846.

WOOLGAR, STEVE (1988), *Science: The Very Idea*. London: Tavistock.

WRIGHT, CRISPIN (1980), *Wittgenstein on the Foundations of Mathematics*. London: Duckworth.

WRIGHT, LARRY (1976), *Teleological Explanations: An Etiological Analysis of Goals and Functions*. Berkeley and Los Angeles: University of California Press.

ZARIN, DEBORAH A. and STEPHEN G. PAUKER (1984), "Decision Analysis as a Basis for Medical Decision Making: The Tree of Hippocrates," *The Journal of Medicine and Philosophy 9:* 181–213.

INDEX

A priori, 65
  knowledge, 109–10, 155
  presupposition, 264
  proof, 369
  reasoning, 56, 111
  rule, 62
Abelson, Robert P., 357
Acceleration , 234, 245
Ackermann, Thomas, 113, 119
Actant, 173, 176
Action, 404, 408, 411–12, 425
Activity, social, 165
Adam, 36
Adaptation, 269, 279, 287–88, 292–
  94, 296–99, 305–06
  abstract theory of, 293
Adaptationist
  assumption, 281
  program, 286
Adaptedness of interactors, 295
Adaptive
  advantage, 284, 304–05
  character change, 281, 303
  opportunity, 281
  problem, 284
  success, 301
Adorno, Theodor, 421
Affect. See Emotion
Affine structure, 213
Affirming the consequent, 62
Age of the earth, 106
AIDS, 34, 338, 340
Airpump, 165
ALARM network, 397, 400–01
Algebra, 365
Algorithm, 365–72, 375, 377–78,
  384, 401–402
  for theory-choice, nonexistence of,
  371
Allele, 276, 282, 300
Alpha particle, 100
Altruism, 295
Ambiguity of I-S explanation, prob-
  lem of, 26
Analytic truth, 110, 118
Anarchism, 157–60
Ancestor-descendant history, 271
And. See Conjunction
Anisotropy, 202
Anomalous advance of the perihelion
  of Mercury, 47, 98–99
Anomaly, 155–56
Anthropology, 405, 408
  cognitive, 410
  cultural, 126

of science, 134–35, 163, 167–73
  symbolic, 410
Anthropomorphism, 13, 298–99, 348
Antibiotics, 311, 313
Antifoundationalism, 151, 157
Anti-inductivism, 92
Antirealism, 170, 179, 184, 189
Antireductionism, 315
Antinomics of reason, 110–12, 119–
  20, 128
Anything goes, 157–60
Apache basketry example, 18, 40
Apel, Karl-Otto, 421
Aplysia, 323–30, 335–37, 339,
  343–44
Apodictic certainty, 155
Applicability, 79
Arabic mathematicians, 365
Argument, 10–12
  deductive, 11–12, 15
  empirical, 109
  general, 63
  inductive, 11–12
Aristarchus, 146
Aristotle, 12, 77, 296–99, 346
Arithmetic, 109–10, 365, 368–71
Arntzenius, Frank, 246
Artifact, 165–66, 170
Artificial intelligence, 328, 362,
  364–03
Ascertainability, 79
Astrology, 39, 58, 126
Astronomy, 296, 406
Athletic scholarship, 413–414
Atomic
  fact, 136, 139
  proposition, 139–40
  theory, 122
    Bohr's, 125
    Dalton's, 125–26, 319
  weight, theory of, 53–54, 104–05,
    118, 124, 391
Atomism, logical, 135, 138–42, 146
Attneave, Fred, 356
Automobile
  accident example, 37
  that won't start example, 400, 402
Autonomy, 405, 408, 425
Autonomous scientific knower, 160
Avogadro's law, 53
Axiom, 114–15, 369
Axiomatization of special relativity,
  190–91
Ayala, Francisco J., 283, 288, 301
Ayer, A. J., 60, 130, 143

Bachelard, Gaston, 135, 147, 173
Back cross experiment, 276
BACON program, 387–92, 402
Balme, David, 297
Balzar, W., 316
Banana example, 111, 129
Bandwagon effect, 406
Barnes, Barry, 161
Barnett, G. Octo, 395–96
Barometer, 22, 34
Bateson, William, 272
Bayes's rule, 71–74, 80, 82–84, 89,
  91, 99–101, 117, 394, 397
  proof of, 72
Baycsian, 82, 144. See also Probability, interpretation of, personalist
  confirmation theory, 89–99
  networks, 393–399
  reasoning, 311
Beatty, John, 290–91, 301
Beauchamp, Tom, 342, 344
Beginning of world, 111
Behavior, 122, 424–25
  causes of, 406–12
  linguistic, 35, 420
  physical, 420
  prediction of, 405–06
  rational, 412–15
  rules of, 409, 415–19
  social dimension of, 351
  voluntary. See Free will
Behaviorism, 349, 351–53, 358, 362–
  63, 420–22
Belief, 128, 133, 404, 407
  causes of, 160–76, 420
  consistent, 82
  justification of, 129
Bell's inequalities, 255–60, 262
Bending of starlight passing close to
  the sun, 99
Berger, Peter L., 171
Berlin group, 2, 142
Berofsky, Bernard, 263
Betting quotient. See Dutch book
Bever, Thomas, 353
Billiard ball, 35, 56
Bioassay. See Inscription device
Biological science, 34, 79, 141, 168,
  311–45, 351
  biochemistry, 300, 305, 329, 332,
    343–44
  bioenergetics, 294, 298
Biomedical science, 312–16, 338, 343
  biopsychosocial model, 313–15,
    344

Biosynthetic pathway, 298
Bits of information, 136
Black holes, 249–50
Blite, 55, 63
Block, Ned, 349
Bloor, David, 161–64
Boag, Peter T., 286–87
Boden, Margaret, 363
Bohr, Niels, 127
Boolean formula, 348, 377, 384
Bootstrapping, 52–54, 102, 389, 392
  macho, 54, 102
Bourdieu, Pierre, 173
Bowler, Peter J., 309
Boyle's law, 16, 44–6, 48, 54, 99,
  102, 387–89, 417
Boyle, Robert , 165, 387
Bradshaw, Gary L., 403
Brain, 349, 380, 401
  in a vat, 106–07, 130
Braithwaite, R. B., 13
Brandon, Robert N., 271, 289–91,
  293, 295, 298, 309
Braybrooke, David, 425
Brentano, Franz, 366
Brewster's law, 322
Bridge collapse example, 411
Bridgeman, Percy, 105
Bright spot, 47
Broad, C. D., 58
Bromberger, Sylvain, 21
Brooks, Daniel R., 294
Brouwer, L. E., 141
Brucke, Ernst, 366
Bruner, Jerome, 353–55
Buchanan, Bruce, 394
Burian, Richard M., 290, 293, 309
Burks, Arthur W., 89
Burnet, MacFarlane, 340

Cahn, Steven M., 130
Cairns, Huntington, 130
Calculus, 78, 366
Caloric, theory of, 126, 172
Cancer, 340–41
Cannizzaro, Stanislao, 391
Cantor's diagonalization, 370
Cantor, Georg, 368, 377
Carbon-14, 23–24, 31, 58, 79
Carnap, Rudolf, 62, 85–90, 94, 113,
  115–18, 123, 128, 135–38, 140–
  43, 179, 334, 367
Carousel example, 181–81
Cartwright, Nancy, 136
Cauchy surface, 247–50
Causal
  antecedent, 299
  connection, 35, 56–57, 96, 300
    necessary, 107
  curve, 247, 262
  explanation. See Explanation,
    causal
  generalization, 330
  gloss, 322
  glue, 96
  interaction, interlevel, 314
  law. See Law, causal
  mechanism, 304

model, 271, 399
regularity, 110
relation, 21–23, 34, 39, 55–57,
  292, 306, 319, 400, 404, 425
relevance, 22–23, 33, 298
role, 123
structure, 400
  of Minkowski spacetime, 216–18
  of spacetime, 247
theory, 276, 290
  of time, 179–80, 190–194, 222,
    230
  See also Determinism
Causality, 172, 414
  Einstein, 258, 260
  Hume's problem of, 35
  Kant's Law of Universal, 110–111,
    233–34
  Maxwell's maxims of, 237
  of beliefs, 160–76, 420
  social, 162–63, 418
Causey, Robert L., 318
Ceteris paribus, 416–17
Challenger space shuttle, 34
Chance, 236–40, 288, 299–02
  recombination, 302
  set-up, 80
Change
  evolutionary, 278, 287–88, 295–96,
    307
    cause of, 287–88, 307
  social, 422
Chaos theory, 236–41, 244, 259
Chemistry, 34, 269, 311–12, 315,
  318
  protein, 231
Chernobyl, 8–9
Chess tournament example, 90
Child abuse, 350
Childress, James, 342, 344
Chisholm, Roderick M., 261
Chomsky, Noam, 352
Christ's virtues, 366
Christensen, David, 54
Chromosome, 272–73
Church's thesis, 243, 375–77
Church, Alonzo, 143, 376
Churchland, Patricia, 319–20, 343–
  44, 363
Churchland, Paul, 319, 343
Circular reasoning, 57
Clark, Gregory A., 329
Clark, Samuel, 228
Class, 136, 166
Clock, 191–93, 292. See also Syn-
  chronized clocks
Coefficient, expansion, 251
Cognition, 292, 354, 362–63
  social dimension of, 122, 351
Cognitive
  aims, 164
  anthropology. See Anthropology,
    cognitive
  community, 162
  goals of science, 155
  goodness, 150
  meaningfulness, 142–45
  order, 163

psychology. See Psychology, cogni-
  tive
  stance toward of theory, 155
Cognitivism, 351, 353–58, 362–63
Coherence of degrees of conviction,
  82
Coin
  biased, 29
  fair, 75, 77, 79, 83
Collectivism, 164, 404, 419–20
Collingwood, R. G., 408, 421–22
Comet, 14
Common cause, Reichenbach's principle of, 257–58
Communication, 164, 166, 169
Compatibilism, 419–20
Competition among populations,
  217
Complementarity, 127
Completeness, 62, 113, 252–53, 255,
  370–71, 378
Complexity, 380, 384–86
  expected, 386
  exponential, 385
  logical, 120
  polynomial, 385
  worst case bound on, 383, 385
  worst case, 386
Computability
  analogue, 243
  computational cost, 381
  digital, 243
  effective, 243–44
  computational system, 380, 385
  theory of, 242, 378
Computer, 364–65, 401
  analogy with mind, 348
  digital, 353, 357, 367, 373, 386
  science, 365
  vision, 386
Condition
  anticorrelation, 262
  completeness, 253
  conjunction, 90
  consistency, 51
  converse consequence, 51, 101
  entailment, 50–51, 100
  equivalence, 50
  Markov, 399–402
  of adequacy, 16
  of generalized
    connectibility, 318, 321, 332–33,
      342
    derivability, 318, 321, 333
    special consequence, 51, 90–91,
      101
  universal, 49
Confirmation, 42–103, 114, 116–19,
  123, 125, 129, 133, 144–45,
  155, 275–76, 302–07, 389, 392,
  416
  confirmational adequacy, 156
  Bayesian. See Bayesian confirma-
    tion
  bootstrapping theory of. See Boot-
    strapping
  function, C*, C◇, M*, M◇, 88–89
  high-probability conception of, 89

hypothetico-deductive, 44–55, 64, 66, 91, 97–99, 394
incremental, 89–91, 97–98
independent, 306
instance, 89, 99, 144
Nicod criterion of, 50
probabilistic, 91
qualitative, 43–55, 66, 89–90
quantitative, 66
Congruence of intervals, 187, 189
Conjecture, 142–45
and refutation, 64, 144
Conjunction, 32, 50–51, 67, 85, 139
constant, 56
irrelevant, 32, 50–51, 67, 85, 91, 139
Connectionist machine, 380
Consequence
etiologies, 298
observational, 44, 183, 185, 188–89
Conservation of angular momentum, law of, 15, 36, 58–59
Conservative dynamical system, 239–40
Consensus, 129, 151. See also Intersubjective agreement
social, 162, 171, 173
Consilience of inductions, 272
Consistency, 62, 82, 86, 107, 158, 369, 370, 401
Construct, social, 151, 161–62, 169
Constructionism, 114, 135–38
Constructive axiomatization, of relativity theory, 190
Constructivism, social, 134, 141, 167–73. See also Cognitivism
Content, empirical, 13, 26, 146, 154–56, 414
Context, 163, 165, 174
historical, 150
mental, 163–64
of discovery, 143, 145, 147
of justification, 143–45, 147
of theory-evaluation, 156
social, 150, 165
Contract, social, 165
Contradiction, 84, 86
Contraposition, 65
Contrast class, 36
Convention, 179, 368
conventionalist strategy, 153
geometrical, 105, 184–89
meaning, 118
measurement, 369
of simultaneity in special relativity, 180, 190–94, 214, 222
Convergence, 79, 111
Converse, Philip, 415, 418
Cooper, John, 297
Coordinate system, 195–205
Cartesian, 205, 207
general, 199–202, 218
standard, 196–98, 201, 203, 205, 207, 218
transformation, 196–207
Coordinative definition, in geometry, 185–87, 190

Copernicus, 104, 146, 164, 181
Corpuscular theory, 47
Corroboration, 64, 96, 133
Cosmic censorship hypothesis, 249–50
Cosmology, 122, 421
Cost, social 419
Cotman, Carl W., 340
Coulomb's law, 188–89
Countable additivity, 94
Counterexample, 94
Counterfactual
conditional, 19, 137, 416–17
definiteness, 260
degrees of belief, 102
logic, 259
Counterinductive rule, 63, 89
Covariance, 194, 196–98, 221–27
group, 196, 198, 207
general, 199–203, 208, 222
Lorentz, 222
Covariant quantity, 201
Covector, 201–02
Covering-law model. See, Explanation, covering-law
Creation science, 13, 39, 289
Creativity, 158
Crime rate in high-rise buildings, 415, 418
Criterion of reality, 250–55, 260
Critical theory, 407, 419–23, 425
Culp, Sylvia, 335
Cultural imperialism, 168
Culture, 128, 410
Cumulative
retention, 145, 156, 158–59
progress, 135, 144
Curves, lightlike, timelike, and spacelike, 216
Curvefitting, 387
Cushing, James T., 263
Custom, 57–58, 126, 128
Cutland, Nigel, 242
Cyclic AMP, 326–29

Dalton's atomic theory. See Atomic, theory, Dalton's
Darwinism, 296, 300, 307
Darwin, Charles, 270–74, 278–79, 286, 288, 290–91, 293, 296, 299–300, 302–03, 307–08
Darwinian theory of evolution. See Evolution, theory of, Darwinian
Darwin's finches, 286–87, 291, 294, 298
Dasein, 147
Davidson, Donald, 120, 411–12
Dawe, C. M., 316
Dawkins, Richard, 295
De Finetti, 94
De Vries, Hugo, 273
De Witt, Bryce S., 252
De-contextualized property, 145–46, 160
Decision
problems, 376–78
theoretic justification, 65
theory, 386, 395, 400, 412–15
Deduction, 10–12, 26, 49, 61–62

Deductive
countersupport, 98
logic, 61, 82, 85, 331
nomological explanation (D-N). See Explanation, deductive nomological (D-N)
reasoning, 334, 402
relationship, 86
support, 98
statistical explanation (D-S). See Explanation, deductive statistical (D-S)
validity, 12, 47, 61
Deductivism, 63–64
Definition, constitutional, 137
Degenerating research program. See Research program, degenerating
Degree. See also Confirmation; Induction
of confirmation, 85–89, 142
of conviction, 81–84
of inductive support, 12
DELPHI, 392
Demon
deceptive, 107–08, 111, 119, 123, 129
Laplace's. See Laplace's demon
Demarcation criterion, 3, 138
DENDRAL, 393–94
Dennett, Daniel C., 348
Depew, David J., 294
Depth perception, 361–63
Derived rules, 68–74
Descartes' theory of motion, 151
Descartes, Rene, 151, 348
Design, 269, 284, 300, 306
Desires, 407
Determinism, 29–31, 232–68
and the absolute vs. relational conceptions of space and time, 245
and chaotic behavior, 236–40
in classical physics, 232, 236–40, 245–46
and computability, 242–43
and free will, 233, 260–62
hard, 260
in general relativity theory, 232, 247–50
Laplacian, 235, 246–47,
local, 247
and open systems, 243–45
and prediction, 240–41, 247, 262
in quantum mechanics, 232, 250–60, 263–68
in special relativity theory, 232, 247
Development, 354
dev(H), of an hypothesis, 51
psychosexual, 350
Devitt, Michael, 131
Diachronic
perspective, 134, 152, 157, 177–78
phenomenon, 145, 177
Diagnosis, 311. See also Medical diagnostic system
Dialectic, 170

Dice, biased, 80
Differential
    adaptation, 281, 288, 290, 293
        principle of, 293–94
    preservation, 288
    equation, 392
    transmission, 281
Directed graph, 399–400
Directly-Hempel-confirm, 52
Disconfirm, 44, 123
Discourse, social, 410
Discovery, 104, 129, 152, 171, 364,
    391–92, 401
Discrimination, 352
Disease, 22, 312, 341
Disjunction, 51, 85, 136
    inclusive, 51
Dissipative dynamical systems, 239
Divine foreknowledge, 241
DNA, 34, 58, 318, 338
    recombinant, 316
Dobzhansky, Theodosius, 300
Domain
    of a model, 113. See also Model
        theory
    of a theory, 321
Dominant gene, 275–76, 295
Doppler effect, 8
Double blind experiment, 27–28,
    49
Dot,(.), 32, 85. See Conjunction
Draw, with and without replacement,
    68–69
Dreyfus, Hubert L., 349
Driesch, Hans, 315
Drugs, 423
Dualism. See also Reduction
    mind/universe, 134
    nature/society, 161
    Cartesian, 252, 345
Duhem, Pierre, 125, 140
Duhem-Quine thesis, 125–26, 189
Dulong, P. L., 104, 118, 124
Dumas, Jean Marie, 104–05
Dummett, Michael, 116, 130
Dunn, J. Michael, 98
Durkheim, Emile, 418–19
Dutch book, 82, 99

Earman, John, 4, 5, 42, 227, 230,
    232, 246, 263
Eclipse, 7–8, 20, 22
Ecology, 405
Economic infrastructure, 163
Economics, 284, 286, 398, 408,
    424
Eden, Garden of, 36
Edidin, Aron, 54
Edinburgh School, 160–61, 167,
    175
Education, 398–99
Eells, Ellery, 99
Effect, 172
Ego, 350–51
Eigenvalues and eigenvectors, 251–
    52, 255, 264, 266–68
Einstein, Albert, 98, 115, 179–80,
    182–83, 185, 187, 191–92, 219–

21, 227, 229–30, 252, 254–55,
    263, 270, 368
degrees of belief in 1915, 98
general theory of relativity. See
    Relativity, general theory of
principle of equivalence, 219–20
relativistic dynamics, 159
special theory of relativity. See Rel-
    ativity, special theory of
Einsteinian framework, 150
Elasticity, 185, 290–91
Electromagnetism, 125, 182
    Maxwell's equations of, 236, 244
Electronics, 125
Elementary experiences, 367
Emergentism, 315, 343
Emotion, 340, 344, 348, 350–51, 362
    causal theory of, 350
Empirical support, 156–57, 270, 286–
    87. See also Confirmation
Empiricism, 140–42, 173
    classical, 347
    logical, 2–3, 134–35, 138–43, 179,
        425
    radical, 158–60
Endler, John A., 293, 306
Endocrine system, 315. See also
    Neuro-immune-endocrine connec-
    tion
Engel, George, 313–14, 339, 344
Engineering analysis, 281, 284, 286,
    294, 304–05
Englehardt, Jr., H. Tristram, 310–11,
    341
Entailment,
    logical, 51, 67, 86
    partial 85
Entelechy, 13
Entity, theoretical, 44, 172
Entrenchment, 101
Entropy, 320
Environment, 271, 279, 281, 290,
    292, 294–95, 297–98, 303–04
    drought, 287, 298
    interaction, 294
    malarial, 284, 293. See also Ma-
        laria
    role in perception, 358–62
Epidemiology, 282, 304
Epistemic commitment, 138
Epistemology, 128–29, 347
    Kantian, 114, 127. See also Tran-
        scendentalism
EPR (Einstein-Podolsky-Rosen) para-
    dox, 252–57, 263–68
EPR-Bell experiment, 254, 259
Equivalence,
    functional, 348
    Leibniz, 228
    logical, 50, 86
    material, 50
Erlangen program, 198
Error, sampling, 302
Essentialism, 144, 177. See also
    Meaning essentialism
Ether, 182–83
Ethics, 422
    medical. See Medical ethics

Ethnography, 165–67, 410
Euclid, 184
Euclidean. See also Geometry, Euclid-
    ean
    space, 37, 203–07, 209, 211–12,
        250–51
    two-dimensional, 203–04, 207
    three-dimensional, 208, 240
Euthanasia, 341
Event,
    factual, 166
    inside vs. outside of, 408
Evidence, 59–60, 72, 88, 123–24,
    173, 175, 281, 303, 349, 401
    clinical, 350
    empirical, 43–44, 142
    expectedness of, 73
    inductive, 59
    observational, 84
    old, problem of, 98–99, 102
    surprisingness of, 73, 91, 98
Evolution, 270–309, 339, 421
    challenges to, 282, 288–302
    Darwinian, 269–309
    Lamarckian, 300
    mutationist vs. selectionist, 273
    neo-Darwinian. See Neo-Darwinism
Exchangeability, 95–96
Exemplar, 147
Expansion pattern, 361
Expected utility, 396, 413–15
Experiment, 166–72, 174–76, 319,
    364, 407
Expert systems, 393–99
    rule-based, 394
Explanandum, 10, 16–17, 32, 331,
    337
Explanans, 10, 13, 16, 26, 32, 42,
    331
Explanation, 7–41, 114, 151, 162,
    392
    as answer to question, 9–10, 303
    as argument, 10–12, 32
    as reduction to the familiar, 14
    behaviorist, 352. See also Behavior-
        ism
    biochemical, 297
    causal, 34, 39, 162, 296, 298
        mechanical, 34–35, 38, 40, 105,
            334, 338
    covering-law, 17, 40, 410, 412–15,
        421, 425
    deductive-nomological, 14–17, 20–
        31, 35–36, 40, 318, 338,
        394
        problems for, 20–23
    deductive-statistical, 17, 23–24, 40
    developmental, 297
    epistemic approach to, 333
    evolutionary, 175, 269–309. See
        also Evolution
    ideal, 271, 291
    explanatory information, 37
    explanatory power, 158
    genetic, 297
    global conception of, 34
    historical, 269, 297, 303
    ideal explanatory text, 37–38

inductive-statistical, 17, 24–31, 36, 39–40
  criticisms of, 27–9
  in social science, 39, 405, 407, 410, 412–15, 422, 425
  lawful, 410
  molecular biological, 331–32
  of laws, 16, 33–34
  of particular facts, 16, 33–35
  ontic approach to, 335, 338
  pattern of, 269–70
  post hoc, 153
  potential, 16
  pragmatics of, 35–38
  -prediction symmetry thesis, 25, 39
  received view of, 27, 33–35, 39
  selectionist, 293, 297–98. *See also* Explanation, evolutionary
  statistical, 23–33
  relevance, 32–33, 40
  teleological, 13, 269, 308
  unification approach to, 33–35, 38–40, 151, 171, 334, 338–39
Extrapolation, 78–79

Fact, 139, 169–71
  irreducibly general, 140
  novel, 154–55
  particular, 16
  social, 126, 161, 167, 172–73
    management of, 164–67
Falsifiability, 144–45, 154, 302, 349
Family complex, 350
Feigl, Herbert, 60, 233
Fermat, Pierre de, 67
Feyerabend, Paul, 135, 145, 147, 157–60, 177–78, 318–19, 347
Feynman, Richard, 121
Field, Hartry, 124, 130
Field, 136
  gravitational, 219–21
  electromagnetic, 236
Figure skater example, 15, 36
Fine, Arthur, 130, 252–53, 261
Finite state automaton, 379, 380, 402
  stochastic, 379
First-order logic, 62, 331
Fisher, Ronald, 401
Fitness, 278–79, 287–96, 307
  as relative survival frequency, 278
  definition of, 289
Flagpole example, 21–22, 34
Flock, Howard R., 360
Fodor, Jerry, 353, 363
Folk psychology, 319, 343, 349
Force, universal, 185–88
Formalism, logical, 317
Forms
  of intuition of space and time, 109–10
  of life, 147, 164–67
Forstrom, Lee A., 311
Fortune teller example, 58–59
Fossil evidence, 106, 272, 279, 284–85, 303
  punctuated, 287

Foucault pendulum, 182
Foundationalism, 146
Founder principle, 301
Frame of reference, 209
  inertial, 209, 222, 244–5
Free will, 404–06, 408, 419–21, 425
  and determinism and indeterminism, 260–62
  and forensics, 420
  incompatibilism (hard determinism), 260
  supercompatibilism (hard indeterminism), 261
Free-floating intellectual, 163
Frege, Gottlob, 112–13, 118, 141, 367–69
Frequency, relative 77–79, 85, 143, 255
  ascertainability of limit of, 79
Freud, Sigmund, 347–51, 366–67
Friedman, Michael, 33–34, 40, 221, 230, 338
Frisbee factory example, 70–72, 80
Frohlich, Cliff, 9
Frost, William N., 328–29
Fruit flies, 273
Function, 365
  computable, 357, 371–73, 375–76, 384
  in exponential time, 384
  in polynomial time, 383–85
  exponential, 385–86
  exponentially bounded, 382
  functional advantage, 296, 298
  functional property, 294–99
  functional relationship, 291
  functional value, 304
  measure, 88
  number theoretic, 376
    Turing, 378–79
  polynomially bounded, 382–83
  recursive, 372, 376
  wave, 304
Functionalism, 348–49

Gaifman, Haim, 92, 97
Galapagos Islands. *See* Darwin's finches
Galilean
  transformation, 245
  velocity addition law, 244
Galileo, 151, 219
Galileo's law, 20, 33
Game
  of chance, 67
  theory, 286, 294, 305
Galison, P.L., 174–75
Gametes, 272
Garber, Daniel, 98–99
Gardner, Martin, 403
Geertz, Clifford, 410
Generalization,
  accidental, 19, 39, 234, 422
  empirical, 234
  inductive, 55
  interlevel, 330
  law-like, 19, 39, 138, 413, 415, 424

statistical, 100
  universal, 17–8, 44, 94, 96, 416
Genes, 273, 298, 318, 322
  recombinant, 301
Genesereth, Michael R., 332–33
Genetic drift, 269, 277, 301–02
Genetics, 269, 282, 303–06, 316, 322
Genotype, 273, 276, 282, 288–89, 304
  frequencies of, 276–78, 290, 293–94, 301
Geographic
  distribution, 302
  isolation, peripheral importance of, 307
Geology, 307, 421
Geometric object, 201–05
Geometry, 78, 110, 120, 187, 189, 317, 368
  Euclidean, 59, 378, 105, 109, 112, 184–86, 208, 230, 365
  of motion, 151
  of space, 179, 230. *See also* Coordinate systems
  non-Euclidean, 112–13, 184–85
Geophysics, 79
Geospiza fortis. See Darwin's finches
Germ theory of disease, 310, 319
Geroch, Robert, 230
Gerver, Joseph L., 244
Gestalt psychology, 347
Ghiselin, Michael T., 299
Gibbs, Josiah Willard, 110
Gibson, J. J., 358–61, 363
Giere, Ronald N., 122
Gill-siphon withdrawal reflex, 325
Gillies, Donald, 97–98
Gilligan, Carol, 350
Gleick, James, 239
Globalism, 145–46, 157, 177
Glymour, Clark, 4–5, 52, 54, 102, 104, 120, 130, 364, 403
God's eye view, 134
Gold sphere, 18–19, 39, 417
Goodenough, Ward, 410
Goodman's paradox, 54–55, 91
Goodmanized predicate, 95–96
Goodman, Nelson, 55, 61–63, 93, 100–02, 131
Gorry, G. Anthony, 395–96
Gosse, Phillip, 106
Gotthelf, Allan, 297
Gödel incompleteness, 370, 371, 401
Gödel, Kurt, 113, 370–72, 377, 401–02
Graham, Neill, 252
Grant, Peter R., 287
Graph, 384
Gray, Asa, 299
Greeks, ancient, 132, 184, 346
Gregory, Richard, 355–56
Grobstein, Clifford, 315
Groups, social, 425
Grue, 63
Grünbaum, Adolf, 120, 189, 191, 226, 230, 311, 421, 349
Guessing, 59, 65
Guyer, Paul, 130

Haber, Ralph, 354
Habermas, Jürgen, 349, 421
Habit, 57–58
Hacking, Ian, 67, 152, 154–55, 173,
 178
Halley, Edmund, 14
Halting problem, 377
Hamilton, Edith, 130
Hamiltonian operator, 250
Hanson, Earl D., 308
Hanson, N.R., 147, 347
Hard core of a research tradition,
 153–54
Hardy, G.H., 274, 277, 401
Hardy-Weinberg equilibrium (HWE),
 277, 300–01, 307
 proof of, 277
Harmonic oscillator, 121
Heart attack, 32–33, 314
Heat
 capacity, 104–05
 diffusion, Fourier equation of, 245
Heated metal plate, 185–87
Heberman, Ronald, 340
Heidegger, Martin, 147, 173
Heisenberg uncertainty, 127. See also
 Quantum mechanics measurement
Helium–4 atoms, 75
Helium balloon example, 37–38
Hellman, 98, 258, 260
Hemoglobin, 282, 297–98
Hempel, Carl G., 3, 13–17, 20–21,
 23–28, 34–36, 38–40, 49–50,
 52, 54, 89–91, 99–100, 103,
 117, 392, 407, 410–14, 417,
 425
Hempel-confirms, 52, 101–02. See
 also Confirmation, hypothetico-
 deductive
Heritability, 271, 281, 285
Hermeneutics, 171, 349
Herschel, John, 272, 299
Hertz, H. R., 182
Heterosis, 282
Heterozygote, 276, 292
 advantage, 282, 304–05
Heuristics, 389, 392, 398, 401–02
 negative and positive, 152–53
Hilbert program, 370–71, 401
Hilbert, David, 112–14, 118–19, 141,
 368–71
Historicism, 134–35, 150–51, 157,
 163
History, 405, 408
 evolutionary, 279
 of science, 104, 154, 157, 177
 of a system, 235
 physically possible, 235
Hobart, R. E. 261
Hobbes, Thomas, 141, 165, 366
Hochberg, Julian, 356
Hole argument, 180, 227–29, 250
Holism, 123–27, 419. See also Col-
 lectivism
Holmes, Frederic L., 174–75
Homans, George, 407
Hooker, C. A., 319
Horkheimer, Max, 421

Horse
 evolution of teeth, 279–81, 291,
  294, 298, 303–04
 evolution of hoofs, 284–86, 292,
  294, 296, 298, 304–06
Horseshoe, 49. See also Implication,
 material
Horwich, Paul, 92, 168, 230
Howson, Colin, 99
Hughes, R. I. G., 263
Hull, Clark, 352
Hull, David L., 157, 174–75, 269,
 295, 299, 302, 316, 319, 321,
 332, 334, 344
Human
 activity, 133, 166
 need, 164
 rights, 423
Humanistic discipline, 151
Hume projectability, 95–96
Hume's problem of induction. See
 Induction, problem of, Hume's
Hume, David, 35, 55–59, 61, 63–66,
 100, 102–03, 107–08, 112, 140,
 230, 342–44, 408
Humean
 regularity, 162
 skeptic, 94
Humphreys, Paul, 40
Huygens, Christian, 67
Hybrids, 275–78
Hypersurface of simultaneity, 209,
 211, 214, 219, 223, 225, 247
Hypothesis, 44–55, 72, 83, 87–88,
 99
 admissible, 101
 alternative, problem of, 48
 auxiliary, 46, 48, 52, 54, 99, 416
 past success of, 64
 statistical, 49, 66
 test, 48, 302

Idealization, 78
 mathematical, 77
Ideas, 136
 world of, 106
Individualism, 419, 425
Immunological
 system, 313, 315, 341
 surveillance, 340
Implication, 136
 material, 49
Impartiality, 162
Incoherence, 82–83, 98
Incommensurability, 124, 132, 146–
 51, 154, 159, 164
Incompatibilism, 260
Inconsistency, 159
Independence, 68–69, 286
 of outcome, 257–58
 of setting, 257–58
Indeterminism, 29–1, 180, 229–30,
 260–62
 in hole argument. See Hole argument
 and Quantum mechanics. See Quantum mechanics and indeterminism

Induction, 10–12, 26, 42, 49, 142–45
 by enumeration, 62–63, 93, 95,
  99–100
 general, 93–94
 justification of, 58, 61
  pragmatic, 65, 143
 new riddle of, 55, 61, 63, 100
 particular, 93
 pessimistic, 125
 problem of, 63, 93
  Hume's, 42, 55–66, 79, 100,
   102, 107, 111
 validation of, 60
 vindication of, 60, 65–66
Inductive
 correctness, 12
 extrapolation, 100
 intuition, 61–63
 practice, 62
 relationship, 86
 support, 25, 42, 46, 86, 97–98,
  142, 304. See also Confirmation
Inductivism, 96–97
Inertia, law of, 109–10
Infallibility, 146
Inference, 11–12
 ampliative, 55, 58
 deductive, 58–59, 394
 erosion-proof, 26
 inductive, 62, 64, 66
  from observed to unobserved, 58
 necessarily truth-preserving, 55, 62,
  333
 nonampliative, 55
 not-erosion-proof, 36
Infinite
 divisibility, 111–12
 sets, countable and uncountable,
  377
Influence, 100, 414
Information processing theory,
 353–58
Inheritance, 272, 274, 279, 307
 blending, 273–74
 polygenic, 273
Initial conditions, 45, 330, 332
Inkstain example, 23, 34
Inscription device, 169–70, 172
Instinct, 351
Institutionalism, 159
Instrumentalism, 104, 121
Insulin, 313
Intention, 348, 408–09
Intentional state, 164
Interest, 161
Internalization, 171
Interpretivism, 407–15, 421, 423, 425
Intersubjective agreement, 141, 144,
 170
Intertranslatability, 145
Intrinsic metrical properties, 189
Intuition, 109, 149, 194, 196–98,
 204, 244
 of time, 109
Invariant duration, definition of, 201
Irrelevance, explanatory, 22–23, 28,
 31–3

Isomers, 393–94
Is/ought distinction, 342–44

Jarret, Jon P., 258
Jealousy, 351, 363
Jeffrey, Richard, C., 40, 93, 96–99
Jeffrey's limit theorem, 96
Jenkin, Fleeming, 273–74, 302–03, 305–07
Johannsen, W. L., 272
Johnson-Laird, P. N., 363
Joint occurrence, 67
Judgement, 100

Kahneman, Daniel, 415
Kandel, Eric R., 315, 317, 323–29, 331, 336–38, 344
Kant, Immanuel, 108–12, 114, 116, 119, 127–28, 130, 184, 233, 342, 368
Kelly, Kevin, 129–30
Kemeny, John G., 316
Kepler's laws, 16–17, 20, 24, 33, 109, 151, 158, 387–88, 402
Kepler, Johannes, 151, 387
Kiger, Jr., John A., 282
Kin selection, 295
Kitcher, Patricia, 130
Kitcher, Philip, 34, 271, 302, 306, 316, 319, 322–23, 332, 335, 338, 344
Klein, Felix, 198
Knowledge
  as socially determined, 160–73
  background, 72–73, 83, 122–23, 129, 305
  by acquaintance and description, 139
  empirical, 114
  growth of, 151–54
  infinite, 108
  necessary condition for, 109
  normative, 160
  of things themselves, 106, 108–09
  representation, 386
  structure of, 347
Kosslyn, Stephen Michael, 363
Kretschmann, Erich, 191
Kronecker, Leopold, 368, 370
Kuhn, Thomas, 3–4, 100, 124–25, 135, 145–55, 159–61, 318–19, 347

La Mettrie, Julien de, 366
Laboratory, 165, 167–70, 173, 175–76
Lacan, Jacques, 349
Lakatos, Imre, 135, 137, 145, 151–55, 157, 160–62
Lakoff, George, 363
Lamarck, Jean Baptiste, 300
Land snail example, 278
Langley, Pat, 387, 403
Language, 165
  arithmetizing, 370
  canonical, 136
  formal, 112–13, 119. See also Logic

fundamental, 140
games, 163–64
inner Russellian, 139, 164
natural, 129, 352
neutral observation, 148, 150
of a mathematical theory, 369–370
observation, 117. See also Vocabulary, observational
ordinary language dissolution, 59–61
protocol, 141
Laplace's demon, 241
Laplacian form of determinism, 235, 245–47. See also Determinism
Laplace, Pierre Simon de, 74–75, 235–37, 240–41
Larmor, Joseph, 241
Latour, Bruno, 151, 165, 168, 170–76, 178
Laudan, Larry, 145, 155–57, 159–60, 162, 177–78
Lavoisier, Antoine, 175
Lavoisier's oxidation theory, 319
Law, 17–20, 37, 188–89, 317, 321–22, 392, 405
  basic, 20
  causal, 23, 161
  derived, 20
  explananda, 17
  general, 16, 26, 116, 337
  of chance, 272
  of cultural diffusion, 418
  of evolutionary potential, 418
  of gravitation, 109. See also Newton's law of gravitation
  of Higgledy-Piggledy, 299
  of human actions, 407, 415–19, 422
  of nature, 234, 241, 262
    causal, 409, 411, 421
    psychophysical, 411
  of thought, 419
  physical, 186, 188, 263, 329, 366, 417, 421
  statement, 17
  statistical, 17, 24, 26, 79, 416
  stochastic, 301, 261
  universal, 17, 24, 26, 95, 112
Learning, 317
  anti-inductive, 89
  behaviorist, 352
  computational, 387
  from experience, 86
  in Aplysia, 323–29
  theory, 337, 343
Lehninger, Albert L., 294
Leibniz, Gottfried, 141, 228–29, 233, 366, 371, 377
Lennox, James G., 5, 13, 24, 269, 296
Lerman, Hannah, 350
Level aggregation, 337
Leviathan, 165
Levins, Richard, 294
Levy, Sandra, 340
Lewin, Benjamin, 332
Lewis, C. I., 114
Lexicon, 146–51

Light, 47
  cone, 216–18, 223–26, 246, 258
  consists of
    particles, 47, 247
    waves, 47, 182
  one-way vs. round trip velocity, 194
  postulate, 213
  signal, 180, 192
  speed of, 19, 58, 216
Likelihood, 73–74, 83–84
Limit, definition of, 93. See Frequency, relative
Lineage, historical, 306
Linguistic
  framework, 150
  network, 148
  practice, 169
  representation, 171
Linguistics, structural, 148
LISP, 378
Little, Daniel, 425
Locality, 257–58
  in physical theories, 254–55
  Einstein, 258–59. See also Quantum mechanics
Locke, John, 57, 106, 108, 114–15, 124, 128
Logic, 100, 136, 138, 144, 365, 368
  inductive, 62, 85, 94
  mathematical, 367
  of degrees of conviction, 82
  of science, 99–100, 112–13
Logical
  correctness, 61
  inconsistency, 159
  linguistic approach to knowledge, 115
  reconstruction, 115, 123
  simplicity, 144
  symbol, 114
  truth, 114
Logically possible state of this universe, 85
Logicism, 114
Long run, 77–78, 120, 143
Long-term memory. See Aplysia
Looming phenomena, 360–63
Lorentz invariance system transformation, 214–15
Lorentz, H. A., 182, 215, 239–40
Luckman, Thomas, 171
Lull, Ramon, 366

Mach, Ernst, 140, 230
Machamer, Peter, 5, 346
Machine learning, 387
MacIntyre, Alasdair, 415–17, 422
Mackie, J. L., 334
Magnetism, theory of, 417
Malament, David, 120, 180, 194, 222–23, 225–26
Malaria, 282–84, 297
Malcolm, Janet, 350
Manifold, 203–04, 206, 209, 211, 216, 221, 228–29
Mannheim, Karl, 162–63
Marcuse, H., 421

Mark theory, 191
Markov condition, 399–402
Martini example, 75–76, 82
Marxism, 163
Mass, 150, 159
  spectrometry, 393–94
Masson, Jeffrey Moussaieff, 350
Material dictionary, 171–72
Materialism, 235–36
Matters of fact and existence, 55
Maxwell's theory, 33
Maxwell, James Clerk, 182, 236–37, 241
Maynard Smith, John, 280, 284, 294, 304–05, 308
Mayr, Ernst, 301, 308, 335
McGuire, J. E., 4, 132
McMillan, Erwin, 263
Meaning, 123–25, 127, 138, 171–72, 408
  essentialism, 164
  holism, 124–26
  invariance, 145, 158–59
  postulate, 118–19, 123, 125
  variance, 124–26, 128
Measure of the ranges, 88
Measurement,
  standard of, 368
  theory of, 391
Mechanism, 235–36, 331
  deterministic, 259. See also Determinism
  evolutionary, 287
  mental, 350
Medical diagnostic system, 395–98
Medicine, 351
  aim of, 310–11
  as a healing art, 314–15, 341
  as a social science, 311–13, 343
  ethical dimension of, 341–42, 423
  philosophy of, 310–45
  scientific status of, 310, 342, 344.
    See also Reduction
Medin, Douglas L., 357
Meiosis, 273
Mendel, Gregor, 272, 274–76
Mendel's
  experiment, 274–76
  factors, 272–73, 276
  genes, 272, 274, 276, 332
  laws, 322
Meno, 107
Mentalism, 356
Mercury, 47, 98–99
Metalanguage, 141, 169
Metamathematics, 370
Metaphysics, 13, 39, 140–41, 172, 234, 346, 368
Method, statistical, 306
Methodological ecumenicalism, 153
Metric tensor, 204
Metric, 188–89, 199, 201, 206–208, 211, 216–18, 220, 228–29
  Euclidean, 206–08, 211, 216
  Minkowski, 216–18, 220
  spatial, 188–89
  temporal, 199, 201, 208, 213
Méré, Chevalier de, 67, 69

Michelson-Morely experiment, 183
Microscope, 46
Microsociology, 165, 174–75
Migration, 277, 301
Mill, John Stuart, 342, 406–07, 410, 412, 414–16, 418–19, 424
Miller, David, 97
Miller, Randolph A., 311
Miller, Richard W., 131
Mills, 290
Mind, 307, 346
  mechanical view of, 349, 366
  philosophy of, 317, 348–49
Mind-independent world, 146
Minkowski spacetime. See Spacetime, Minkowski
Minkowski, Hermann, 214–15, 218–19, 221
Mitchell, Sandra D., 294
Modality, 172
  historical, 133
  physical, 18
Model
  of a theory, 121–22
  of perception, 355
  of speciation, 306
  theory, 113, 286, 318, 320
Modus ponens, 62
Modus tollens, 62, 64
Molecular
  and atomic weights. See Atomic weight, theory of
  biological explanation. See Explanation, molecular biological
  biology, 313, 317, 322, 332, 336–37
  genetics, 282, 305, 332
  kinetic theory of gases, 34
  neurobiology, 343
  proposition, 139
Moliere, Jean Baptiste, 352, 356
Monism, theoretical, 158
Moore, George Edward, 343
Morgan, T.H., 273, 322
Morowitz, 329
Moscatelli, Anthony, 360
Motion, inertial, 181–82, 209
Muller-Lyer illusion, 355–56
Munson, Ronald G., 311–12, 315, 339, 341, 344
Murdock, George P., 417–18
Murphy, Gregory, 357
Musgrave, Alan, 152–54
Mutation, 277, 297, 303
  genetic, 273, 295
Mutually exclusive, 67, 69–70
Myers, Jack D., 311

Nagel, Ernest, 13, 318–20, 330, 343–44
Naked singularity, 247–49
Narrative, historical, 151
Natural
  killer, 340
  logarithm, 392
  selection, 270, 272–73, 277, 279, 282, 284–99, 308

Darwin's principle of, 24, 271
  genetic model of, 279
  mechanics of, 286
Naturalism or Naturalistic tradition, 163–64, 343, 405–08, 410, 412–15, 419, 423
Naturalistic fallacy, 343
Naturalized epistemology, 122–23
Nature, uniformity of, 57, 65
Need, 161
Negation, 32, 50
Neisser, Ulric, 357
Neo-Darwinism, 274, 295, 307
  and evolution, 279
  and explanation, 281
Neo-Darwinian synthesis, 307
Neo-Kantian tradition, 142
Neptune, 47
Neumann, J. von, 251
Neural network, 337, 366–67
Neurath, Otto, 138, 140
Neuro-immune-endocrine connection, 339–41
Neurobiology, 313, 317, 329–32, 337, 339, 362
Neurological system, 315
Neurosis, 350
Neuroscience, 316
Neurotransmitter, 325–26, 329–30
Neutrinos emitted by the sun, 100
New philosophy of science. See Philosophy, of science, new
New riddle of induction. See Induction, new riddle of
Newman, Oscar, 415, 418
Newton, Sir Isaac, 14, 112, 151, 180, 183, 194, 209, 214–15, 227–28, 272, 406
  laws, 17, 20, 153, 157, 180–82, 236, 240, 320, 330,
  of motion, 16, 234, 393
  of gravitation, 125, 233–35, 243–44, 392
Newtonian
  framework, 150
  mass, 105, 150, 159, 234
  mechanics, 16, 64, 105, 110, 121–22, 181, 234–36
  particle, 34, 234–36, 239, 243–44, 250, 262
  spacetime. See Spacetime, Newtonian
  synthesis, 33, 151
  theory, 158, 194
Niche, 269, 303
Nicod, Jean, 49–50
Nilsson, Nils J., 332–33
Nomological or Nomothetic skepticism, 407, 410, 415–19
Non-Euclidean geometry. See Geometry, non-Euclidean
Non-inductive method, 65
Noncumulative theory change, 155
Nonequilibrium thermodynamics, 294
Nonlinear definitional relationship, 76
Norm. See Rules and norms

Normal science. *See* Science, normal
Norton, John D., 4, 179, 227, 392
Nozick, Robert, 261
NSV, 253
Number
  definition of, 136
  natural, 112, 243, 368–72, 376
  prime, 371
  real, 195–96, 199, 202–04
    anisotropy of, 196
  theory, 368

Object,
  language, 141
  ordinary physical, 136–37
Objectivity, 133–34, 144, 164,
  169–70
Observation, 43, 49, 57, 100, 105,
  123–24, 347
Occurrent objects, 133
Ohm's law, 389, 391
Olber's paradox, 14
Olby, Robert C., 322
Omniscience, logical, 98
One-form, 201–02
Ontology, 141, 172
Open systems, 243–45
Operant conditioning, 352–53
Operationalism, 105, 352
Operationalization, 288
Opinion, 161
  poll, 406
Oppenheim, Paul, 14–17, 20–21, 25–
  26, 34–35, 38, 40, 316
Optics, 125, 322
  wave vs. geometrical, 158
Optimality, 284, 293, 304–06, 308
Or, inclusive, 69–70, 139
Oral contraceptive, 22–23, 28
Ornithology, 54, 92, 100
Orthogonal, 214, 223, 225
Osherson, Daniel N., 130
Osiander, A., 116
Outcomes, 74, 77, 80
Overdetermination, 167
Overdominance, 282. *See also*
  Church's thesis; Turing machine;
  Universal register machine; Com-
  putability, effective

Paleoecological data, 303
Papineau, Daivd, 172, 413–14, 419,
  425
Paradigm, 124, 128–29, 146–51, 158,
  160, 173, 287, 323, 337
  experimental, 35, 352
  incommensurability of, 148
  shift, 147–49
Paradox
  of self membership, 136
  of the ravens, 49, 54, 91, 93, 99
  of time travel, 248
Parallel processing, 328–30, 380. *See
  also* Aplysia
Paranoia, 351
Paresis, 28–30
Parkinson's disease, 316
Parole officer, 398

Particle
  accelerator, 46, 124
  fundamental, 121
Partition, 32–33
PASCAL, 378
Pascal, Blaise, 67
Pauker, Stephen G., 311
Pauli, Wolfgang, 231
Pauling, Linus, 27
Peano's axioms, 370–71, 377, 401
Peano, Guiseppe, 114, 371
Peirce, Charles Saunders, 80, 152
Pellegrino, 310, 312, 315
Perception, 109, 134, 351, 354–56,
  358–62
  distance, 360
Perfect gas law. *See* Boyle's law
Personality, Freud's theory of, 350
Persuasion, 100, 149, 158, 169–70,
  175
Petit, A. T., 104, 118, 124
Phases of the moon, 406
Phenomenalism, 137, 140–41, 143
Phenotype, 273, 284–85, 295, 304
Philosophy
  of mathematics, 365, 367
  of mind. *See* Mind, philosophy of
  of science, 1, 4, 128–29, 154, 157,
    401
    new, 4, 133, 135, 145–46, 173
    normative, 123, 138
    standard view of, 3–4, 135, 140,
      142, 144–46, 173, 177,
      330
Phlogiston, 319
Physicalism, 135, 140–41
Physics, 52, 79, 106, 120, 125, 311–
  12, 315, 331, 351, 391, 392,
  412
  atomic, 269. *See also* Particle, fun-
    damental; Atomic weight, the-
    ory of
  classical, 236, 243–44. *See also*
    Newtonian mechanics
Physiology, 305, 349, 353, 362
Piaget, Jean, 354
Pianka, Eric R., 294
Pickering, Andrew, 165, 174–75
Pinch, Trevor J., 165
Pitt, Joseph C., 41
Pitts, Walter, 367
Planck, Max, 234
Planck's constant, 265
Plasmodium falciparum, 282–83
Plato, 107–08, 116, 119, 141, 296,
  299, 346
Platonic form, 136
Pluralism, theoretical, 158–60
Pneumatics, 165
Podolsky, Boris, 252, 263
Poincare, Henri, 105, 112, 184–85
Political science, 398, 405, 408
Politics, 124
Polity, 164–66
Polyani, Michael, 173
Polygraph operators, 398
Pople, Jr., Harry E., 311

Popper, Sir Karl, 13, 63–64, 80, 92,
  95–97, 99, 103, 135, 142, 144,
  161, 241, 302–03, 318–19, 349,
  414
Population genetics, 274, 282, 290,
  293, 295–96, 300–01, 304,
  306–07
Posit, 143
Positivism, logical, 2–3, 112–15,
  134–35, 141–43, 177, 179, 228,
  347
Possible world, 113–14, 233. *See also*
  Logically possible state of this
  universe
Postmodernism, 164
Potentially infinite sequence, 77, 80
Potter, Mary C., 355
Power, 165
Pragmatics, 35, 150, 168
  of social consensus, 162
Prediction, 55, 58, 65, 87, 114, 122,
  241, 405–07, 410, 419, 422
  observational, 45–47, 66, 99
  of novel facts, 154–55
  of single events, 79
Predictive success, 64
Price, Barbara J., 418
Primary qualities, 105–06, 474
Principle. *See also* individual names
  of principles
  of acquaintance, 138
  of indifference, 75–77, 102
  of sufficient reason, Leibniz's,
    233
  of tolerance, Carnap's, 137
Priority of simples, 145
Probability, 31, 42, 49, 66–102, 117,
  143, 236, 386, 398, 400
  absolute, 92
  axioms of, 67–68, 74, 79, 94, 99,
    277, 301
  betting quotient definition of. *See*
    Dutch book
  calculus, 66, 80, 101, 348
  causal tendency, 80
  distribution, 379
  incremental, 92
  inductive, 85, 142
  interpretation of, 101
    admissible, 74–89
    classical, 74–77, 80
    frequency, 77–81, 143
    logical, 85–89
    personalist, 82–84
    propensity, 80–81
    subjective, 81–84
  logical, 85
  numerical value of a, 25, 27
  personal, 81–84
  physical, 81
  posterior, 32, 73–74, 83–84,
  prior, 31–33, 73, 83–84, 88, 91,
    101
  a priori, 88
  nonzero, 93–97
  zero, 121
Problem-individuation, transparadigm
  criteria of, 150

Problem-solving, 155–58, 160
  effectiveness, 155
  model of scientific advance, 147–
    51, 156
Program,
  computer, 373, 378
  research. *See* Research program
Progress. *See* Scientific progress
Projectability, 55, 93–94
Proof theory, 369
Propaganda, 158
Propensity. *See* Probability, interpreta-
  tion of, propensity
Propositional calculus, 348
Protective belt of auxiliary hypothe-
  ses, 153–54
Protagoras, 164
Provine, William B., 274
Proximity, 56
Pseudoscience, 39
Psyche, 346
Psychiatry, 311, 317
Psychoanalysis, 349–51
Psycholinguistics, 362
Psychological
  and sociological influence, 100
  expectation, 35
  makeup, 57
Psychology, 141, 312, 318, 405, 408
  cognitive, 328, 347, 364–65, 398,
Psychotherapy, efficacy of, 27–28,
    31–33
Ptolemaic astronomy, 146
Punnett square, 276
Punnett, R.C., 274, 276
Putman, Hilary, 119, 130–31
Puzzlesolving, 147–49
Pythagorean theorem, 59, 185, 204

Quality classes, 367
Quantum mechanics, 24, 30, 127–28,
    250–60, 263–68, 399
  and Bell's inequalities, 256–60
  and chaos, 250
  and indeterminism, 127, 250–52
  collapse of a superposition, 252
  completeness and incompleteness
    of, 253–57
  equation of temporal evolution
    (Schrödinger equation),
    250–52
  Hilbert space formalism, 250–53
  and locality, 254–55, 257–58
  many worlds interpretation, 252
  measurement problem, 127, 252,
    254–55
  statistical character of, 416
  testable statistical prediction of, 255,
    257. *See also* EPR paradox
Quantifier, 119–20
  existential (∃), 51, 85
  universal (∀), 16, 49, 51, 85
Quark, 52
Question
  explanation-seeking, 9, 36
  how, what, why, 8–9, 36, 296
  metaphysical, 138
Quine, W. V. O., 120, 122–25, 347

Radioactive decay, 23–24, 30
Radiologist, 398
Railton, Peter, 37, 41
Range
  of a statement, 86
  of evidence, 87
Rational
  being, humans as, 405
  reconstruction, 137–38, 154, 336,
    343
Rationality, 59–61, 164, 367
  theory of, 384
  standard of, 424
Rawlsian social contract, 342
Realism, 104–31, 177, 179–80, 184
  and space and time, 227–30
  local, 172
  Platonic, 164
  psychological, 351, 358–63
Reality, 167–68
  artificial, 169
  cultural, 133
  extralinguistic, 152
  simultaneous, 263–68
  social, 166
Reason. *See* Rationality
Reasons, 407–09, 411–12, 425
Received view
  of philosophy of science. *See* Phi-
    losophy, of science, standard
    view of
  of scientific explanation. *See* Expla-
    nation, received view of
Recessive gene, 295
Recidivism, 398
Recognition, 354
Recursive function. *See* Function, recursive
Red shift, 8–9
Reduction, 137, 149, 163, 310–45,
    419
  and explanation, 322, 337–39
  branch, 320
  causal-mechanical account of,
    332–37
  epistemic, 333
  general reduction-replacement
    model, GRR, 319–21, 329,
    333–37, 339, 343–44
  historical account of, 317–23
  in neurobiology, 329–32
  intertheoretic, 316, 335
  methodological, 315, 343
  Nagel model of, 318–20, 343
  of classical genetics to molecular
    genetics, 316
  of mind to brain, 172, 316–17,
    348
  of predictive import, 87
  of scientific fact to social fact, 172–
    73, 177
  of time to causation, 179, 190–91
  ontological, 166, 315, 343
  partial model, 331, 337
  solipsistic, 137
  to the familiar. *See* Explanation, as
    reduction to the familiar
Redundancy theory of truth, 129

Reference class, 32
Reflexive criterion, 162
Refutation. *See* Conjecture, Corrobo-
  ration, Popper, Sir Karl
Register machine, 242–43. *See also*,
  Universal register machine
  (URM)
Reichenbach, Hans, 56, 61, 64–65,
    117, 120, 123, 135, 142–44,
    179, 184–85, 187–88, 190–92,
    231, 258, 399
Reification, 133, 171
Reinforcement, 352–53, 420
Relativism, epistemelogical, 125–29,
    164, 363
Relativistic quantum field theory,
    258
Relativity, 181–83, 208–13, 227–30
  of simultaneity, 213–14
  general theory of, 47, 98, 179,
    190–91, 219–21, 227–28,
    247–50, 368
    Einstein's field equations of,
      219, 221–22, 247–48
  special theory of, 125, 179, 190–
    91, 213–22, 227, 246–47
  standard formulation of, 215
Relevance, 32–33, 125, 133
Representation
  mental, 134
  of reality, 151–52
  theorem, 94
Representative sample, 63, 301
Repression, 350
Reproductive success, 288, 290–93,
    295, 301
  actual, 290–92, 295
  differential, 288, 290, 293
Requirement of maximal specificity,
  (RMS), 26
Research
  program, 151–58, 160, 302, 323,
    332
  degenerating, 152, 307
  tradition, 155–58, 323
Rest, absolute, 209, 230
Retrodiction, 55. *See also* Prediction
Revolution, 3, 124, 129, 132, 144,
    146, 148–52, 177, 237
Rhetoric, 124, 134, 149, 151, 158–
    59, 170
Ricoeur, Paul, 349
Rigid rod, 185–87
RNA, 335, 339
Robb, Alfred R., 191
Rorty, Richard, 127
Rosen, Nathan, 252, 263
Rosenberg, Alexander, 269, 282,
    292, 298, 309, 316, 319, 332,
    425
Rosinski, Richard R., 359, 361
Ruben, David-Hillel, 172–73
Rule,
  general addition, 69
  general multiplication, 68
  inductive, 143
  in perception, 355
  negation, 69, 73, 82

special addition, 67, 69, 277
special multiplication, 68, 277
of total probability, 70
Rule-governed activity, 143
Rules and norms, 158–59, 409–10, 421
cultural, 133, 409, 414
Ruse, Michael, 309, 319
Russell, Bertrand, 93–94, 112–14, 118, 122, 135–40, 142–43
Rutherford, E., 100
Rutherford-Soddy law of radioactive decay, 387–88
Ryle, Gilbert, 410

Sahlins, Marshall D., 418
Salience, 37
Salmon, Merrilee H., 5, 404
Salmon, Wesley C., 4, 7, 34, 41, 42, 58, 64, 66, 77, 89–90, 100, 103, 120, 194, 231, 321–22, 333–35, 338–99, 344
Sample, unrepresentative, 63
Sampling, 68–69, 301
Sanders, William T., 418
Sarkar, Sahotra, 335
Satellite, 64
Saturn, 296
Scale, 199–202, 204, 225
Schaffer, Simon, 164–68
Schaffner, Kenneth F., 5, 269, 310, 316, 318–21, 330–31, 334, 336, 339, 345
Scheines, Richard, 403
Schema, 356–58, 363
Schiff, William, 361
Schlick, Moritz, 179
Schrödinger equation, 250–52
Schwartz, James H., 323–25, 344
Science
as a cognitive activity, 392
as a goal oriented activity, 176
as a human activity, 160–76
as a rational enterprise, 157–58
as a rule-bound activity, 147, 157, 167
as universal, 144
crisis within, 146–47, 151
historical, 421
inexact vs. exact, 406, 419
normal, 147, 149, 152
physical, 368, 406–08, 415–17, 421, 423–25
puzzle-solving potential of, 147–48, 151
social, 298, 398
fallibility of, 423
success of, 58–59
Scientific
change, 132–78, 308, 330
community, 121, 146–48, 168, 177
culture, 132, 164, 168
discovery. See Discovery
growth, 155
knowledge, limits of, 119–23
practice, 142, 146, 158, 167–68, 171, 173–76
autonomy of, 177

progress, 133–34, 149, 151–60, 178, 298, 307
pre-Kuhnian accounts of, 135–46
Revolution, the, 132
theory. See Theory
understanding, 7, 38–39
Scientology, 39
Scope modifier, 416–17
Script. See Schema
Scriven, Michael, 23, 34, 41, 417
Sea hare. See Aplysia
Secret power, 56
Selection. See Natural selection
Self-adjoint operators, 251
Self-determination, 343
Semantic conception of theories. See Theory, semantic conception of
Semantics, 35–36, 138
Sensation, 358–62
Sense-data, 136–37, 145
Sensitization, 324–30, 334, 344
Sentence,
lawlike, 194
observational, 120, 125, 140
novel, 352–53
protocol, 136, 140, 146
Service, Elman R., 418
Set theory, 368–69
Shannon-Weaver mathematical information theory, 348
Shapere, Dudley, 320
Shapin, Steven, 164–68
Sheppard, P. M., 293, 308
Sickle cell anemia, 282–84, 292–94, 316
Simon, Herbert, 349, 367, 387, 403
Simplicity, 49
Simpson, George G., 280, 284, 304, 315
Simultaneity, 179, 224–26, 230. See also Hypersurface of simultaneity
in special relativity, 230
of distant events, 179
standard, 224–25
topological, 226
Single case, problem of, 79–80
Skepticism, 57, 65, 110–12, 134
metaphysical, 105–06, 108–110, 114–17, 119, 123
nomological. See Nomological or Nomothetic skepticism
Skinner, B. F., 352, 356, 420
Sklar, Lawrence, 231, 318
Sloppy bartender example, 75, 82
Smith, James D., 294
Sneed, Joseph D., 122
Snell's law, 322
Snir, Marc, 97
Sober, Elliott, 288–90, 292, 295, 298, 306, 309
Social phenomenon, 151, 160, 162–64, 166–67, 404, 422
Society, 122, 125–27, 165
Sociology, 398, 405, 408
of knowledge, 126, 134–35, 160–63
Socrates, 107
Soul, 346

Source law, 295
Space, 109
absolute, 180–83, 227
homogeneous, 205–06
isotropic, 206
inertial, 181–82, 199, 214
relative, 180–82
Space-time matrix, 134
Spacetime, 205, 208–16, 227–30, 245–46, 262
curved, 221
Minkowski, 214–17, 219–21, 224, 226, 246–62
Newtonian, 179, 208–16, 222, 227–30
Species, 298
Spirtes, Peter, 225, 403
Spontaneity, human, 405
State
description, 85–87, 89
of a system, 234–35, 264
Statistics, 424
Stigler, Stephen M., 67
Stimulus, 352–58, 361, 420
Stob, Michael, 130
Strange attractors, 239
Strawson, P. F., 60, 103
Strep infection, 24–26, 29
Strong program, 126, 149, 160–63, 167
Structure description, 88–89
Subjectivism, 128
Substantivalism in spacetime theories, 180, 227–30
Success,
reproductive, 288, 290–93, 295, 301
adaptive, 301
Suicide, 418–19
Summum bonum, 342
Superconductor, 100
Superstructure, cultural, 163
Supervenience, 292, 332
Suppe, Frederick, 122
Suppes, Patrick, 92, 101, 122, 318, 399
Suppes-Adams model, 320
Support,
empirical, 156–57, 270, 286–87.
See also Confirmation
inductive, 25, 42, 46, 86, 97–98, 142, 304. See also Confirmation
partial, 85
Swamping of the priors, 84, 97
Symbiosis of practice and theory, 174–75
Symmetry, 205–06, 209, 223
principle, 162, 205–08, 211, 244
transformation, 205–07, 213, 222
Symptom of illness, 22
Synchronic perspective, 134
Synchronicity, 214
Synchronized clocks, 193, 214, 219
Synonymy, 117
Syntax, 35–36, 136, 138, 141, 318
Synthetic a priori, 110–12, 118, 142, 318

Syphilis, 28–30
System,
  constitutional, 137
  theoretical, 159

Tachyons, 346–47
Tardyons, 246
Tautology, 289, 308
Taylor, Edwin F., 231
Technician, 174
Technique, 404, 422, 424
Teleology, 296–89
Teleonomic, 299
Telescope, 43, 46, 379, 402
Tell, William, 101
Temporal priority, 56
Test of hypothesis, empirical, 16, 308
Testability, 349, 405
TETRAD, 399–401
Text, 153, 177
Texture gradient, 359–62
Theology, 13, 39
Theory, 114, 172, 289, 316–17, 321,
  401
  alternate, 105
  axiomatizable, 53, 330, 371,
    377–78
  choice, 135, 142, 150
  competing, 127, 152
  complete, 252–53, 255, 370–71,
    378
  content, 105
  decidable, 377–78
  evaluation, 157
  incompatible, 129
  -independent entities, 151, 170, 172
  laden, 124, 148, 173
  proliferation, 159
  reduction, 330. See also Reduction
  semantic view of, 121–23, 194,
    318, 334
  social, 135, 160, 162, 177
  standard conception of, 116–20,
    123–28, 135
  statistical, 300
  structure of, 108
  success of, 302
  vs. observation, 147. See also Vo-
    cabulary, theoretical, and ob-
    servational
Thermodynamics, laws of, 34, 110,
  158
Thick description, 410
Thinking machine, 353
Thomasma, 310, 312, 315
Thought experiment, 219, 232–33,
  244, 253, 302
Three Mile Island, 34
Tidology, 406
Tilde (~), 32, 50. See Negation
Tile floor, 359–60
Time,
  isotropic, 193, 195, 205
  linear theory of, 195–203, 205, 208
  reversal invariance, 235

slice, 247
  to collision, (TC), 361–62
Topology of time, space and space-
  time theories, 203
Toulmin, Stephen, 116
Tradition vs. innovation, 176
Trajectory, 64
Transcendentalism, 109, 142
Transformation,
  inertial, 211
  operation, 137
Translation, piecemeal, 148
Trial and error, 64
Triple bar (≡), 50. See Equivalence
Truth, 108, 152, 155, 160, 168
  function, 62, 139–40
  timeless, 133
Turing machine, 242, 372–86, 402
Turing, Alan, 242, 372, 375–76
Turnstile (⊢), 50
Tversky, Amos, 415
Two-headed coin, 83–84

Unconscious, 349–50
Underdetermination, 120, 122, 129
Understanding, 13
  concepts of, 110
Unification. See Explanation, unifica-
  tion approach to
Uniformity of nature, 57, 65
Unity of science, 138, 141
Unit of scientific achievement, 149
Universals, 139
Universe, universal, 89
Unlimited Register Machine (URM),
  375–76, 378, 385
Unobservables, 44
Uranium sphere, 18–19, 39
Uranus, 47
Utilitarianism, 342

Validity, deductive. See Deductive
  validity
Value judgment, 423
Van Fraassen, Bas C., 36, 41, 54, 99,
  122, 130, 136
Van Heijenoort, Jean, 130
Van Herik, Judith, 350
Van Inwagen, Peter, 263
Vapor density, 104–05
Variation, 271–74, 279, 285, 287,
  307
  advantageous, 271, 281–82, 295,
    298
  chance, 300
  heritable, 281, 284, 287, 300
  random, 288, 300, 307
Velocity, 234
Venn, John, 77
Verifiability, 133, 143
  criterion, 143, 179, 228, 230
  empirical, 346–47
Verification principle of meaning,
  115–18, 183
Verification, 142–45, 181–82

vs. conjecture and refutation, 147
Verisimilitude, 160
Vienna Circle, 2, 138, 141–42, 179,
  347
Vindication of induction. See Induc-
  tion, vindication of
Virchow, R., 311, 313
Virtus dormitiva, 352, 356
Vision, Gerald, 131
Vitalism, 13, 315, 343
Vitamin C, 27–28, 31, 49
Vividness of specific information,
  415
Vocabulary,
  observational, 44, 118–19, 125–26,
    136, 140
  theoretical, 44, 118–19, 121, 123–
    26, 140, 159

Wainwright, S. A., 294
Wald, Robert M., 231
Watkins, 318
Watson, Gary, 263
Watson, James D., 322, 332, 336
Watson, John W., 351
Weber, Bruce H., 294
Wedge (∨), 69, 85. See Or, inclusive
Weight, 87, 88, 368. See also Atomic
  weight, theory of
Weinberg, W., 277
Weinstein, Scott, 130
Weiss, Paul, 315
Weizenbaum, Joseph, 349
Well-formed formula, 369
Weyl, Hermann, 117, 191
Wheeler, John Archibald, 231,
  252
Whewell, William, 272
Whitehead, Alfred North, 114, 136
Wigner, Eugene P., 257
Wiley, E. O., 294
Wilf, Herbert S., 403
Williams, Donald Cary, 63
Williams, George C., 294, 308
Williams, Juanita, 350
Wilson, David Sloan, 288, 293,
  308
Wimsatt, William C., 319, 321–22,
  331, 334–35
Winch, Peter, 409–10, 412
Winnie, John A., 191, 222
Wittgenstein, Ludwig, 86–88, 137,
  147, 163–64, 347–48, 409
Wohwill, Joachim F., 360
Woolgar, Steve, 151, 168, 170–75,
  178
World, physical, 78, 134
Worldline, 209, 216, 219, 223, 225,
  246
Wright, Crispin, 116
Wright, Larry, 41, 296–98

Zarin, Deborah A., 311
Zurek, Wojciech Hubert, 252
Zytkow, Jan, 392, 403